Physical, Chemical, and Thermal Technologies

Remediation of Chlorinated and Recalcitrant Compounds

Editors

Godage B. Wickramanayake
Battelle

Robert E. Hinchee
Parsons Engineering Science, Inc.

The First International Conference
on Remediation of Chlorinated and
Recalcitrant Compounds

Monterey, California, May 18–21, 1998

 BATTELLE PRESS

Columbus • Richland

Library of Congress Cataloging-in-Publication Data

International Conference on Remediation of Chlorinated and Recalcitrant Compounds
    (1st : 1998 : Monterey, Calif.)
        Physical, chemical, and thermal technologies : remediation of chlorinated and
    recalcitrant compounds / editors, Godage B. Wickramanayake, Robert E. Hinchee.
            p.   cm.
        "First International Conference on Remediation of Chlorinated and Recalcitrant
    Compounds, Monterey, California, May 18–21, 1998."
        Includes bibliographical references and index.
        ISBN 1-57477-060-8
        1. Organochlorine compounds--Biodegradation--Congresses. 2. Hazardous
    waste site remediation--Congresses. I. Wickramanayake, Godage B., 1953–   .
    II. Hinchee, Robert E. III. Title.

    TD1066.O73I58   1998e
    628.5′2—dc21   98-25125
                                                                       CIP

Printed in the United States of America

Battelle Press
505 King Avenue
Columbus, Ohio 43201, USA
614-424-6393 or 1-800-451-3543
Fax: 1-614-424-3819
Internet: press@battelle.org
Website: www.battelle.org/bookstore

For information on future symposia and conference programs, write to:
    Bioremediation Symposium
    Battelle
    505 King Avenue
    Columbus, Ohio 43201-2693
    Fax: 614-424-3667

CONTENTS

Air Sparging

Dehalogenation Technologies

Oxidation Technologies

Photo-, Electro-, and Catalytic Mechanisms

FOREWORD

Physical, chemical, and thermal technologies represent the mainstream remediation approaches for sites contaminated by chlorinated and recalcitrant compounds worldwide. Recently, new technology developments and significant advancements have emerged for some of the more traditional remediation technologies. *Physical, Chemical, and Thermal Technologies: Remediation of Chlorinated and Recalcitrant Compounds* offers a comprehensive overview of the remediation technology spectrum, with chapters covering such diverse topics as soil vapor extraction, multiphase extraction, enhancements to pump-and-treat systems, groundwater circulation wells, thermal desorption, in situ thermal technologies, plasma technologies, in situ steam processes, dehalogenation technologies, air sparging, oxidation technologies, and photo-, electro-, and catalytic mechanisms.

This is one of six volumes published in connection with the First International Conference on Remediation of Chlorinated and Recalcitrant Compounds, held in May 1998 in Monterey, California. The 1998 Conference was the first in a series of biennial conferences focusing on the more problematic substances—chlorinated solvents, pesticides/herbicides, PCBs/dioxins, MTBE, DNAPLs, and explosives residues—in all environmental media. Physical, chemical, biological, thermal, and combined technologies for dealing with these compounds were discussed. Several sessions dealt with natural attenuation, site characterization, and monitoring technologies. Pilot- and field-scale studies were presented, plus the latest research data from the laboratory. Other sessions focused on human health and ecological risk assessment, regulatory issues, technology acceptance, and resource allocation and cost issues. The conference was attended by scientists, engineers, managers, consultants, and other environmental professionals representing universities, government, site management and regulatory agencies, remediation companies, and research and development firms from around the world.

The inspiration for this Conference first came to Karl Nehring of Battelle, who recognized the opportunity to organize an international meeting that would focus on chlorinated and recalcitrant compounds and cover the range of remediation technologies to encompass physical, chemical, thermal, and biological approaches. The Conference would complement Battelle's other biennial remediation meeting, the In Situ and On-Site Bioremediation Symposium. Jeff Means of Battelle championed the idea of the conference and made available the resources to help turn the idea into reality. As plans progressed, a Conference Steering Committee was formed at Battelle to help plan the technical program. Committee members Abe Chen, Tad Fox, Arun Gavaskar, Neeraj Gupta, Phil Jagucki, Dan Janke, Mark Kelley, Victor Magar, Bob Olfenbuttel, and Bruce Sass communicated with potential session chairs to begin the process of soliciting papers and organizing the technical sessions that eventually were presented in Monterey. Throughout the process of organizing the Conference, Carol Young of

Battelle worked tirelessly to keep track of the stream of details, documents, and deadlines involved in an undertaking of this magnitude.

Each section in this and the other five volumes corresponds to a technical session at the Conference. The author of each presentation accepted for the Conference was invited to prepare a short paper formatted according to the specifications provided. Papers were submitted for approximately 60% of the presentations accepted for the conference program. To complete publication shortly after the Conference, no peer review, copy-editing, or typesetting was performed. Thus, the papers within these volumes are printed as submitted by the authors. Because the papers were published as received, differences in national convention and personal style led to variations in such matters as word usage, spelling, abbreviation, the manner in which numbers and measurements are presented, and type style and size.

We would like to thank the Battelle staff who assembled this book and its companion volumes and prepared them for printing. Carol Young, Christina Peterson, Janetta Place, Loretta Bahn, Lynn Copley-Graves, Timothy Lundgren, and Gina Melaragno spent many hours on production tasks. They developed the detailed format specifications sent to each author, tracked papers as received, and examined each to ensure that it met basic page layout requirements, making adjustments when necessary. Then they assembled the volumes, applied headers and page numbers, compiled tables of contents and author and keyword indices, and performed a final page check before submitting the volumes to the publisher. Joseph Sheldrick, manager of Battelle Press, provided valuable production-planning advice and coordinated with the printer; he and Gar Dingess designed the volume covers.

Neither Battelle nor the Conference co-sponsors or supporting organizations reviewed the materials published in these volumes, and their support for the Conference should not be construed as an endorsement of the content.

Godage B. Wickramanayake and Robert E. Hinchee
Conference Chairman and Co-Chairman

FULL–SCALE THERMAL DESORPTION OF SOIL CONTAMINATED WITH CHLORINATED COMPOUNDS

Pim I.M. Vis (Ecotechniek bv, Maarssen, The Netherlands)
Piet Krijger (Ecotechniek Bodem bv, Maarssen, The Netherlands)

ABSTRACT: Ecotechniek successfully treated several types of soils contaminated with HCH, drins, PCDD/F and PCB. In total 6,500 tons of heavily contaminated soils were treated with the thermal desorption plant of Ecotechniek during several full–scale test treatments. The treated soils were cleaned to not detectable levels of contamination for pesticides (HCH and drins) and below the target levels for PCB and PCDD/F. The average PCDD/F emissions (0.035 ng TEQ/m^3_0) in the smoke-stack were obtained far below the target value of 0.1 ng TEQ/m^3_0.

INTRODUCTION

Since the eighties, soil contaminated with hydrocarbons (BTEX, TPH and PAH) and cyanides has been treated in thermal desorption plants. These contaminants are completely removed from the soil by thermal treatment (Koopmans and Reintjes, 1988). Many batches of soil are also contaminated with halogenated hydrocarbons like pesticides (HCH, drins), PCB, dioxins and furans (PCDD/F). Compared to contaminants like PAH and cyanides chlorinated hydrocarbons have a relatively low boiling piont, so removal of the chlorinated contaminants from the soil is feasible with a thermal desorption process. However, because of the formation of dioxins and furans in the off–gases, due to the socalled Denovo-synthese process, thermal treatment of soils contaminated hydrocarbons was not allowed before the start of the test treatments.

The integration of an adsorbent unit for the removal of chlorinated hydro-carbons, in particular PCDD/F, in the existing off–gas treatment system could be a possible solution to this problem. Literature research on the various types of adsorbent for the removal PCDD/F from the off–gases results in a mixture of lime and active carbon as adsorbent.

The objective of the test treatment was to apply technology for treating soil contaminated with halogenated hydrocarbons in an environmentally responsible way and on an industrial scale. The main criteria in this context are the end con-centrations in the treated soil and the emission into the air of the chlorinated components (in particular PCDD/F).

PROCESS DESCRIPTION OF SOIL TREATMENT FACILITY

The thermal soil treatment plant of Ecotechniek BV was converted in 1993, in particular, the afterburner and the related off–gas treatment technology were adapted especially to remove chlorinated hydrocarbons from the off–gases. Figure 1 shows the process diagram of the soil treatment plant of Ecotechniek BV.

Soil flow. The contaminated soil is discharged from a storage site into a dosaging hopper. The soil is transported on conveyer belts to a sifter which is equipped with a magnetic separator for removing any iron particles from the soil. A crusher may be used to reduce the coarse rubble into smaller parts. The sifted soil is transported on conveyer belts from a feeding screw into the rotary kiln. In the rotary kiln, the soil is heated indirectly and directly. In the first part of the rotary kiln the soil is indirectly preheated to approx. 300°C by the hot exhaust gases. As a result, water and volatile components will evaporate from the soil. In the second part of the rotary kiln the soil is heated directly to a maximum of 600°C by a gas burner, so that less volatile components will evaporate from the soil. The soil is removed and cooled with watersludge to 75–90°C in a soil cooler. The treated soil is removed on a conveyer belt to a storage site for treated soil.

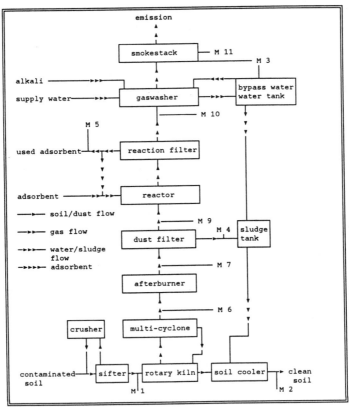

FIGURE 1. Process flow diagram, including sampling points during test periods, of the soil treatment plant.

Process gas flow. The gases which develop in the rotary kiln are aspirated by a ventilator to the off–gas treatment. The gas flow consists mainly of water, the burner air of the gas burner and the evaporated contaminants from the soil. The process gas flow is led through a series of multi–cyclones for dust removal, pre–heated in a heat exchanger and fed to an afterburner. In the afterburner the waste gas contaminants are destructed into CO_2, H_2O and HCl. At the end of the after–burner the hot gas flow heats the incoming gas flow of the afterburner by means of a heat exchanger. The process gas heats indirectly the first part of the rotary kiln, following which the burner air of the gas burners is heated by means of a second heat exchanger. Next, more dust is removed from the gas flow by means of a dust filter. From the dust filter the gas flow is fed to the adsorbent unit. The adsorbent unit consists of a reactor and a reaction filter. In the reactor, an adsorbent is added to the gas flow. The adsorbent consist of a mixture of lime and carbon. The lime removes the acid exhaust gases (SO_2, HCl), while the carbon removes any organic components which are present in the gas flow such as dioxins and furans. The adsorbent with the contaminants is captured in the reaction filter and largely recirculated. Accordingly, a mixture of "fresh" adsorbent and recirculated adsorbent is added to the reactor. A small fraction of the adsorbent is discharged. Subse–quently, the gas flow is led through a quench gas washer and than exhausted through a smokestack.

TEST AND MEASUREMENT PROGRAM

Ecotechniek has in several periods tested the removal of chlorinated hydrocarbons from contaminated soil in the soil treatment facility in Rotterdam, The Netherlands (Noorman en Vis, 1995).

Test program: the following batches of soil were involved in the research pro–gramm:

* 3,000 tons of soil contaminated with drins (1993)
* 1,200 tons of soil contaminated with HCH, dioxins and furans (1993)
* 800 tons of soil/rubble contaminated with dioxins and furans (1996)
* 1,500 tons of soil contimatined with PCB (1997)

Before the test treatment with contaminated soil was executed the perfor–mance of the installed afterburner and absorbent unit were tested by the addition of tracers in the off–gases. The tracers used for testing the afterburner were trichlo–roethylene (tri), tetrachloromethane (tetra) and monochlorobenzene (MCB) based on the Guidance Handbook of the EPA (EPA, 1989). The tracer used for testing the adsorbent unit was naftalene, based on the chemical/physical property of naftalene compared to PCDD/F.

Measurement program: the measurement program was based on the Dutch and German requirements. The measurements were performed by two laboratories, i.e., the Dutch company TAUW Infra Consult Deventer and the German RWTÜV (Rheinisch–Westfälischer Technische ÜberwachungsVerein) of Essen. Both companies operate according to established regulations and standards and they have sufficient expertise to monitor the process data, and to process the measure–ment results. Samples were taken of all the relevant flows (see figure 1) in the soil treatment facility. The soil flow was measured at M1 and M2, the water flow at M3, the dust flow at M4, the adsorbent flow at M5 and the process gas flow at M6, M7, M9, M10 and the stack–emission at M11.

RESULTS AND DISCUSSION

Tracer tests: The preconditions set for a good performance of the installed afterburner and adsorbtion unit were a destruction efficiency of the afterburner higher than 99,99% for all tested tracers and a removal efficiency for the adsorbent unit of at least 95%. The result of the tracer tests, shown in table 1, show that with a temperature in the afterburner of 1050 °C and the use of a mixture of carbon and lime as adsorbent the preconditions set for the performance of the afterburner and adsorbtion unit, based on the emission target values for PCDD/F, were fulfilled.

TABLE 1. Tracer test results

AFTERBURNER				ADSORPTION UNIT		
	Tracer				Adsorbent	
Temperature	tri	tetra	mcb	Tracer	lime/carbon	lime
	destruction efficiency (%)				removal efficiency (%)	
850 °C	99,99	99,98	99,99	Naftalene	> 99,5	90
1050 °C	> 99,99	> 99,99	> 99,99			
1200 °C	> 99,99	> 99,99	> 99,99			

Soil treatment: In total 6500 tons of soil contaminated with chlorinated hydrocarbons were treated in the thermal desorption process during the test treatment periods. The following process conditions were maintained during the test days:
* temperature of rotary kiln: 600 °C
* temperature of afterburner: 1050 °C
* treatment capacity: 25 tons/hour

TABLE 2. Average composition of contaminated and treated soil for soils contaminated with chlorinated pesticides (HCH and drins)

contaminant	DRINS (pesticide)		contaminant	Hexachlorohexane (HCH)	
quantaty of soil tested	3,000 ton		quantaty of soil tested	1,200 ton	
concentration	input	output	concentration	input	output
	mg/kg ds	mg/kg ds		mg/kg ds	mg/kg ds
aldrin	70	< 0.1	alpha-HCH	2300	< 0.1
dieldrin	82	< 0.1	beta-HCH	550	< 0.1
endrin	320	< 0.1	gamma-HCH	< 1	< 0.1
telodrin	24	< 0.2	delta-HCH	15	< 0.1
endrin aldehyde	510	< 0.2			

Table 2 and table 3 show the treatment results of four batches of soils contaminated with chlorinated compounds. The results show, as expected, a very high removal efficiency of the chlorinated compounds. The concentration of the contamination in the treated soil is below the detection limit for HCH and drins. For PCB and PCDD/F the end concentration is relatively low and far below the Dutch target values for reuse of soil at civil works. The overall removal efficiency with the thermal desorption process of Ecotechniek for the removal of chlorinated compounds from contaminated soil is higher than 99,8%.

TABLE 3. Average composition of contaminated and treated soil for soils contaminated with PCB and dioxins ande furans (PCDD/F)

contaminant	PCB		contaminant	dioxins and furans (PCDD/F)	
quantaty of soil tested	1,500 ton		quantaty of soil tested	800 ton	
concentration	input	output	concentration	input	output
	mg/kg ds	mg/kg ds		ng TEQ/kg ds	ng TEQ/kg ds
sum PCB (7 Ballschmitter)	1250	0.02	dioxins and furans (PCDD/F)	30,280	60
EOCl*	7200	0.2			

total extractable organic chlorinated compounds expressed in therms of Cl⁻

Gas treatment: The gas flow was sampled at 5 check points in the soil treatment facility. In this paper only the emission values in the smokestack are shown. Table 4 shows the process conditions and concentrations in the smokestack.

TABLE 4. Process conditions and concentrations of components in smoke-stack.

	drins	HCH	dioxins /furans	PCB	emission target values
gasflow (Nm³/hour)	26100	24500	22600	23000	-
temp.(°C)	76	75	76	75	-
O_2 (vol.%)	11.3	12.1	10.9	12.2	-
CO (mg/m³)	25	12	18	20	50
SO_2 (mg/m³)	‹5	‹5	‹5	‹5	40
HCl (mg/m³)	0.4	0.3	0.3	3.4	10
org. C (mg/m³)	5.8	3.6	2.9	5.7	10
dust content (mg/m³)	‹5	‹5	‹5	‹5	5
PCDD/F* (ng TEQ/m³)	0.03	0.02	0.01	0.07	0.1

all concentrations have been converted to dry in normal conditions at 11% O_2
*: the concentration PCDD/F is given for the "dirty seventeen" (ng TEQ/m³).

The main research object of the test treatment was to achieve PCDD/F concentrations in the smokestack below the Dutch emission target values. The results show that the average emission values of dioxins and furans remains far below the target value of 0.1 ng TEQ/m³.

Furthermore can be concluded (results not shown) that the afterburner has a

very high destruction efficiency, higher than 99,99%, for chlorinated compounds. The adsorbent unit removes any dioxins and furanes formed after the afterburner due to the "Denovo–synthese" with an efficiency higher than 95%. HCl, formed after the destruction of the chlorinated compounds is removed with a high degree of efficiency in the adsorbent–reactor and the gas washer.
For all of the other gaseous components, e.g. CO, SO_2, HF, organic C and dust the values measured are lower than the Dutch emission target values.

CONCLUSIONS

Soil contaminated with halogenated hydrocarbons (drins, HCH, PCB, dioxins, furans, etc.) can be treated in an environmentally responsible way with the thermal treatment plant of Ecotechniek BV.

Soil contaminated with halogenated hydrocarbons can be cleaned to values lower than the laboratory detection limit for HCH and drins and to extremely low end concentrations for PCB and PCDD/F, even when the levels of halogenated hydrocarbons of the contaminated soil is very high.

The emission of dioxins and furans is considerably lower than the target value of 0.1 ng TEQ/m_0^3. The average emission was 0.035 ng TEQ/m_0^3. Low values were also measured for other components in the gas flow (CO, SO_2, HCl, HF, organic C and dust).

Ecotechniek has, based on the test results from the first test treatment with drins, a permit to treat soil contaminated with halogenated hydrocarbons. So far already 300,000 tons of soil contaminated with halogenated hydrocarbons are treated succesfully with the thermal treatment plant of Ecotechniek. In total Ecotechniek has already treated more than 3,000,000 tons of contaminated soils over the past 17 years with one mobile and two stationary thermal treatment plants.

REFERENCES

EPA, 1989, Handbook Guidance on Setting Permit Conditions and Reporting Conditions and Reporting Trail Burn Results, *Volume II of the Hazourdous Waste Inceration Guidance Series*. 1989.

Koopmans W.F., Reintjes R.C. 1988."Six years of experience in thermal soil cleaning", in K. Wolf et al. (eds.) *Contaminated Soil '88*, Kluwer Academic Publishers, Dordrecht, pp 787–796.

Noorman F, Vis P.I.M. 1995. "Succesful soil purification test by Ecotechniek", in W.J. van den Brink et al. (eds) *Contaminated Soil '95*, Kluwer Academic Publishers, Dordrecht, pp 1013–1022.

THERMAL DESORPTION TREATABILITY STUDIES: REMOVING CHLORINATED ORGANIC COMPOUNDS FROM SOILS

Jerome P. Downey, *Lawrence D. May*, and Kari D. Moore
Hazen Research, Inc., Golden, Colorado, USA

ABSTRACT: Hazen Research, Inc. has developed a bench-scale apparatus and methodology especially suited to thermal desorption treatability studies of media contaminated with chlorinated and recalcitrant compounds. A batch rotary kiln system is used to mix the media while maintaining it at relatively uniform temperature. Desorption characteristics of organic contaminants such as polychlorinated biphenyls (PCBs), dioxins, furans, petroleum-based organic compounds, and other volatile (VOC) and semivolatile (SVOC) organic compounds have been examined. Data show that most organic compounds can be desorbed from soils and sludges at temperatures ranging from 100 to 650°C and retention times of 5 to 30 minutes. Hazen's experience in performing thermal desorption studies on materials contaminated with chlorinated compounds is discussed. The experimental apparatus and methodology are disclosed, along with a discussion of the relationships between desorption efficiency and the pertinent process parameters.

INTRODUCTION

Technology. Thermal desorption technologies use direct or indirect heat to vaporize and remove organic compounds from soils, sludges, and other solid materials. Whereas incineration is intended to fully combust organic compounds, thermal desorption processes physically separate the contaminants from the media, while minimizing organic decomposition. Air or inert gas is normally used to convey the vaporized organic compounds from the contaminated media, but recycled process gas can also be used. Process gases containing vaporized organic compounds can be treated by a number of secondary treatment processes, including thermal oxidation, condensation, carbon adsorption, or chemical neutralization.

Objective. The main objective of most batch kiln thermal desorption test programs is to assess whether the cleanup criteria can be met; if so, the optimization of the process operating parameters becomes the focus of the test work. Cleanup standards for most sites are determined by the appropriate federal, state, or local regulations, or may even be determined on a site-by-site basis. Therefore, the cleanup goal may not be consistent from one site to the next. As a general guideline, the Universal Treatment Standard (40 CFR sec. 268.48) is often quoted.

Testing. Since 1992, Hazen has performed more than 40 studies on materials contaminated with various volatile and semivolatile organic compounds. These studies were conducted using representative samples of soils, sediments, and sludges from RCRA and CERCLA sites throughout the U.S. In many cases, the media tested contained more than a single contaminant.

THERMAL DESORPTION TESTING

Media and Contaminants. Soils and sludges are the most common media treated by thermal desorption technology. These often come from areas around historical chemical processing plants, drainage basins downstream of such plants, tailing ponds, and even from river dredgings. Contaminants can include inorganic species, organic species, and radionuclides. The organic compounds are classified as either volatile or semivolatile, depending on the boiling point. Generally, compounds that boil

TABLE 1. Typical boiling point ranges for common contaminants.

Contaminant Category	Boiling Point Range, °C
VOCs	<205
SVOCs	>205
2,3,7,8 TCDD	500d
PCBs	275 - 385

below 205°C are considered volatile while those that boil above 205°C are classified as semivolatile. Boiling points for the contaminants of concern are key information when considering the application of thermal desorption; Table 1 summarizes the boiling point ranges for common types of contaminants. Most troublesome organic compounds are amenable to thermal desorption in the range of 100 to 650°C. Some of the media tested, the contaminants of concern, and their concentrations in the untreated media are summarized in Table 2.

TABLE 2. Contaminant concentrations in untreated media.

Media	Contaminant	mg/kg
Soil/sludge	Bis(2-chloroethyl)ether	6.04 - 6.56
Soil/sludge	1,2-Dichlorobenzene	0.38 - 0.42
Soil/sludge	1,2-Bis(2-chloroethoxy)ethane	15.2 - 15.8
Soil	Pentachlorophenol	27.5 - 46.4
Soil	Total dioxins	0.35 - 0.54
Soil	Total furans	0.023 - 0.040
Soil	PCBs, Aroclor 1248	6.3 - 26,300
Soil/humus	PCBs, Aroclor 1248	20,000
Soil/clay	PCBs, Aroclor 1248	800
Sludge	PCBs, Aroclor 1248	280,000 - 340,000
Sediment	PCBs, Aroclor 1248	260

Apparatus. A 4-inch-diameter batch quartz kiln system (Figure 1) is used for bench-scale thermal desorption testing. Operating temperatures up to 1,000°C are attainable by indirectly heating the kiln in an insulated clamshell furnace. Raised dimples act as lifters to enhance the mixing and tumbling of the sample as the kiln rotates. Typical sample charges range from 300 to 1,000 grams, depending on the material to be tested and the planned operating conditions. Control parameters include temperature, pressure, kiln rotational speed, sweep gas composition, and gas flow rate. Process exhaust gases can be treated using condensers, carbon columns, or a thermal oxidizer. Alternatively, the exhaust gases can pass through an emission sampling train to quantify volatile and semivolatile organics, including PCBs, dioxins, and furans. Additionally, a portion of the exhaust gas can be analyzed for concentrations of O_2, CO_2, CO, SO_2, NO_x, and THC using continuous emissions monitors (CEM).

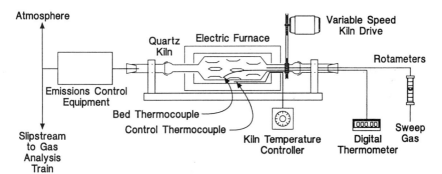

FIGURE 1. Batch rotary kiln system.

Methodology. For a typical thermal desorption test, a known mass of a contaminated soil or sludge is added to the kiln. The kiln is placed in the clamshell furnace and a thermocouple is positioned in the media to measure the temperature. Sweep gas (nitrogen or a blend of nitrogen and air) and the kiln rotation are started. In some tests, the time required for the media in the kiln to reach the designated temperature is defined as the retention time, at which point the heat is turned off and the kiln is removed from the furnace. In other applications, the media are maintained at the designated temperature for a set period of time. During a test, selected data such as temperatures and gas composition are continuously recorded by a data acquisition system. Data not electronically recorded (such as pressures and flow rates) are manually entered onto operational data sheets.

Following a test, the system is disassembled and the products recovered. The mass and/or volume of each product stream is quantified. General physical characteristics of each sample are recorded and chemical and physical analyses may be performed. Representative splits of the test products are packaged and saved for analyses according to the designated protocols for the specific program.

Advantages. The batch kiln system and test methodology offer distinct advantages over other practices. Only a small sample mass is needed to quantify the desorption characteristics of a contaminated soil or sludge. The actual temperature of the media is measured, providing more accurate information about the process requirements. The rotating kiln provides mixing not available in static applications, improving the potential for physically separating contaminants from the media. In addition, the potential for "clinkers" (agglomerates of material that become very hard on the outside and may not be sufficiently treated on the inside) to develop can be identified. On-line gas analysis can be performed and problems with plugging of the gas handling system can be seen. Finally, the methodology is economical; several tests can be run to assess organic removal as a function of time and temperature at a relatively low cost.

Quality Assurance. Several measures are employed to ensure that the data generated from a desorption test are consistently of high quality. The following protocols are followed:
- Representative portions of contaminated media and test products are analyzed according to recommended protocols (EPA SW-846).
- At least one replicate test is performed per program.
- Routine equipment calibration is conducted, including:
  - Verification of gauge, thermocouple, and flowmeter readings.
  - Confirmation of CEM measurements against certified span gas.
  - Verification of scale accuracy using calibrated weights.
- Equipment is precleaned and triple rinsed.
- Sample blanks are taken when emission sampling is performed.
- Analytical samples are collected and stored in precleaned amber glass jars with Teflon-lined lids, and refrigerated if appropriate.

For all tests, data are recorded both electronically and manually to document and verify the important parameters. In addition, a project journal is maintained to record aspects of a program not covered by routine data collection. All data and results are reviewed by senior members of Hazen's technical staff to ensure accuracy and completeness.

Results. Thermal desorption studies have been conducted with a variety of contaminant types and concentrations in many types of media. Some representative results are summarized in Table 3. Except where noted, retention time is defined as the period of time that the sample was held at the stated temperature.

The first three entries in Table 3 demonstrate the effectiveness of thermal desorption in removing organic compounds with relatively low boiling points, such as bis(2-chloroethyl)ether, 1,2-bis(2-chloroethoxy)ethane, and 1,2-dichlorobenzene. Nearly complete removal of each compound was achieved by processing the samples under relatively mild conditions, i.e., 10 minutes at 230°C.

The next three examples in Table 3 illustrate the effect of temperature on the removal of pentachlorophenol from samples of contaminated soil. Pentachlorophenol proved somewhat more difficult to remove, as relatively high concentrations of the compound remained with the solids after processing for 20 minutes at 340°C. Greater than 99% removal was obtained by processing the

TABLE 3. Summary of typical results for thermal desorption studies.

Number of Tests Summarized	Media	Time, min.	Temp., °C	Contaminant of Interest	Untreated, mg/kg	Treated, mg/kg	% Removal Efficiency
2	Soil/sludge	10	230	Bis(2-chloroethyl)ether	6.04 - 6.56	<0.011	99.81 - >99.99
				1,2-Dichlorobenzene	0.38 - 0.42	<0.043	88.68 - >99.99
				1,2-Bis(2-chloroethoxy)ethane	15.2 - 15.8	<0.014 - 0.036	99.76 - >99.99
3	Soil	10 - 30	340	Pentachlorophenol	27.5 - 46.4	0.99 - 1.39	96.40 - 97.00
3			455			<0.130	99.53 - >99.99
3			595			<0.130	99.53 - >99.99
3	Soil	10 - 30	340	Total dioxins	0.35 - 0.54	0.35	0.00 - 35.19
3			455			0.0007	99.80 - 99.87
3			595			0.0	>99.99
3	Soil	10 - 30	340	Total furans	0.023 - 0.040	0.006	73.91 - 85.00
3			455			0.0	>99.99
3			595			0.0	>99.99
1	Soil	GEC	540 - 595	PCBs, Aroclor 1248	26,300	<1	>99.99
5	Soil				6.3 - 120	<1	>99.99
2	Soil/humus				20,000	<1	>99.99
2	Soil/clay				800	<1	>99.99
2	Sludge				280,000 - 340,000	<1	>99.99
2	Sediment				260	<1	>99.99

Note: GEC - tests were run until gas evolution ceased.

material at 455°C; no measurable improvement was realized by increasing the temperature to 595°C.

The test series conducted with soil samples contaminated with dioxins and furans showed that greater than 99.99% removal of each compound was possible. As expected, the furans were more easily desorbed, and greater than 99.99% removal was achieved when the samples were processed at or above 455°C. For the dioxins, 595°C was required to exceed four nines removal efficiency.

In contrast to the other tests summarized in Table 3, the PCB-bearing media were typically processed at temperature until evidence of gas evolution had virtually ceased. This mode of operation was initially selected because the as-received samples had high initial moisture content and/or high levels of other, more volatile contaminants relative to the PCB concentrations (measured as Aroclor 1248). This methodology has proven exceptionally successful for PCB removal. Regardless of the media type or the initial PCB concentration, every sample that was processed within the temperature range of 540 to 595°C analyzed less than 1 mg/kg PCB.

Conclusions. The removal efficiency of any given contaminant will be affected by the type of matrix (sand, clay, soil, sludge, or sediment). A well-designed test program and experimental matrix are essential to determine the feasibility of applying thermal desorption technology. The batch kiln system and methodology can be used to establish the efficiency at which various organic compounds can be desorbed from a representative sample of media. Also, the requisite solids temperature and retention time can be expediently determined from batch kiln test results. However, it is important to understand the limitations of conducting small-scale tests in batch mode and the risks involved in extrapolating laboratory data to a commercial scale operation. Before implementing any thermal desorption process, it is advisable to conduct confirmatory tests in continuous mode using a larger, pilot scale system.

REFERENCES

Regulations for Implementing the Procedural Provisions of the National Environmental Policy Act. 40 CFR sec. 268.48.

U.S. Environmental Protection Agency. *Test Methods for Evaluating Solid Waste, Physical/Chemical Methods.* SW-846.

REMEDIATION OF CONTAMINATED SOILS USING THERMAL DESORPTION

Christine A. Nardini (Environmental Soil Management, Inc., Loudon, NH)
James C. O'Shaughnessy (Worcester Polytechnic Institute, Worcester, MA)
Robert D. Manz (Environmental Soil Management, Inc., Loudon, NH)

ABSTRACT: A three year research and development (R&D) project was undertaken from 1994 to 1997 demonstrating both pilot and full scale thermal desorption of soils contaminated with numerous compounds such as manufactured gas plant (MGP) wastes, low level PCBs and chlorinated solvents. The ability of a bench-scale test unit to predict the performance of full scale thermal desorption for treating contaminated soils was researched. The measure of treatability used for the pilot scale tests was removal efficiency of the contaminant(s) of concern. Upon demonstration of contaminant treatability in the pilot scale, soils were then treated in full scale thermal desorption units, both fixed and mobil facilities. Consistent successful treatment in the pilot scale unit, allowed many types of contaminated soil to be treated in the full scale plants. Removal efficiencies for coal tar contaminated soils were measured as reduction in total petroleum hydrocarbon (TPH) and polycyclic aromatic hydrocarbons (PAHs) concentrations. TPH and PAH removal efficiencies ranged from 98.97 to 99.99% and 95.44 to 99.99%, respectively. All soils contaminated with low level PCBs (<50 parts per million (ppm)) were remediated to below treatment standard of 1 ppm with removal efficiencies ranging from 90.11 to 99.47%. Soils contaminated with chlorinated solvents were successfully treated with removal efficiencies of 95.8 to 99.98% for trichloroethylene (TCE) and 93.5 to 99.98% for perchloroethylene (PCE).

INTRODUCTION

Environmental Soil Management, Incorporated (ESMI) began full scale thermal desorption remediation of soils contaminated with virgin petroleum fuels in 1992. In 1994, research began to evaluate the effectiveness of a bench-scale treatability test to predict the performance of full scale thermal desorption process for treating soils contaminated with a variety of contaminants. When treatability was predicted to be effective in the bench scale unit, full scale thermal desorption was performed on the same soil stream to ensure that treatment standards for the contaminant(s) of concern could be met. Original research focused on remediation of soils contaminated with used oils, followed by low level PCBs and coal tars. When the versatility and effectiveness of treatment was established, full scale technology was extended to remediation of chlorinated solvents.

A wide variety of soil origins and contaminant scenarios were considered under the R&D project. Soils originated from active and inactive industrial sites, former manufactured gas plants, abandoned illegal disposal sites and sites in the Comprehensive Environmental Response, Compensation and Liability Information System (CERCLIS) database. R&D pilot scale treatment was conducted at Environmental Soil Management, Incorporated in Loudon, New Hampshire. Full scale soil treatment was completed at ESMI facilities located in both New Hampshire and New York. Ultimately, the full scale technology was applied to successfully remediate soils contaminated with chlorinated solvents at a Superfund site in New York.

RESEARCH BACKGROUND

Based on original successful treatment of virgin petroleum products, ESMI applied for and was granted a R&D permit from New Hampshire Department of Environmental Services (NHDES). The research was overseen by Worcester Polytechnic Institute and NHDES. Initial stages of R&D work investigated the treatability of soils contaminated with used oil products, including those contaminated with low level (<50 ppm) polychlorinated biphenyls (PCBs)[1]. The objectives of the initial research on used oil contaminated soils were to ascertain the ability of the bench scale plant to predict full scale plant performance and to define optimum full scale plant operating conditions.(O'Shaughnessy, 1997)

As research continued, new soil contaminants were considered as candidates for thermal treatment. The scope of the R&D was expanded to study MGP waste contaminated soils, including a wide variety of waste sources such as coal tar, water-gas tar, oil-gas tar and tar emulsions. Upon completion of coal tar contaminated soil studies, research began on the treatability of other contaminants such as chlorinated solvents, waxes and greases. In New Hampshire, non-hazardous solvent contaminated soils were treated at a fixed base facility while in New York, treatment of chlorinated solvent contaminated soils was conduct onsite at a Superfund site using a transportable thermal desorption plant.

Bench Scale Research . A propane-fired furnace was designed and constructed for use as a static tray bench scale unit to test the treatability of contaminated soils (Figure 1). The bench scale unit was designed as a conservative predictor of the

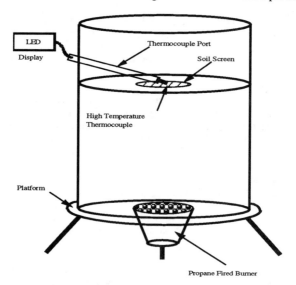

FIGURE 1. Schematic of the Bench Scale Dryer Unit

full scale plant's rotary dryer. Soil samples of approximately 50 grams were placed on the soil screen inside the bench scale unit. Treatment temperatures in the bench

---

[1] Soil contaminated with low level PCBs (<50 ppm) are not regulated by the Toxic Substance Control Act (TSCA).

scale unit were set based on the boiling points of the contaminant(s) of concern. TPH and other contaminants of concern were measured before and after test burning. The bench scale unit was used to predict the soil exit temperature required for effective plant scale treatment.

THERMAL DESORPTION TREATMENT PROCESS

Plant scale remediation entailed heating contaminated soils in a rotary dryer at temperatures between 450 and 900+ °F. The contaminants are driven from the soil and destroyed in a thermal oxidizer with a hydrocarbon destruction efficiency of greater than 99.6%. At fixed facilities, soils are stored and pre-processed in an enclosed storage building. Pre-processing includes blending, screening, crushing and removal of untreatable debris (wood, plastic, metal). Pre-processed soils are feed via conveyors to the thermal desorption plant. Soils are fed to the rotary dryer and tumble through for a designated residence time. The burner is fired to achieve a desired soil exit temperature. Treated soils exit the burner end of the dryer and are rehydrated in a pugmill. Treated soils are stockpiled, sampled and segregated awaiting post-treatment analytical results (Figure 2).

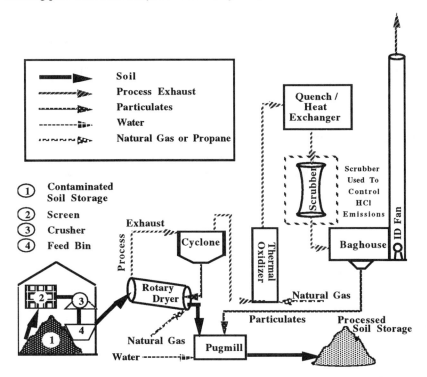

FIGURE 2. Thermal Desorption Plant Process Schematic

Soil contaminants volatilize from the soil in the dryer and exit with the flue gas stream at the feed end of the dryer. Flue gases flow from the dryer to a cyclone where initial particulate removal takes place. Particulate matter collected in the cyclone is returned to the burner end of the dryer for treatment and exits with the dryer discharge soils. Flue gases exit the cyclone and are treated for contaminant removal in a refractory lined thermal oxidizer. Thermal oxidizer temperatures are set to ensure complete combustion of the contaminants entering and range from 1500 to 1800 °F. Oxidizer exit gases are cooled in a quench/heat exchanger system followed by a baghouse for final particulate removal. If soils containing high concentrations of chlorinated solvents are to be treated, a wet scrubber is added to the thermal desorption train to control hydrochloric (HCl) acid gas emissions.

RESULTS

Bench Scale Treatment. The results of the bench and plant size thermal desorption runs were compared and the bench scale test methodology was found to be a conservative predictor of full plant size results. The conservative nature of the bench scale apparatus is explained by the static tray treatment area versus the turbulence of the full scale rotary dryer (O'Shaughnessy & Nardini, April, 1997). The plant scale dryer rotates as soil is fed with internal flights enhancing soil agitation. This mixing increases the surface area exposure of the soil particles, allowing for more effective mass transfer of contaminants from an adsorbed liquid phase on the soil particles to a vapor phase in the flue gas stream.

MGP Waste Contaminated Soils. MGP waste contaminated soils originating from a variety of sites were treated during the research period. TPH and total PAHs removal were used as the measures of treatment effectiveness. The pre- and post-treatment TPH and total PAHs of 11 batches of MGP contaminated soils that were tested are listed in Table 1. Treatment conditions (temperature, residence time and soil feed rate) were varied to evaluate plant scale treatability and optimum operating conditions.

TABLE 1: Pre and post treatment TPH and total PAH concentrations

Batch #	Pretreatment (mg/kg)		Post Treatment (mg/kg)	
	TPH	Total PAH's	TPH	Total PAH's
CT 1	1450	186	15	8.5
CT 2	555	229	<5	<0.4
CT 3.2	14475	2574	14	10.9
CT 3.3	14475	2574	<5	<0.7
CT 4.1	13308	2734	2	14.4
CT 4.2	13308	2734	17	1.3
CT 4.3	13308	2734	<5	0.2
CT 4.4	13308	2734	14	5.4
CT 5.1	9753	1249	6	6.2
CT 5.2	9753	1249	7	5.1
CT 5.3	9753	1249	<5	<0.2

For all batches, the TPH treatment standard was 25 ppm and the PAH treatment standards varied by individual compound from 0.7 ppm [e.g., benzo(a)pyrene] to 1,000 ppm [e.g., anthracene]. The resulting treatment efficiencies varied from a low of 98.97% to 99.99% for TPH removals. Total PAH removals varied from a low of 95.44% to 99.99%, under the various operating conditions. The removal of individual PAH compounds to below treatment standards was successful a total of 98.6% of the time. Of the 216 PAH measurements only three were at concentrations greater than the specific treatment standard.

PCB Contaminated Soils. PCB contaminated soils from a number of different industrial sites were treated during the research period. Under existing regulations only soils with less than 50 ppm PCB were eligible to be remediated at the existing fixed facilities. Dryer temperatures were in excess of 850 °F, and dryer residence times were 10 minutes. In all cases the soils were treated to PCB contamination levels of less than 1.0 ppm. The treatment standard for PCB contaminated soils in both New Hampshire and New York is 1.0 ppm. Table 2 features treatment results recorded during a thermal desorption plant performance test at the New York fixed facility. Soils used in this performance test originated from a former transformer maintenance area at a utility company. Table 2 data and other data collected during the research period supports the conclusion that a treatment standard of 1 ppm residual PCBs in treated soil is consistently achievable.

TABLE 2: Pre- and Post-treatment PCB concentrations

Batch #	Pre-treatment (mg/kg)	Post-Treatment (mg/kg)	% Removal
P 1.1	11.8	0.32	99.29
P 2.1	11.2	0.45	95.98
P 3.1	11.3	0.06	99.49
P 4.1	7.9	0.63	92.03
P 2.1	11.6	0.39	96.64
P 2.2	9.4	0.39	95.85
P 2.3	9.6	0.32	96.67
P 2.4	12.3	0.11	99.11
P 3.1	31.2	0.88	97.18
P 3.2	12.4	0.91	92.66
P 3.3	8.9	0.88	90.11
P 3.4	6.9	0.29	95.80

Chlorinated Solvent Contaminated Soil. A transportable thermal desorption plant was used to remediated solvent contaminated soil at the American Thermostat Superfund site in South Cairo, New York. The contaminants of concern at the site were TCE and PCE. Lower concentrations of TCE and PCE, as well as other chlorinated solvents, have been successfully remediated during the research period at the New Hampshire facility. Chlorinated solvent contaminated soils are well suited for thermal desorption. The lower temperatures needed for desorption insure that soil remediation standards can be readily be achieved. At the American Thermostat site, the average dryer soil exit temperature was 475 °F. The TCE and PE soil concentrations at this site were high enough to require the addition of a wet scrubber to the thermal desorption train for control of acid gases. Thermal oxidizer temperatures averaged 1800 °F. Lower dryer soil exit temperatures reduce fuel costs for soil treatment, however, the cost savings are offset by the higher temperature requirements in the oxidizer. The treatment standard for TCE and PCE contaminated soils were 400 ppb and 1,000 ppb, respectively. Table 3 presents some of the treatment results for the American Thermostat site.

TABLE 3: Pre- and Post-treatment TCE and PCE concentrations

Batch #	Pre-treatment (µg/kg)		Post-treatment (µg/kg)	
	TCE	PCE	TCE	PCE
C 25.1A	14000	140000	22	290
C 25.5A	2000	660000	4	160
C 27.1A	27000	80000	4	220
C 27.2A	6000	34000	55	15
C 27.3A	5800	43000	55	55
C 27. 4A	5400	41000	54	44
C27.1B	2700	11000	55	160
C27.2B	5300	5300	3	140
C 27.3B	1300	2000	55	130

CONCLUSIONS
The following conclusions were reached based on the analyses of data collected in this study.

- The bench scale dryer unit is an effective indicator of full plant scale performance.
- Thermal desorption was effective in remediating the MGP waste contaminated soils, PCB contaminated soils, and chlorinated solvent contaminated soils.
- Thermal desorption temperatures between 800°F and 900°F were sufficient to reduce TPH and PAH concentrations in MGP contaminated soils to below treatment standards.
- Thermal Desorption of PCB contaminated soils required temperatures in excess of 825 °F.
- Soils having low concentrations of PCBs (<50 ppm) were remediated to less 1 ppm .
- Thermal Desorption of TCE and PCE contaminated soils required temperatures of 475 °F to be remediated to below standards of 400 ppb and 1000 ppb, respectively.

REFERENCES
O'Shaughnessy, J.C. 1997. *Thermal Desorption of Used Oil Contaminated Soils (Final Report)*. Department of Civil & Environmental Engineering, Worcester Polytechnic Institute, Worcester, MA.

O'Shaughnessy, J.C., and Nardini, C. June, 1997. "Heating the Tar Out", *Soil & Groundwater Cleanup*. June, 1997, pp. 7-13.

O'Shaughnessy, J.C., and Nardini, C. April, 1997. "Remediation of Coal Tar Contaminated Soils by Thermal Desorption." In D. K. Moon (Program Chair), *Proceedings of the Ninth Annual EnviroExpo - New England Conference,* pp. 206-215. Longwood Environmental Management, Belmont, MA.

Farmer, J.K. 1996. *American Thermostat LTEVF Performance Test.* O'Brien & Gere Technical Services, Inc., Cheshire, CT.

A COMPARISON OF INSITU SOIL MIXING TREATMENTS

Steven R. Day (Inquip Associates, Denver, Colorado)
Larry Moos (Argonne National Laboratory-East, Chicago, Illinois)

ABSTRACT: A series of bench-scale tests were performed to determine the suitability of insitu soil mixing to treat clayey soils contaminated with chlorinated compounds and solvents at a site also used as a radioactive storage yard at Argonne National Laboratory-East near Chicago, Illinois. Soil mixing with hot air injection was proposed as the initial treatment and as the delivery system for a number of polishing techniques including zero valent metals, biological agents, potassium permanganate, humic acid, and soil vapor extraction. Batch tests were performed on soils samples using simulated soil mixing and hot air injection. Additional tests were performed on several polishing technique, after a portion of the volatile contaminates were removed by the hot air injection and soil mixing. The results were compared for removal efficiency and cost.

It was found that about 50 to 80% of the contaminate mass was removed by hot air injection. The polishing techniques were used to increase treatment efficiencies to 70 to 99%. Soil vapor extraction and zero valent metals (iron filings) treatments provided the greatest short- term treatment efficiency. Proprietary combinations of treatments such as biodegradation/zero valent metals and humic acid/biodegradation produced treatment efficiencies near 90% in about two weeks with the promise of continued improvement with additional time.

INTRODUCTION

The 317 Area French Drain had been used as a disposable area for waste fluids at the Argonne National Laboratory-East during the late 1950's, before current waste management practices were instituted. Workers previously poured liquid wastes into gravel filled trenches or into tanks where they slowly leaked into the soil. As a result, total volatile organic chemical (VOC) contamination in the soil was recorded as high as 3,500 mg/kg. The contamination was found up to 30 feet (9 m) deep over an area of about 1 acre (4000 m^2). The soils at the site are hard and impermeable glacial clays and till. The site is currently used as a staging area to temporarily store radioactive wastes prior to off-site disposal. A voluntary cleanup was designed to remediate the 317 Area.

Insitu soil mixing is a construction technique which has increasingly been relied upon for the remediation of contaminated soils. Depending on the application, large or small (4 to 0.3 m) mixing augers can be used to inject cement, bentonite or other reagents to modify soil properties and thereby remediate contaminated soils and sludges. A major advantage of the method is the capability to inject a variety of reagents and treat soils at depth (up to 35 m deep) without excavation, shoring or dewatering. Advantages of soil mixing over alternative technologies include lower cost, less exposure of wastes to the surface environment and eliminating off-site disposal. These advantages convinced Argonne National Laboratory-East to use a system of insitu soil mixing coupled with hot air injection, as shown in the sketch for the 317 Area soil remediation.

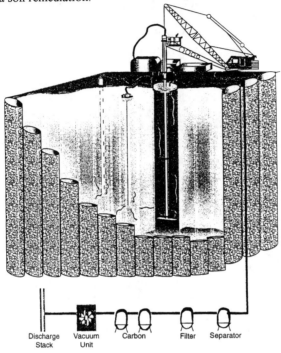

Discharge   Vacuum   Carbon   Filter   Separator
  Stack   Unit

Process Diagram for Insitu Soil Mixing

In order to properly evaluate the potential success of soil mixing at a particular site it is generally advisable to conduct a pre-construction laboratory testing program to establish treatment capabilities, optimize reagent usage, and better define costs. Such a study was performed for the 317 Area project. The purpose of this study was to determine the total percentage of contaminates which could be removed.

METHODS AND MATERIALS

The methods and materials tested were selected to mimic field construction based on the author's previous experience, economic considerations, and local availability. The following materials were included in the testing:

MATERIAL	SOURCE
contaminated soil	samples from the site obtained via drilling
hot air (temperature = 100^0C)	laboratory generator
potassium permanganate	Carus Chemicals Co.
humic acid (Humasorb)	Arctech, Inc. (proprietary formulation)

iron filings	Peerless Metals, Inc.
biological/metals (Daramend)	W.R. Grace (proprietary formulation)
guar gum polymer	Rantec, Inc.
methanol	laboratory stock
formate	laboratory stock
phenol	laboratory stock

The testing was performed in a series of steps 1) first the contaminated materials were tested for total VOCs, 2) the contaminated soils were next treated with hot air injection and mixing for 30, 60, 90, and 120 minutes and retested, and 3) each of the polishing techniques was mixed with identical samples of the hot air treated soil and retested at various proportions and intervals.

The soils were tested for 28 different volatile compounds in accordance with EPA Method 8260A. Of the 28 volatile compounds only six had significant concentrations and these were carbon tetrachloride, chloroform, MIBK (4-methyl-2-pentanone), tetrachloroethene, 1,1,1 trichloroethane, and trichloroethene. Two other compounds, methyl chloride and vinyl chloride were also tracked during the testing due to a concern that some of the treatments could create these compounds if misapplied.

The efficiency of treatment was calculated for each compound and for the sum of the compounds as a "treatment efficiency". The specific compound treatment efficiency was calculated as follows:

$$\text{Specific Compound Treatment Efficiency} = 1 - \frac{\text{post-treatment concentration (mg/kg)}}{\text{pre-treatment concentration (mg/kg)}}$$

For the eight compounds previously discussed, an average treatment efficiency was calculated as follows:

$$\text{Average Treatment Efficiency} = 1 - \frac{\text{sum post-treatment concentration (mg/kg)}}{\text{sum pre-treatment concentration (mg/kg)}}$$

The higher the treatment efficiency the more contamination removed by the treatment. A method with a treatment efficiency of 100% removed all of the contamination.

RESULTS AND DISCUSSION

Soil Characterization. The untreated site soils exhibited significant contamination, especially with trichloroethene. A summary of test results on the untreated soil is reproduced in the table below.

Chemical Compound	Concentration (mg/kg)	Physical Properties	Result
Carbon Tetrachloride	22	PH	7.6
Chloroform	5.2	Moisture Content (%)	19
Methylene Chloride	0.4	Loss on Ignition (%)	2
MIBK	6.2	Unit Weight (kN/m^3)	21.0
Tetrachloroethene	160	% Gravel	3
1,1,1 Trichloroethane	7.6	% Sand	15
Trichloroethene	16	% Silt	38
Vinyl Chloride	0	% Clay	44
Total of 8 VOCs	217.4	Plastic Index (%)	16

Hot Air Treatment. The soils were treated by mixing and injecting hot air in an open container. About 2500 gm of soil were mixed in a modified laboratory-type blender at 60 rpm with hot air (100^0C) injected at 10 liters per minute. Samples of the treated soils were taken at intervals to evaluate VOC removal efficiency. After 90 minutes of mixing in the laboratory, a treatment efficiency of about 70% was obtained which is consistent with previous field experience for penetration and initial mixing of a large diameter soil column. Six duplicate batches were mixed and tested with typical results as shown in the table below.

Compound	Batch 2 (mg/kg)				Batch 4 (mg/kg)	Batch 5 (mg/kg)
Mixing Minutes→	30	60	90	120	90	90
Carbon Tetrachloride	21	22	18	18	13	12
Chloroform	1.4	1.4	0.81	0.74	0.7	0.57
Methylene Chloride	0	0	0	0	0	0
MIBK	4.3	2.9	2.6	2.2	2.3	1.8
Tetrachloroethene	58	48	39	38	33	30
1,1,1 Trichloroethane	1.7	1.8	1.4	1.7	1	1.2
Trichloroethene	8.8	8.1	5.1	4.6	4.4	3.2
Vinyl Chloride	0	0	0	0	0	0
Total of 8 VOCs	95.2	84.2	66.9	65.2	54.4	48.8
Treatment Efficency (%)	56	61	69	70	75	77

Notes: 1) 0 = not detected
         2) Pre-treatment concentration = 217 mg/kg

Polishing Techniques. Six polishing techniques were used to treat the contaminated soil after hot air mixing and injection. A number of vendors participated in the study by providing materials, expertise, and in the case of W.R. Grace, active involvement in the laboratory mixing. The iron filings and Daramend materials are solids and therefore, were blended into a slurry using a water-guar gum solution to suspend and carry the solids for mixing with the soil. For the iron filings, a series of trial mixtures were made with progressively finer filings until the workable slurry was made composed of iron filings (finer than #50 sieve size), guar gum and water. Potassium permanagate (KMn04) was mixed into a solution with water at the solubility limit at about 5%. The Humasorb product is a thick solution and was usable without amendments. A formate solution was mixed with the soils to enchance biological activity. The tests were performed on mixtures which were blended and then stored in a dark, controlled atmosphere for 2 days to 2 weeks to model insitu conditions.

A summary of the results of the testing are presented in the table below:

Compound	KMn04	SVE	Iron Filings	Humic Acid	DARA-MEND	Bio-Treat-ment
Additive→	10%	7 Days	5% & Guar	L-3	D92 & Guar	Formate Solution
Carbon Tetrachloride	12	0	0.028	1.9	0.4	5.4
Chloroform	0.371	0.043	0.3	0.50	3.4	1.3
Methylene Chloride	0	0.15	0.063	0	0.041	0.058
MIBK	1.8	0.63	1.1	0.61	0.33	0.31
Tetrachloroethene	27	2.7	1.3	9.5	18	23
1,1,1 Trichloroethane	0.87	0.52	0	0.24	0.5	0.51
Trichloroethene	2.3	0.11	0.37	1.00	5	4.4
Vinyl Chloride	0	0	0	0	0	0
Total of 8 VOCs	44.6	4.15	3.16	13.75	27.67	34.98
Treatment Efficency (%)	79	98	98	94	87	84

Notes: 1) SVE = soil vapor extraction

The results shown above indicate that insitu soil mixing with hot air can be significantly improved by the use of polishing techniques. Workability concerns can be addressed by making slurries with water and guar gum that enhance the process and the treatment. While these results are encouraging, several of the techniques should show continued improvement as biological organisms continue to degrade the contaminates and these techniques include Daramend and humic acid.

Specific Compound Treatment Efficiency. In general, all of the compounds were reduced by the hot air and polishing treatments. Different treatments produced differing treatment efficiencies with specific compounds. For example, carbon tetrachloride was more difficult to treat with hot air injection and potassium permanganate but easily treated with soil vapor extraction and iron filings. Trichloroethene was the found in the largest concentration of the contaminants, but was reduced to low levels by all of the treatments. While not detected in the characterization of the untreated soils, very small concentrations (< 0.2 mg/kg) of methylene chloride were detected in all of the polishing treatments except potassium permanganate. No vinyl chloride was detected in any treatment. A summary of specific compound treatment efficiencies are shown in the table below.

Compound	Specific Compound Treatment Efficiency (%)						
	Hot Air Only	KMn04	SVE	Iron Filings	Humic Acid	DARA-MEND	Bio-Treat-ment
Carbon Tetrachloride	23	45	100	100	91	98	75
Chloroform	85	86	99	94	90	35	75
Methylene Chloride	0	0	0	0	0	0	0
MIBK	65	71	90	82	90	95	95
Tetrachloroethene	80	83	98	99	94	89	86
1,1,1 Trichloroethane	86	89	93	100	97	93	93
Trichloroethene	71	86	99	98	94	69	73
Vinyl Chloride	0	0	0	0	0	0	0

Notes: 1) 0% = no change in concentration
       2) 100% = complete removal of compound

Comparative Costs. A cost analysis was performed using the data obtained, above, to estimate the cost of each of the different techniques. The cost comparison was based on treating 20,000 cubic yards (15,000 m^3) of contaminated soils. Each vendor was contacted and allowed to optimize the expected reagent usage and cost. A summary of the estimated costs are shown in the following table.

Technique	Initial Mixing with Hot Air Cost (\$/cy)	Polishing Cost (\$/cy)	Total Cost (\$/cy)
Soil Vapor Extraction	30	45	75
Zero Valent Metals	30	65	95
Daramend	30	31	61
Humic Acid	30	58	88
KMn04	30	105	135

CONCLUSIONS

Insitu soil mixing can be used to remediate clayey soils with relatively high levels of VOC contamination. Hot air injection can be augmented with a number of highly effective polishing techniques to provide greater than 90% removal of volatile compounds.   This type of remediation can be relatively inexpensive when compared to other methods.

In Situ Thermal Desorption Using Thermal Wells and Blankets

H. J. Vinegar, E. P. de Rouffignac, Shell E&P Technology Company
G. L. Stegemeier, GLS Engineering
J. M. Hirsch and F. G. Carl, TerraTherm Environmental Services

ABSTRACT:
    A new in situ thermal desorption (ISTD) process has been demonstrated to remove high-concentration PCB contaminants such as Aroclor 1260 even from tight clay soils. The ISTD technology uses thermal blankets for surficial contamination and thermal wells for deeper contaminated soils.

INTRODUCTION

    The difficulty in remediating the large number of sites contaminated by toxic, carcinogenic, or radioactive chemicals has generated interest in developing improved processes for cleaning these sites. *In-situ* processes, which either destroy contaminants in place or remove them without disturbing the soil, offer distinct advantages over those requiring excavation: they minimize exposure to workers and the public and they can reduce costs of full remediation.

    One of the most versatile and effective of these *in-situ* processes is *In-Situ* Thermal Desorption (ISTD), in which heat and vacuum are applied simultaneously to subsurface soils. For shallow contamination (less than 2 ft depth) heat and vacuum are applied by thermal blankets [1-3] and for deeper contamination thermal wells are used. [4-5]

    The ISTD process has a high degree of social acceptance because it is a clean, closed system that is reliable and fast. It destroys a wide range of VOC and SVOC pollutants in place without having to move the soil. Thermal wells can be used under roads, foundations and other fixed structures. If required, the thermal wells can be slanted or drilled horizontally. The operations are low profile, quiet, and cause little disruption of adjoining neighborhoods. There are no odors.

    The ISTD process possesses a high removal efficiency because the narrow range of soil thermal conductivities provides excellent sweep efficiency and because its high operating temperature assures complete displacement efficiency of contaminants in the gas phase. Unlike fluid injection processes, ISTD is applicable to tight soils, clay layers, or in soils with wide variations in permeability and water content.

DESCRIPTION OF PROCESS

    For ISTD-Thermal Blankets heat is supplied to the soil by downward conduction from a surface heater and vaporized products are collected under an impermeable sheet into a vacuum system (**Figure 1a**). The ISTD-thermal blanket is an 8-ft by 20-ft stainless steel box covering 160 square feet. Each blanket has 33 heating rods which together apply 100 Kw of radiant energy to the soil surface. The near surface soil is raised to about 1600

°F by the radiant energy and the heat front then propagates downward into the soil by thermal conduction. A 12-inch thick insulation layer in the stainless steel box covers the heating elements, reducing heat losses out the top to less than 10%. Multiple blankets are installed next to one another along the 20 ft edges to treat large areas. Finally, a silicone rubber vapor barrier is installed over the multiple blanket set and vapors are collected by the vacuum system. As the contaminants in the soil are drawn upwards, the high temperatures in the soil near the heating elements convert the majority of contaminants (typically 98-99%) to CO_2 and water vapor.

For ISTD-Thermal Wells, an array of heater/vacuum wells is placed vertically in the ground in triangular patterns (**Figure 1b**). The wells are equipped with high-temperature electric heaters (1700°F) and connected to a vacuum blower. As heat is injected and soil temperatures rise, the vaporized products are drawn into the wells by the applied vacuum. Just like for the Thermal Blanket, contaminants are mostly converted in the soil near the heater well to CO_2 and water vapor.

For both ISTD-Thermal Blankets and ISTD-Thermal Wells, produced vapors are treated further in Mobile Vapor Treatment System consisting of a flameless thermal oxidizer with >99.99% Destruction Removal Efficiency (DRE), followed by two carbon beds in series. The overall DRE in the hot soil and the mobile vapor treatment system is extremely high. A continuous emission monitoring system in the stack measures total hydrocarbons, wet and dry oxygen, carbon monoxide and carbon dioxide.

DEMONSTRATIONS

Both Thermal Wells and Thermal Blankets have been demonstrated to be highly effective in removing PCBs, pesticides, and chlorinated solvents from soils. Commercial remediation services are now available through TerraTherm Environmental Services, a wholly owned subsidiary of Shell Technology Ventures.[6]

A recent field demonstration is described in which ISTD–Thermal Blankets and Wells were shown to remediate high concentration PCB contamination from shallow and deep clay soils.[7-8] The demonstrations were conducted at the Missouri Electric Works (MEW) Superfund site in Cape Girardeau, Missouri. The MEW site was contaminated with PCBs in both shallow and deeper soils during past operations including selling, servicing, and remanufacturing transformers, electric motors and electrical equipment controls, and recycling dielectric fluids containing PCBs. The site clean up level specified in the ROD was 2 ppm total PCBs. The field demonstration was carried out in an area devoid of underground gas, water, or electric utilities. The natural stratigraphy is brown clay soil; the water table is located about 40 feet below ground surface.

The objectives of the MEW field test included (1) clean up clay soils contaminated with high concentrations of the highest boiling point PCB, Aroclor 1260, to less than 2 ppm, (2) demonstrate that stack discharges were in compliance with state and federal standards for PCBs and polychlorinated dibenzodioxins/polychlorinated dibenzofurans (PCDDs/PCDFs), and (3) obtain a system DRE for PCBs greater than 99.9999%. The demonstration was conducted in support of TerraTherm's application for a modification of the TSCA permit for alternate PCB treatment. EPA Region VII and the Missouri Department of Natural Resources (MODNR) monitored the demonstration.

For the blanket demonstration, two heater blankets were placed side-by-side in an area where PCB concentrations had averaged 510 ppm near the surface and 2.7 ppm at 12-18 inches. The target treatment depth was 18 inches.

For the well demonstration, twelve heater/vacuum wells were completed in a multiple triangular array with a 5-foot well spacing to a depth of 12 feet. As shown in **Table 1**, the area chosen had PCB contamination as high as 19,900 ppm near surface and still above 2 ppm at the target depth of 10 ft. **Figure 2** shows contour values of initial PCB concentration. **Figure 3** shows the well test pattern and post treatment sampling locations. During the demonstration, electrical resistance heating and vacuum were applied to the wells for a period of 42 days. Soil temperatures were monitored throughout the experiment, and soil samples were taken with a split-spoon sampler fitted with six-inch brass coring sleeves to verify the removal of contaminants.

RESULTS AND DISCUSSION

In the blanket demonstration, the soil was successfully remediated to a depth of 18 inches. The upper one foot of soil was non-detect for PCBs (i.e. < 33 ppb) and averages at all depths met the remedial objective of < 2 ppm.

In the well demonstration, temperatures above 1000°F were achieved in the interwell regions. The temperature history at the centers of the triangles near the middle of the heated interval (depth 0-6 ft) is shown in **Figure 4**. There are three distinct phases in the heating process. During the first phase, the soil temperature rose nearly to the boiling point of water in about 250 hours from the start of heating. During the second phase, water boiling occurred and the temperature remained near the boiling point of water. The duration of this phase was dependent on the pore water content and the water inflow. This phase ended at between 560 and 630 hours, with the center and adjoining triangles drying first and the outer triangles later. During the third (superheating) phase, soil temperatures rose rapidly until the heaters were turned off on day 42. Maximum temperatures over 1000°F were reached at the center of the triangles, and about 50% of the volume was over 1100°F. **Figure 5** shows the maximum temperatures reached along profile 7-G.

As shown in **Table 1**, sampling after 42 days showed complete clean up of all contaminants to levels below 1 ppm to a depth of 10 ft below ground surface. Eighty-one samples in the treatment zone were non-detect (<33 ppb) by EPA Method 8080. Sampling down to 15 ft in the center of the treated zone showed that no vertical migration of contamination had occurred.

Emission stack sampling by EPA methods demonstrated that the discharge of PCBs and combustion byproducts complied with state and federal ambient air requirements. Stack testing of emissions from the process indicated 99.9999998% DRE of the PCBs by combined in-situ and surface treatment. The sampling and analysis results of the EPA Method 680 analysis performed on the stack samples indicates that a total of 0.10 mg of PCB was emitted from the stack from a conservative estimate of 40 kilograms of PCB in the treated area.

Post-treatment soil samples composited vertically and areally from the treated zone were analyzed for PCDD and PCDF and exhibited TEQ levels from non-detect to 0.00684 ppb, with an average of 0.003 ppb. This is below the background level of 8 ppt for uncontaminated soil in North America.

CONCLUSIONS

In summary, the ISTD-Thermal Blanket and Thermal Well technologies was effective in achieving the site remediation goals of <2.0 ppm at all locations sampled within the treatment zone. The ISTD technologies volatilized, extracted, and effectively treated high concentrations of the highest boiling point PCBs from dense clay overburden soils without excavation. The discharge of PCBs and combustion by-products detected during stack testing activities conducted on the mobile vapor treatment system confirmed that the ISTD process did not adversely impact ambient air quality.

REFERENCES

1. Stegemeier, G. L. and Vinegar, H. J. (1995), "Soil Remediation By Surface Heating and Vacuum Extraction," SPE 29771, SPE/EPA Exploration and Production Environmental Conference, Houston, Texas, March 27-29.

2. Iben, I.E.T. et al. (1995), "Thermal Blanket for In-Situ Remediation of Surficial Contamination: A Pilot Test," Environmental Science and Technology, Vol 30, No. 11, Nov, 1996.

3. Sheldon, R.B. et al. (1996), "Field Demonstration of a Full-Scale In Situ Thermal Desorption System for the Remediation of Soil Containing PCBs and Other Hydrocarbons," HAZWaste World Superfund XVII, Washington, D.C., October 15-17.

4. Vinegar, H. J., Stegemeier, G. L., de Rouffignac, E. P., and Chou, C.C. "Vacuum Method for Removing Soil Contaminants Utilizing Thermal Conduction Heating," U.S. Patent Nos. 5,190,405, issued March 2, 1993, and 5,318,116, issued June 7, 1994.

5. Vinegar, H. J et al. (1997), "Remediation of Deep Soil Contamination Using Thermal Vacuum Wells, " SPE 39291, 1997 Soc. Pet. Eng. Annual Technical Conference, San Antonio, Texas, October 5-8.

6. "Low-Cost Solutions to Difficult Cleaning Problems," TerraTherm Environmental Services, Inc., an affiliate of Shell Technology Ventures, Inc., Houston, Tx.

7. Vinegar, H. J. et al. (1997), "In Situ Thermal Desorption (ISTD) of PCBs," HAZWaste World Superfund XVIII, Washington, D.C., December 2-4.

8. France-Isetts, P. (1998), "In Situ Thermal Blankets and Wells for PCB Removal in Tight Clay Soils," CLU-IN: Tech Trends, EPA Region 7, February.

Table 1. Thermal Wells Pre-Demo Soil Sampling Results

Boring ID	Sample #	Depth (ft)	ATAS Lab Result PCB Concentration (ppm)
TW-1	S1-A	0.0-2.0	1590
	S1-B	2.0-3.4	357
	S2-A	3.4-5.4	<0.5
	S2-B	5.4-8.1	<0.5
	S5	8.2-10.0	NA
	S6	10.0-12.0	13.5*
TW-3	S1-A	0.2-2.2	2190
	S1-B	2.2-4.2	59.5
	S2-A	4.2-6.2	ND
	S2-B	6.2-8.2	ND
	S5	8.2-10.0	6.37*
	S6	10.0-12.0	4.34*
TW-3T	S1	0.0-0.5	614
	S2	0.5-1.0	2970
	S3	1.0-2.0	16.5
	S4	2.0-4.0	0.694
	S5	4.0-6.0	4.42
	S6	6.0-8.0	2.32
	S7	8.0-10.0	0.084
	S8	10.0-12.0	<0.033
	S9	12.0-14.0	<0.033
	S10	14.0-16.0	<0.033
TW-4	S1-A	0.2-2.2	3030/8030
	S1-B	2.2-4.2	NA
	S2-A	4.2-6.2	0.913
	S2-B	6.2-8.2	<0.50
	S5	8.2-10.0	0.418
	S6	10.0-12.0	3.63*
TW-6	S1-A	0.2-2.2	299
	S1-B	2.2-4.2	393
	S2-A	4.2-6.2	342
	S2-B	6.2-8.2	114
	S3-A	8.2-10.2	<0.50
	S3-B	10.2-12.2	0.973
TW-6T	S1	0.0-0.5	19900
	S2	0.5-1.0	2190
	S3	1.0-2.0	885
	S4	2.0-4.0	234
	S5	4.0-6.0	46.2
	S6	6.0-8.0	5.33
	S7	8.0-10.0	0.061
	S8	10.0-12.0	0.158
	S9	12.0-14.0	0.22
	S10	14.0-16.0	0.043
TW-7	S1-A	0.2-2.2	25.7
	S1-B	2.2-4.2	<0.50
	S2-A	4.2-6.2	11.4
	S2-B	6.2-8.2	<0.50
	S3-A	8.2-10.2	<0.50
	S3-B	10.2-12.2	<0.50
TW-10	S1-A	0.2-2.2	2.39
	S1-B	2.2-4.2	<0.50
	S2-A	4.2-6.2	<0.50
	S2-B	6.2-8.2	<0.50
	S5	8.2-10.0	0.475
	S6	10.0-12.0	<0.50

Boring ID	Sample #	Depth (ft)	ATAS Lab Result PCB Concentration (ppm)
TW-13	S1	0.2-2.2	253
	S2	2.2-4.2	2.23
	S3	4.2-6.2	0.099
	S4	6.2-8.2	NA
	S5	8.2-10.2	<0.50
	S6	10.2-12.2	<0.50
TW-14	S1	0.2-2.2	4100
	S2	2.2-4.2	1060
	S3	4.2-6.2	276
	S4	6.2-8.2	67.5
	S5	8.2-10.2	3.98
	S6	10.2-12.2	<0.50
TW-14T	S1	0.0-0.5	9210
	S2	0.5-1.0	1450
	S3	1.0-2.0	984
	S4	2.0-4.0	1470
	S5	4.0-6.0	134
	S6	6.0-8.0	11.8
	S7	8.0-10.0	<0.033
	S8	10.0-12.0	<0.033
	S9	12.0-14.0	<0.033
	S10	14.0-16.0	<0.033
TW-15	S1	0.2-2.2	93.8
	S2	2.2-4.2	5.3
	S3	4.2-6.2	NA
	S4	6.2-8.2	2.03
	S5	8.2-10.2	NA
	S6	10.2-12.2	8.35*
TW-16	S1	0.2-2.2	61.8
	S2	2.2-4.2	NA
	S3	4.2-6.2	1.14
	S4	6.2-8.2	NA
	S5	8.2-10.2	3.11
	S6	10.0-12.0	1.22 (10.2)*
TW-17	S1	0.0-0.5	93.7
	S2	0.5-1.0	2530
	S3	1.0-2.0	<0.50
	S4	2.0-4.0	1.66
	S5	4.0-6.0	<0.50
	S6	6.0-8.0	<0.033
	S7	8.0-10.0	0.146
	S8	10.0-12.0	<0.033
	S9	12.0-14.0	1.27
	S10	14.0-16.0	0.395
TW-18	S1	0.0-0.5	9090
	S2	0.5-1.0	1690
	S3	1.0-2.0	762
	S4	2.0-4.0	450
	S5	4.0-6.0	293
	S6	6.0-8.0	1.53
	S7	8.0-10.0	0.421
	S8	10.0-12.0	0.136
	S9	12.0-14.0	0.051
	S10	14.0-16.0	<0.033

NOTES:
1. NA denotes that sample analysis results are not available at this time.
2. NS indicates no sample was collected.
3. Samples taken at locations of thermal wells, e.g., TW-1 as shown on Figure 3.
4. "T" denotes twinned geoprobe location.
5. * Split spoon sample, possible contamination from shallow cavings
6. PTW-8 samples were collected adjacent to the PTW-1 location.

Table 2. Thermal Wells Post-Demo Soil Sampling Results

Boring ID	Sample #	Depth (ft)	ATAS Lab Result PCB Concentration (ppm)	Boring ID	Sample #	Depth (ft)	ATAS Lab Result PCB Concentration (ppm)
PTW-1	S1	0.0-0.5	<0.033	PTW-8	S1	0.0-0.5	<0.033
	S2	0.5-1.0	<0.033		S2	0.5-1.0	<0.033
	S3	1.0-1.5	<0.033		S3	1.0-2.0	<0.033
	S4	1.5-2.0	<0.033		S4	2.0-4.0	<0.033
	S5	2.0-2.5	<0.033		S5	4.0-6.0	0.036
PTW-2	S1	0.0-0.5	<0.033	PTW-9	S1	0.0-0.5	<0.033
	S2	0.5-1.0	<0.033		S2	0.5-1.0	<0.033
	S3	1.0-2.0	<0.033		S3	1.0-2.0	<0.033
	S4	2.0-4.0	<0.033		S4	2.0-4.0	<0.033
	S5	4.0-6.0	<0.033		S5	4.0-6.0	<0.033
	S6	6.0-8.0	<0.033		S6	6.0-8.0	<0.033
	S7	8.0-9.9	<0.033		S7	8.0-9.9	<0.033
PTW-3	S1	0.0-0.5	<0.033	PTW-10	S1	0.0-0.5	<0.033
	S2	0.5-1.0	<0.033		S2	0.5-1.0	<0.033
	S3	1.0-2.0	<0.033		S3	1.0-2.0	<0.033
	S4	2.0-4.0	<0.033		S4	1.0-2.0	<0.033
	S5	4.0-6.0	<0.033		S5	2.0-4.0	<0.033
	S6	6.0-8.0	<0.033		S6	4.0-6.0	<0.033
	S7	8.0-9.9	<0.033		S7	6.0-8.0	<0.033
PTW-4	S1	0.0-0.5	<0.033		S8	8.0-9.9	0.302
	S2	0.5-1.0	<0.033	PTW-11	S1	0.0-0.5	<0.033
	S3	1.0-2.0	<0.033		S2	0.5-1.0	<0.033
	S4	2.0-4.0	NS		S3	1.0-2.0	<0.033
PTW-6	S1	0.0-0.5	<0.033		S4	1.0-2.0	<0.033
	S2	0.5-1.0	<0.033		S5	2.0-4.0	<0.033
	S3	1.0-2.0	<0.033		S6	4.0-6.0	<0.033
	S3 DUP	1.0-2.0	<0.033		S7	6.0-8.0	<0.033
	S4	2.0-4.0	<0.033		S8	8.0-9.0	<0.033
	S5	4.0-6.0	<0.033		S9	9.0-9.9	<0.033
	S6	6.0-8.0	<0.033	TW-12	S1	0.0-0.5	<0.033
	S7	8.0-10.0	<0.033		S2	0.5-1.0	<0.033
	S8	10.0-12.0	<0.033		S3	1.0-2.0	<0.033
	S9	12.0-13.5	<0.033		S4	1.0-2.0	<0.033
	S10	13.5-14.0	0.072		S5	2.0-4.0	<0.033
	S11	14.0-15.5	<0.033		S6	4.0-6.0	<0.033
PTW-7	S1	0.0-0.5	<0.033		S7	6.0-8.0	<0.033
	S2	0.5-1.0	<0.033		S8	8.0-9.9	<0.033
	S3	1.0-2.0	<0.033	TW-13	S1	0.0-0.5	0.045
	S4	2.0-4.0	<0.033		S2	0.5-1.0	0.045
	S5	4.0-6.0	<0.033		S3	1.0-2.0	0.042
	S6	6.0-8.0	<0.033		S4	2.0-4.0	<0.033
	S7	8.0-9.9	0.168		S5	4.0-6.0	<0.033
					S6	6.0-8.0	<0.033
					S7	8.0-9.9	<0.033
				PTW-14	S1	0.0-0.5	<0.033
					S2	0.5-1.0	<0.033
					S3	1.0-2.0	<0.033
					S4	1.0-2.0	<0.033
					S5	2.0-4.0	<0.033
					S6	4.0-6.0	<0.033
					S7	6.0-8.0	<0.033
					S8	8.0-9.9	<0.033

NOTES:
1. NA denotes that sample analysis results are not available at this time.
2. NS indicates no sample was collected.
3. Samples taken at locations of thermal wells, e.g., TW-1 as shown on Figure 3.
4. "T" denotes twinned geoprobe location.
5. * Split spoon sample, possible contamination from shallow cavings
6. PTW-8 samples were collected adjacent to the PTW-1 location.

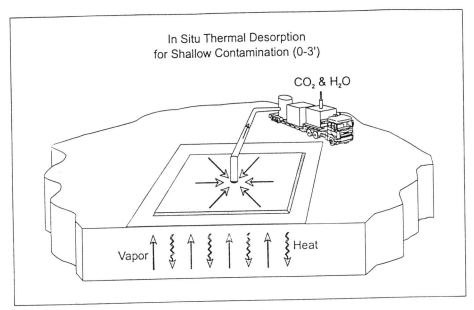

Figure 1a

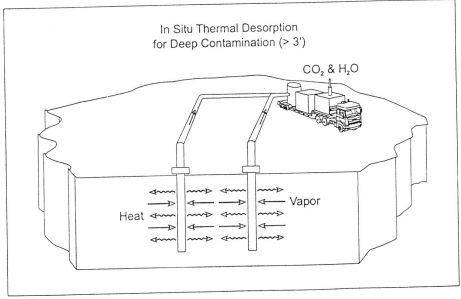

97/0161/06 Figure 1b

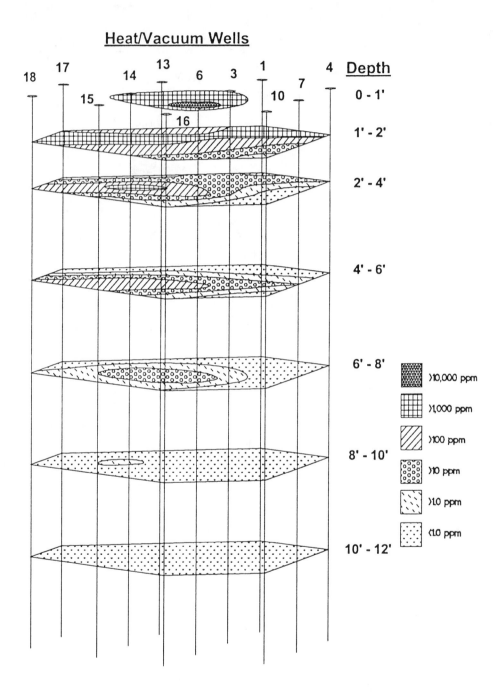

Figure 2.Contour values of initial PCB concentration

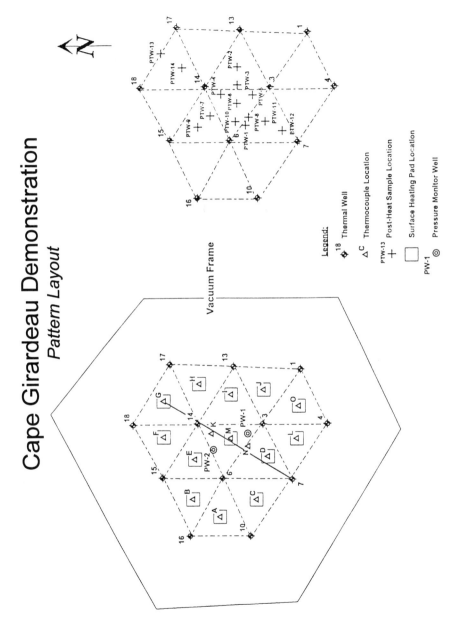

Figure 3. Left, position of thermal wells and thermocouples. Right, post-heat soil sample locations (Table 2.).

Soil Temperature History at 6-Foot Depth

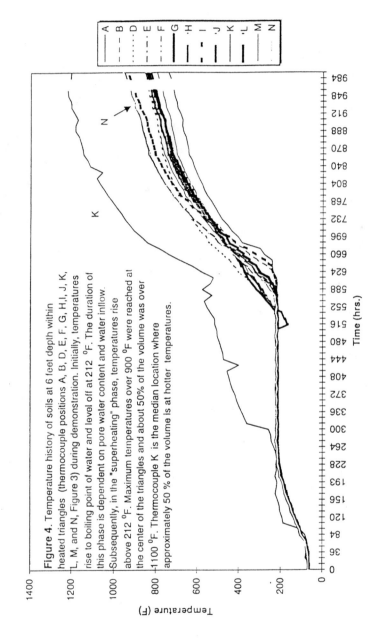

Figure 4. Temperature history of soils at 6 feet depth within heated triangles (thermocouple positions A, B, D, E, F, G, H,I, J, K, L, M, and N, Figure 3) during demonstration. Initially, temperatures rise to boiling point of water and level off at 212 °F. The duration of this phase is dependent on pore water content and water inflow. Subsequently, in the "superheating" phase, temperatures rise above 212 °F. Maximum temperatures over 900 °F were reached at the center of the triangles and about 50% of the volume was over ~1100 °F. Thermocouple K is the median location where approximately 50 % of the volume is at hotter temperatures.

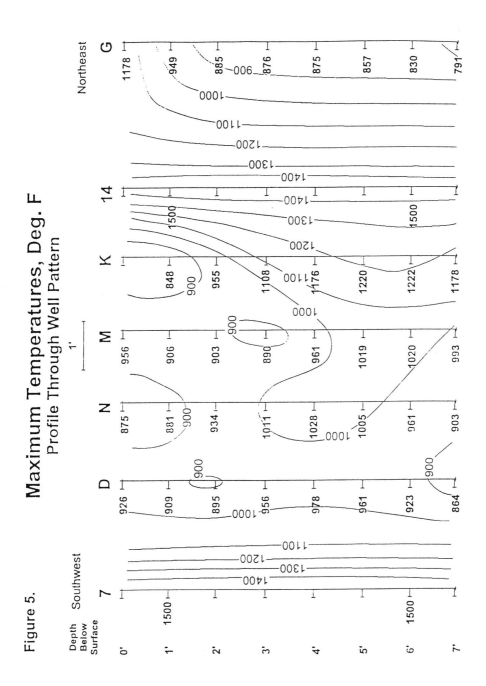

Figure 5.

Maximum Temperatures, Deg. F
Profile Through Well Pattern

SOIL HEATING FOR REMEDIATION OF DISSOLVED TRICHLOROETHYLENE IN LOW-PERMEABLE SOIL

Gorm Heron and Thomas H. Christensen (Technical University of Denmark, Lyngby, Denmark)
Marcus Van Zutphen (Netherlands Institute of Applied Geoscience TNO, Delft, The Netherlands)
Carl G. Enfield (US EPA, Cincinnati, OH)

ABSTRACT: Electrical heating of a low-permeable silt layer contaminated with dissolved trichloroethylene (TCE) was simulated in a laboratory tank. Heating to 85 and 100 °C led to dramatically increased mass fluxes in the soil vapor extraction system used for capture of the effluent. Within 45 days, soil concentrations were reduced by 99.8%, without significant reduction in the soil moisture content. This study showed that thermal techniques are not only efficient in removing liquid contaminants, also dissolved and adsorbed contaminants are removed rapidly due to in-situ steaming, dramatic increases in the volatility at higher temperatures, and reduced sorption coefficients. Clean-up criteria for groundwater and soil concentrations can be met by ensuring good hydraulic control and capture of all the evolved gases. Thermal in-situ treatment appears economically attractive, since treatment times are in the order of months, low residual contaminant levels can be achieved, and the energy consumption is not cost-prohibitive.

INTRODUCTION

Soil heating using steam injection and electrical heating has shown great promise for rapid removal of organic contaminants from soils and groundwater (Udell and Stewart 1989; Gauglitz et al. 1994; Newmark and Aines 1997, Phelan et al. 1997; Udell et al. 1997). The major mechanisms are currently studied in the laboratory, showing that mass-transfer limitations are practically eliminated when the subsurface reaches boiling temperatures. Non-aqueous phase liquids (NAPL) are removed rapidly due to preferential boiling, steam distillation and volume expansion when pore water is converted into steam (Udell 1996).

Research and field demonstrations have shown that NAPL contaminants may be removed, but the question still remains whether the soil and groundwater will retain contaminants as dissolved or adsorbed phases, and whether clean-up criteria can be met. Especially low-permeable layers are of concern, since these are in poor contact with circulating fluids, and typically are rich in organic carbon and clay minerals (leading to more adsorption). A recent review on mass-transfer limitations indicated by theoretical calculations, that contaminant left in the dissolved state may also be removed rapidly when in-situ boiling of the pore fluids is induced, and that very low residual concentrations are achievable (Udell

1996). This can be accomplished either by cyclic steam injection combined with vacuum extraction, where pore water boils during depressurization (Itamura & Udell 1995), or by electrical heating of the soil layers leading to pore water boiling and partial desaturation.

This paper presents a laboratory study on electrical heating of a low-permeable silty soil contaminated with trichloroethylene (TCE) at concentrations close to the solubility limit, simulating the worst case after removal of the NAPL phases. In addition, we briefly discuss the energy demands and cost of thermal remediation.

LABORATORY JOULE HEATING SIMULATION

An overview of the laboratory setup is given in Figure 1. The silty soil was 50 cm high, 120 cm wide and 12 cm deep. Soil vapor extraction was driven by a vacuum pump, leading to air flow through the entire system from the air inlet ports, sweeping across the top sand layer in the box, through the central vent into the vapor treatment line, where excess moisture was removed by condensation, and TCE concentrations were measured by gas sensors placed in a GC oven (Heron et al 1998c). Finally, the air was passed through a drierite trap for removal of the last moisture and on to an exhaust vent.

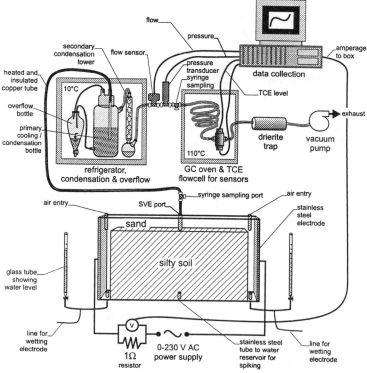

Figure 1. Laboratory setup for simulation of electric heating and soil vapor extraction in a silty TCE-contaminated soil (from Heron et al. 1998b).

The 2-dimensional box was spiked with water (1120 mg/L TCE) and then left to equilibrate for 8 days prior to onset of soil vapor extraction. The dissolved TCE concentration distribution shape was dominated by the fact that water and TCE was added from the bottom central screen, and kinetic adsorption of TCE onto the soil (Heron et al., 1998b).

Soil vapor extraction alone (days 0-8) removed 2.4 g of TCE from the box (7 % of the added TCE). Clean air entered through the top inlet ports and quickly concentrations in the outlet port dropped to very low levels with a flux of about 0.13 g/d. This flux indicated that it would take at least 250 days to clean the silty layer, assuming that this flux could be maintained.

At day 8, the electrical power was switched on, and the box was heated at a rate of 8 °C/day until 80 °C was reached on day 18, followed by slower heating to 85 °C on day 21 (Figure 2). This resulted in dramatically increased fluxes of TCE out of the central vent between days 10 and 20, dropping off towards a steady flux of about 0.35 g/d. The temperature was kept almost constant until day 35, where a total of 20 grams had been removed, and 15 gram of TCE remained in the box. The steady state flux of 0.35 g/d reached indicated that at least another 50 days would be required for complete removal of the TCE. In reality, more time would be needed, since this flux would level off with soil concentrations approaching low levels.

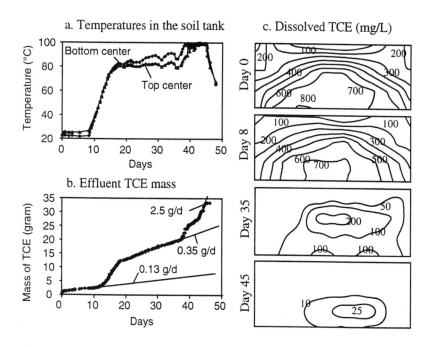

Figure 2. Temperature evolution (a), TCE mass recovery (b), and dissolved TCE distribution at selected times during electrical heating and vapor extraction (c).

Then the power was increased, and 99-100 °C was reached and maintained through days 39-45. This led to another peak in the flux and later to stabilization around 2.5 g/d. The power was shut off on day 45, since the mass balance indicated that only trace amounts of TCE remained in the soil. The box was left to cool down slowly over a 2 week period. The soil vapor extraction was discontinued on day 48 after the flux of TCE out of the box had dropped to below detection limits, and finally soil samples were collected for determination of TCE levels in the treated soil.

The TCE mass balance indicated an overall recovery of 94.3 % by the SVE system, and a remediation efficiency of 99.8% with a residual TCE content of 70 mg in the soil compared to the initial 35.5 g on day 0.

DISCUSSION AND PERSPECTIVES

Heating the soil favored the transport of TCE to the gaseous phase, and reduced the importance of adsorption onto the soil. However, even at 99 °C more TCE was found in the adsorbed state than dissolved (on a mixed media basis; Heron et al. 1998b). This showed that sorption was still a dominant mechanism, even when the volatility was increased 9 fold (Henrys law constant; Heron et al. 1998a). At the end of the heating period, only traces of TCE were found in the vapor, water and soil. Overall, all three phases were reduced simultaneously, even though TCE was only removed as vapor.

Recent results indicate that very little drying out is required for almost complete removal of organics from soil (Udell 1996). A theoretical study showed that the concentration (C) of contaminants in the dissolved phase is reduced by orders of magnitude when the soil dries from a water mass m_o to m_1 following

$$C_1/C_o = (m_1/m_o)^{(H \rho_{water}/\rho_{vapor} - 1)}$$

where H is the dimensionless Henry's law constant, and ρ is density (Udell, 1996). At 99 °C, the exponent may be estimated to 5640 for TCE, assuming that the vapor is pure steam and performs as an ideal gas (H value from Heron et al. 1998a). Drying out of 1 % of the water in the soil would lead to a concentration drop to well below detection limits in the water. Although a very simplified calculation, this supported our observation that very little mass-transfer limitations occurred in the later stages when the soil was heated to 99 °C.

In this study, steam production greatly enhanced the recovery of TCE from the soil. When the temperatures dropped slightly, and steam production ceased, the TCE flux would decrease similarly. An important mechanism for the TCE movement may be the expansion of water as it evaporates. One gram of liquid water occupies 1 mL. Once evaporated at 100 °C, the same gram of water behaves as an ideal gas and thus occupies 1.7 L as steam at 1 atm pressure, which is a 1700-fold volume expansion. This expansion may push out large amounts of soil vapor in the surrounding areas, and thus be a very important mechanism for the enhanced contaminant recovery at steam temperatures (Itamura & Udell 1995).

The enhancement of the TCE recovery by heating does not directly depend on exceeding the boiling point of TCE (87 °C), since there was no free phase TCE and thus not a direct boiling of such a phase. Actually, Figure 2 shows that the TCE fluxes picked up dramatically long before 87 °C was reached. This indicates that similar enhancement factors can be achieved for other compounds with higher boiling points. Independent of the boiling point, properties such as the Henry's constant, vapor pressure and diffusion coefficients will increase with temperature in a manner similar to those of TCE. Adsorption onto the soil will decrease, also leading to faster removal. Actually, the stronger the contaminant sorbs (high $-\Delta H_s$), the stronger the effect of heating up on K_d (Heron 1997). Thus, electrical heating is very promising for accelerating removal of a wide range of organic compounds, including semi-volatile compounds.

Drying out was not necessary for almost complete TCE removal from the silty soil. In fact, even the 100% saturated bottom parts of the silty layer were cleaned to very low residual TCE levels. This is encouraging for the application to groundwater plumes, where desaturation would be hard to achieve without lowering the water table hydraulically to below the target area. Resistive heating may thus be feasible in the upper part of aquifers, overlayered by an unsaturated zone suited for soil vapor extraction.

A typical concern on thermal remediation is the cost of heating soil. However, simple calculations show that approximately 80 kWh is needed for heating one cubic meter of water-saturated soil from 10 to 100 °C (heat capacity of soil and water of 1,000 and 4,200 J/(kg K), respectively). This equals a cost of $8/m^3$ when heated using electricity, and a cost of $3/m^3$ when the heat is supplied as steam. When significant evaporation of water occurs, the heat of vaporization may lead to greater energy consumption. Field scale applications show that thermal cleanup can be done with a total treatment cost in the range of $45 to $150 per cubic meter, including the cost of research and development (Phelan et al. 1997; Newmark & Aines 1997). A recent study on the cost of thermal methods concluded that in all 5 scenarios presented, thermal remediation was cheaper than the best comparable technology (Bremser & Booth 1996). It appears that not only are thermal methods efficient for removal of organics from low-permeable soils that presently cannot be remediated by other techniques, they are also economically very competitive. The main reason for this is the low treatment time needed, when elevated temperatures speed up mass removal, leading to much lower long-term expenses.

ACKNOWLEDGMENTS. The U.S. Environmental Protection Agency through its Office of Research and Development (funded and managed or partially funded and collaborated in) the research described here under Cooperative Agreement CR 823908-01-0 entitled Joule Heating for DNAPL Remediation. It has not been subjected to Agency review and therefore does not necessarily reflect the views of the Agency, and no official endorsement should be inferred.

42

REFERENCES

Bremser, J., and S.R. Booth. 1996. *Cost Studies of Thermally Enhanced In Situ Soil Remediation Technologies.* Report LA-UR-96-1683. Los Alamos National Laboratories, Los Alamos, NM.

Davis, E. L. 1997. *How heat can enhance in-situ soil and aquifer remediation: Important chemical properties and guidance on choosing the appropriate technique.* US EPA Issue paper EPA/540/S-97/502, 1997.

Gauglitz, P., J. Roberts, T. Bergsman, R. Schalla, S. Caley, M. Schlender, W. Heath, T. Jarosch, M. Miller, C. Eddy-Dilek, R. Moss, and B. Looney. 1994. *Six-phase soil heating for enhanced removal of contaminants: Volatile organic compounds in non-arid soils. Integrated demonstration, Savannah River Site.* Report No. PNL-10184, UC-406. Pacific Northwest Laboratory, California, USA.

Heron, G. 1997. *Using elevated temperatures to enhance in-situ remediation in low-permeable soils and groundwater.* ATV Meeting on Groundwater Contamination, March 11-12,Vingstedcentret, Denmark, pp. 257-271.

Heron, G., T.H. Christensen, and C.G. Enfield. 1998a. Henry's Law Constant for Trichloroethylene between 10 and 95°C. *Revised for publication in Environ. Sci. Technol.*

Heron, G, M. van Zutphen, T.H. Christensen, and C.G. Enfield. 1998b. Soil heating for enhanced remediation of chlorinated solvents: A laboratory study on resistive heating and vapor extraction in a silty, low-permeable soil contaminated with trichloroethylene. *Revised for publication in Environ. Sci. Technol.*

Heron, G., M. van Zutphen, and C.G Enfield. 1998c. In-line measurement of trichloroethylene vapors using tin dioxide sensors. Accepted for publication in Ground Water Monitoring and Remediation.

Itamura, M.T., and K.S. Udell. 1995. *An analysis of optimal cycling time and ultimate chlorinated hydrocarbon removal from heterogeneous media using cyclic steam injection.* Proceedings of the ASME Heat Transfer and Fluids Engineering Divisions, ASME, HTD-Vol. 321/FED-Vol. 233.

Newmark, R.L., and R.D. Aines. 1997. *Dumping Pump and Treat: Rapid Cleanups Using Thermal Technology.* Report UCRL-JC-126637, Lawrence Livermore National Laboratory, Livermore, CA.

Phelan, J., Reavis, B., Swanson, J., Cheng, Wu-Ching, Dev, H., and L. Enk. 1997. *Design, Demonstration and Evaluation of a Thermal Enhanced Vapor Extraction System.* Report SAND97-1251 UC 2010,Sandia National Laboratory, Albuquerque, NM.

Udell, K.S., and L. Stewart. 1989. *Field study of in situ steam injection and vacuum extraction for recovery of volatile organic solvents.* UCB-SEEHRL Report No. 89-2. University of California, Berkeley.

Udell,K.S. 1996. *Heat and mass transfer in clean-up of underground toxic wastes.* In Chang-Lin Tien (Ed.): Annual Reviews of Heat Transfer, Vol. 7, Begell House, Inc.: New York, Wallingford, UK, 333-405.

SOIL VAPOR EXTRACTION ENHANCED BY MICROWAVE ENERGY

Zdzisław Kawala (Inst. Chem. Eng., Wrocław, Poland)
Tomasz Atamańczuk (Inst. Chem. Eng., Wrocław, Poland)

ABSTRACT: Microwave energy supplied to the contaminated soil is released in form of heat and allows for significant temperature rise of the soil bed. In connection with soil vapor extraction it can be used to remove contaminants of various volatility from soil in a short time. A large-scale laboratory experiment was performed, in which 550 kg of sand contaminated with chlorinated hydrocarbons of various volatilities (TCE, PCE and o-dichlorobenzene) was treated. Microwave energy was supplied to the soil bed by means of an antenna placed in the axis of a vapor extraction well, powered by a 750 W / 2450 Mhz magnetron. During four days of the experiment TCE and PCE were removed completely, and o-dichlorobenzene was cleaned up in 97 percent. In course of the remediation process two contaminant removal fronts moving in opposite directions were observed within the range of influence of the extraction well. The first of them, moving radially from the well outside, resulted from the intensive evaporation of contaminants around the antenna brought about by the microwave radiation. The other one, headed inwards, was the effect of the suction of contaminant laden soil air and can be described in terms of a standard SVE process.

INTRODUCTION

Soil vapor extraction (SVE) is one of the most widespread in-situ soil remediation techniques. Low costs and uncomplicated equipment are its unquestioned advantages. The process is carried out in ambient temperature and the removal rate depends to the greatest extent on the contaminant concentration in the extracted soil air. The applicability of SVE is thus limited to sites contaminated with volatile fuels and solvents exhibiting high vapor pressures in temperatures ranging from 10 to 20°. Treatment of contaminants with boiling points exceeding 150° is therefore impractical from the economic point of view, as the time needed to achieve acceptable concentration levels would amount up to several years.

The above mentioned reasons give rise to the efforts aiming at the reduction of the duration of SVE treatment and exceeding its applicability to higher boiling contaminants. This can be achieved by the temperature rise of the treated soil volume. Higher ground temperature would result in the rise of the equilibrium gas phase contaminant concentrations, and thus in the increase of the removal rates.

A medium that allows for the in-situ supply of heat to the ground is electromagnetic radiation. High frequency electric currents may be induced in soil by means of electrodes placed in bores drilled in a contaminated soil layer. The electromagnetic energy is absorbed by water molecules, soil particles and polar contaminants, whereby a corresponding amount of heat is being released. A primary mechanism of this process is dielectric heating.

Techniques utilizing radio frequencies for soil heating have been used in the 1980s in pilot-scale experiments for the remediation of airfield soils contaminated by petroleum products (Edelstein, 1994, Downey, 1990).

Due to the higher frequency the heat release rate achieved by microwave heating is much higher than in the case of radio frequency radiation. The laboratory scale experiments performed by the authors (Kawala, 1996) proved a high efficiency of the technique, especially in the case of polar contaminants. The removal efficiency was close to 100 percent. When soils with a high moisture content are treated, the primary mechanism of the contaminant removal is steam distillation. This may be of importance in the case of humus soils with a high content of the organic material, where too high a temperature could burn the humus matter.

EXPERIMENTAL APPARATUS AND PROCEDURES

A model cell of a microwave soil treatment system was built in a laboratory scale. The contaminated soil is heated by means of a custom-built microwave-propagating stub antenna.

The microwave heating system consisted of three elements: a microwave generator, a power supply system and the stub antenna. The shell of the generator housed a typical commercial magnetron of the type Toshiba 2M240. The output power of the system ranges between 550 and 650 W and the operating frequency is 2450 MHz. The watt-hour efficiency of the system was in excess of 60 percent.

The design of the stub antenna was based on the work by Pushner (1966). The antenna was 0.7 m long, but the microwaves are emitted only through slots cut in its 0.3 m long lower section.

The design of the large scale experimental apparatus is shown in Figure 1. Contaminated soil was placed in a 1 m i.d. cylindrical container. The side wall of the container was perforated in order to enable fresh air flow from outside. Upper and lower surfaces of the soil bed were sealed and insulated thermally. In this way air was forced towards the extraction well in radial direction only. The contaminated sand layer representing a typical sandy soil was 0.45 m high. A 35 mm i.d. perforated PTFE tube was fixed vertically in the axis of the container and the stub antenna was placed inside it. The tube was connected to a vacuum pump and acted as a vapor extraction well. Vapors generated in the heated soil were transported under a negative pressure to the condenser, where they were condensed and collected as distillate in scaled receivers. The remainder of uncondensed vapors was sorbed on activated carbon placed in two adsorbers working alternately. Activated carbon bed was regenerated periodically with steam, whereby the adsorbed organic phase was recovered. The side wall of the container was provided with an array of openings for the introduction of a thermocouple probe and for the collection of soil samples to be analyzed by chromatography. In course of the experiment soil samples were collected at 5 positions along the container radius and analyzed by HPLC for the contaminant content. Changes in the moisture content were also monitored.

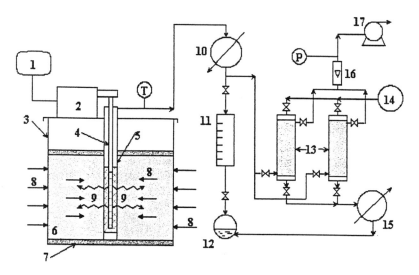

FIGURE 1. Schematic view of the large scale laboratory microwave apparatus. 1.Power supply, 2.Microwave generator, 3.Cylindrical container, 4.Antenna with a microwave propagation section, 5.Perforated PTFE vapor extraction tube, 6.Contaminated soil, 7.Sealing and thermal insulation, 8.Movement direction of the extracted vapors, 9.Propagation of microwaves, 10,15.Condensers, 11,12.Organic phase receivers, 13.Activated carbon filters, 14.Superheated steam generator, 16.Rotameter, 17.Vacuum pump

550 kg of sand with a porosity of 0.28 was uniformly contaminated with a mixture of three chlorinated hydrocarbons differing in volatility. The initial concentrations of trichloroethylene, tetrachloroethylene and o-dichlorobenzene along with their boiling points are shown in Table 1. All these substances are frequently identified at contaminated sites, where they can persist for a long time due to their low biodegradability. Because of their high toxicity a soil with TCE or PCE concentration higher than 50 ppm (10 ppm for o-dichlorobenzene) is regarded as contaminated and subject to remediation according to the Dutch List. The initial soil moisture content was 4.7 percent by weight.

TABLE 1. Contaminant initial concentrations and boiling points.

Substance	Initial concentration [mg/kg]	Initial content in soil [kg]	Boiling point [°C]
Trichloroethylene	3620	1.991	86.7
Tetrachloroethylene	5729	3.151	120.8
o-Dichlorobenzene	4676	2.572	179
Total	14025	7.714	

RESULTS AND DISCUSSION

The contaminant removal process is based primarily on the equilibrium between the organic liquid phase, contaminants adsorbed on sand and soil air. The microwave energy supplied by the antenna is initially absorbed in the vicinity of the antenna, mostly by the soil moisture contained therein. A temperature rise in this zone results in a rapid evaporation of water and the contaminants. The arising vapors are removed through the extraction well. The zone around the antenna is thus cleaned up and a removal front between the "clean" and contaminated zone appears. As the remediated zones dry up their transmissivity for microwave radiation increases (dry sand absorbs microwaves much weaker than water) and the electromagnetic energy can penetrate farther into more distant zones along the container radius. This mechanism, but also conductive heat transfer, is responsible for the gradual temperature rise in zones more distant from the container axis (Figures 2 and 3). The radius of the "cleaned" zone increases and the contaminant removal front moves in radial direction outwards towards the container side wall.

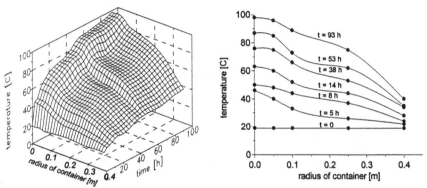

FIGURE 2. Temperature distribution in soil bed

FIGURE 3. Temperature profiles in soil in course of the remediation

Fresh air flowing into the container through the perforated side wall gets into contact with the organic phase as soon as it passes the wall. It instantly becomes saturated with contaminant vapors according to the gas-liquid equilibrium conditions. In this way contaminants are initially removed from a zone most distant from the extraction well. A second contaminant removal front emerges, for which the mechanisms ruling the standard soil vapor extraction are responsible. As the remediation proceeds, this front moves inwards towards the extraction well, in the direction of the air flow.

On its movement towards the extraction well, the contaminant laden air flows through the higher temperature zones that lie more closely to the extraction well, where additional amounts of contaminants and water can evaporate.

Concentration changes of the organic phase constituents at various distances from the container axis in course of the remediation are shown in Fig.4. After a period of 5 and 8 hours a maximal TCE and PCE concentrations are observed at the radius of 0.35 m from the antenna. Both at the container side wall and in the immediate vicinity of the antenna the concentrations are lower, what can be attributed to the two-side mechanism described above. The same can be observed for o-dichlorobenzene after 38 and 56 hours.

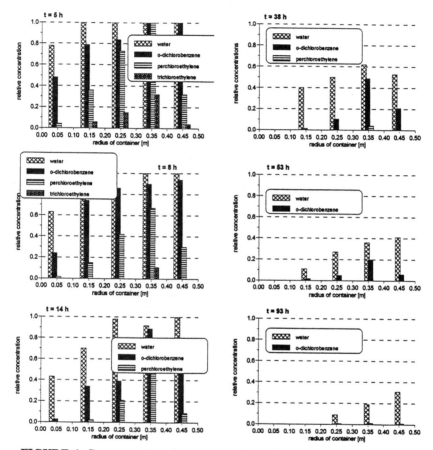

FIGURE 4. Concentration changes at various distances from the extraction well indicating the existence of two contaminant removal fronts. The concentration is shown in relation to the initial value.

Mean values of the contaminant concentrations in sand and the rate of contaminant and water removal are shown in Figures 5 and 6. It may be observed that the removal rate of a particular substance is directly related to its volatility. The most volatile constituent - TCE - was completely removed from soil after 14 hours, while a semivolatile compound - PCE - after 45 hours. In the case of a low volatile compound - o-dichlorobenzene - after 93 hours of the experiment about 3 percent of the initial content remained in the sand. Results analogous to the above described were achieved in other experiments, also with the use of dry sand.

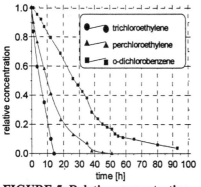

FIGURE 5. Relative concentration changes of contaminants in sand

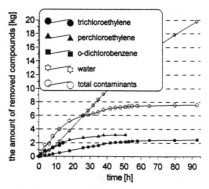

FIGURE 6. Removal rate of the contaminants and water

According to the authors' conception of a large scale remediation system the supply of microwave energy to the contaminated soil would be achieved through an array of independent antennas, each powered by a separate microwave generator. A custom-built high power generator can thus be replaced by low-cost commercially available 750 W generators used in microwave ovens, working at a frequency of 2450 MHz. The wholesale cost of these generators does not exceed 15$, whereas an eight times more powerful 6 kW magnetron costs as much as 3300$ (Schiffmann, 1990). A system of low power generators is very flexible and it imposes no restrictions on the number and the arrangement of the antennas. Thanks to the individual power supply a break-down of a single antenna-generator cell can be dealt with without a need to shut off the entire system.

REFERENCES

Edelstein W.A.. 1994. Radiofrequency ground heating for soil remediation: science and engineering. Env. Prog., 13(4): 247-252.

Downey D.C., Elliot M.C. 1990. Performance of selected in-situ decontamination technologies. An Air Force perspective. Env. Prog., 9(3): 169-173.

Kawala Z., Atamańczuk T. 1996. Application of microwave heating for the remediation of soil contaminated with hydrocarbons and fuels. in Proc. Third Int. Symp. on Environmental Contamination in Central and Eastern Europe, Warsaw.

Schiffmann R.F. 1990. Microwave and dielectric drying in Handbook of industrial drying, Mujumdar A.S. ed., pp. 327-356. Marcel Dekker Inc., New York

Pushner H. 1966. Heating with microwaves. Cleaver-Hume Press Ltd..

RESEARCH ISSUES FOR THERMAL REMEDIATION

Eva L. Davis (U.S. EPA, Ada, Oklahoma)
Gorm Heron (Technical University of Denmark, Lyngby, Denmark)

ABSTRACT: In order to optimize thermal remediation techniques, all of the effects of heat on the subsurface system must be understood and taken into consideration during the remediation. Research is needed to provide a better understanding of the effects on temperature on capillarity in soils. This should include laboratory data on the effect of temperature on displacement pressures which is needed to determine the potential for downward movement of DNAPLS. Solubility data as a function of temperature over the temperature range of interest in thermal remediation is lacking for many organic contaminants and thus temperature effects on Henry's constant is not known. Movement of vapors and liquids in response to thermal gradients also requires additional research to understand the implications for thermal remediation.

INTRODUCTION

The in-situ thermal remediation methods currently being developed, which include steam injection, hot water injection, and electrical heating (both powerline frequency and radio frequency), were first developed for enhanced oil recovery. When the adaption of these methods for remediation purposes was initiated, the technical problems of application of these techniques in the field had already been devised. Adaption of the techniques to environmental problems quickly proceeded with pilot scale field demonstrations. However, the objectives of applying thermal techniques to oil recovery and for remediation purposes are significantly different. Complete or near-complete recovery of organic contaminants from the relatively near-subsurface soil and aquifer environment requires a greater level of understanding of the physical and chemical interactions between the heat, soil, water, and contaminant than does the application of heat to recover mobile oil, with the recovery operation terminated at some economic endpoint. Also, although oils are generally lighter than water, many remediation challenges we face are contaminants that are more dense than water. Because thermal remediation is applied to the shallow subsurface, there is a greater risk associated with the use of these technologies, which also demands that the processes which occur in the subsurface be better understood, A thorough understanding of these processes will also allow optimization of these remediation techniques.

Many physical properties of chemicals are a function of temperature, and the temperature-sensitivity of the these properties, such as density, viscosity, and vapor pressure, are different for different organic chemicals (Davis, 1997; Heron, 1997). Also, many of the physical properties of these contaminants are affected by the capillary-size pores in which the contaminant resides in the subsurface, such as the effect that a curved interface has on vapor pressures. The effects of the porous media on the fluids themselves, on the heating process, and on the resulting fluid flow at higher temperatures and in the presence of a temperature gradient must be fully considered. Thus, laboratory research on the effects of heat on contaminants in the subsurface, and under the conditions that would be created with the application of these thermal remediation processes is needed.

MOBILIZATION OF LIQUID PHASE CONTAMINANTS

The primary risk associated with the use of thermal methods that physically displace liquid contaminants, such as hot water and steam injection, is the potential for mobilization of contaminants to areas other than the recovery points. One concern that has often been associated with steam injection is the possibility for downward mobilization of DNAPLs (Hunt et al., 1988; Udell, 1996; Sleep and Ma, 1997; Heron et al., 1998c). Mobilization of separate phase contaminants as heat is added to the subsurface may occur through expansion of the fluid or reduction in capillary forces, which may occur through decreases in the surface and/or interfacial tension and the contact angle. For the case of a DNAPL pool on top of a capillary barrier, the question arises that if the temperature is raised, and thus the interfacial tension is lowered, will the DNAPL now penetrate the fine layer? Measurements of interfacial tension as a function of temperature for a variety of pure organics shows that the interfacial tension decreases linearly as a the temperature is increased (Jasper, 1972). The few measurements of contact angle with temperature that have been made show in general a decrease in contact angle with increasing temperature, but the relationship was not always linear (Poston et al., 1970; McCaffery, 1972). With the combination of these two reductions, the capillary forces in a soil system will decrease as the temperature increases.

Capillary pressure-saturation curves are normally used to measure capillary forces in soil systems, and where these curves have been measured as a function of temperature, they indicate a more complex system than is predicted by interfacial tension and contact angle measurements on pure systems. The effect of temperature on the relationship is greater than would be predicted based on surface tension reductions alone, and the extent of the effect is also dependent on the grain size of the sand (Wilkinson and Klute, 1962). For one oil/water system studied, although the water/oil interfacial tension changed more with temperature than did the water/ air surface tension, the oil/water capillary pressure-saturation curves showed less change with temperature than did the water/air curves (Davis, 1994). As pointed out in Davis (1994), there are also many other factors in the soil system that can effect both interfacial tension and contact angle, making for a complex relationship between soils, the chemicals contained in the soils, and the temperature. Laboratory measurements of displacement pressures at various temperatures for the soil/contaminant system of interest are the best way to determine if displacement pressure will change with temperature and downward movement will occur.

This reduction in capillary forces with increasing temperature may also reduce the residual saturation of a separate phase liquid. Often the capillary number, defined as $N_c = q\mu/\sigma$, where N_c is the capillary number, q is the specific discharge, μ is the viscosity of water, and σ is the interfacial tension, is used to determine the potential to physically displace residual NAPL (Wilson and Conrad, 1984). Because the viscosity of water decreases more than the interfacial tension as temperature is increased, this relationship would predict that the residual saturation of a system will increase as the temperature is increased. However, this is contrary to the experimental data, both measured in the laboratory in the form of capillary pressure-saturation curves (Poston et al, 1970; Davis, 1994) and in displacement experiments (Davis and Lien, 1993; Pytte, 1997), and in the field where a hot water flushing was done after an ambient temperature water flush (Michalski et al., 1995). Thus, research is needed to provide an understanding of the effects of temperature on capillarity in soils.

Another complicating factor may be the effect that temperature has on the permeability and relative permeability of porous media. Experimental data shows changes in permeability in various types of porous materials with temperature for various fluids (see for example Casse and Ramey, 1979), and the implications of these results for thermal remediation methods need to be assessed.

PARTITIONING BETWEEN PHASES

Vaporization is the dominant process during most thermal remediations (Udell, 1996), thus the partitioning of contaminants between the vapor, aqueous, and solid phases will have a strong effect on the recovery of the contaminant. For a volatile contaminant, recovery in the vapor phase is often preferred, however, at ambient temperatures the contaminant may be concentrated in the aqueous or solid phase. Vapor pressure, aqueous solubility, and adsorption on the soil surfaces are all generally temperature dependent, thus the partitioning between phases will change as the temperature is increased (Davis, 1997). The distribution of the contaminant between the phases as a function of temperature and the process of phase changes in pores will determine the trade off between temperature and time for remediation, and perhaps the point where thermal remediation is no longer cost effective and a polishing step should be initiated.

Vapor pressures increase monotonically with temperature and are fairly well described (Reid et. al, 1987). However, the effect that the capillary pores of the soil will have on vaporization as a function of temperature has received only a limited amount of attention (Fares et. al, 1995). The solubility of many organic contaminants as a function of temperature over the temperature range of interest has not been determined, but the data available shows that some organic chemicals have a maximum or minimum solubility at a temperature around ambient temperature. Thus, the Henry's constants (the ratio of the vapor phase concentration to the solubility) at the temperatures reached in thermal remediation systems are not generally known. For many sparingly soluble contaminants, such as TCE, the Henry's constant will increase with temperature to the point that heating will allow the removal of the dissolved contaminant to very low concentrations (Heron et al., 1998a&b). However, for more soluble contaminants, such as ketones, it is not known if the chemical can be readily shifted to the vapor phase by raising the temperature. Highly water soluble contaminants may require a thermal technique that does not leave a residual water saturation.

The vaporization of contaminants in porous media requires vapor formation in a small pore which is sometimes mostly water-filled. Thus the processes that control nucleation phenomena, which will prevent the formation of superheated contaminant states and encourage bubble production through vaporization, must be understood (Udell, 1996).

Adsorption onto soil surfaces and partitioning into soil organic matter as a function of temperature also have not been greatly studied. It has been shown that over limited temperature ranges, adsorption can sometimes increase as the temperature increases. Because adsorption is an exothermic reaction, this would not be expected to be true over greater temperature ranges. However, sometimes the desorption of organic chemicals from clays requires much higher temperatures than the boiling temperature of the compound (EPA, 1995a; Tognotti et. al, 1991). Initial studies by Heron et al. (1997) on TCE indicate that solid-liquid partition coefficients may decrease by 40 to 50 percent between 20°C and 90°C, while vapor adsorption coefficients may decrease by orders of magnitude over the same temperature range. This will lead to strongly reduced adsorption. Research is needed to determine if the temperature effects on the octanol-water partition coefficient can be used to describe the partitioning the occurs in the soil as a function of temperature. A clear understanding of the lower level of remaining contaminants in the soil that can be economically achieved using thermal methods will require an understanding of the partitioning of the chemicals between the phases as a function of temperature.

THERMAL DESTRUCTION OF CONTAMINANTS

The literature contains many examples where not all of the contaminants identified in the soil after thermal treatment had been present before the thermal treatment (EPA, 1991; EPA, 1995c), or some particular volatile contaminants increased in concentration during thermal treatment rather than decreased (EPA, 1995b; Swanstrom and Besmer, 1995). Other in situ thermal techniques claim the ability to thermally destroy contaminants in situ (Knauss et. al, 1997; Vinegar et al., 1997), and indeed the above evidence appears to point to thermal degradation of some contaminants during in situ thermal processes. Clays are used as catalysts in some processes for the thermal cracking of petroleum products, it may be reasonable to assume that the high temperatures in situ will cause degradation of contaminants. It may be beneficial to promote the in situ destruction of contaminants, as this would eliminate the need for storage and treatment on the surface (Knauss et al., 1997).

Considerable research has been done under conditions applicable to incinerators to determine the temperature needed for the thermal destruction of numerous hazardous chemicals in the vapor phase and on the formation of hazardous byproducts such as chlorinated dioxins and furans. However, very little research has been done on the destruction of contaminants during an in situ thermal remediation. Research is needed to understand how temperature, composition of the wastes, and the soil environment - with and without water present - will affect the thermal destruction of contaminants. There are some apparent inconsistencies in the data reported thus far in the literature. An example is the fact that thermal degradation of naphthalene was found to be very difficult under incineration conditions (Taylor et al., 1990) but readily achievable at much lower temperatures in the presence of water and oxygen (Knauss et. al, 1997), despite the fact that both degradations are thought to be via free radical substitutions.

FLUID MOVEMENT IN RESPONSE TO THERMAL GRADIENTS

Most movement of contaminants by nonthermal remediation methods is brought about by pressure gradients in each of the phases. When thermal methods are applied there is an additional gradient that can alter both the flow of liquids and vapors - the temperature gradient. The movement of water in soil due to thermal gradients has been studied extensively, however, the theoretical and experimental studies done to date have not been reconciled. The classical theoretical work by Philip and deVries (1957) predicted that temperature will affect water flow in unsaturated porous media only through the effect of temperature on the capillary pressure. However, experimental work shows that water vapor movement in the presence of a temperature gradient is greater than can be accounted for by diffusion (Ho and Webb, 1996). Bear and Bensabat (1989) developed a theory on the simultaneous movement of heat and multiple fluid phases in porous media which considered the transfer of momentum through the air-water interface, and found that it led to a coupling between the capillary pressure and temperature gradients, which could cause air flow even in the absence of pressure gradients and may explain the experimental data. Ho and Webb (1996) also developed a theory that could account for enhanced vapor flow; theirs is based on evaporation occurring on one side of a water droplet held between sand grains and condensation on the other side in the presence of a temperature gradient. However, there is not adequate experimental data to test either of these theoretical developments.

Much less is known about how temperature gradients affect solute movement or the movement of free phase volatile organics in the subsurface. Prunty (1992) studied the movement of water and octane in soil in the presence of a temperature gradient, and found that although the octane originally moved away from the heat source, at later times water and octane moved in opposite directions. At equilibrium, very little water was contained at the warm end of the column, but the octane was more equally distributed through the column's length. A full understanding of the movement of water and organic contaminants in the presence of a temperature gradient will require understanding of the interactions between evaporation and vapor phase movement with capillary forces and liquid phase movement, and the relative importance of each of these for different conditions for organic contaminants with different physical properties.

ACKNOWLEDGEMENTS
Although this research has been supported by the U.S. Environmental Protection Agency, it has not been subjected to the Agency review and therefore does not necessarily reflect the views of the Agency. The mention of trade names or commercial products does not constitute endorsement.

REFERENCES
Bear, J., and J. Bensabat. 1989. "Advective Fluxes in Multiphase Porous Media under Nonisothermal Conditions." *Trans. Porous Media*, 4: 423-448.

Casse, F. J., and H. J. Ramey Jr. 1979. "The Effect of Temperature and Confining Pressure on Single-Phase Flow in Consolidated Rocks." *J. Petrol. Technol.*, 1051-1059.

Davis, E. L. 1994. "Effect of Temperature and Pore Size on the Hydraulic Properties and Flow of a Hydrocarbon Oil in the Subsurface." *J. Contam. Hydrol.*, 16: 55-86.

Davis, E. L., and B. K. Lien. February 1993. *Laboratory Study on the Use of Hot Water to Recover Light Oily Wastes from Sands*, EPA/600/R-93/021, Robert S. Kerr Environmental Research Laboratory, Ada, OK.

Davis, E. L. April 1997. *How Heat Can Enhance In-Situ Soil and Aquifer Remediation: Important Chemical Properties and Guidance on Choosing the Appropriate Technique*, US Environmental Protection Agency, Ground Water Issue Paper, EPA/540/S-97/502.

Environmental Protection Agency. March 1991. *Toxic Treatments, In Situ Steam/Hot-Air Stripping Technology*, Applications Analysis Report, EPA/540/A5-90/008, Risk Reduction Engineering Laboratory, Cincinnati, OH.

Environmental Protection Agency. March 1995a. *Thermal Desorption at the McKin Company Superfund Site, Gray, Maine*, Cost and Performance Report, Office of Solid and Emergency Response, TIO.

Environmental Protection Agency. April 1995b. *Radio Frequency Heating, KAI Technologies, Inc.*, Innovative Technology Evaluation Report, EPA/540/R-94/528.

Environmental Protection Agency. July 1995c. *Low Temperature Thermal Aeration (LTTA) Process Canonie Environmental Services, Inc.*, Applications Analysis Report, EPA/540/AR-93/504.

Fares, A., B. Kindt, R. Lapuma, and G. P. Perram. 1995. "Desorption Kinetics of Trichloroethylene from Powdered Soils." *Environ. Sci. Technol.*, 29(6): 1564-1568.

Heron, G. March 11-12, 1997. "Using Elevated Temperatures to Enhance In-Situ Remediation in Low-permeable Soils and Groundwater." ATV Meeting on Groundwater Contamination, Vingstedcentret.

Heron, G., M. van Zutphen, T.H. Christensen, and C. G. Enfield. 1998a. "Henry's Law Constant for Trichloroethylene between 10 and 95°C, Revised for publication in *Environ. Sci. Technol.*, January.

Heron, G., M. Van Zutphen, D. J. La Brecque, G. Morelli, T. H. Christensen, and C. G. Enfield. 1998b. "Soil Heating for Enhanced Remediation of Chlorinated Solvents: a Laboratory Study on Resistive Heating and Vapor Extraction in a Silty, Low-permeable Soil Contaminated with Trichloroethylene." Revised for publication in *Environ. Sci. Technol.*, January.

Heron, G., T.H. Christensen, T. Heron, and T.H. Larson. 1998c. "Thermally Enhanced Remediation at DNAPL Sites: The Competition between Downward Mobilization and Upward Volatilization." *First International Conference on Remediation of Chlorinated and Recalcitrant Compounds*, May 18-21, Monterey, CA.

Ho, C. K., and S. W. Webb. May 1996. *A Review of Porous Media Enhanced Vapor-Phase Diffusion Mechanisms, Models, and Data - Does Enhanced Vapor-Phase Diffusion Exist?* SAND96-1198, Sandia National Laboratories.

Hunt, J. R., N. Sitar, and K. S. Udell. 1988. "Nonaqueous Phase Liquid Transport and Cleanup, 1. Analysis of Mechanisms." *Water Res. Res.*, 24(8): 1247-1258.

Jasper, J. J. 1972. "The Surface Tension of Pure Liquid Compounds." *J. Phys. Chem. Ref. Data*, 1(4): 841-1009.

Knauss, K. G., R. D. Aines, M. J. Dibley, R. N. Leif, and D. A. Mew. March 10-12, 1997. "Hydrous Pyrolysis/oxidation: In-Ground Thermal Destruction of Organic Contaminants." *In Situ Remediation of Soil and Ground Water*, Houston, Texas.

McCaffery, F. G. May, 1972. "Measurement of Interfacial Tensions and Contact Angles at High Temperature and Pressure." *J. Canad. Petrol. Technol.*, 26-32.

Michalski, A., M. N. Metlitz, and I. L. Whitman. Winter, 1995. "A Field Study of Enhanced Recovery of DNAPL Pooled Below the Water Table." *Ground Water Monitor. Rev.*, 90-100.

Philip, J. R., and D. A. Vries. April 1957. "Moisture Movement in Porous Materials Under Temperature Gradients." *The American Geophysical Union*, 38: 222-232.

Poston, S. W., S. Ysrael, A. K. M. S. Hossain, E. F. Montgomery, and H. J. Ramey, Jr. June 1970. "The Effect of Temperature on Irreducible Water Saturation and Relative Permeability of Unconsolidated Sands." *J. Petrol. Tech.*, 171-180.

Prunty, L. 1992. "Thermally Driven Water and Octane Redistribution in Unsaturated, Closed Soil Cells." *Soil Sci. Soc. Am. J.*, 56: 707-714

Pytte, K. O. 1993. *Dimensionality and Hetergeneous Effects on Enhanced LNAPL Recovery Using Hot Waterflooding,* M.S. Thesis, University of Colorado, Boulder, CO.

Reid, R., J. Prausnitz, and B. Poling. 1987. *The Properties of Gases and Liquids,* McGraw-Hill Book Company, NY.

Sleep, B. E., and Y. Ma. 1997. "Thermal Variation of Organic Fluid Properties and Impact on Thermal Remediation Feasibility." *J. Soil Contamin.,* 6(3): 281-306.

Swanstrom, C. P., and M. Besmer. March 9, 1995. "Bench- and Pilot-Scale Thermal Desorption Treatability Studies on Pesticide-Contaminated Soils from Rocky Mountain Arsenal." *Seventh Annual Gulf Coast Environmental Conference,* Houston, TX.

Taylor, P. H., B. Dellinger, and C. C. Lee. 1990. "Development of a Thermal Stability Based Ranking of Hazardous Organic Compound Incinerability." *Environ. Sci. Technol.,* 24(3): 316-328.

Tognotti, L., M. Flytzani-Stephanopoulos, A. F. Sarofim, H. Kopsinis, and M. Stoukides. 1991. "Study of Adsorption-Desorption of Contaminants on Single Soil Particles Using the Electrodynamic Thermogravimetric Analyzer." *Environ. Sci. Technol.,* 25(1): 104-109.

Udell, K. S. 1996. "Heat and Mass Transfer in Clean-up of Underground Toxic Wastes." *Annual Review of Heat Transfer,* Vol. 7, Chap. 6, pgs. 333-405.

Vinegar, H. J., E. P. de Rouffignac, G. L. Stegemeier, M. M. Bonn, D. M. Conley, S. H. Phillips, J. M. Hirsch, F. G. Carl, J. R. Steed, D. H. Arrington, P. T. Brunette. 1997. In "Situ Thermal Desorption (ISTD) of PCBs." Submitted to *Hazwaste World Superfund Conference XVIII,* Washington, DC.

Wilkinson, G. E., and A. Klute. 1962. "The Temperature Effect on the Equilibrium Energy Status of Water Held by Porous Media." *Soil Sci. Soc. of Am. Proc.,* 26: 326-329.

Wilson, J. L., and S. H. Conrad. November 5-7, 1984. "Is Physical Displacement of Residual Hydrocarbons a Realistic Possibility in Aquifer Restoration?" *NWWA/API Conference on Petroleum Hydrocarbons and Organic Chemicals in Ground Water - Prevention, Detection and Restoration,* Houston, Texas.

REMOVAL OF DISSOLVED SOLVENTS FROM HEATED HETEROGENEOUS SOILS DURING DEPRESSURIZATION

Kent S. Udell, University of California, Berkeley, CA, USA
Michael T. Itamura, Sandia National Laboratories, Albuquerque, NM, USA

ABSTRACT: Prior research has shown that chemicals that tend to partition into the vapor phase can be quickly removed from a heated porous media during a depressurization cycle. In the work presented here, one-dimensional simulations were performed using M2NOTS to investigate the rate of removal of dissolved TCE from uniform and dual permeability configurations during depressurization. Only one depressurization cycle was simulated in order to determine the concentration reduction of TCE per cycle and the time frame of the removal. The simulation domain was four meters long with a uniform initial temperature, water saturation, and contaminant concentrations. Simulations showed effective reduction of TCE concentrations from near its solubility limit to below the drinking water standards in high permeability soils. The total clean up times for a 10^{-12} m^2 permeability material with an initial temperature of 100 °C was on the order of a few hours. If the permeability is reduced two orders of magnitude to 10^{-14} m^2, the clean up time increases significantly. Simulations with heterogeneous media showed similar limits, but more rapid vaporization rates were observed at the interface between high and low permeability zones.

INTRODUCTION

Research has shown that the removal of volatile and semi-volatile DNAPLs from unsaturated soils can be achieved by simple heating to temperatures near 100°C (DeVoe and Udell, 1998). However, the presence of VOCs and SVOCs that are dissolved in the aqueous phase can still pose a significant problem to the environmental clean up of heterogeneous media since the post-heating aqueous phase concentrations may be nearly six orders of magnitude greater than the drinking water standards. Zero-dimensional, thermodynamic modeling (Itamura, 1996, Udell, 1996) has shown that chemicals that tend to partition into the vapor phase can be quickly removed from a porous media during a depressurization cycle (Udell, et al., 1991). During that depressurization process, the temperature must drop to satisfy thermodynamic equilibrium constraints, releasing the energy from the porous structure to vaporize interstitial water. According to that theory, the reduction of the water mass due to that vaporization process from the initial value, m_{wo}, to the final value, m_w, results in the reduction of the concentration of species i in the water phase from the initial value, $C_{i,wo}$, to the final value, $C_{i,w}$, as a function of the mass fraction ratio, Γ, as quantified by the following equation:

$$\frac{C_{i,w}}{C_{i,w0}} = \left(\frac{m_w}{m_{w0}}\right)^{\Gamma-1}.$$ (1)

The mass fraction ratio is defined by $\Gamma = H \rho_w/\rho_v$ where H is the dimensionless Henry's constant, ρ_w is the water density, and ρ_v is the vapor density. Equation 1 is plotted in figure 1 for different values of Γ and water mass fraction vaporized $(1-m_w/m_{wo})$. As shown in figure 1, aqueous phase concentrations can change by several orders of magnitude for even small amounts of water evaporation for large values of Γ. Thus, once separate phase volatile and semivolatile contaminants with high values of Γ have been removed by heating, aqueous contaminant concentrations may be reduced to drinking water standards very quickly by depressurization.

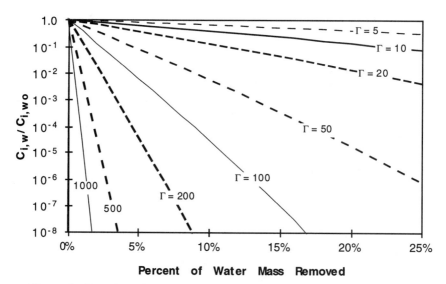

Figure 1. Fraction of contaminant remaining as a function of Γ and the percentage of water mass removed during depressurization.

The value of Γ for several different chemicals at 20 °C are listed in Table 1. Also listed in the table are the mass fraction of the contaminant in the water and in the liquid phase. Large values of the mass fraction of the contaminant in the vapor phase are realized for many of the contaminants because of the large ratios of the molecular weight of the contaminant compared with that of the water vapor or air. Combining this effect with the low solubility of many of the contaminants in water leads to the high values of Γ which range from a low of 41 for Dichloromethane to a high of 900 for PCE.

TABLE 1. Values for the Mass Fraction Ratio Γ for various chemicals at 20°C

	Molecular Weight [g/mole]	Vapor Pressure [kPa]	Water Solubility [gm/l]	Mass Percentage Water Phase	Mass Percentage Vapor Phase	Γ
Benzene	78.11	12.7	1780	0.18%	28%	156
Toluene	92.13	3.80	510	0.05%	11%	216
Ethylbenzene	106.2	1.27	160	0.02%	4%	278
p-Xylene	106.2	1.17	190	0.02%	4%	216
TCE	131.40	10.0	1100	0.11%	33%	302
PCE	165.83	2.5	140	0.01%	13%	896
Carbon Tetrachloride	153.80	15.1	1160	0.12%	48%	415
Chloroethane	64.90	100.7	5710	0.57%	100%	175
Dichloromethane	84.90	58.4	19400	1.94%	80%	41

While these equilibrium results imply a simple means to meet drinking water standards, it is not clear how actual dimensional effects and heterogeneities will affect those predictions. The simulations reported herein, performed using M2NOTS (Adenekan et al., 1993), investigate the time-scale of removal of dissolved TCE in a uniform and dual permeability systems and to determine how much TCE can be removed in one cycle.

SYSTEM MODELING

The timescale of depressurization of a steam saturated porous media is expressed as the Fourier number, Fo - the generalized timescale of phenomena that are governed by diffusive rates, such as heat conduction in the energy equation and mass transfer in mass conservation equations (Itamura, 1996):

$$Fo = \frac{\left(\lambda + \frac{kk_r}{\mu_g T} \rho_g^2 h_{fg}^2 \right)}{\overline{\rho C_p}} \frac{t}{L^2} . \tag{2}$$

where L is the domain length, t is time, λ is the thermal conductivity, $\overline{\rho C_p}$ is the overall heat capacity of the soil-fluid system, k is the permeability, k_r is the relative permeability, μ_g is the gas dynamic viscosity, T is the absolute temperature, ρ_g is the gas density, and h_{fg} is the enthalpy of vaporization. Using the properties of the simulations, the value contribution of the enthalpy flux in equation (2), $\frac{kk_r}{\mu_g T} \rho_g^2 h_{fg}^2$, equals the thermal diffusion, λ when the permeability is 1.6×10^{-15} m^2. Thus for media of lower permeability, the timescale of thermal response remains essentially unchanged.

It has been shown (Itamura and Udell, 1995, Itamura 1996) that $Fo = 1$ defines the time at which most of the mass that will be removed. The characteristic time for the three 4 meter cases examined are thus 6 hours (2×10^4 sec) for the 10^{-12} m^2 case, 65 days (5×10^6 sec) for the 10^{-14} m^2 case, and 460 days (4×10^7 sec) for 10^{-16} m^2 and lower cases.

Simulation Set-up A system of forty elements was used with each element being a cube 10 cm on each side. The extraction pressure was set at the top element at 7.3 psia (50.5 kPa). All other boundaries were closed and treated as adiabatic. Thus, these calculations represent the extreme condition of a 26.25 foot (8 m) thick zone heated by steam flowing in bounding permeable zones. Gravitational effects were ignored. Uniform permeability simulations were performed for permeabilities of 10^{-12}, 10^{-14} and 10^{-16} m^2.

The dual permeability simulations consisted of a system with the twenty elements nearest the extraction point having a permeability of 10^{-12} m^2 and the bottom twenty elements having a permeability of either 10^{-14} or 10^{-16} m^2. The initial temperature of the system was 100°C and the initial partial pressure of the TCE in the vapor phase was set at 30 kPa. The initial mass fraction of TCE in the water was 225 ppm. The initial water saturation was set at the irreducible saturation of 10% to eliminate migration of liquid water during the simulations.

RESULTS
The temperature profiles for one example simulation are shown in figure 2. The temperature of the system drops from the initial value of 100 °C to the final temperature of 82 °C, the steam-water equilibrium value at the pressure of the boundary (1/2 atmosphere). For the k = 10^{-12} m^2 simulation, the system reached overall thermal equilibrium after 300 hours. For the 10^{-14} m^2 permeability simulation, it took 100 hours for the last element to drop 1 °C, and 10,000 hours for the system to reach equilibrium.

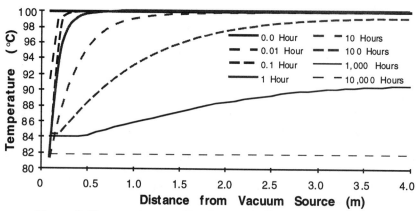

Figure 2. Temperature profiles for various times for k = 10^{-14} m^2.

Two additional simulations were performed to determine the effect of a permeability discontinuity on the temperature and TCE concentration profiles. Simulations were again performed with an initial temperature of 100 °C. The 20 elements closest to the extraction point were given a permeability of 10^{-12} m^2 with the other 20 elements given permeabilities of either 10^{-14} m^2 or 10^{-16} m^2. The masses remaining in the system for all of the simulations are presented in figure 3.

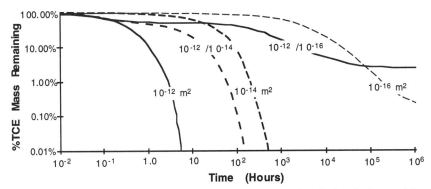

Figure 3. TCE remaining in system for a 1-Dimensional simulations with an initial temperature of 100˙C.

For the 100 °C initial condition case, it took 1 hour to remove 90% of the TCE for the 10^{-12} m^2 case. After 3 hours, only 0.7% of the TCE remained in place and after 6 hours, less than 0.01% of the TCE remained. For a permeability of 10^{-14} m^2, 90% of the TCE was removed in the first 100 hours, 1% remained after 200 hours, and less than 0.01% remained in place after 500 hours. For the case where the permeability was 10^{-16} m^2, 10% of the TCE remained in place after 80 days (2,000 hours) and 1% remained in place after 17 years (150,000 hours).

Discussion These simulations show that the time to remove TCE from the aqueous phase in heterogeneous systems heated to 100 °C is reasonable for all but very thick, low permeability regions. The total clean up times for a 10^{-12} m^2 and eight meter thick material with an initial temperature of 100 °C and decreased pressures on either side, was on the order of a few hours. For permeabilities two orders of magnitude lower (10^{-14} m^2), the clean up time increases to several months. Dropping the permeability two more orders of magnitude to 10^{-16} m^2 will slow down the clean-up time to decades. However, a permeability of 10^{-16} m^2 is lower than the EPA guidelines (40CFR258.40b) for the maximum allowable permeability for the clay liners used to contain the migration of hazardous wastes from landfills, and thus represents a nearly impermeable medium. Deep penetration of aqueous phase contamination distances of the order of 4 meters into 10^{-16} m^2 media is not expected due to negligible convection and diffusion limitations. Thus, those time restrictions are not seen as prohibitive. Since the time period for cleanup of media of such low permeability is scaled with respect to the length to the second power (Itamura, 1996), these simulations show that cyclic steam injection is also applicable to the cleanup of thin strata of low permeabilities within the time scale appropriate for the thicker regions of higher permeability media.

ACKNOWLEDGMENTS

This study was performed with the financial support from NIEHS Grant No. 5 P42 E504705-11, Project #11.

REFERENCES

Adenekan, A. E., T. W. Patzek, and K. Pruess. 1993 "Modeling of Multiphase Transport of Multicomponent Organic Contaminants and Heat in the Subsurface: Numerical Model Formulation", *Water Resources Research*, Vol. 29, No. 11, pp. 3727-3740.

DeVoe, C. and K. S. Udell. 1998. "Thermodynamic and Hydrodynamic Behavior of Water and DNAPLs During Heating", *Proceedings of First International Conference on Remediation of Chlorinated and Recalcitrant Compounds*.

Itamura, M. T., and K. S. Udell. 1995. "An Analysis of Optimal Cycling Time and Ultimate Chlorinated Hydrocarbon Removal from Heterogeneous Media Using Cyclic Steam Injection," *Proc. Heat Transfer and Fluids Engineering Division*, ASME HTD-vol. 321 and FED-vol. 233, pp. 651-660.

Itamura, M. T. 1996. "Removal of Chlorinated Hydrocarbons from Homogeneous and Heterogeneous Porous Media Using Steam," Ph. D. dissertation, University of California, Berkeley, CA.

Udell, K. S., J. R. Hunt, N. Sitar, and L. D. Stewart, Jr. 1991. "Process for *In Situ* Decontamination of Subsurface Soil and Ground water," U. S. Patent # 5,018,576.

Udell, K. S. 1996. "Heat and Mass Transfer in Clean-up of Underground Toxic Wastes", *Annual Review of Heat Transfer*, C-. L. Tien (ed.), Begell House, Inc., New York, NY, 7: 330-405.

SIX-PHASE SOIL HEATING OF THE SATURATED ZONE

Loni M. Peurrung, Theresa M. Bergsman, Thomas D. Powell, Janet S. Roberts, and Ronald Schalla (Pacific Northwest National Laboratory, Richland, WA)

ABSTRACT: Although Six-Phase Soil Heating (SPSH) has been demonstrated to remove recalcitrant compounds such as trichloroethylene (TCE) and perchloroethylene (PCE) from the unsaturated zone of soils (Bergsman et al., 1995; Gauglitz et al., 1994), it had never been applied to a permeable saturated zone. In early 1997, a SPSH field demonstration was successfully conducted in an uncontaminated aquifer at Dover Air Force Base using tracer compounds to mimic Dense Nonaqueous Phase Liquids (DNAPLs) such as TCE and PCE commonly found at Air Force sites. A zone roughly 42 feet (13 m) in diameter and 15 feet (3 m) thick was heated for a total of 30 days. Boiling occurred throughout the aquifer. A total of 50,000 gallons (200 m^3) of condensate was ultimately removed from the site during operations, an amount approximately equal to all the subsurface moisture initially present in the heated region. The energy used was 200,000 kW-hrs. The tracer results showed no significant migration of tracers in the groundwater, some migration of tracers in the unsaturated zone, full recovery of one tracer in the extracted off gas, and 35% recovery of the other.

BACKGROUND

In August 1995, the Air Force Research Laboratory, Airbase and Environmental Technology Division, Tyndall AFB Florida, selected Six-Phase Soil Heating as part of their program to identify technologies for treating DNAPLs in the saturated zone. An expert panel reviewed various technologies, and SPSH was identified as a promising technology for further evaluation. Six-Phase Soil Heating uses electrical resistive heating to raise the temperature of the subsurface to the point that its moisture boils, creating an in-situ source of steam to strip contaminants. A field test was performed at the Groundwater Remediation Field Laboratory (GRFL) at Dover Air Force Base, Delaware, to determine the effectiveness of SPSH for heating the aquifer sufficiently to remove target DNAPL contaminants. This field test was conducted in an uncontaminated aquifer using tracer compounds to mimic DNAPLs commonly found at Air Force sites.

DESCRIPTION OF DEMONSTRATION

An array consisting of six electrodes was installed into the subsurface at the GRFL site as shown in Figure 1. The stratigraphy at the site consisted of layers of sand and gravel with thin clay layers and silt to a depth of 33.5 to 34 feet (10.2 to 10.4 m) below ground surface (bgs) and an underlayer of dense clay containing thin laminations of silt and fine sand. The water table was located at approximately 25 feet (7.6 m) bgs and extended to the clay layer. However,

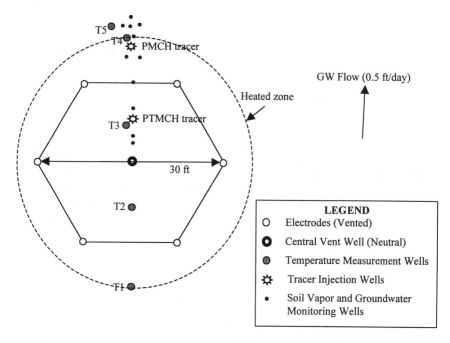

FIGURE 1. Schematic layout of Six-Phase Soil Heating array and injection and monitoring wells.

immediately above the clay was a 2- to 4-ft (0.6- to 1.2-m) thick low-permeability layer of silty and clayey sandy gravel or sands and gravel with interbedded clay layers. Therefore, the upper, high permeability region of the saturated zone was only 5 to 7 feet (1.5 to 2 m) thick. Groundwater flow through the saturated zone was about 0.5 ft/day (0.15 m/day).

Electrodes were installed to a depth of 35 feet (10.7 m) bgs, and the active heated zone extended from 20 feet (6.1 m) bgs to 35 feet (10.7 m) bgs. This design allowed heating of not only the saturated zone but also approximately 5 feet (2 m) of the unsaturated zone above the aquifer to assist in steam collection. The diameter of the electrode array was 30 feet (9 m), creating a heated zone roughly 42 feet (13 m) in diameter and 15 feet (5 m) thick for a total heated volume of about 800 yd^3 (600 m^3). The vapor extraction system used for this demonstration was designed as part of the electrode array to collect soil vapor, steam, and mobilized contaminants (hereafter referred to as "off gas") from each electrode and from a central vent. The vent well was screened from 10 to 20 ft (3.0 to 6.1 m) bgs, and the electrodes were screened from 15 to 20 ft (4.6 to 6.1 m) bgs.

Other below-ground installations included five temperature monitoring wells, tracer injection wells, and soil vapor and groundwater monitoring wells. Within the temperature monitoring wells, thermocouples were spaced vertically at 5-ft (1.5-m) intervals, starting at 12.5 ft (3.8 m) bgs and extending into the aquitard to 42.5 ft (13.0 m), for a total of seven vertical locations.

To study the potential for DNAPL migration and test the effectiveness of the soil vapor extraction system for removing DNAPL mobilized by SPSH, two nonhazardous, low-solubility organic tracers, perfluoromethylcyclohexane (PMCH) and perfluorotrimethylcyclohexane (PTMCH), were selected that mimic TCE and PCE, respectively. Table 1 shows that the tracers have roughly equivalent volatility and density as the contaminants they simulate but the tracers' solubility is much lower. Approximately 1 kilogram of each tracer was injected via screened wells approximately 1 ft (0.3 m) above the bottom of the saturated zone. As shown in Figure 1, the PTMCH was injected inside the array, while the PMCH was injected 1.4 array radii from the array center, which is the edge of effective heating. Injecting a single tracer at two different points essentially "labeled" these locations for the purpose of studying DNAPL migration and removal. Since the originating point of each tracer was known, subsequent sampling and analysis of the groundwater and soil vapor for both tracers during and after heating revealed the extent and direction of migration. Likewise, off-gas sampling for both tracers revealed when each location reached the temperature necessary to treat the mock contaminant. These samples also quantified the amount of each compound extracted from the subsurface.

The soil vapor and groundwater monitoring points were aligned roughly with the injection points and the natural groundwater flow. Four monitoring wells for extracting soil vapor samples were installed to a depth of 20 ft (6.1 m), with a screened interval from 17 to 20 ft (5.1 to 6.1 m) bgs. In addition, soil vacuum was monitored from two of the wells. Six groundwater monitoring wells with 3-ft (1-m) screened intervals were also installed, five for collecting samples near the bottom of the saturated zone and one for sampling near the top to assess vertical tracer migration.

The above-surface equipment included a transformer to convert standard three-phase line power into six phases, a collection header, a vacuum blower, a condenser and knockout box, and granulated activated carbon drums to treat both the off gas and condensate.

TABLE 1. Physical properties of TCE, PCE, and surrogate tracer compounds.

Physical Properties:	TCE	PMCH	PCE	PTMCH
Boiling point, °C	87	76	121	127
Density, g/cm^3	1.46	1.79	1.62	1.89
Water solubility, mg/kg	1100	< 1	150	< 1

RESULTS

Power was applied to the array beginning on February 7, 1997. Over 12-17 days, temperatures in the saturated zone within the array rose to 100°C. Figure 2 shows temperatures during heating at about 27 ft and 32 ft (8 and 10 m) bgs as measured at four temperature wells, T1-T4, and the neutral electrode (i.e., the central vent). As shown in Figure 1, temperature wells T2 and T3 were directly within the heated region. Temperature well T1 represents temperatures at the fringe of expected heating hydrologically up gradient of the array. Temperature well T4 is the same distance from the heated region as T1, but down gradient. Heating and boiling of the saturated zone continued for another 13 days while sampling for the tracers proceeded. The total duration of the heating operation was 30 days, during which 50,000 gallons (200 m³) of condensate were removed from the site, an amount roughly equal to all the subsurface moisture initially within the heated zone. The energy used over 30 days was 200,000 kW-hrs. The bulk of the tracer was removed over the first 21 days. The energy used up to that time was 136,000 kW-hrs, and the condensate removed was 29,000 gallons (110 m³).

Saturated Zone Temperatures

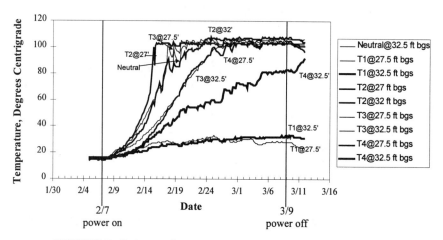

FIGURE 2. Saturated zone temperatures during soil heating.

Tracer sampling results showed no significant migration of tracers in the groundwater, due in part to their low solubility. However, soil vapor samples taken within the heated zone showed evidence of some outward migration of both tracers in the unsaturated zone. The vacuum applied to the soil was therefore increased to improve off-gas capture.

Figure 3 shows the cumulative amount of each tracer recovered by the soil vapor extraction system as calculated from off-gas concentrations and flow rates. Data prior to 2/16 were lost due to a breakdown of the off-gas analysis instrument. The data show the lower-volatility PTMCH coming out more rapidly than the PMCH since the center of the array heated more quickly. A small jump in the total amount of collected tracer on 3/8 was the result of one high-concentration sample that may or may not have been an artifact.

The data show full recovery of the PMCH but only 35% recovery of the PTMCH. The fate of the remaining PTMCH is uncertain. Its appearance in the off gas may have been missed during the analytical system outage during the first

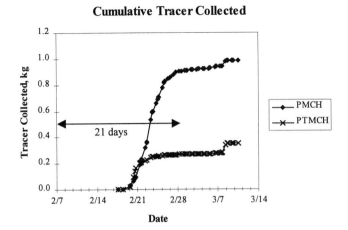

**FIGURE 3. Cumulative removal of the two tracer compounds,
PMCH and PTMCH, in the extracted soil vapor.**

week of heating or, conceivably, in a sudden surge between sampling events. Soil
vapor and off-gas analyses at the end of the operation did not show evidence of
residual tracer. That is, if two-thirds of the PTMCH were still in the subsurface,
one would expect continued, slow evolution, which was not seen. It is possible
that the PTMCH somehow migrated out of the heated zone, but it would have had
to do so from the center of the array (while the PMCH, which was out at the
fringe of the heated zone, was completely captured). None of these theories is
completely consistent with the data and our understanding of how SPSH would
affect DNAPLs. We suspect that some tracer may have been lost by migration
but that more was removed than the data show.

CONCLUSIONS

Six-Phase Soil Heating was successful in heating the saturated zone to
temperatures sufficient to boil the groundwater and remove target DNAPL
contaminants.

A significant portion of the injected tracers was removed during treatment,
indicating that SPSH has the potential to volatilize and remove DNAPL from the
subsurface. The apparent lower recovery of PTMCH may be due to lost or
incomplete data, migration out of the heated zone, or incomplete removal. Soil
vapor samples suggest that the tracer compounds did migrate outward through the
vadose zone when steam was first generated, indicating incomplete control of
vapor by the off-gas collection system. However, the high recovery of PMCH
(the tracer placed at the edge of heating) indicates that an increase in the vacuum
applied to the soil during the operation enabled an overall high capture efficiency
for the system.

Energy requirements for SPSH treatment of an aquifer were roughly as
predicted. At 20 percent of the total cost, energy costs are an important part, but
not a majority, of the overall treatment cost. For the 30-day test, 200,000 kW-hrs

were used and 50,000 gallons (200 m³) of condensate were collected. Most of the tracer was removed during the first 21 days of heating. During that period, 136,000 kW-hrs were used and 29,000 gallons (110 m³) of condensate were collected. At $0.07 per kW-hr, this represents an energy cost of $9,500 or approximately $12/yd³ ($16/m³) heated.

RECOMMENDATIONS

Six-Phase Soil Heating is applicable for full-scale deployment at a DNAPL site. The GRFL demonstration was successful at showing that the technology can be used to heat a flowing aquifer to temperatures sufficient to remove targeted DNAPL compounds. The technology has also been deployed, full scale, at a saturated, tight-soil DNAPL site at Ft. Richardson, Alaska, where it was successful in removing over 90% of the contaminants in six weeks. The success of the GRFL demonstration and the Chicago deployment support moving forward with a full-scale demonstration or deployment of this technology.

REFERENCES

Bergsman, T. M., P. A. Gauglitz, J. S. Roberts, and M. H. Schlender. 1995. "Soil-Heating Technology Shown to Accelerate the Removal of Volatile Organic Compounds from Clay Soils," *Federal Facilities Environmental Journal, Winter 1995/96*, 69-79.

Gauglitz, P. A., J. S. Roberts, T. M. Bergsman, S. M. Caley, W. O. Heath, M. C. Miller, R. W. Moss, and R. Schalla. 1994. "Six-Phase Soil Heating Accelerates VOC Extraction from Clay Soil." In the proceedings of *Spectrum '94: International Nuclear and Hazardous Waste Management* (Atlanta, Georgia, August 14-18, 1994), pp. 2081-2089. American Nuclear Society, La Grange Park, IL.

PLASMA REMEDIATION TECHNOLOGY FOR CHLORINE, SULFUR AND PHOSPHORUS CONTAINING WASTES

Darryl Freed and *Laszlo Heredy, Ph.D.*
(Plasma EnvironmentalTechnologies, Inc., Toronto, ON)
Ferenc Pocsy (Hungaroplazma KFT., Budapest, Hungary)
Kevin R. Bruce and Abderrahame Touati, Ph.D.
(ARCADIS Geraghty & Miller, Research Triangle Park, NC)

ABSTRACT: Tests were conducted evaluating a plasma-based non-incineration technology for the destruction of hazardous liquid organic wastes. In tests performed at the Environmental Research Center of EPA's National Risk Management Research Laboratory the following destruction/removal efficiencies (DREs) were achieved: (a) for polychlorinated biphenyl (PCB) surrogates containing chlorobenzenes and biphenyl: greater than 99.9999%; and (b) for simulants of chemical warfare agents and energetics: greater than 99.99999%. Minimal formation of polychlorinated dibenzo-p-dioxins and furans (PCDD/PCDFs) was found in the PCB surrogate tests: the toxicity equivalency (TEQ) value was 0.03 ng/m^3, nearly one order of magnitude below regulatory limit.

INTRODUCTION

As incineration of hazardous wastes comes under increased scrutiny from the public and from regulatory agencies, non-incineration remediation technologies provide effective alternatives, particularly for certain waste streams such as polychlorinated biphenyls (PCBs), chlorofluorocarbons and chemical warfare agents. Plasma Environmental Technologies, Inc. (PET) has arranged to test a non-incineration thermal treatment technology, Plasma Arcing Conversion (PARCON), to determine its efficiency at destroying PCB-like wastes and simulants for chemical warfare agents.

The PARCON technology was developed in Hungary by the Research Institute for the Electrical Industry (VKI), and a commercial demonstration unit (PARCON 125) was produced by Hungaroplazma Ltd, a privately owned company in Budapest. PET contracted with ARCADIS Geraghty & Miller (formerly Acurex Environmental Corporation) to perform a technology evaluation of the PARCON 125. The device was installed at the Environmental Research Center managed by the Air Pollution Prevention and Control Division of the Environmental Protection Agency's National Risk Management Research Laboratory (EPA/NRMRL) to test the destruction/removal efficiencies with surrogate mixtures simulating the characteristics of (a) PCBs and (b) various chemical warfare agents.

Operation of the PARCON unit was performed by engineers from Hungaroplazma, gas analyses to determine DRE values and gas compositions including the concentrations of trace contaminants were performed by ARCADIS Geraghty & Miller in accordance with applicable EPA testing protocols. This paper presents a summary of the testing results.

DESCRIPTION OF THE PARCON UNIT:

The PARCON 125 components are mounted inside two standard 20 ft. long cargo containers which are stacked to form a two-story installation in the final assembly. A schematic diagram of the system is shown in Figure 1.

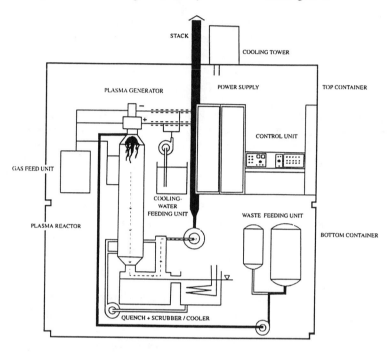

FIGURE 1. Flow diagram of PARCON 125.

The main system components are as follows:

1. a 487V, 60 Hz power supply which provides AC power for the system and 1,000V DC power for the plasma generator;
2. a gas supply system to provide compressed air for the plasma torch and the reactor; and argon gas for start-up;
3. the ceramic lined stainless steel reactor including the air plasma generator in the upper section of the reactor;
4. the quenching and neutralization system which includes the lower section of the reactor and a scrubber column; and
5. the cooling water system which provides for intensive cooling to the electrodes of the plasma generator and supplies cooling water to the heat exchanger of the neutralization system; the water system includes a cooling tower located at the top of the upper container.

Operation of the PARCON machine begins by starting the non-transferred arc type plasma generator with argon gas and then switching the gas supply to air. After feeding clean fuel oil into the air plasma until the reactor reaches the desired operating temperature, feeding of the waste begins. The finely dispersed liquid waste is injected directly into the air plasma torch which enters the reactor at a temperature of 3,000-5,000 °C. Destruction of the waste begins in the top section of the reactor, the plasma chamber, by thermal disintegration to its atomic constituents. Partial oxidation of the disintegrated components also takes place in this reactor section. Full oxidation to CO_2 and H_2O occurs in the central section of the reactor by air introduced radially in several steps. The chlorine content of the waste is converted to hydrochloric acid, the sulfur to sulfur oxides and the phosphorus to phosphorus pentoxide.

In the lower section of the reactor the hot oxidized gases are quenched by a spray of circulating alkaline solution (aqueous sodium carbonate or calcium hydroxide) to prevent harmful recombination reactions which otherwise would lead to the formation of highly toxic PCDD/PCDF derivatives. The cooled gas enters the bottom of a washing tower where its acid content is fully neutralized by the circulating alkaline solution. A water-cooled heat exchanger and a cooling tower are used to reject the heat from the alkaline washing solution.

RESULTS AND DISCUSSION

Testing of PCB Surrogate Mixture. Pending the receipt of a Toxic Substance Control Act (TSCA) permit for testing PCBs and PCB-containing transformer oils, a surrogate mixture of the following composition was used in the tests:

- chlorobenzene (30 percent)
- 1, 2-dichlorobenzene (30 percent)
- biphenyl (40 percent)

The chlorine content of the mixture was approximately 27 weight percent which closely approximates actual PCB waste streams.

The principal objectives of the tests were (a) to determine destruction/ removal efficiencies; and (b) to investigate the possibility of the formation of PCDD/PCDF under typical operating conditions. To this end, exit gas samples were withdrawn over extended periods (up to four hours) to measure the amount of undestroyed surrogate mixture components. A separate exit gas sampling stream was used for investigating the possible presence of PCDDs and PCDFs. Other gaseous reaction products were measured using continuous emission monitors.

EPA Method 0010 for collecting semi-volatile organic compounds (SVOCs) was used in the determination of DREs, with one exception: since no particulate formation or particulates passing through the scrubber was anticipated, the filter was eliminated from the train. The sample trains were extracted using Method 3542, and the extracts analyzed by Method 8270C. Samples for PCDD/PCDF determination were collected using EPA Method 23, again without using filters.

DRE (in percent) was calculated as follows:

$$\frac{\left(POHC_{input} - POHC_{output}\right)}{POHC_{input}} \times 100, \text{ where}$$

$POHC_{input}$ = Mass per unit time of the principle organic hazardous constituent (POHC) fed into the system (kg/h)
$POHC_{output}$ = Mass per unit time of POHC found in the flue gas (kg/h)

The results of DRE tests demonstrated that the PARCON unit is capable of treating surrogates for PCBs to greater than 99.9999% destruction/removal efficiency. A test conducted using a mixture of chlorobenzene and o-dichlorobenzene yielded DREs of 99.99993 percent and 99.99995 percent, respectively, while testing of a 1: 1 mixture of the ternary PCB surrogate and No. 2 fuel oil yielded DREs over 99.9999 percent. Minimal formation of PCDD/PCDF occurred in the system. No measurable PCDDs were found; the value of the toxicity equivalency quotient (TEQ) was 0.03 ng/m^3, nearly one order of magnitude below the regulatory limit. Details of the experimental conditions are shown in Table 1.

TABLE 1. Summary of DRE Tests for PCB Surrogates.

Parameter	Test 1	Test 2
POHC	chlorobenzene 1,2-dichlorobenzene	standard mix in No. 2 fuel oil
Waste Feed Rate (kg/h)	4.68	4.58
Total Air (Nm^3/h)	86	70
Injection Air (Nm^3/h)	11	10
Reactor Temperature (°C)		
T1	991	894
T2	1369	1353
T3	1219	1185
DRE (%)		
chlorobenzene	99.99993	99.99990
1,2-dichlorobenzene	99.99995	99.99993
biphenyl	NA	99.9995

NA = not applicable

Testing of Simulants for Chemical Warfare Agents and Energetics. These tests were conducted using similar general operating conditions and analytical procedures as in the tests with PCB surrogates, except for the following important difference: the gas samples for DRE determination were withdrawn from the lower part of the plasma chamber using a quartz sampling tube. The results demonstrate the destruction efficiency of the ionized, hot plasma stream in a very short residence time of less than 150 ms. The simulants used in these tests were taken from a list developed by the U.S. Army's Chemical and Biological Defense Command. The test results are summarized in Table 2.

TABLE 2. Test results using simulants of chemical warfare agents and energetics.

Parameter	Test 1	Test 1	Test 1	Test 2
Simulant	Thioanisole	Dimethyl methylphos-phonate	Tributyl phosphonate	2,4-Dinitro-toluene in toluene
Material simulated	HD, HT	GB	VX	TNT, Tetrytol
Waste Feed Rate (kg/h)	2.18^a	0.736^a	2.01^a	0.45^b
Total Air (Nm^3/h)	40	40	40	40
Reactor Temperature (°C)				
T1	850	850	850	1270
T2	1350	1350	1350	1350
T3	570	570	570	1200
DRE (%)	>99.999992	>99.999994	>99.999998	>99.999996

[a]Total simulant feed rate was 4.9 kg/h
[b]Total simulant/toluene solution feed rate was 5.1 kg/h

DESTRUCTION OF HAZARDOUS MILITARY WASTES USING PLASMA ARC TECHNOLOGY

Mahmood A. Qazi (Concurrent Technologies Corp., Johnstown, Pennsylvania)
Louis Kanaras (US Army Environmental Center, Aberdeen Proving Ground, MD)

ABSTRACT: A Plasma Arc Technology (PAT) system treats hazardous wastes in a furnace, at temperatures of 2000° C, or higher, using a plasma torch. The organic components vaporize, decompose or oxidize. The off-gases consist of hydrogen, carbon monoxide, carbon dioxide and nitric oxides. A wet air scrubber is used to remove most of these gases. The scrubber water is treated and recycled. Metal-bearing solids are melted or vaporized. The solids are usually recovered as molten metal, or as non-leachable vitrified slag, suitable for disposal in a landfill.

The evaluation, a two-phase program, included determining the process capability of PAT for the ultimate destruction of hazardous components of nine military hazardous wastes. Goals included verifying the suitability of the resulting slag for regular landfill disposal; identifying potential hazards associated with the process emissions; and developing qualified cost estimates for the future utilization of the process on large scale operations. Process variables such as feed rate, composition of off-gases, run duration, percentage of oxygen used in the secondary chamber, and chamber temperature were controlled. Material balance, destruction and removal efficiency (DRE), air emission quality, wastewater quality, and suitability of slag for landfill were determined. The results from these test runs and analyses are presented in this paper. The ultimate success of PAT for destroying hazardous wastes depends on achieving a slag that is non-leachable and meets Resource Conservation and Recovery Act (RCRA) requirements, and off-gases and scrubber water that meet Clean Air and Clean Water Act standards. Results of Phases 1 and 2 show that all slags passed Toxicity Characteristics Leaching Procedure (TCLP) tests, for land disposal requirements; air quality meets California standards, and scrubber water can be treated and recycled.

INTRODUCTION

The National Defense Center for Environmental Excellence (NDCEE), operated by Concurrent Technologies Corporation (*CTC*), was tasked to evaluate the applicability of plasma arc technology (PAT) to treat complex military wastes. The U.S. Army Environmental Center (AEC), sponsor of this task, established a statement of work (SOW) containing two phases to accomplish the evaluation.

Phase I. Identification and selection of candidate waste materials, identification and selection of a suitable plasma waste treatment system, and conduct of testing.

Phase II. Identification and selection of additional candidate waste materials, conduct of testing, documentation of results, preparation of a procurement/design/ fabrication guidance manual, and production of a video presentation.

BACKGROUND

This paper reports the results of Phases I and II pilot-scale testing of candidate waste materials using PAT. PAT can be used to convert complex hydrocarbon molecules into common, simple molecules. PAT processing temperatures of 3632°F (2000°C) and higher lead to breakdown of the organic components of the waste into smaller molecules, while the non-volatile inorganic components are vitrified in the resulting slag. Plasma arc furnaces are currently used commercially for industrial and manufacturing applications, including iron ore reduction, titanium scrap melting, platinum and aluminum metal recovery, and steel ladle or tundish heating. During the past two decades, plasma arc technology has emerged as a very effective method of converting hazardous waste materials into recyclable, non-hazardous by-products. Plasma arc furnaces have the potential to stabilize weak foundation soils in-situ, produce energy, and destroy or immobilize waste materials in a cost-effective and environmentally safe manner. The system was used to evaluate this technology for destruction of nine military hazardous wastes, as follows:

- sludge from Longhorn Army Ammunition Plant, TX;
- agriculture blast media from Letterkenny Army Depot;
- glass and plastic glass media from Letterkenny Army Depot;
- medical incineration ash from Aberdeen Proving Ground, MD; and
- contaminated soil from open burning/open detonation ground at Picatinny Arsenal, NJ.
- waste paint from a U.S. Navy facility;
- garnet blast media from a U.S. Air Force Base;
- simulated oil-contaminated sorbent, used by Tri-Services and private industry; and
- Mendocino soil spiked with dichlorobenzene, a surrogate for PCB.

Through a competitive bidding process, *CTC* subcontracted the pilot-scale testing to Retech, Inc. of Ukiah, CA. Retech proposed a Plasma Arc Centrifugal Treatment (PACT) system for this task. Retech designated their system the PACT-2, where 2 designates a furnace that is two feet in diameter. At the heart of the PACT-2 system is the RP-75T Plasma Torch, which is capable of a sustained output of 160 kilowatts (kW). This torch operates in the transferred arc mode, in which the rear electrode is the positive attachment point and the negative attachment point is the workpiece or molten material.

RESULTS

The PAT testing program was divided into Phase I (tests 1-18) and Phase II (tests 19-31). Test #19 was a repeat of test #1 for medical ash because of operational problems with the latter. Actual number of valid tests performed per the SOW was 30. Phase I was divided into Phase IA (tests 1-6) and Phase IB (tests 7-18). Equipment modifications were completed after tests 6 and 18. Nine different waste materials, listed in Table 1, were treated during Phase I and II testing.

TABLE 1. Treated Waste Materials

Test Series	Waste Type	Source
Phase I	Longhorn Sludge (LS)	Longhorn Army Ammunition Plant, TX
	Open Burning Ground Soil (OBG)	Picatinny Arsenal, NJ
	Agricultural Blast Media (ABM)	Letterkenny Army Depot, PA
	Plastic and Glass Blast Media (PGBM)	Letterkenny Army Depot, PA
Phase I & II	Medical Incinerator Ash (MIA)	Aberdeen Proving Ground, MA
Phase II	Garnet Blast Media (GBM)	McClellan Air Force Base, CA
	Surrogate Absorbent Material (SAM)	Retech, CA
	Waste Paint (WP)	Norfolk Naval Base, VA
	Dichlorobenzene spiked Mendocino Soil (MS/DCB)	Mendocino County, CA

Feedstock materials were selected, based on existing waste disposal requirements at military installations. The selection criteria include: their hazardous waste status; the lack of satisfactory treatment and disposal procedures; the presence of energetics or heavy metals; potential cost benefit to treating the waste with a PAT; and the existing inventory of a particular waste. A total of 5203.52 lb. of waste was treated in the PACT-2 during Phases I and II. The results are summarized in Tables 2 and 3.

TABLE 2. Summary of Phase I PAT Testing Results

Retech Test No.	Waste Type	DRE	TCLP	CO	Amount Processed (lb)	Rate (lb/h)	Scrubber Water[2]	Air Particulate Emissions (g/hr)
1R	MIA	99.9999	Pass	Pass	162.5	31.0	T.A.	Fail (8.46)
2	MIA	99.9999	Pass	Pass	49.3	13.0	T.A.	Fail (26.08)
3	MIA	N/A	Pass	Fail	48.6	13.0	T.A.	Fail (4.31)
4	MIA	99.9999	Pass	Pass	48.6	16.0	T.A.	Fail (21.59)
5	LS	N/A	Pass	Pass	58.8	34.0	T.A.	Fail (31.52)
6	LS	99.9999	Pass	Pass	96.5	25.0	T.A.	Fail (28.17)
7	OBG	N/A	Pass	Pass	152.4	83.0	T.A.	Pass (0.09)
8R	OBG	99.9999	Pass	Pass	105.8	90.0	T.A.	Fail (6.94)
9	OBG	N/A	Pass	Pass	303.98	134.0	T.A.	Pass (0.01)
10	OBG	99.9999	Pass	Pass	318.63	113.0	T.A.	Pass (0.09)
11	LS	N/A	Pass	Pass	244.07	82.0	T.A.	Pass(0.11)
12	LS	99.9999	Pass	Pass	177.19	59.0	T.A.	Fail (0.17)
13	ABM	N/A	Pass	Fail	138.8	36.0	N.D.	Pass (0.09)
14	ABM	N/A	Pass	Fail	166.36	32.0	N.D.	Fail (0.92)
15	PGBM	N/A	Pass	Fail	47.5	28.0	N.D.	Fail (0.32)
15R	PGBM	N/A	Pass	Fail	101.0	34.0	N.D.	N/A
16	PGBM	N/A	Pass	Fail	108.02	30.0	N.D.	Fail (0.14)
17	PGBM	N/A	Pass	Fail	122.47	40.0	T.A.	Fail (5.22)
18	PGBM	N/A	Pass	Pass	137.4	36.0	T.A.	Fail (4.90)

Because of the potentially large variation in the chemistry of materials processed in PAT systems, feedstock material was thoroughly analyzed prior to processing. This analysis yields important information regarding composition, which in turn influences process parameters (i.e. oxygen, gas flow values, potential metals emissions, etc.). Pre-qualification testing was performed to test and adjust system parameters prior to long term processing. Stack emissions were monitored closely, when increasing feed rates, to avoid overloading the off gas cleaning train.

One of the major objectives of any PAT system is to process hazardous wastes in a manner that allows the PAT solid effluent to be taken to a non-hazardous landfill. Based on the parameters measured for waste slag from a PAT system, the slag could be used for roadbed materials, aggregate for hot and cold mix asphalt, blast media, abrasives, or filter media. In order to reduce the overall waste from the PAT System, the wet scrubber solution in Phase II was not changed after every test, only the solid sludge was removed. This solid sludge can be taken to a non-hazardous landfill.

TABLE 3. Summary of Phase II PAT Testing

Retech Test No.	Waste type	DRE	TCLP	NO CO HCl	Amount Processed (lb)	Rate (lb/h)	Scrubber Water[2]	Air Particulate Emissions (g/h)
19	MIA	99.9999	Pass	Pass	598	124.6	N.D.	Pass (0.47)
20,21,22	SAM	99.9999	Pass	Pass	240	32.7	N.D.	Pass (0.154)
23,24,25	GBM	N/A	Pass	Pass	772	85.3	N.D.	Pass (0.032)
26,27,28	MS/D CB	99.991	Pass	Pass	654	68.3	N.D.	Pass (0.067)
29,30,31	WP	99.9999	Pass	Pass	351.6	34.9	N.D.	Fail (0.334)

[1]Dry weight
[2] Scrubber water is hazardous waste and should be treated and recycled
N.D.= Non-Detection of SVOCs (tested for)
DRE = Destruction and Removal Efficiency
CO = Carbon monoxide
TCLP = Toxicity Characterization Leaching Procedure
T.A. = Trace Amounts of either SVOCs or VOCs (tested for).
SVOC = Semivolatile organic compounds
VOC = Volatile organic compounds

LESSONS LEARNED

Equipment Changes
Several equipment changes were implemented after Phase I testing.

Throat Design. The PC throat was fabricated from a high alumina casta-ble, Novacon 95®, which replaced the silicon carbide and Ruby plastic used previously. This material produced a 10-fold improvement in throat life, thereby decreasing overall refractory maintenance.

Geometry. Throat geometry was changed from a cylindrical to a parabolic cross-section. This geometry change aided in reducing throat closure during treatment as well as the quantity of unprocessed material falling through the throat.

Tub Materials. The tub bottom refractory was changed from a graphite ring to Nuline *RS20*® (magnesia) conductive refractory brick. This change facilitated hot and cold torch starting and produced a better conductive path for both heat and electricity at the plasma-slag interface.

Auxiliary Torch Starter (ATS). The auxiliary torch starter (ATS) is a slag heater used in the PACT-2 unit to transfer the primary torch when the slag skull is not molten. This system was used because chipping the tub bottom damaged the tub refractory, and direct transfer of the arc to the tub bottom severely eroded the refractory near the throat. In operation, the primary torch transfers to the ATS, effectively heating the relatively constant slag layer using the evolved hot gases. The plasma arc transfers once the slag layer reaches approximately 3,000 °F (1649°C). After plasma arc transfer occurs, the ATS is removed vertically from the tub by a hydraulic manipulator. A typical slag skull preheat cycle was about thirty minutes.

Tub Drive Upgrades. The tub hydraulic drive was upgraded to produce higher torque during normal tub rotation. This upgrade reduced tub seizure due to refractory melt from the spool section, or from feed material depositing in the gap between the spool and tub. The mold extractor interconnect arm was redesigned to ensure a constant grip around the mold. This redesign also reduced fatigue failure induced by cyclic heat and mechanical stresses. Mold alignment software was debugged so that the mold consistently lined up with the mold extractor. An Aerojet designed, platelet-cooled electrode replaced the standard Retech electrode to increase normal operating life by using diffusion-bonded, directionally-cooled surfaces and argon surface-arc diffusion at the arc termination-electrode interface. The SCC baffle was removed to enhance combustion efficiency and to reduce the amount of refractory maintenance required during Phase I. The inlet and outlet of the SCC were changed to increase the off-gas residence time.

A feed rate calibration system was installed to properly determine the weight of material at a given feed rate, as determined by the requirements of an individual test. Feeder can flight geometry was modified to produce a more constant feed to the primary chamber. The new flight geometry eliminated "spiked" cyclic feed behavior, and permitted better control of the combustion process. A liquid lance with pumping unit was designed to feed waste paint and other liquids.

The off-gas filtering system was modified to provide two filtering stages. The first stage was a sintered, commercially pure (CP) titanium (Ti*)* metal filter placed at the exhaust of the wet scrubber. The filter had a nominal pore size of 10 micrometers, and acted as a coarse prefilter for the second stage. It was expected that the majority of the particulates would be kept in the scrubber water. The second stage was a 0.1 micrometer, sintered CP Ti filter located downstream from the 10 micron filter. The sintered metal filters were intended to be more resistant to moisture in the gas stream. These filters also can be cleaned and reused, thereby eliminating the disposal of a HEPA filter secondary waste stream.

Wet scrubber liquor was sampled after every test but was not changed. The liquor was pumped into a circulating holding tank, from which representative samples were taken. The liquor remained in the tank (not circulating) overnight to allow for settling of the particulate material. The supernatant liquor was then pumped back into the scrubber. The sludge was placed in sludge bins to be reintroduced into the system at a later date.

NOx and THC measurements were conducted before and after the wet scrubber (off-gas treatment system) in order to determine NOx and THC removal efficiencies.

Method and Procedural Modifications After Phase I Testing

The technique of adding hexachlorobenzene (HCB) seemed to work well as the principal, organic, hazardous constituent (POHC). Being a SVOC, it makes transportation to the laboratory for analysis much less critical than when using xylene. The spiked feed also can be stored without evaporation of the HCB. The HCB is easy to handle while weighing, because it is a granular solid.

The metal spikes did not work well for the fine, fluffy type wastes treated (such as MIA and ABM) because the metal spikes themselves were powders. The powders tended to be blown around in the system. Under ideal conditions, the metal spikes should coat the waste particles uniformly. This would better mimic the behavior of the actual waste and eliminate scatter in the elemental analyses of the feed.

A reliable and accurate analytical method for total metals is required for glassy silicate slags, because EPA Method 6010 is not adequate. Slag chemistry analyses show that analyses using EPA 6010 consistently report lower values than those concentrations actually present. The glassy slags require a digestion. The boiling nitric acid digestion called for in Method 6010 is not suitable.

The sampling procedures for heterogeneous wastes need to be re-evaluated in order to compensate for potential sampling inaccuracies. There is so much scatter inherently generated by the present techniques that a reliable, representative analysis cannot be obtained.

Chemical analyses on the as-received wastes are needed. Representative analyses should be requested from the waste suppliers, but analyses of the as-received wastes are essential in order to eliminate confusion caused by different parties performing different analyses at different laboratories. An accurate assessment of the waste to be treated is required before testing begins.

The standard metals, organics, and semi-volatile organic compounds analyses performed do not provide sufficient information from an operational standpoint. A moisture analysis and a method that determines the total organics available (such as loss on ignition or total organic content) are critical for controlling the treatment process. Any future testing, involving a chloride containing waste or a chloride spike, should allow for a chloride mass balance. A better method to determine organic content of the waste, including free carbon, needs to be identified and used.

Two important process parameters are oxygen and carbon monoxide content in the PC off-gas pipe. These parameters are critical in assuring the correct combustion atmosphere (reducing or oxidizing). Gas sampling in this area is difficult because of the high temperature, the acid gases present, and the amount of particulates present. A sampling technique needs to be developed to solve this problem.

Equipment Modifications During Phase II Testing

The only design change implemented during Phase II testing was to increase the surface area of the CP Ti filters in the off-gas filtering system. It was doubled prior to test 22 (2.3 ft^2 to 4.6 ft^2) to increase run time when processing materials that generate large amount of particulate material.

CONCLUSIONS AND RECOMMENDATIONS

The PAT system tested successfully treated the wide variety of complex military wastes selected by the Army/*CTC* team. Nearly all tests met or exceeded the design feed rate recommended by the equipment supplier. Lower feed rates were used for the organic wastes because of their high volatile content, and higher feed rates were used for the inorganic wastes because of their low volatile content. For all wastes treated, the Destruction and Removal Efficiencies (DREs) were in excess of the Environmental Protection Agency (EPA) requirement 99.99%.

All slag materials formed passed Toxicity Characteristics Leaching Procedure (TCLP) testing, which demonstrates that PAT can be used to produce a product with commercial value, or one which can be landfilled. Gas emissions for nitrous oxide, carbon monoxide, (except for test 13-17) hydrochloric acid, or metals were below permit levels. Particulate emissions were below permit level for all wastes tested, with the exception of the paint waste, for which there were problems with soot build-up. Scrubber water, containing metals, particulates, and dissolved solids can be treated, for example, by processes such as reverse osmosis, to remove these impurities. The water then can be recycled.

For future applications, the following recommendations should be considered:

- A portable PACT-5 or fixed PACT-8 system is recommended. The system should be installed at a regional location to treat wastes from surrounding bases which have sufficient materials for treatment and disposal.
- The system should be located in those states where permits are easily obtainable.
- For efficiency of the operation, the system should be run on a 24-hour basis, seven days per week.
- More pilot scale studies should be performed to evaluate the applicability of a PAT system to other hazardous wastes, e.g. chemical weapons.

REFERENCES

1. D.J. Freeman and H.H. Zaghloul, "Evaluation of Plasma Arc Pyrolysis for the Destruction of Hazardous Military Wastes," presented at the 86th Annual Meeting of the Air and Waste Management Association, Denver, CO, June 15 - 17, 1993.
2. D. J. Freeman and H.H. Zaghloul, "Evaluation of Plasma Arc Pyrolysis for the Destruction of Thermal Batteries and Proximity Fuzes", presented at the 1993 JANNAF Environmental Protection Committee Meeting in Las Cruces, NM, August 1-4, 1993.
3. D.J. Freeman, R.E. Haun, and H.H. Zaghloul, "Destruction of Pyrotechnics Manufacture Wastewater Treatment Sludge by Plasma Arc Pyrolysis," presented at the 1994 JANNAF Environmental Protection Committee Meeting, San Diego, CA, August 1 - 4, 1994.
4. Louis Kanaras, R.J. Patun, M.A. Qazi, "Destruction of Hazardous Waste Materials using Plasma Arc Treatment Technology," presented at the Interna-

tional Symposium on Environmental Technologies, Plasma Systems and Application, October 8 - 11, 1995, Atlanta, GA.

5. Retech Reports

    Retech 1600-9796-002 Rev. 2; Interim Report - Plasma Arc Centrifugal Treatment Tests; 12 DEC 95 (Phase I)

    Retech 1600-9796-002 Rev. 3; Interim Report - Plasma Arc Centrifugal Treatment Tests; 12 FEB 96 (Phase I)

    Retech 1600-9796-005 Rev. 0; *CTC* test 19 Report; Medical Incinerator Ash (MIA) Treatability Study (Phase II); 11 JUN 96

    Retech 1600-9796-009 Rev. 0; *CTC* Tests 20, 21, & 22 Report, Surrogate Absorbent Material (SAM) Treatability Study (Phase II); JAN 97.

    Retech 1600-9796-010 Rev. 0; *CTC* Tests 23, 24, and 25 Report, Garnet Blast Media (GBM) Treatability Study (Phase II), 6 FEB 97.

    Retech 1600-9796-011 Rev. 0; *CTC* Tests 26, 27, & 28 Report, Mendocino Soil Spiked (MSS) with Dichlorobenzene Treatability Study (Phase II), 6 FEB 97.

    Retech 1600-9796-012 Rev. 0; *CTC* Tests 29,20, and 31 Report, Waste Paint (WP) Treatability Study (Phase II); 6 FEB 97.

    Retech 1600-9796-015 Rev. 0; *CTC* Summary Report (Phase II); 6 FEB 97.

    Retech 1600-9796-016 Rev. 0; *CTC* Technical Report - Phase I and II Test Results; 6 FEB 97.

6. Concurrent Technologies Corporation (*CTC*) Reports

    Plasma Arc Technology Evaluation Task, Waste Stream Material Report, *CTC*, December 16, 1994

    Plasma Arc Technology Evaluation Task, Activity Log, V. Saccone, *CTC*

    Plasma Arc Technology Evaluation Task, Test Plan, *CTC*, February 2, 1995

    Plasma Arc Technology Evaluation Task, Quality Assurance/Quality Control Plan, *CTC*, January 4, 1995

    Plasma Arc Technology Evaluation Task, Equipment Selection Report, *CTC*, July 11, 1995

    Plasma Arc Technology Evaluation Task, Resource Utilization Plan, *CTC*, November 18, 1994

    Plasma Arc Technology Evaluation Task, Safety Plan, *CTC*, June 19, 1995

CATALYTIC COMBUSTION OF CARBONTETRACHLORIDE

Süheyda ATALAY (Ege University, Bornova, Izmir, Turkey)
Tamer Tanılmış, H.Erden Alpay and Ferhan S. Atalay (Ege University, Bornova,
Izmir, Turkey)

ABSTRACT: The catalytic combustion of carbontetrachloride was investigated on metal oxide catalysts coated on the monolith support. The prepared catalysts were tested at the different temperatures between 300 and 800°C and the varying GHSV values with an excess air ratio of 3100%. The catalyst, having the composition of 18% Cr_2O_3, 2% Ce_2O_3 and 80% γ-Al_2O_3, was found to be suitable for the almost complete destruction of carbontetrachloride. The operating conditions were proposed as 5702 h^{-1} for GHSV, 3100% for the excess air. The temperature should be slightly higher than 800°C. The reaction rate expression was found to be independent of oxygen partial pressure and strongly dependent on carbontetrachloride partial pressure.

INTRODUCTION

Chlorinated hydrocarbons are widely used chemicals in industry and the wastes of them are classified as hazardous and toxic. The simplest subgroup of the chlorinated hydrocarbons family is the chlorinated methanes, which are generally used as industrial solvents including methylchloride (CH_3Cl), methylenechloride (CH_2Cl_2), chloroform ($CHCl_3$) and carbontetrachloride (CCl_4). Incinerating difficulties of these compounds are ranked from the most to the least as CCl_4, $CHCl_3$, CH_2Cl_2 and CH_3Cl (Miller et al., 1984).

Although the incineration of the gas effluent containing these compounds presents several complications, the catalytic incineration technique used successfully in the destruction of volatile organic compounds may overcome these difficulties. In the studies on the chlorinated hydrocarbons both in open and patent literature it is seen that more severe conditions than that of VOCs are required (Subbanna et al., 1988 ; Weldon and Senkan, 1986 ; Sare and Lavanish, 1977; Yang et al., 1977). The thermal incineration of chlorinated hydrocarbons requires the temperatures up to 1200°C involving a high fuel oil consumption (Müller et al., 1993). However, in the catalytic combustion of chlorinated hydrocarbons, mild conditions are required and these conditions change according to the hydrocarbon studied.

The catalysts often used are the active catalysts prepared by the nobble metals such as platinum, rhodium, and palladium or the metaloxide catalysts containing the oxides of cobalt, copper, manganese or chromium resistant to chlorinated compounds. Since the washcoat monolith carrier has some advantages to a conventional carrier, it is preferred in the combustion of chlorinated hydrocarbons (Barresi et al., 1992 ; Tichenor, 1987 ; Friedel et al., 1993).

In the study on combustion of waste gases containing chlorinated hydrocarbons, a bimetallic Pd, Pt / Al_2O_3 catalyst has been used with an each pollutant concentration of 0.1 vol% (Müller et al., 1993). For a GHSV value of

15 000 h^{-1} and for a temperature range of 350-680°C, it was found that the waste gas could be oxidized totally and that the catalytic combustion had a significant economical and technical benefits compared to the conventional thermal incineration.

In our earlier study catalytic combustion of methylenechloride was searched and a temperature range between 550 and 600°C and a GHSV value of 78 000 h^{-1} were found to be suitable conditions to destruct methylenechloride completely on bimetallic Cr, Ce / Al_2O_3 monolith catalyst (Ballıkaya et al., 1996).

A few of the studies on catalytic combustion could explore the kinetics of chlorinated hydrocarbons. The kinetics of catalytic oxidation of CH_3Cl on Cr_2O_3 catalyst obeyed a nonlinear second order rate model and the reaction rate was independent of the concentration of oxygen (Weldon and Senkan, 1986). In the combustion of chlorinated hydrocarbons from soil venting and ground water remediation on the commercial HDC catalyst developed by Allied Signal a negative order with respect to 1,1,2-trichloroethane was reported. In our earlier work the kinetics of the catalytic combustion of methylenechloride was searched by using the model proposed by Downie based on Mars and van Krevelen mechanism and it was found that the reaction was dependent weakly on methylenechloride partial pressure and strongly on oxygen partial pressure (Ballıkaya et al., 1996).

Objective. The objective of this study is to determine the process conditions satisfying the total combustion of carbontetrachloride on the prepared catalysts and to explore the kinetics of the reaction. This study should be evaluated in the frame of preparation of the data base on catalytic combustion of the chlorinated methanes. In the earlier study the catalytic combustion of methylenechloride was searched.

MATERIALS AND METHODS

The experimental system consisting primarily of a vaporiser, a preheater and a fixed bed reactor and the procedure applied were presented in detail elsewhere (Ballıkaya et al.,1996). The two catalysts used in this study were prepared as mixed metal oxides on a cordierite type ceramic monolith. The compositions of the catalysts are given in Table 1.

TABLE 1. Compositions of the catalysts.

Catalyst No	Components (w%)		
	γ-Al_2O_3	Cr_2O_3	Ce_2O_3
1	80	15	5
2	80	18	2

The experiments were performed at atmospheric pressure and using very lean hydrocarbon concentrations changing from 1418 to 7374 vpm. A few runs were conducted to have an idea about the possibility of the reactions between the gas reactants in the absence of the catalyst. In these experiments the reactor temperature range was between 300 and 700°C and excess air was 3100% (4357 vpm). Uncoated monolith pieces were placed into the reactor and the GHSV value was kept at 7603 h^{-1}. The performances of the prepared catalysts were searched carrying out the experiments at different temperatures changing from 300 to 800°C. Then the first set of the kinetic experiments was carried out with constant partial pressure of oxygen and the second set was carried out with constant partial pressure of carbontetrachloride to investigate the dependency of the reaction rate on partial pressures of the reactants.

RESULTS AND DISCUSSION

The homogeneous reactions between 300 and 700°C were evaluated by calculating the conversion of carbontetrachloride, which was the measure of combustion efficiency. The calibration experiments performed at 100 °C by changing flow rate of carbontetrachloride with an excess air of 3100% were used in this evaluation. The gas phase reactions without need a catalyst were in substantial amount. Even at a temperature of 300 °C, approximately 30% of carbontetrachloride was converted.

The Catalyst1 was tested at different preheater (250 °C and 350 °C) and reactor temperatures(between 300 °C and 800 °C). The preheater temperature and the reactor temperature are proportionally effective on the combustion efficiency. At a preheater temperature of 350 °C and at the reactor temperature of 800 °C and with a GHSV value of 7603 h^{-1} the combustion efficiency was increased up to 81 %. To be able to increase the conversion GHSV values were changed down to 4701 h^{-1}. As it is expected, the combustion efficiency increased as GHSV value was decreasing. The maximum conversion be obtained was 84%. Due to setup limitations, it was not possible to decrease the GHSV to amuch lower value.

The combustion efficiencies at the studied conditions were not satisfactory, and it was decided to prepare the second catalyst containing more chromium. The second catalyst was tested at a preheater temperature of 350°C and at a GHSV value of 5702 h^{-1} taking the experiences obtained with the Catalyst 1 into consideration.Thecombustion efficiencies obtained by using this catalyst are shown

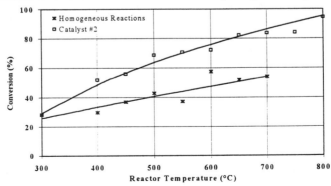

FIGURE 1. The performance experiments with the Catalyst 2.

in Figure 1.The benefit of using the catalyst can be seen easily from the figure. The second catalyst gave us more efficiency than the first catalyst; it enabled at least 25% extra efficiency. It satisfied 94.5% conversion at 800 °C. Although this conversion could be increased by increasing the temperature or by decreasing GHSV value, no further experiment was performed and it was decided that the catalyst having the given composition could success the total combustion by arranging the operating conditions.

The kinetic experiments were studied on the second catalyst and temperature of the reactor was changed from 350 °C to 500 °C. The conversion values in the kinetic experiments were greater than 10%. It was not possible to operate the reactor in differential mode at the studied conditions. For that reason, the reaction rates could not be calculated directly. The conversion values in the experiments with variable partial pressure of carbontetrachloride were plotted versus space time values at each reactor temperature studied(Figure 2). And the equations describing conversion vs space time were derived and the reaction rates at the different temperatures were calculated by differentiating these equations.

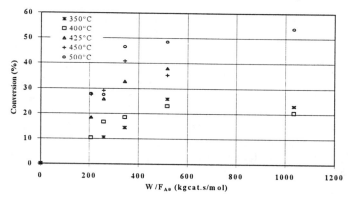

FIGURE 2. Coversion values vs space time at constant oxygen pressures.

In the experiments carried out at variable partial pressure of oxygen and nearly constant partial pressure of carbontetrachloride. The oxygen conversions were calculated and plotted versus space time values (Figure 3). Although the conversions calculated according to the carbontetrachloride was very higher than 10%, the oxygen conversions were calculated to be less than 10%. The experimental points in Figure 3 could be expressed by line equations. This finding was explained by the reaction rate was not affected by oxygen partial pressure at studied conditions. By using the least squares method, equation of each line was derived and reaction rates were calculated by slopes of these lines. The calculated reaction rates are given in Table 2.

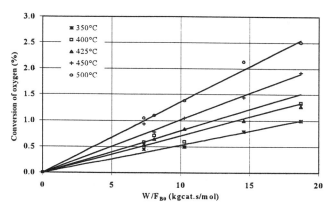

FIGURE 3. **Coversion values vs space time at constant carbontetrachloride pressures.**

TABLE 2. **Reaction rates at different temperatures in the experiments with the variable oxygen partial pressure.**

Reactor Temperature (°C)	Log Mean P_{O2} (Pa)	Reaction Rate (mol/kgcat.s)$\times 10^5$
350	7350 – 18 772	53.453
400	7337 – 18 757	71.483
425	7340 – 18 762	80.988
450	7314 – 18 719	101.915
500	7290 – 18 706	136.335

The kinetic model proposed by Mars and van Krevelen based on the reaction of the hydrocarbon from gas phase with adsorbed oxygen was assumed and it was simplified by assuming the oxygen adsorption was a rapid process compared to the surface reactions. At the same time a kinetic expression was derived according to mass of actions and remembering that the finding on the dependencies of the reaction rates on partial pressures of the species it was reduced to

$$-r_{CCl_4} = kP_{CCl_4}^m \qquad (1)$$

The values of reaction rate constant, k, at different temperatures and reaction order satisfactorily representing the experimental data were determined. The reaction order was found to be 1.14. The change of k values with respect to temperature tested by Arrhenius equation and the following expression was found:

$$k = 8.75 \times 10^{-6} \exp\left(-\frac{18,930}{RT}\right) \qquad (2)$$

CONCLUSIONS

A bimetallic catalyst having Cr, Ce / Al_2O_3 on monolith support was able to combust carbontetrachloride almost completely. The temperature slightly higher than 800°C and a GHSV value of 6000 h^{-1} were recommended.

The reaction rate was dependent on the hydrocarbon partial pressure and it could be taken independent of partial pressure of the oxygen. Starting the Mars and van Krevelen mechanism, the reaction rate was simplified to a nonlinear equation containing only carbontetrachloride partial pressure. Order was found 1.14. The activation energy for the surface reaction was found 18 930 J/mol.

ACKNOWLEDGEMENT

The authors would like to acknowledge the support of this work by Turkish Scientific and Technical Council.

REFERENCES

Ballıkaya, M.M., S. Atalay, H.E. Alpay, and F.S. Atalay. 1996. "Catalytic Combustion of Methylenechloride." *Combust.Sci. an Tech.* 120 : 169-183.

Barresi, A.A., I Mazzarino, and G. Baldi. 1992. "Gas Phase Complete Catalytic Oxidation of Aromatic Hydrocarbon Mixtures." *The Canadian Journal of Chemical Engineering.* 70 : 286-293.

Friedel, I.M., A.C.Frost, K.J.Herbert, F.J.Meyer, and J.C.Summers. 1993. "New Catalyst Technologies for the destruction of Halogenated Hydrocarbons and Volatile Organics." *Catalysis Today.* 17(1-2): 1-18.

Miller, D.L., W.S. Dwight, V.A. Cundy, and R.A.Matula.1984."Chemical Considerations in the Incineration of Chlorinated Methanes. I-Methylchlorie." *Hazardous Waste.* 1(1). 1-18.

Müller, H., K. Deller, B. Despeyrouk, and E. Peldssuz. 1993."Catalytic Purification of Waste Gases Containing Chlorinated Hydrocarbons with Precious Metal Catalysts." *Catalysis Today.* 17: 383-390.

Sare, E.J., and J.M. Lavanish, 1977. "Catalytic Oxidation of C_2-C_4 Halogenated Hydrocarbons." *U.S. Patent.* 4, 065, 543.

Subbanna, P., H. Greene, and F. Desal. 1988. "Catalytic Oxidation of Polychlorinated Biphenyls in a Monolithic Reactor System. "*Environ. Sci. Technol.* 22(5): 557-561.

Tichenor, B. A, 1987. "Destruction of Volatile Organic Compounds via Catalytic Combustion." *Environmental Progress.* 6(3): 172-176.

Weldon, J., and S. M. Senkan. 1986. "Catalytic Oxidation of CH_3Cl by Cr_2O_3. "*Combust. Sci. and Tech.* 47: 229-237.

Yang, K., J. D. Reedy, and J.F. Scamehorn. 1977. "Decomposition of Halogenated Organic Compunds." *U.S. Patent.* 4, 0509, 675.

IN-PLACE VOLATILIZATION OF CHLORINATED SOLVENTS USING STEAM AND COMPRESSED AIR

John G. Funk, P.E. (Rust Environment & Infrastructure, Raleigh, NC)
Michael Taylor, P.G. (Rust Environment & Infrastructure, Raleigh, NC)

ABSTRACT: As a source removal technology, an auger 10 feet (3 meters) in diameter was used to inject steam and heated compressed air into the soil and groundwater beneath a pit used for evaporation of spent solvents. The solvents consisted primarily of methylene chloride, trichloroethene, Freon 113, acetone, and methyl ethyl ketone (2-butanone). The pit was located in a geologic feature known as a Carolina Bay, which is locally characterized by discontinuous lenses of sands, silts, and clays. During site assessment, a dense non-aqueous phase liquid (DNAPL) consisting of methylene chloride and trichloroethene was encountered above a clay confining layer. Prior to implementing the volatilization process at the site, treatability testing was conducted to compare the removal efficiency of air injection versus steam and air injection. During implementation, the auger was driven to a depth of 20 to 22 feet (6.1 to 6.7 meters) below ground surface and 16 feet (4.9 meters) into the groundwater table, while injecting steam which heated the subsurface soils in the auger column to a temperature of 60 degrees C. At the end of the first auger cycle, the steam injection was stopped, and compressed air at a temperature of 150 degrees C was injected at a rate of 1,000 to 1,400 standard cubic feet per minute (scfm) (40 cubic meters per minute). The compressed air volatilized the solvents, and the off-gases were collected in a shroud placed over the auger column. Off-gas samples were collected and analyzed using an organic vapor analyzer (OVA) and field gas chromatograph (GC), and the results were used to direct movement of the auger up through the column. A total area of 5,000 square feet (464.5 square meters) was treated to a depth of 20 feet (6.1 meters) using the process, and based upon off-gas analysis it is estimated that 1,900 pounds (862 kilograms) of solvents were removed from the subsurface.

SITE HISTORY

A manufacturer of medical products manufactured from polyvinyl chloride and other plastics discharged solvents from its manufacturing operations into an unlined pit. The solvents included methylene chloride, trichloroethene, methyl ethyl ketone, cyclohexane, toluene, acetone and Freon 113. The manufacturing facility and solvent pit are located in a geologic feature known as a Carolina Bay. The strata of these features are characterized by discontinuous layers of sands and clays. The groundwater depth in the former pit area is 3 to 4 feet (0.9 to 1.2 meters) below land surface. During site investigations a dense non-aqueous phase liquid (DNAPL) consisting of methylene chloride and trichlorethene was recovered from a monitoring well immediately downgradient of the former pit area. The site investigation also located a confining clay layer at a depth of 20 to 22 feet (6.1 to 6.7 meters) below land surface. The thickness of the layer varied from 1 to 2 feet (0.3 to 0.6 meters), and a steep gradient of contaminant concentration was exhibited across the layer.

TREATMENT METHOD SELECTION

A feasibility study was prepared for the site, and the client selected in-place volatilization for implementation. The process consists of using a 10-foot (3.05 meter) diameter auger to advance into the subsurface soil while injecting steam and/ or compressed air through nozzles in the auger blades. The compressed air volatilizes the solvents, which are collected in a shroud placed over the auger. The shroud is placed under vacuum, and the vapors which collect in the shroud can be treated and discharged to the atmosphere. The system was exempted from air permitting and treatment of air emissions.

The client selected this remedial approach because it promised to remove a large amount of contaminants in a short period of time, compared to other alternatives such as air sparging/vacuum extraction, pump and treat, or recirculation. The auger would move down through the intervening clay and sand layers to the confining layer, and would break up preferred pathways to liquid or vapor flow formed by the clay lenses.

Prior to implementation, treatability testing was conducted to compare ambient air injection versus steam and air injection to determine which process to implement at the site. The tests also evaluated operational parameters such as steam and air flow rates, and were used to determine treatment times and to estimate contaminant removal.

TREATABILITY TEST RESULTS

Rust conducted the volatilization treatability tests at the WMX Clemson Technical Center using a 1/16 scale Millgard MecTool Auger. Representative clayey soils from the site were spiked with the four primary solvents of concern: methylene chloride, trichlorethene, acetone, and methyl ethyl ketone. The soils were spiked to ensure adequate mass of solvents to obtain contaminant concentrations throughout the planned 8 hours of testing. Two treatability tests were performed. The first test consisted of running the auger through the soil column while injecting ambient compressed air at a temperature of 21 degrees C and a flow rate of 0.5 standard cubic feet per minute (scfm) (0.014 standard cubic meters per minute). The second test consisted of injecting saturated steam into the test container and then injecting compressed air. Steam was injected at a pressure of 25 pounds per square inch gauge (psig) (1.7 atm) at a rate of 3.09 pounds per hour (1.4 kilograms) for 40 minutes, then compressed air was injected for the remainder of the test.

The results of the first test using ambient compressed air for the removal of acetone, methyl ethyl ketone, trichloroethene, and methylene chloride are provided in Figure 1. The results of the second test, using steam to raise the soil temperature followed by heated compressed air, are provided in Figure 2. Analysis of the results indicated that the effectiveness of methylene chloride removal was not enhanced by increasing the soil temperature, but removal of trichloroethene, methyl ethyl ketone and acetone were improved.

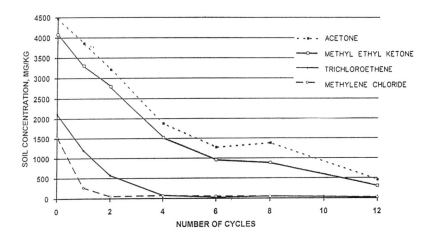

FIGURE 1. Treatability test results of ambient air.

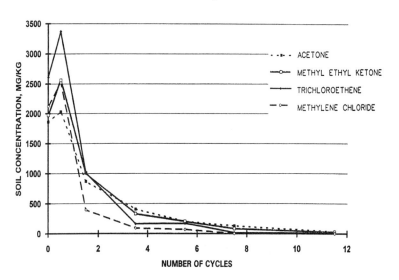

FIGURE 2. Soil VOC concentrations for steam/hot air stripping.

Contaminant concentrations from the test data in Figures 1 and 2 were plotted as log concentration versus a ratio of the soil volume in the test chamber to the total air flow injected into the soil (air:soil ratio). The plot of the data indicated a straight-line correlation between log solvent concentration and air:soil ratio. The plots of the test data for trichloroethene and methyl ethyl ketone are provided in Figures 3 and 4, respectively.

Figures 3 and 4 were used to develop air flow rates and treatment times for the full-scale process. In both figures, the initial concentration and 90% removal lines are shown. The difference between the air:soil ratio for the initial concentration and final target concentration (90% removal) times the soil volume in the auger column of 1570 cubic feet (44.46 cubic meters) (10 foot [3 m] diameter by 20 feet [6.1 m] deep for full-scale system) provided the total volume of air injection required to achieve the target concentration in the soil. Knowing the air injection rate then provided an estimate of the total time required to achieve the target concentration. The procedure was used to estimate the remedial cost, and to compare the cost of ambient air injection versus steam and air injection, as described below.

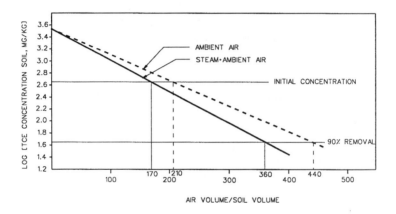

FIGURE 3. Trichloroethylene 90% removal ambient air vs. steam and ambient air.

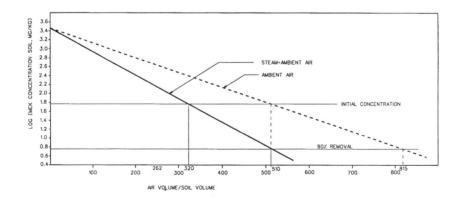

FIGURE 4. Methyl ethly ketone 90% removal ambient air vs. steam and ambient air.

Cost Comparison. Table 1 compares the cost of ambient air injection versus steam injection for both trichloroethene and methyl ethyl ketone. The cost estimates are based on $ 0.53/cubic yard/minute for ambient air injection and $ 0.62/cubic yard/ minute for steam and air injection, as provided by a remedial contractor. The volume of soil treated in one auger cylinder 10 feet (3 meters) in diameter and 20 feet (6.1 meters) deep is 1570 cubic feet or 58 cubic yards (44.46 cubic meters). The planned air injection rate for the full-scale system was 1400 scfm (40 cmm). Refer to Figures 3 and 4 for estimated treatment air/soil ratios for 90% removal of both contaminants. For the estimate 40 minutes was assumed for steam injection; that time was added to the total treatment time.

TABLE 1. Cost comparison ambient air injection versus steam and air injection for trichloroethene and methyl ethyl ketone.

	Tricloroethene	
	Ambient Air Injection	**Steam & Air Injection**
Air:Soil	440-210 = 230	360-170=190+40=230
Soil Volume	1570 cf	1570 cf
Air Volume	361,100 cf	361,100 cf
Min./1400 cfm	258	258
Cost/cy/min.	$ 0.53	$ 0.62
Cost/cubic yard	$ 138	$ 160
	Methyl ethyl ketone	
	Ambient Air Injection	**Steam & Air Injection**
Air:Soil	815-510 = 305	510-320=190+40=230
Soil Volume	1570 cf	1570 cf
Air Volume	478,850 cf	361,100 cf
Min./1400 cfm	342	258
Cost/cy/min.	$ 0.53	$ 0.62
Cost/cubic yard	$ 181	$ 160

The treatability test results concluded that for highly volatile compounds such as methylene chloride and trichloroethene, there is no advantage in using steam injection prior to air injection. If the site contains less strippable compounds such as methyl ethyl ketone, then there is a slight cost advantage to using steam injection prior to air injection.

PROCESS IMPLEMENTATION

After performance of the treatability tests and review of the test results, the client decided to implement steam injection followed by heated air injection at the site. Soil data developed during the assessment phase was used to delineate the horizontal and vertical extent of solvent contamination and the extent of the remediation area. The soil data and removal graphs (Figures 3 and 4) were used to develop treatment times for different areas of the site. The following describes the effect of steam injection on subsurface temperature, monitoring parameters and performance evaluation of the process.

Steam Injection and Heat Loss. During implementation, saturated steam at a pressure of 110 to 140 psi (7.5 to 9.5 atm) was injected into the subsurface at a rate of 4,500 pounds/hour (2,041 kg). Using an average pressure of 120 psi (8.16 atm) and an injection time of 40 minutes yields a total heat input into the system of 2,787,000 BTU (702,324 kg-cal). The total heat input assumes that all the steam condenses to vapor and the resultant liquid remains at 100 degrees C.

During steam injection, heat is lost from the system through conduction from the sides of the auger cylinder into the surrounding soils. This amount of heat loss was calculated to be approximately 7,100 BTU (1,789 kg-cal) for a 40-minute period. Heat can also be lost through convection out of the top of the auger cylinder, but the top is covered by the shroud. Neglecting heat loss out the top of the cylinder, the total heat into the system then is approximately 2,780,000 BTU (700,560 kg-cal) for a 40-minute injection period. During implementation, soil temperatures were measured after steam injection, and the temperature in the auger column was raised from 13 degrees C to 60 degrees C, which indicates a temperature rise of 47 degrees C. The field measures of soil temperature indicate that the heat transfer from the steam to the subsurface soils and groundwater was approximately 70% of the theoretical maximum. Therefore, the results confirmed the initial assumption that very little heat would be lost due to transfer to the surrounding soils or to the air above the auger column.

Volatilization Treatment Time. As shown in Figure 1 and 2, the removal of the contaminants approaching asymptotic levels after two cycles for methylene chloride and after 4 to 6 cycles for the other compounds. Because removal becomes asymptotic, the technology reaches a point where additional expenditure of time does not produce appreciable additional contaminant removal.

In addition, the solvents were not uniformly distributed throughout the 20- to 22-foot treatment depth. During implementation, organic vapor analyzer (OVA) measurements of the system off-gas provided a real-time indication of contaminant removal. The measurements were also used to vertically profile the contaminant mass with depth. An example of vertical profiling from the site is provided in Figure 5. Profiling the vertical distribution of contaminants on the first two cycles indicated that the highest concentration of solvents were in the 8- to 11-foot (2.4 to 3.3 m) depth for this particular auger location. Using the profiling data, the auger was cycled through the 8- to 11-foot (2.4 to 3.3 m) depth rather than through the full 20-foot (6.1 m) soil column. In that manner, the effectiveness of the volatilization process was optimized, since the auger operated for the majority of time at the depths with the highest contaminant concentration. The auger was operated in a column until the VOC concentration in the off-gas did not appreciably change, as shown in Figure 5. The use of vertical profiling and treating to asymptotic off-gas OVA concentrations reduced the treatment times anticipated by the treatability test.

Performance Evaluation. The results of the treatability tests indicated a direct correlation between the reduction in the off-gas concentration of VOCs and the remaining concentrations of VOCs in the soil. Soil sampling was conducted before and after the volatilization work, but there was very poor correlation between the before and after analytical data. The poor correlation could be attributed to several causes. Spreading of contaminant was indicated by the sampling, and that could be caused by the steam and air injection. Also, the soil was mixed during auguring and the resultant mixing caused a vertical redistribution of contaminants.

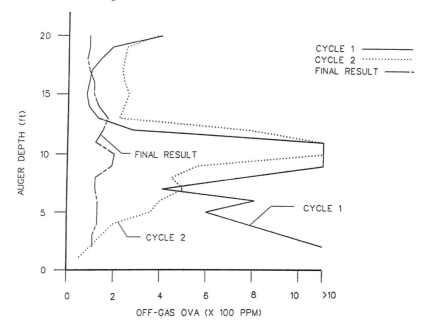

FIGURE 5. Off-gas measurement vs. time and depth.

It was found that the effectiveness of the volatilization treatment based upon the concentration of VOCs in the off-gas measured at the start of the each auger column and at the end of the last cycle provided a more reliable measure of contaminant mass removal than before and after soil data. The field-measured off-gas VOC concentration data and the known air injection flow rate provided an estimate of the mass of contaminants removed from the soil during operation of the MecTool volatilization system. This method provided a estimate of total contaminant mass removal of approximately 1,900 pounds (861.8 kg).

Conclusion. The client selected in-place volatilization as the remedial approach since it was believed it would meet his objective of removing the bulk of contaminant mass from the site in a relatively short period of time. The process met that objective and proved an effective means for removing solvent in free-phase concentrations from a subsurface environment of sands and clays.

CLOSING A DNAPL SITE THROUGH SOURCE REMOVAL AND NATURAL ATTENUATION

Gregory Smith (ENSR Consulting and Engineering, Westmont, Illinois)
Timothy V. Adams (ENSR Consulting and Engineering, Westmont, Illinois)
Valdis Jurka (Lucent Technologies, Inc., Morristown, New Jersey)

ABSTRACT: An Illinois EPA-approved closure of a DNAPL remediation is underway, where trichloroethylene (TCE) and 1,1,1-trichloroethane (TCA) have been remediated from soil and groundwater. These compounds were present in the subsurface in the form of dense non-aqueous phase liquids (DNAPL), forming pools measured as thick as 2.4 m (8 feet) below the water table in the extraction wells. The associated aqueous phase liquid (APL) plume was undergoing natural biotransformation producing lower molecular weight alkenes and alkanes.

In 1991, we initiated a full-scale remediation process using steam flooding and simultaneous enhanced biotransformation[1] to remove the chlorinated solvents from low permeability sediments. The steam flooding vaporizes and flushes the DNAPL for removal through conventional vapor and groundwater extraction. The simultaneous enhanced biotransformation produces lower molecular weight compounds that are more amenable to removal through conventional vapor and groundwater extraction. The process has removed in excess of 14,900 kg (32,800 pounds) of hydrocarbon from low permeability ($k = 1.05$ x 10^{-5} to 1.23 x 10^{-4} cm/sec) saturated fine sands, silts, and clays.

Monitoring of the remedial progress detected previously undiscovered areas of DNAPL in 1991, 1992, 1994, and 1997 resulting in modifications to the treatment system and approach. In most areas undergoing remediation, DNAPL has been removed with residual concentrations below the regulatory criteria (Illinois Class II Groundwater Standards). Due to the discovery of additional DNAPL areas during remediation, we initiated a phased closure of remedial operations under the Illinois Site Remediation Program. In the first area to be closed down, we have received a "no further remediation" (NFR) letter from Illinois EPA. Maximum concentrations in this area have been reduced to below an excess lifetime cancer risk of 1.83 x 10^{-7} for exposure through incidental contact and inadvertent ingestion. Natural attenuation calculations (ASTM 1739-95) show that the maximum concentrations of residual compounds in groundwater will be attenuated to Illinois Class II groundwater standards within 6.7 m (22 feet) of the source areas. This negates the need for long-term monitoring and restrictions impairing land use, facilitating site redevelopment. Most of the site has been redeveloped for retail use. This closure will soon allow for complete site redevelopment.

INTRODUCTION

During the decommissioning of a manufacturing facility in the Chicago, Illinois suburbs, chlorinated solvents in the form of TCE and TCA were discovered in soils and groundwater beneath the plant. In significant areas, the solvents were found to be in an undissolved state (DNAPL). Where the solvents were present in a dissolved state (APL),

---

U.S. Patent No. 5,279,740

they were undergoing a natural transformation, producing di- and mono- chlorinated aliphatics in the soils and groundwater.

The manufacturing facility is located on a lacustrine sequence of sediments, consisting of fine sands, and silts, with some clay lenses to a depth of 5.5 - 6.1 m (18 - 20 feet). The lacustrine sediments have hydraulic conductivities measured from 1.05×10^{-5} to 1.23×10^{-4} cm/sec. The lacustrine sequence is underlain by a dense glacial clay till. The clay till is a silty clay formation known as the Tinley Groundmoraine, present throughout a significant portion of the Chicago area.. The groundmoraine extends to depths of 16.8m (55 feet).

The authors' approach for the groundwater remediation has been to use steam injection to remove the DNAPL, and simultaneous enhanced biotransformation of the dissolved phase. This has been done to modify subsurface conditions and produce compounds more amenable to vapor and groundwater extraction. Further, this approach aids the intrinsic remediation of residual concentrations in the subsurface.

STEAM INJECTION

DNAPL refers to compounds that have measurable aqueous solubilities and are denser than water. As such, when released to the subsurface in sufficient quantity, they have the ability to migrate under gravity, contrary to the prevailing hydraulic gradient. As DNAPL migrates, it leaves behind blobs, or ganglia of material within the soil pores, or if sufficient quantity is present, it may pool on lower permeability strata. These ganglia and pools represent long term sources of groundwater contamination. Conventional pumping techniques are limited by the mass transfer into the surrounding fluid (air or water).

The chlorinated aliphatics identified in the subsurface at the facility have boiling points of less than 87⁰C (188.6⁰F). Mechanisms that promote the recovery of DNAPL compounds during steam flooding include, vaporization of compounds in the hot water zone, viscosity reduction, distillation of the more volatile components at the steam front, and the subsequent movement as a solvent bank (Hunt, et al., 1988).

Steam injection allows heating of the solid phase porous media and displacement of the pore water below the phreatic surface. Residual DNAPL and aqueous- phase solvents are dissolved and volatilized as a result of the pore water displacement. According to Miller (1975), the movement of the steam front is controlled by temperature gradients and the heat capacity of the porous media, and less so by pressure gradients and permeability. Thus, advective steam movement is less important in providing optimal thermal gradients that expedite volatilization of DNAPL from below the water table.

BIOTRANSFORMATION TO ACCELERATE GROUNDWATER CLEANUP

Compounds such as TCE and TCA are readily biodegradable under anaerobic conditions (Vogel and McCarty, 1985), producing di- and mono- chlorinated alkenes and alkanes. The biotransformed compounds have different chemical characteristics than their parent compounds (such as aqueous solubility, vapor pressure octanol-water, and octanol-carbon partition coefficients) and hence have different fate and transport characteristics. These compounds are more mobile than their parent compounds and, as such, are more easily removed through conventional vapor and groundwater extraction techniques. The differing fate and transport characteristics results in a distribution in the subsurface that has been likened to a chromatographic dispersion in an aquifer (Jackson and Patterson, 1989) with the more mobile compounds predominating at the plume front. This behavior has implications in groundwater pump ant treat remediation. Figure 1 presents a model depicting the number of pore volumes to attain a 90% decontamination compared against octanol-water partition coefficients for various compounds (as modified from Jackson and Patterson, 1989). According to this model, it would require 30.5 pore volume flushes of an aquifer to remove 90% of the TCE APL within an aquifer. This compares with 4.4 and 5.3 pore volume flushes for trans and cis-1,2-dichloroethene, respectively.

Figure 1:   **Estimation of Number of Pore Volumes to Attatin 90% Aquifer Decontamination. Modified from Jackson and Patterson (1989).**

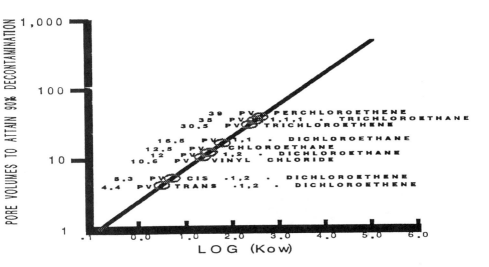

The design of the remediation system to enhance the biotransformation of TCE and TCA to daughter compounds is based on the microbial ecology of anaerobic bacteria. Vogel and McCarty (1985) and Fathepure, et al., (1987) demonstrated that methanogenic bacteria are capable of performing sequential reductive dehalogentation of chlorinated solvents under anaerobic conditions. A suitable donor for energy metabolism must be present, such as hydrogen, acetate, formate, or methanol. Optimizing these factors provide for the enhancement of the natural biotransformation. Data on the transformation of halogenated aliphatic compounds (Vogel and McCarty, 1985), suggest that the cis-monomer is produced in preference to the trans- and 1,1-dichloroethene from TCE. More recent research (McCarty, 1994) has shown that the production of vinyl chloride tends to occur under sulfate reducing and methanogenic conditions, while dichlorinated ethenes are readily produced under nitrate to iron reducing conditions. Therefore, maintaining anaerobic conditions in the range of nitrate to iron reducing conditions provides a means of controlling the production of vinyl chloride.

MONITORING OF REMEDIAL PROGRESS

To evaluate effectiveness of remediation, each extraction well is sampled quarterly.

Samples from the individual extraction wells are analyzed using a head space extraction technique, with the chemical analysis performed using a Hewlett-Packard HP-5980 gas chromatograph equipped with an electron capture detector. Standards for TCE, TCA, cis and trans-1,2-dichloroethene and 1,1-dichloroethane were used to calibrate the instrument. A subset of the sample population was analyzed using a gas chromatograph with a mass spectrometer following EPA Method 8240 procedures. Comparisons of the data showed that the headspace technique identifies the compounds and provides approximate concentrations. The results have been correlated, yielding correlation coefficients ranging from 0.82 to 0.93, indicating a very good fit for the data and high degree of confidence for decision making. We find the headspace analysis technique useful for identifying areas that have achieved cleanup, areas requiring further remedial effort, and general data trends. Figures 2 and 3 show the area above the cleanup criteria for both TCE and TCA as of October 1997, approximately 6 years after treatment began. The central area of the plume which initially had the greatest quanity of DNAPL has also shown the greatest concentration reduction. It was realized later that this area also experienced the highest degree of dewatering during this period and therefore the most efficient heating.

Figure 2: Areas above Cleanup Criteria for TCE

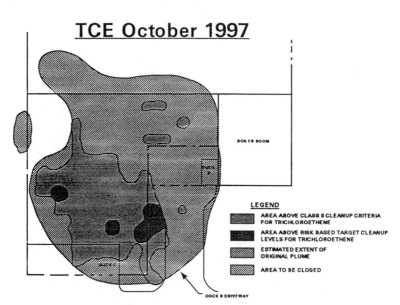

At the start of treatment, an isolated area west of the main treatment area was discovered to have DNAPL. In one extraction well, DNAPL was measured to be 0.3 m (1 foot) thick. The treatment system was extended to this area during the spring and summer of 1991. After the monitoring program was expanded to include the headspace extraction analysis of the extraction wells, areas of suspect DNAPL were discovered in the north and west portions of the treatment area in 1992, in the southwest portion in 1994, and in the south-central portion of the treatment area in 1997. These particular areas showed higher concentrations than were previously detected during the investigation portion of the project. It is believed that disturbing equilibrium conditions through groundwater pumping can mobilize DNAPL, resulting in the observation of its presence where it was previously undetected. The system, or its operations were modified to address these areas.

Figure 3: Areas above Cleanup Criteria for TCA

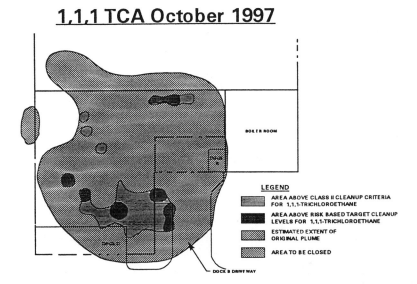

1,1,1 TCA October 1997

LEGEND

AREA ABOVE CLASS II CLEANUP CRITERIA FOR 1,1,1-TRICHLOROETHANE

AREA ABOVE RISK BASED TARGET CLEANUP LEVELS FOR 1,1,1-TRICHLOROETHANE

ESTIMATED EXTENT OF ORIGINAL PLUME

AREA TO BE CLOSED

INITIAL CLOSURE AREA

The area undergoing closure involves the north portion of the treatment area and the separate area west of the main treatment area containing DNAPL discovered in 1991. The area to north had DNAPL present in the structural backfill surrounding the building footings, and in the capillary fringe. Concentrations of TCE and TCA have been reduced to below Illinois Class II Standards (non-potable, general use criteria). Residual concentrations of the di- and mono- chlorinated compounds have been reduced to below an excess lifetime cancer risk of 1.83×10^{-7} for incidental contact and inadvertent ingestion.

The increased rate of biotransformation that has been achieved aids in the rate of natural attenuation. Table 1 presents the literature half-lives (35 Illinois Administrative Code, Part 740) compared to the apparent half-lives currently being observed. Natural attenuation calculations (ASTM 1739-95) show that the maximum residual concentrations in groundwater will be attenuated to Illinois Class II Standards within 6.7 m (22 feet) of the source area . Given that the affected groundwater is not used for drinking water due to its naturally high TDS and low well yield, this indicates that further remedial effort is unwarranted, and that the redevelopment of the area can proceed.

Table 1. Comparison of Literature-Derived Half-Lives and Apparent Half-Lives

Compound	Literature-Derived Half-Life (Days	Apparent Half-Life (Days)
Chloroethane	N/A	22.6 - 155.8
1,1-Dichloroethane	364.7	18.7 - 411.8
1,1-Dichloroethene	N/A	5.5
cis-1,2-Dichloroethene	2887.5	14 - 51
trans-1,2-Dichloroethene	2887.5	20.3 - 84.8
1,1,1-Trichloroethane	533.1	11 - 30.5
Trichloroethylene	1650	7.1 - 29.4
Vinyl chloride	2887.5	32.6 - 87

N/A = not available from IL Admin Code Part 740

CONCLUSIONS

A groundwater remediation process has proven itself capable of removing DNAPL and reducing the risk to public health that will facilitate property redevelopment. This has been demonstrated to the satisfaction of Illinois EPA, who have issued a "no further remediation" letter in accordance with the Illinois Site Remediation Program (35 IL Admin Code Part 740) for an initial area which had DNAPL present, measured as thick as 0.3 m (1 foot). The remediation is continuing for the remainder of the site.

The process consists of simultaneous steam injection coupled with enhanced biotransformation. The process modifies subsurface conditions to utilize natural fate and transport mechanisms to enhance the rate of hydrocarbon removal through conventional vapor and groundwater extraction techniques. This process appears to have enhanced intrinsic remedial processes, reducing the biodegradation half-lives, and provided more optimal conditions for natural attenuation of the chlorinated solvents in groundwater.

REFERENCES

Fathepure, B.Z., Z.P. Nengu, and S.A. Boyd (1987). "Anaerobic Bacteria that Dechlorinate Perchloroethene". *Applied Environ. Micro.*, November, pp. 2671-2674.

Hunt, J.R., N. Sitar, and K.S. Udell (1988). "Nonaqueous Phase Liquid Transport and Cleanup 1. Analysis of Mechanisms". *Water Resources Res.*, Vol 24, No. 8.

Illinois Register (1997). "Illinois Administrative Code Part 740, The Illinois Site Remediation Program".

Jackson, R.A. and R.J. Patterson (1989). "A Remedial Investigation of an Organically Polluted Outwash Aquifer". *Groundwater Monitoring Review* 3(ix).

McCarty, P.L. (1994). "An Overview of Anaerobic Transformation of Chlorinated Solvents". *Symposium on Intrinsic Bioremediation of Ground Water*. Denver, Co, August 30 - September 1, 1994.

Miller, C.A. (1975). "Stability of Moving Surfaces in Fluid Systems with Heat and Mass Transport, III, Stability of displacement Fronts in Porous Media". *AICHE J.* 2163.

Vogel, T.M. and P.L. McCarty (1985). "Biotransformation of Tetrachloroethylene to Trichloroetheylene, Dichloroethylene, Vinyl Chloride and Carbon Dioxide Under Methanogenic Conditions". *Applied Environ. Micro.* 49(No. 5).

DNAPL/LNAPL REMEDIATION IN CLAY TILL USING STEAM-ENHANCED EXTRACTION

Timothy V. Adams (ENSR, Westmont, Illinois, USA)
Gregory J. Smith (ENSR, Westmont, Illinois, USA)

ABSTRACT: Steam-enhanced groundwater and vapor extraction in a dense glacial clay till has successfully removed greater than 11,748 kg (25,900 pounds) of volatile organic hydrocarbons (VOH) during the first 29 months of system operation. VOH consists of trichloroethylene (TCE) and mineral spirit constituents (toluene and xylene). These solvents are present as both dense and light non-aqueous phase liquid (DNAPL/LNAPL). The remediation site is a former telecommunications manufacturing facility in the metropolitan Chicago, Illinois area. The remediation program is being conducted under the Illinois Voluntary Cleanup Program (35 IL Admin Code Part 740). The remediation system objective is to provide for sufficient and expedited mass removal in order to meet IEPA Tier I cleanup goals. Steam is being injected into two highly permeable zones within the dense glacial clay till at depths of 11.6 m (38 feet) and 14 m (46 feet) below grade via 65 injection wells. After 29 months of operation, soil temperatures measured at nested thermocouple strings surrounding selected steam injection wells ranged from 29°C to 60°C (84°F to 140°F). Groundwater temperatures range between 20°C and 74°C (68°F and 165°F). In this temperature range, significant TCE (boiling point 86°C, 188°F) is currently being removed through groundwater and vapor extraction, while the di- and mono- chlorinated daughter products have apparently been removed (boiling point range -4°C to 60°C, 25°F to 140°F,). Mean TCE concentrations have been reduced from 45,000 µg/l to 500 µg/l in 24 months, with Illinois EPA approved shutdown of 20% of the treatment area occurring after15 months of operation.

INTRODUCTION

As part of the decommissioning of a telecommunication manufacturing facility in the suburban Chicago, Illinois area, chlorinated aliphatic hydrocarbons in the form of TCE and cis 1,2-dichloroethene (DCE), and mineral spirit components in the form of xylene and ethylbenzene were discovered in the soil and groundwater in the vicinity of a former underground storage tank farm. Over 57,200 tonnes (63,000 tons) of impacted soil above the water table was excavated from outside the plant and disposed off site. Soil and groundwater beneath the plant were found to contain solvents as immiscible phase (DNAPL) and (LNAPL) as well as in a dissolved phase.

The investigated geologic strata at the site consists of three till layers overlying a dolomite bedrock The site consists of 1.5 to 2.1 m of clayey silt fill material underlain by 9 to 12 m of the Tinley Till, which is composed of dense silty clay with thin discontinuous sand and silt stringers. A fine- to medium-grained sand layer ranging from 0.15 to 0.6 m in thickness underlies the Tinley Till. Well-sorted laminated silt underlies the sand layer. The silt is extremely dense and overconsolidated. The sand layer and silt comprise the Valparaiso Till which is found at a depth of 10.7 to 13.7 m across the site. The Valparaiso Till is underlain by the Lemont Drift, which consists of a 0.3 to 1.2 m thick, coarse-grained sand and gravel underlain by fine-grained dolomitic sand and silt with some gravel fragments. The Lemont Drift unconformably overlies weathered Silurian dolomite bedrock. Competent dolomite bedrock is encountered at depths of 18 to 27 m.

The remedial objective for the site was to dewater the Tinley and Valparaiso Tills to facilitate steam injection and heating of the tills to enhance hydrocarbon removal through vapor and groundwater extraction.

SYSTEM COMPONENTS

The remediation system layout covers approximately 18,580 m^2 and consists of the following components:

- 186 shallow vapor extraction wells located in the Tinley Till
- 76 combination groundwater/vapor extraction wells screened across the Tinley and Valparaiso Tills
- 39 steam injection wells screened across the sand layer at the base of the Tinley Till, and 26 steam injection wells screened across the cobble layer at the base of the Valparaiso Till
- 2 Lemont Drift groundwater extraction wells

Vapor Extraction

Two PEGO® vapor extraction units (VES #1 and VES #2) supply 23.6 to 47.3 kPa (7 to 14 inches of mercury) vacuum while pulling up to 7.8 m^3/min (275 ft^3/min) via polyvinyl chloride header piping. Vapor extraction occurs via 186 shallow vapor extraction wells in the Tinley till and 76 combination groundwater/vapor extraction wells screen across the Valparaiso and Tinley tills.

Hydrocarbon emissions from VES #1 and VES #2 are measured continuously using a TECO® 51 flame ionization detector. The TECO is designed to sequentially sample total hydrocarbon concentrations (THC) in mg/L at VES #1 and #2. THC readings are recorded and averaged using LABTECH® CONTROL software installed at the site computer. Hydrocarbon emissions from the air stripper are measured by a separate channel on the TECO. Total hydrocarbon removal totals (in pounds) per month from system startup in August 1995 through January 1998 are presented in Table 1.

Groundwater Extraction and Treatment

Groundwater extraction occurs via the 76 combination groundwater/vapor extraction wells, 2 Lemont Drift recovery wells, and 1 excavation dewatering well. The combination wells provide for vacuum enhanced groundwater recovery while removing vaporized VOH.

Groundwater treatment consists of a stainless steel shallow tray air stripper with a 900 m³/min blower followed by secondary treatment through two 4500 kg (1,000-pound) activated carbon vessels before discharge under the facility's NPDES permit. Groundwater discharge averages approximately 1,890,000 l/month (500,000 gallons) per month.

TABLE 1. Hydrocarbon Removal Totals

MONTH	AIR STRIPPER DISCHARGE (kg)	VES#1 (kg)	VES#2 (kg)	MONTHLY TOTAL (kg)
Aug 95	0.00	0.00	222.39	222.39
Sep 95	147.92	64.28	152.59	364.79
Oct 95	110.36	114.17	198.58	423.10
Nov 95	82.62	319.67	190.22	592.51
Dec 95	113.55	247.21	185.78	546.54
Jan 96	139.62	193.09	228.65	561.36
Feb 96	101.78	107.53	106.65	315.96
Mar 96	131.29	400.42	160.92	692.63
Apr 96	181.89	331.48	133.60	646.97
May 96	262.76	298.28	145.42	706.46
Jun 96	225.22	128.59	109.89	463.69
Jul 96	122.83	243.84	72.92	439.59
Aug 96	118.74	202.34	119.80	440.89
Sep 96	127.49	114.43	90.68	332.61
Oct 96	145.63	107.65	98.21	351.50
Nov 96	97.75	128.64	104.07	330.45
Dec 96	86.75	148.49	93.15	328.39
Jan 97	81.57	131.12	82.77	295.46
Feb 97	72.22	71.41	42.96	186.59
Mar 97	87.19	144.67	105.72	337.58
Apr 97	89.57	161.97	86.36	337.90
May 97	98.59	136.17	68.44	303.20
Jun 97	69.95	60.57	44.58	175.11
Jul 97	50.26	28.41	32.67	111.35
Aug 97	132.18	41.75	204.29	378.23
Sep 97	126.55	40.81	164.82	332.18
Oct 97	94.99	87.66	182.30	364.94
Nov 97	224.39	243.76	35.10	503.25
Dec 97	84.58	213.63	46.02	344.24
Jan 98	90.98	121.15	87.67	299.80
TOTALS:	**3,499.23**	**4,633.20**	**3,597.25**	**11,729.68**

Steam Injection

Steam is injected via 65 steam injection wells into the sand and cobble layer which occur at the base of the Tinley and Valparaiso Tills, respectively. Steam is supplied by a 294 kilowatts (30 boiler horsepower) series HF Scotch-Box boiler at pressures ranging from 20.6 to 48.3 kPa (3 to 7 psi). Temperature thermocouples are installed around two of the deep steam injection wells and one shallow steam injection well to measure heat propagation in the subsurface. Temperature readings are averaged daily and recorded by the LABTECH software program. Through January 1998, soil temperatures ranges from 29°C to 60°C and groundwater temperatures ranged from 20°C to 74°C. The subsurface heating has enhanced TCE (boiling point 86°C) removal through groundwater and vapor extraction while the di- and mono- chlorinated daughter products have been removed with the groundwater or potentially vaporized. Steam front movement through the clay till is controlled by temperature gradients and heat capacity of the media and not necessarily through advective transport (Miller, 1975).

We are currently evaluating cycling the steam injection to improve the rate of hydrocarbon removed. This is analogous to oil industry practice using steam drive for enhanced oil recovery. In the oil industry, this is known as "huff and puff". Turning the steam off appears to allow soil pressures to dissipate, allowing vapors to escape. We experimented with this in August 1997 (See Figure 1) where VES #2 saw a dramatic increase in hydrocarbon removal following steam shutdown. We are currently evaluating appropriate frequencies for the steam cycle.

TCE REDUCTION

Quarterly groundwater sampling at the 76 combination extraction wells is conducted to evaluate system performance. Groundwater samples are analyzed using a head space extraction technique using a Hewlett-Packard HP-5890 gas chromatograph equipped with an electron capture detector. Approximately 10% of the samples are split for laboratory analysis by EPA Method 8240.

TCE is the predominant chlorinated alkene present in the remediation area. Figure 1 shows the area where TCE is above 1,000 ug/L in December 1995 (4 months following system startup) and October 1997 (26 months following system startup). Table 2 presents percent reductions TCE and cis 1,2-DCE at 17 of the 76 combination extraction wells from December 1995 to October 1997.

TABLE 2. % Reduction TCE and cis 1,2-DCE

Well Location	Dec-95 TCE	1,2-DCE	Oct-97 TCE	1,2-DCE	%Reduction Dec95 to Oct 97	%Reduction Dec95 to Oct 97
200n230e	94166	2311	74	0	100%	100%
220n210e	3007	1	212	17	93%	+1600%
220n250e	337	29	28	0	92%	100%
240n190e	431318	168	2890	101	99%	40%
260n250e	161	11	33	7	80%	36%
276n110e	7615	74	342	0	96%	100%
276n230e	1336589	437	4488	80	100%	82%
276n270e	164764	478	140	0	100%	100%
276n310e	190527	467	4700	4	98%	99%
300n270e	46743	1	1941	10	96%	+900%
300n290e	189610	456	1466	13	99%	97%
320n110e	352639	47	39	0	100%	100%
320n220e	266	22	599	34	+125%	+55%
320n290e	341207	259	10526	73	97%	72%
340n270e	75213	228	270	0	100%	100%
360n180e	86	33	28	14	67%	58%
360n240e	954	423	497	122	48%	71%

Figure 1. TCE in Groundwater (> 1,000 ug/L)

October 1997

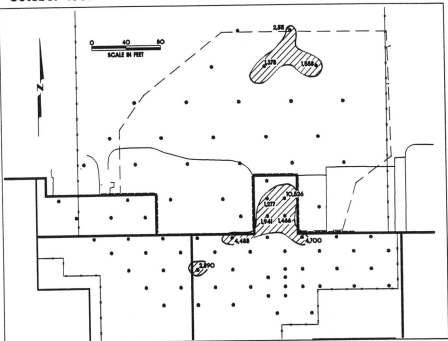

December 1995

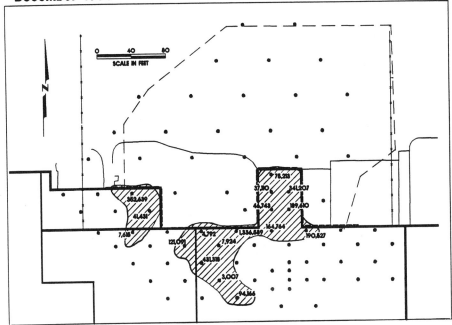

Figure 2 presents mean TCE concentrations during quarterly sampling for the first 2 years of system operation. Mean TCE concentrations decreased from over 45,000 ug/L in December 1995, to less than 600 ug/L in September1997.

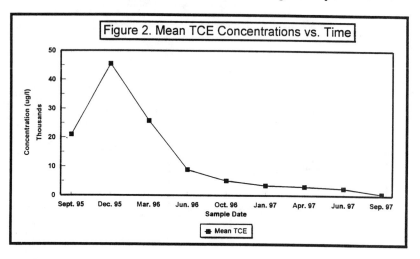

CONCLUSIONS

Steam-enhanced vapor and groundwater extraction has been successful in removing over 11,700 kg of hydrocarbon consisting mainly of TCE from a low-permeability glacial till in a 29-month period. Over 2,000 m^2 of remediation area undergoing vapor extraction has been approved for closure by Illinois EPA after only 14 months of system operation. Mean TCE concentrations have been reduced from 44,000 µg/l to 500 µg/l in a period of 24 months. As an indication of the removal of DNAPL, the number of wells above 10 mg/l has been reduced from 13 to 1 in a period of 21 months. Subsurface heating has enhanced mass transfer of DNAPL and LNAPL at the propagating steam front, resulting in vaporization and flushing from the soils of TCE.

REFERENCES

Illinois Register (1997). "Illinois Administrative Code Part 740, The Illinois Site Remediation Program".

Miller, C.A. (1975). "Stability of Moving Surfaces in Fluid Systems with Heat and Mass Transport, III, Stability of Displacement Fronts in Porous Media". *AICHE J* 2163.

THERMAL ENHANCEMENT AND FRACTURING TO REMOVE SEMIVOLATILE ORGANIC COMPOUNDS FROM CLAY

Jay Dablow (Fluor Daniel GTI, Irvine, California)
Dr. Kathy Balshaw-Biddle (Rice University, Houston, Texas)

ABSTRACT: An innovative combination of thermal and physical enhancement technologies has been successfully implemented to remediate clay soils impacted by jet fuel. At a site at Ft. Hood, Texas, tight, clay soils were hydraulically fractured to create discrete, high-permeability flow paths in the clay. Fractures were created at three depths in three separate test cells. The fracture configuration was controlled by the target depth and site lithology. Fractures were consistently smaller than the design radius. Soil heating by steam injection and electro-heating was used to volatilize contaminants and enhance vapor flow along fractures. Soil heating by steam injection achieved soil temperatures of 100°C, while the electro-heating system raised the soil temperatures to 45°C. A significant increase in the biodegradation rates was associated with the electro-heating process. An evaluation of the hydrocarbon range distribution in the vapor extraction system off gas indicates that volatilization of semi volatile compounds took place in both the electro-heating and steam injection test cells.

INTRODUCTION

A research project, funded by the Department of Defense (DOD) through the Advanced Applied Technology Demonstration Facility (AATDF) at Rice University in Houston, Texas, was conducted to investigate hydraulic fracturing and soil heating (steam injection and electro-heating) to remediate semivolatile organic compounds in fine-grained soils. Hydraulic fracturing increases secondary permeability of the clay soils by creating discrete flow paths at several depths in the vadose zone. Soil heating, employing steam injection and electro-heating, raises the contaminant vapor pressure and accelerates mass transfer from separate and adsorbed phases to the vapor phase for eventual removal by standard vapor extraction systems. Creation of horizontal hydraulic fractures also results in shortened oxygen diffusion path lengths, thereby enhancing bioventing to degrade contaminants.

SITE DESCRIPTION

A site at Ft. Hood, Texas, located adjacent to the Rapid Refueling System tank farm of the Robert Gray Army Airfield was used for the demonstration. A series of jet propellant 8 (JP-8) surface spills, which have occurred since the tank farm installation in the early-1980s, has infiltrated the clay soils to the east of the tank farm. One spill in excess of 8,000 gallons, as well as numerous smaller spills, has occurred.

The site is underlain by Lower Cretaceous bedrock consisting of beds of argillaceous limestone and calcareous clay and commonly contains oyster beds. Locally, fill materials, which consist of weathered limestone cobbles, fossil fragments, and silty clay overlie the weathered portion of the bedrock. The JP-8 impact is concentrated in the weathered bedrock layer at depths of 3 to 6 meters. The fill materials, which have been impacted in isolated areas, extend from the ground surface to a depth of approximately 2 to 3.5 m.

Groundwater levels at the site have ranged from about 2.9 to 5.5 m below the ground surface. Product thickness as large as 1.5 m has been recorded in wells at the site. The shallow groundwater is unconfined and generally moves eastward in the weathered shales and limestones as a subdued replica of the topography. The groundwater flow appears to occur through fractures in the clay. Since the fractures generally occupy less than 1 percent of the formation matrix, effective porosity is very low while the matrix porosity may be as high as 40 percent. The response of the formation to rainfall events significantly impacted the remediation operations and is further evidence of the very low effective porosity.

Total petroleum hydrocarbon (TPH) concentrations as high as 35,500 mg/kg (ppm) were encountered. These high concentrations and areas of SPH were much more prevalent in the fractured limestone beds than the surrounding clays. In general, concentrations greater than 1,000 ppm occur in a zone between 3 and 6 m below the ground surface. Most of the soil above depths of 3 m (fill materials) contains TPH concentrations of less than 100 ppm.

TECHNOLOGY PROCESS DESCRIPTION

The remediation strategy initially selected for the AATDF research included hydraulic fracturing and steam injection in three separate test cells, labeled A, B and C from north to south. After hydraulic fracture creation and subsequent field evaluation of the extent of the fracturing, it was determined that the degree of fracturing in two of the three test cells (A and B) was not sufficient to allow effective steam injection as planned. At that time, it was decided to modify the demonstration program strategy to include an evaluation of electro-heating and limited steam injection in the two moderately fractured cells, while proceeding with the original steam injection strategy in the well-fractured cell (C).

Hydraulic fracturing creates lenses of high permeability, granular material, which readily conducts fluid flow and creates more efficient flow patterns. A sand and guar gel slurry was pumped into a borehole with an initiating notch at the target fracture depth to create hydraulic fractures. This slurry filled the fracture as it propagates away from the borehole. Enzyme additives degraded the liquid portion of the slurry leaving a highly permeable sand pathway.

Hydraulic fractures were installed horizontally in each of the three cells at depths of approximately 3.5, 4.5 and 5.5 meters below the ground surface. A distance of 1 m separated the fractures in order to maintain vertical isolation. The fracture propagation was monitored so that a maximum fracture radius of 4.5 m would not be exceeded. This radius was selected so that each test cell would be separated by at least 3 m of unfractured soil. A single dewatering fracture was installed in each test cell at a depth of 6.5 m.

Steam Injection. Steam injection significantly increases the vapor pressure and volatilization rate of volatile and semi-volatile contaminants, such as JP-8. Since large quantities of vapors and mobile SPH are produced, steam injection is generally applied in conjunction with SVE and groundwater extraction systems for capture of mobilized phases.

Steam injection in hydraulically fractured clay transfers heat to the soil via convection more effectively than conventional injection wells in unfractured soils. Additionally, conductive heating of adjacent clays and convection in vertical discontinuities that intercept the hydraulic fractures will improve soil-heating efficiency. These same fractures and discontinuities can transmit volatilized and mobilized fluids to liquid and vapor extraction wells.

Electro-heating. The electro-heating approach for this site is based on a combination of 3-phase electric heating and SVE techniques. The approach uses the soil to form the resistive heating element of an in-situ, multi-phase (AC) heater system. To achieve this effect, 6 electrodes were placed, in a hexagonal array, directly into the soil to supply electrical energy. Since heat energy is transferred electrically, fluid flow is not required for soil heating. Therefore, the low permeability of the clay does not impede the soil heating efficiency. A comparison of the uniformity of heating between the steam injection and electro-heating techniques in clay soils was conducted in the two moderately fractured cells, A and B.

CONCLUSIONS

Fracture Extent and Configuration. The fracture extent and configuration was determined by compiling continuous logs of the subsurface conditions during the well installation process and noting the depth, thickness and orientation of the encountered fractures. A comparison of the average fracture radius at individual fracture depths indicates that the longest fracture was propagated in Cell C at the shallow depth, where the radius was close to 3 m. The shortest fracture was in Cell B at the deepest fracture depth where the fracture radius was about 1.5 m. In all cases the fracture radius was less than the design radius of 4.5 m. Real time electrical resistivity measurements indicated that fractures had propagated to the design radius. Due to wide variations in the soil resistivity, the electrical resistivity approach was not sufficiently accurate to track fracture propagation and should be augmented with other techniques such as tiltmeters.

All of the fractures, with one exception, were roughly symmetrical or had a strong north-south (NS) elongation. The NS linearity increases significantly from Cell A, the most symmetrical, at the northern end of the test area to Cell C, the most elongated, at the southern end of the area. The fractures in Cell C are consistently elongated in the NS direction and cover a larger areal extent than the fractures in either of the other two cells. The fractures in Cell B are smaller than those fractures in the other cells and are slightly elongated in the NS direction. Although only slightly larger than the fractures in Cell B, the Cell A fractures are significantly more symmetrical than those fractures in either Cells B or C.

In general, the fractures in Cell B are relatively flat, while the fractures in Cells A and C are significantly curved. The greatest curvature is consistently encountered in the fractures at the 3.5-m depth where the differential between the principal stresses is smaller and upward curvature is, therefore, more likely. There is less curvature at the 4.5 and 5.5 m depths, with the greatest amount of curvature in the south and west directions. The flattest fractures were created at the 6.5-m depth, where the differential between the principal stresses is likely to be the greatest, constraining the vertical propagation of fractures.

The differential of the principal stresses also appears to be the controlling factor in the size of the fractures as described above. The smallest fractures were encountered at the greatest depth, where the principal stress differential is greatest. The elongation of the fractures and isolated areas of bifurcation of fractures seems to be related to lithologic factors, such as the presence of fractured limestone lenses and oyster beds.

Soil Heating. The soil heating start-up period coincided with a period of abnormally high rainfall. Rainfall infiltration rapidly filled the naturally occurring fractures, which resulted in high groundwater levels throughout the test area. After upgrading the dewatering system by installing a series of venturi eductors to provide high vacuum dual phase extraction of the groundwater and soil vapors, the formation was drained sufficiently to allow steam flow through the fractures and, consequently, soil heating.

Beginning on July 1, the steam injection system was operated regularly until final shutdown on September 5. A thermocouple network installed at several depths throughout the cell monitored the progress of the steam heating operation. The steam boiler operation history and the temperature response from thermocouples are presented on Figure 1. Several episodes of heating are seen to occur in response to the periodic changes in the nature of the steam injection. Significant events occurred at start-up (July 1-15), at a restart following a cool period (July 26-31), and again in August after a two-week shutdown.

The initial response from injection into the upper and middle fractures shows an abrupt increase in temperature from about 22°C to over 100°C. This response can be attributed to rapid migration of steam along the hydraulic fractures and subsequent heating of the surrounding soils. Cooling due to heat loss can be seen shortly after periods of inactivity. However, nearly immediate temperature increases were observed after restarting of the steam injection. During the later periods of the steam injection process temperatures were maintained above 100°C, even at steam flow rates that were 30 percent to 40 percent of peak rates.

Electro-heating of the soils in Cell A was initiated using the hexagonal electrode array by applying power from a system rated at 75 kVa. As can be seen on Figure 2, the soil-heating rate varied as the power system was adjusted to optimize the power-input parameters. The soil-heating rate varied from approximately 0.48°C/day to 1.03°C/day during the period from June 2 to August 1. Increasing the power and irrigating the soil surrounding the electrodes to maintain good current flow to the soil attained the maximum heating rate. The lower heating rate observed after August 1 is due to reduced soil irrigation that consequently reduced the current flow from the electrodes to the soil.

In contrast to the soil heating history in the steam injection cell, which showed significant fluctuations due to the injection rate and groundwater conditions, the electro-heating rates were consistent and did not reflect any fluctuations due to groundwater level conditions. The electro-heating approach appears to be well suited to sites with tight soils and high groundwater. In addition, enhanced biodegradation, as evidenced by a significant increase in carbon dioxide production in the off gas, was observed throughout the operation of the electro-heating system.

Hydrocarbon Range Distribution. A JP-8 standard was compared to the composition of the vapor phase off gas concentrations to evaluate whether soil heating has impacted soil vapor extraction potential. The comparison of the hydrocarbon ranges in the off gas to the JP-8 standard distribution of these ranges identified a number of trends.

In Cell A, the composition of the vapor stream in the three extraction wells was enriched in the light, volatile $<C_{10}$ hydrocarbons throughout the operational period. As the temperature of the soil increased, the middle range (C_{10} - C_{11}) hydrocarbon percentages in the vapor stream increased, indicating volatilization into the vapor stream. At temperatures of 45°C, it appears that the light and middle range hydrocarbons volatilized, while the heavier hydrocarbons remained in the separate and adsorbed phases.

In Cell C, the $<C_{10}$ and C_{10} - C_{11} compounds were enriched in the vapor stream during the early stages of the operation. In late July, when the soil temperatures were greater than 100°C, the vapor phase composition percentage of these compounds began to decrease. In conjunction, a strong increasing trend in the heavier compounds was observed. These trends indicate that, as in Cell A, the rising soil temperatures caused volatilization of the light fractions early in the heating history. As the soil temperatures rose beyond 45°C, the middle and heavier range compounds volatilized into the vapor stream. The volatilization of all the hydrocarbon ranges is supported by a nearly normal JP-8 distribution in the vapor stream. If volatilization had not occurred, the vapor stream would continue to be depleted in the heavier hydrocarbon ranges (C_{12} - C_{13} and $>C_{13}$).

ACKNOWLEDGEMENTS

Funding for the project was provided by the Advanced Applied Technology Demonstration Facility at Rice University under the direction of Dr. Herb Ward and Dr. Carroll Oubre. Dr. Kathy Balshaw-Biddle managed the research work for AATDF.

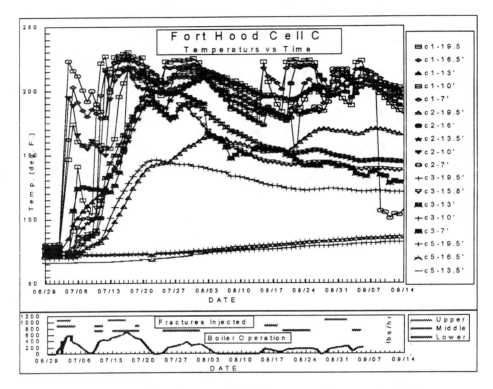

Figure 1. Steam Injection History and Soil Temperature Response in Cell C

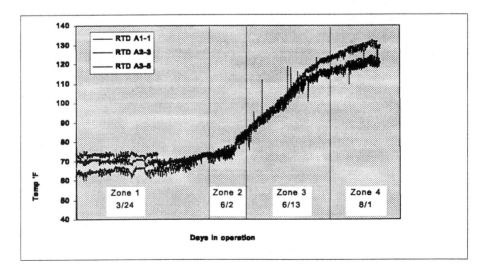

Figure 2. Soil Temperature Response in Cell A due to Electro-heating

HOT WATER INJECTION FOR THE REMEDIATION
OF OILY WASTES

Eva L. Davis (U.S. EPA, Ada, Oklahoma)

ABSTRACT: In recent years a number of sites have been identified where viscous oils of low solubility and volatility are contaminating the subsurface. Raising the temperature of these oils will generally reduce the viscosity by an order of magnitude or greater, which can greatly increase their recovery as a separate phase. One dimensional hot water displacement experiments have shown 15 to 30 percent increases in the recovery of viscous oils after the injection of 10 pore volumes of water by increasing the temperature from 10°C to 50°C. Greater recoveries may be possible with continued hot water injection. The efficiency of the displacement in terms of oil recovery versus pore volumes of water injected is dependent on the oil properties and the grain size distribution of the sand.

INTRODUCTION

Large volume spills of a crude oil (Hess et al., 1992), a naval oil, automatic transmission fluid (Abdul, 1992), creosote (Sale and Applegate, 1997), and a lubricating oil (Fulton et. al, 1992) have been documented. Each of these oils is highly viscous at ambient temperatures, and has a low volatility and solubility. These properties greatly reduce the effectiveness of the commonly-used pump-and-treat or soil vacuum extraction methods for the recovery of these oils. Hot water injection has been used effectively in the petroleum industry to recover viscous oils, and is currently being considered for the remediation of viscous oils which are contaminating the subsurface (Fulton et al., 1991). For these highly viscous oils, moderate temperature increases may significantly reduce their viscosity, as well as shift the relative permeability curves toward increased oil recovery and reduce residual oil saturation.

This laboratory study demonstrated the use of moderately hot water - up to 50°C - to enhance the recovery of waste oils. This limiting temperature was chosen because at industrial sites, waste heat may be able to supply the energy needed to heat the water to this temperature. Two types of displacement experiments have been performed. The constant temperature experiments provide information of the effects of temperature and different fluid properties on the displacement process. The transient temperature experiments more closely model the displacement process that would occur in the field, and help to determine if the benefits observed at higher temperatures in the constant temperature experiments can be achieved under field conditions. To date, these displacement experiments have been performed using three different oils (experiments continue with the transmission fluid) and two quartz sands with different grain sizes. The objective of the continuing research is to correlate the potential recovery with properties of the oil (such as viscosity and interfacial tension), and of the media (such as grain size distribution).

MATERIALS AND METHODS

One dimensional horizontal displacement experiments were performed at constant temperatures in the range of 10°C to 50°C by the method described by Davis and Lien

(1993). The experimental method for the constant temperature displacements basically consisted of first equilibrating the water-saturated sand column to the desired temperature, then displacing the water with oil at the same temperature until residual water saturation is reached. Then, the oil is displaced by water at the same temperature. Throughout the experiment, the column effluent is collected by a fraction collector which allows the proportion of oil and water in the effluent versus the pore volumes of water injected to be determined. The pressure is measured at each end of the column throughout the displacement by pressure transducers. Each experiment was run at least twice and the results were found to be reproducible.

Transient temperature experiments were performed by placing the oil-saturated column in the incubator at 10°C, then using water at 50°C to displace the oil. During these experiments, the temperature was measured at four locations along the column.

The oils used for these experiments were Inland 15 which is a vacuum pump oil, a crude oil obtained from an oil field near Ada, Oklahoma, creosote obtained from the Mississippi Products Lab, and an automatic transmission fluid. Both the crude oil and the creosote were "weathered" in the laboratory to remove the volatile and water soluble fractions. Quartz sands were used as the porous media. The 20/30 sand has a very uniform grain size, with all the grains passing the #20 sieve and retained on the #30 sieve (sand grains are 0.85 to 0.60 mm). The mixed sand has approximately one third of its grains in each of the size ranges: 0.85 to 0.60 mm, 0.60 to 0.25 mm, and 0.25 to 0.15 mm.

RESULTS AND DISCUSSION

Table 1 lists the oil recovery at breakthrough and after the injection of 10 pore volumes of water for the constant temperature displacements and shows that as the temperature is increased, the oil recovery also increases. For the Inland 15, the additional recovery occurs at the early part of the experiment, and is due to a delay in the breakthrough of the water at the far end of the column. The amount of oil recovered between breakthrough and the injection of 10 pore volumes of water is essentially constant. Note that the mixed sand, and its wider distribution of pore sizes, consistently shows greater recovery than the 20/30 sand, and this result was confirmed by simulations using the Buckley-Leverett equation (Davis and Lien, 1993).

The creosote shows increased recoveries as the temperature is increased (at least up to 30°C), but in this case the improvement in recovery occurs during the latter stages of the displacement, and breakthrough recovery changes very little with temperature. Most of the creosote recovery was with the first five pore volumes of injected water (Davis, 1995). The crude oil shows a significant increase in recovery when the temperature is increased from 10 to 20°C, and small additional recoveries are realized with additional increases in the temperature. This oil has an early breakthrough, and significant oil recovery after breakthrough.

'Only preliminary results are available at this time on the displacement of transmission fluid, but the two partial experiments that have been completed definitely how a much greater recovery at breakthrough than was found with the other oils. After breakthrough, the column effluent became an oil/water emulsion. The pressure within the column increased very significantly as the displacement of the emulsion continued. Examination of the sand when the experiment was terminated showed that most of the oil remaining in the column was in the influent end of the column, and the effluent end had only a small amount of the pink-colored emulsion remaining.

Table 2 lists the viscosity as a function of temperature for each of the oils. The creosote has a significantly lower viscosity than the other three oils, and this probably

Table 1. Oil recovery and final oil saturation for each of the experiments.

	10°C	20°C	30°C	40°C	50°C	Trans.
20/30 Sand/ Inland 15 Oil						
Percent recovery at Breakthrough	24.7	27.0	28.9	28.9	29.9	27.5
Percent recovery at 10 PV	60.9	63.0	65.4	71.0	69.9	65.4
Final oil saturation	0.391	0.384	0.344	0.295	0.299	0.297
Mixed Sand/ Inland 15 Oil						
Percent Recovery at Breakthrough	22.9		32.7		33.3	27.9
Percent Recovery at 10 PV	66.8		70.7		75.5	70.4
Final Oil Saturation	0.331		0.291		0.232	0.255
20/30 Sand Creosote						
Percent Recovery at Breakthrough	29.9	31.4	31.4	29.7		30.6
Percent Recovery at 10 PV	63.0	67.4	70.0	67.7		63.7
Final Oil Saturation	0.370	0.326	0.264	0.296		0.363
20/30 Sand/ Crude Oil						
Percent Recovery at Breakthrough	16.0	21.3	24.6	25.9	23.4	17.5
Percent Recovery at 10 PV	58.2	71.1	71.2	72.6	72.2	70.1
Final Oil Saturation	0.413	0.280	0.273	0.259	0.270	0.291

accounts for its later breakthrough than Inland 15 or crude oil. The crude oil has a very high viscosity at low temperatures, and thus an earlier breakthrough. At temperatures of 30° to 50°C the viscosities of the crude oil and Inland 15 are essentially the same, but the crude oil still has a measurably earlier breakthrough. Thus, although viscosity may explain the later breakthrough for some of the oils at higher temperatures, viscosity alone does not account for all of the differences in displacement behavior.

Table 2. Viscosity (centipoise) of the oils as a function of temperature.

	10°C	20°C	30°C	40°C	50°C
Inland 15	111.7	58.7	35.4	23.4	15.4
Crude Oil	149.3	68.2	40.9	25.3	16.9
Creosote	37.9	20.4	13.1	9.0	6.4
Transmission Fluid	125.2	71.8	44.3	29.6	20.4

Interfacial tension may also have a significant affect oil recovery. The surface tension for the four oils are similar (approximately 30 dynes/cm), and decreases as the temperature increases. Their interfacial tensions, however, vary drastically. The Inland 15 has a high interfacial tension that shows a slight decrease with temperature, while the creosote and transmission fluid have very low interfacial tensions that show a slight decrease with temperature. The interfacial tension of the crude oil is intermediate between the others, and does not show a decrease with temperature over this temperature range. Thus, the effect of interfacial tension on the oil recovery curve is not clear, and it appears there are other factors - such as wettability - that must be taken into account to fully explain oil recovery behavior.

Table 1 also lists the oil saturation remaining in the sands after 10 pore volumes of water have been injected at each of the temperatures for which displacements were run. At 10°C, remaining oil saturations ranged from about 0.33 to 0.41, while at the temperature of optimum oil recovery, the remaining oil saturation was 0.23 to 0.30. The reduction in the amount of oil remaining in the pore space was 25% for the Inland 15 in the 20/30 sand, 30% for Inland 15 in the mixed sand, 29% for creosote, and 39% for the crude oil. It should be noted that these remaining oil saturations are after the injection of 10 pore volumes of water,

Table 3. Interfacial tension (dynes/cm) as a function of temperature.

	10°C	20°C	30°C	40°C	50°C
Inland 15	40.0	40.0	40.0	39.7	39.6
Crude Oil	21.2	20.7	20.4	20.7	20.7
Creosote	9.6	7.8	6.2	5.3	5.9
Transmission Fluid	12.5	11.2	10.2	9.5	9.5

but are not necessarily the residual or immobile oil saturations. The creosote at low temperatures had very little additional oil being recovered by the end of the experiment, but the crude oil still had a significant amount of oil in the effluent when the experiment was ended. Thus, lower oil saturations are attainable with this displacement method for some oils if the water injection is continued.

Oil recoveries and the final oil saturation during the transient experiments are also listed in Table 1. During the early part of the displacement, most of the column is at 10°C, so the displacement is similar to the 10°C displacement. The heat does not travel along the column at the same rate as the water; approximately one and half pore volumes of water must be injected before the heat reaches the far end of the column. During the injection of about the next three pore volumes, the temperature all along the column continues to increase, and, as shown in this figure, the displacement then approaches that of the 40 or 50°C constant temperature displacement. The equilibrium temperature in the column ranged from about 38°C at the influent end to 32°C at the effluent end, and the oil recovery after the injection of 10 pore volumes of water was similar to a 40°C constant temperature waterflood.

CONCLUSIONS

Hot water, even in the moderate temperature range of up to 50°C, has been shown to increase the recovery of viscous, nonvolatile oils from porous media. The main mechanism for the increased recovery is the reduction in the viscosity at the higher temperatures, which generally increase the rate at which oil is recovered. Other properties of both the sand and the oil also affect the displacement history. These laboratory experiments have shown that a sand with somewhat smaller grain sizes and a wider distribution of pore sizes has a significantly better recovery of oil. The highly viscous crude oil shows a very significant increase in recovery as the temperature is increased to 20°C and above. The less viscous creosote shows smaller increases in recovery during the latter part of the displacement as the temperature is increased. Based on just the preliminary experiments that have been performed to date for the transmission fluid, it appears that the oil/water is very effective for displacing the oil from the sand. However, the pressure drop during the displacement of the emulsion is very high. The transient temperature experiments show that the benefits of the increased temperatures are achieved under conditions more representative of a hot water displacement in the field.

After the injection of 10 pore volumes of water, the remaining oil saturations in the columns ranged from 0.24 to 0.30. The crude oil experiments in particular were still recovering significant amounts of oil when the displacements were terminated, so it may be possible to reduce its saturation significantly below this amount by hot water injection. However, it is not likely that conditions can be met in a hot water displacement that would remove all of the oil; a residual oil saturation will be left in the pores. Thus a process such as bioremediation will likely be required as a "polishing" step. The moderate temperatures used here should leave the subsurface system amenable to biodegradation of the remaining oil. The removal of a large percentage of the oil will reduce toxicity effects for the subsurface microbial populations, and increased temperatures may increase the solubilization of the oil components so that they are available for biodegradation. The bioremediation is likely to proceed at enhanced rates and the overall time for remediation should be greatly reduced due to the reduction in the amount of oil to be biodegraded.

ACKNOWLEDGEMENTS

Although this research has been supported by the U.S. Environmental Protection Agency, it has not been subjected to the Agency review and therefore does not necessarily reflect the views of the Agency. The mention of trade names or commercial products does not constitute endorsement.

REFERENCES

Abdul, A. S. 1992. "A New Pumping Strategy for Petroleum Product Recovery from Contaminated Hydrogeologic Systems: Laboratory and Field Evaluations." *Groundwater Monitoring Review,* Winter, 105-114.

Davis, E. L., and B. K. Lien. February 1993. *Laboratory Study on the Use of Hot Water to Recover Light Oily Wastes from Sands,* EPA/600/R-93/021, Robert S. Kerr Environmental Research Laboratory, Ada, OK.

Davis, E. L. 1995. "Hot Water Enhanced Remediation of Hydrocarbon Spills." In D. W. Tedder and F. R. Pohland (Eds.), *Emerging Technologies in Hazardous Waste Management V,* pp. 237-250. American Chemical Society, Washington, DC.

Fulton, D. E., G. J. Reuter, and T. E. Buscheck. November 20-22, 1991. "Hot Water Enhanced Recovery of Phase Separated Lubricating Oil." *Petroleum Hydrocarbons and Organic Chemicals in Ground Water: Prevention, Detection, and Restoration Conference and Exposition,* 143-156.

Hess, K. M., W. N. Herkelrath, and H. I. Essaid. 1992. "Determination of Subsurface Fluid Contents at a Crude-Oil Spill Site." *J. Contamin. Hydrol.,* 10: 75-96.

Sale, T., and D. Applegate. 1997. "Mobile NAPL Recovery: Conceptual, Field, and Mathematical Considerations." *Ground Water,* 35(3): 418-426.

STEAM ENHANCED EXTRACTION OF WOOD TREATMENT CHEMICALS FROM SOILS

Kent S. Udell, University of California, Berkeley, California
Robert L. McCarter, University of California, Berkeley, California

ABSTRACT: Experiments were conducted to determine the applicability of Steam Enhanced Extraction (SEE) to remove wood treatment chemicals from soil. In all experiments, the soil was mixed with a viscous hydrocarbon mixture to a concentration of 10,000 gm/kg. The first two one-dimensional experiments demonstrated that over 99% of the hydrocarbon mass was removed after a throughput of about 180,000 pore volumes of steam. The third one-dimensional experiment was conducted with additional Pentachlorophenol (PCP) added to the hydrocarbon mixture obtained from the site subsurface. A significantly shorter cyclic steam injection period (9000 pore volumes of steam) was selected for this third run to more appropriately scale the steam use to field conditions. While the percentage of total hydrocarbon mass removed was unknown for the third experiment, over 80% (probably 99%) of the initial PCP was removed.

INTRODUCTION

The effectiveness of steam enhanced extraction of toxic compounds from soils and groundwater has been demonstrated in laboratory studies (Hunt et al, 1988) and field demonstrations (Udell, 1996, 1997). The operation of steam injection in a cyclic manner after the soils reach steam temperatures has been shown to enhance mass transfer rates and accelerate clean-up (Udell, et al, 1991, and Itamura and Udell, 1995). The applicability of SEE to soils containing low-volatility wood treating chemicals has not been previously studied. This report summarizes such findings.

EXPERIMENTAL APPARATUS AND PROCEDURES

The apparatus for the one-dimensional experiments consisted of a pair of metering pumps to deliver a constant flow rate of distilled water, a steam generator to vaporize the water into steam, a stainless steel pipe to hold the soil sample, and an effluent collection jar maintained in an ice bath. To provide an adiabatic experiment, the stainless steel sample holder was wrapped with heater tape and insulated. Teflon tubing was used to carry all fluids in the system. A condenser stage could be added to the outlet tubing between the outlet of the stainless steel sample holder and the effluent collection jar when the effluent flow rate was high.

Refrigerated, separated, and filtered NAPL pumped from the site subsurface was mixed with the sand at a concentration of one percent by mass

(10,000 mg/kg). This soil-hydrocarbon mixture was packed into the sample holder which was a 9 inches (0.23 m) long, 1.5 in. (3.8 cm) i.d. cylinder for experiments 1 and 2, and a 13 inch (0.33 m) long holder for experiment 3.

Throughout each experiment, the metering pumps delivered a constant water flow rate of 38 ml/hr. All of the effluent coming from the sample holder was collected into sample bottles. To minimize sample vaporization, an ice bath surrounded each bottle while collecting the effluent. As each bottle filled, it was sealed with a Teflon cap and placed it in a refrigerator at 4 °C (± 2 °C) or in an ice bath.

Three 1-D experiments were run. The first experiment consisted of steady steam injection over a long duration (with several short periods of pressure cycling), conducted with a fairly coarse clean sand approximating the most porous soils expected in field operations at the site of interest. Experiment 1 was run with clean, washed sand passing a 40 mesh screen, but not a 50 mesh screen. Pressure was cycled by closing the valve at the outlet end of the sample holder while maintaining steam flow into the inlet. The valve was closed until steam pressures rose to about 30 psig (207 kPa gage). The valve was then opened to depressurize the system over about two minutes.

The second experiment was conducted with finer silty sands taken from the field site. The measured permeability of the sand pack was about three darcy. The experiment was operated using cyclical pressurization of the test cell since this had been found to effectively increase removal rates during the first experiment. The valve was closed for periods of one hour and opened for periods of thirty minutes while maintaining steady steam flow into the inlet. With the valve closed steam pressures rose to about 30 psig (207 kPa gage).

For the third experiment the sample holder was packed with a sand and silt mixture from the site. This mixture was sieved with a 325 mesh screen to eliminate the finer grains that would pass through the screens at the ends of the test section. The resulting permeability of this sand when packed into the test section was about 2.5 darcy. This third experiment was designed to investigate the removal rate of Pentachlorophenol (PCP, formula C_6HCl_5O) from the NAPL during a relatively short injection period. One percent by mass of PCP was added to the NAPL. The NAPL was then added at one percent by mass (10,000 mg/kg) washed and sieved sand to give a 100 mg/kg PCP concentration in the soil. One inch (2.5 cm) of clean sand was placed at each end of the sample holder to minimize end effects such as condensation near the end caps. Pressure was cycled in the sample holder throughout the Experiment 3 as in Experiment 2.

Soil and effluent samples were also analyzed for poly-nuclear aromatics (PNA) by EPA Method 8270. From these analyses, the following chemicals of interest, listed in order of decreasing vapor pressures, were detected in the soil and effluent samples: Naphthalene, 2-Methylnaphthalene, Acenaphthylene, Acenaphthene, Dibenzofuran, Fluorene, Phenanthrene, Anthracene, Fluoranthene, Pyrene, Benzo (a) Anthracene, Chrysene, Benzo (b and k) Fluoranthenes, and

Benzo (a) Pyrene. At 100 °C, their vapor pressures range from about 0.36 psia (2.5 kPa) for Naphthalene to 0.005 psia (0.035 kPa) for Phenanthrene, to still lower values for chemicals shown in this report beyond Phenanthrene.

RESULTS

In all experiments, the nature of the effluent changed visibly with time. Samples collected prior to steam break-through contained water tinted light brown with no visible free product. Samples collected during the next few hours of injection contained free product in the form of light brown DNAPL, with the relative volume of DNAPL to water decreasing with successive samples. Later samples contained waxy solids that floated on the water in the effluent sample bottles. All samples collected while cycling pressure contained noticeably increased levels of NAPL compared to samples collected during steady injection.

The analytical reports from Experiment 1 indicated the total hydrocarbon concentration in the soil was reduced from 6620 mg/kg to 39 mg/kg as determined from EPA Method 8015M. This represents a reduction of hydrocarbon mass of 99.4%. The soil analyses for Experiment 2 indicated a reduction in the total hydrocarbon concentration in the soil from 6870 mg/kg to 19 mg/kg: a mass reduction of 99.7%. In both cases, the chromatogram displayed a dramatic reduction of all compounds but those with the highest column retention times. Hydrocarbons concentrations were not determined for soils of Experiment 3.

Figures 1, 2, and 3 show the concentrations of the PNAs in soil samples before and after steam injection for Experiments 1, 2, and 3, respectively. Comparison of Figures 1 and 2 indicates that compounds were more completely removed in Experiment 2 than in Experiment 1, likely because of the cyclic steam injection employed throughout that experiment and the greater number of pore volumes of steam injected. For both experiments, the data indicate that the more volatile compounds were removed to below detection limits and concentrations of the even less volatile compounds were also reduced by at least an order of magnitude. Figure 3 also indicates the general trend of significantly reduced concentrations of all compounds up through Phenanthrene and an order of magnitude decrease in concentrations of compounds of lesser volatility.

PCP concentrations in the Experiment 3 soil before and after treatment were below the detection limits of 200 mg/kg and 20 mg/kg, respectively. Since the soil was mixed with 100 mg/kg PCP prior to treatment, at least 80% of the PCP was removed from all soil samples. Since Phenanthrene is a less volatile than PCP, PCP reductions should be greater than those of Phenanthrene. Pre-steaming Phenanthrene concentrations in the soil were 560 mg/kg and were reduced to below its detection limit of 2 mg/kg, implying over 99.6% removal. Thus, it is expected that PCP concentrations were also reduced by over 99.6%.

Effluent concentrations of Pentachlorophenol, Phenanthrene, Anthracene, and Floranthene are shown as a function of time (pore volumes effluent condensate) in Figures 4, 5, and 6.

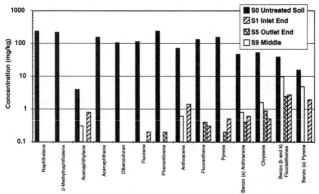

Figure 1: Soil Concentrations Before and After Steam Treatment - Experiment 1.

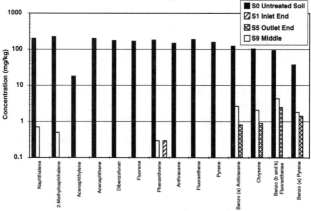

Figure 2: Soil Concentrations Before and After Steam Treatment - Experiment 2.

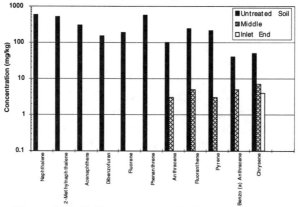

Figure 3: Soil Concentrations Before and After Steam Treatment - Experiment 3.

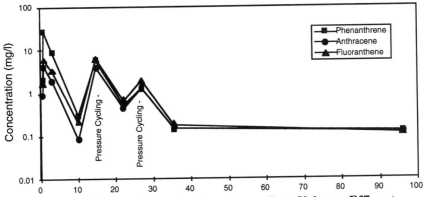

Figure 4: Effluent Concentrations Versus Pore Volumes Effluent Condensed - Experiment 1.

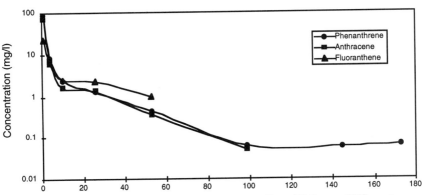

Figure 5: Effluent Concentrations Versus Pore Volumes Effluent Condensed - Experiment 2.

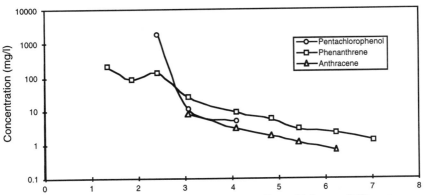

Figure 6: Effluent Concentrations Versus Pore Volumes Effluent Condensed - Experiment 3.

Comparison of Figures 4 and 5 shows the effluent concentrations during cyclic extraction were significantly higher than for steady steam injection, and when cyclic steam was used throughout, the concentration dropped to lower values much quicker (less pore volumes condensate). All other compounds showed the same trend. Figure 6 shows the data for PCP in the effluent of the third experiment. PCP was only detected in three samples. PCP concentrations were at 1830 mg/l after about two pore volumes of condensate and fell to 5.6 mg/l after four pore volumes. Beyond that, the PCP was below the 4 mg/l detection limit. From thermodynamic equilibrium considerations, such a reduction implies a decrease of about 99.8% in PCP mass in the hydrocarbon mixture.

CONCLUSIONS

The overall conclusions that can be drawn from this series of experiments regarding the effectiveness of steam injection to restore the site are:

- Steam will remove most of the hydrocarbon phase, including PCP, from the zones that are to be contacted by significant flowing steam.
- Cyclic steam injection was found to increase the recovery rates of the NAPL as compared to steady steam injection.

ACKNOWLEDGMENT

The support of Southern California Edision and the NIEHS, Grant No. 5 P42 E504705-11, Project # 11, is gratefully acknowledged..

REFERENCES

Itamura, M. T., and K. S. Udell. 1995. "An Analysis of Optimal Cycling Time and Ultimate Chlorinated Hydrocarbon Removal from Heterogeneous Media Using Cyclic Steam Injection," *Proc. Heat Transfer and Fluids Engineering Division*, ASME HTD-vol. 321 and FED-vol. 233, pp. 651-660.

Hunt, J. R., N. Sitar, and K. S. Udell. 1988. "Non-Aqueous Phase Liquid Transport and Cleanup, Part II. Experimental Studies," *Water Resources Research*. 24, No. 8, pp. 1259-1269.

Udell, K. S., J. R. Hunt, N. Sitar, and L. D. Stewart, Jr. 1991. "Process for *In Situ* Decontamination of Subsurface Soil and Ground water," U. S. Patent # 5,018,576.

Udell, K. S. 1996. "Heat and Mass Transfer in Clean-up of Underground Toxic Wastes", *Annual Review of Heat Transfer*. C-. L. Tien (ed.), Begell House, Inc., New York, NY, 7: 330-405.

Udell, K. S. 1997. "Thermally Enhanced Removal of Liquid Hydrocarbon Contaminants From Soils And Groundwater," *Subsurface Restoration*.. Ward, et. al. (ed.), Ann Arbor Press.

REMEDIATION DEMONSTRATION OF
DUAL-AUGER ROTARY STEAM STRIPPING

Barry Rice
Roy F. Weston, Inc.*
U.S. Department of Energy
Grand Junction Office
Grand Junction, Colorado

ABSTRACT: A field-scale remediation demonstration using dual-auger rotary steam stripping was performed at the former U.S. Department of Energy (DOE) Pinellas Plant in Largo, Florida, from January through April 1997. The goal of the remediation demonstration was to reduce areas of high contaminant concentration for the potential future application of in situ anaerobic bioremediation, while collecting cost and performance information on the rotary steam-stripping technology. The demonstration provided adequate analytical and operational data to evaluate the cost and performance of the rotary steam-stripping technology. Initial operational tests indicated that contaminant removal could be accomplished; however, several issues complicated the evaluation of the technology and slowed the progress of remediation. Following numerous system and operational changes, the system demonstrated liberation of contaminants from the site and, under controlled operations, destroyed recovered contaminant vapors with an off-gas treatment system.

INTRODUCTION

Site Description. The former U.S. Department of Energy (DOE) Pinellas Plant, now the Pinellas Science, Technology, and Research (STAR) Center and owned by a branch of the county government, occupies approximately 100 acres (40.5 hectares) in Pinellas County, Florida. The Pinellas Plant operated from 1956 to 1994, manufacturing components for nuclear weapons under contract to DOE. The site Management and Operating contractor at the time of the demonstration project was Lockheed Martin Specialty Components (LMSC), Inc. The portion of the plant site where the demonstration took place, the Northeast Site, was formerly a waste solvent staging and storage area. This area was also used to dispose of drums of waste and construction debris. A debris removal activity at the site removed multiple buried drums, many of which were empty but contained solvent residue.

---

*Work performed under DOE Contract No. DE–AC13–96GJ8335 for the U.S. Department of Energy.

Subsurface Conditions. The Northeast Site consists of a shallow surficial aquifer contaminated with a variety of volatile organic compounds (VOCs), including chlorinated solvents such as trichloroethene (TCE), methylene chloride, dichloroethene (DCE), and vinyl chloride. Estimates of the mass of contaminants suggested that approximately 9,000 lbs (4,082 kg) of VOCs were in the planned treatment area.

The water table is generally 3 to 4 ft (0.9–1.2 m) below ground surface. The base of the aquifer ranges from 25 to 35 ft (7.6–10.7 m) below ground surface and is composed primarily of fine sand. Permeabilities range between 10^{-3} to 10^{-5} cm/s with a variable presence of silt and clay. The Hawthorn Group (composed primarily of clay) underlies the surficial aquifer. The thickness of the Hawthorn Group ranges from 60 to 70 ft (18–21 m).

MATERIALS AND METHODS

Rotary Steam-Stripping Method. The technology evaluated in this field demonstration was rotary steam stripping for the in situ removal of high concentrations of VOCs from the soil and groundwater. This system is based on rotary drilling technology and consists of a drill tower attached to a mobile platform (Milgard, 1993; U.S. EPA, 1991). In most applications, the drill tower supports one or two drill blades or augers designed to inject hot air or steam into the subsurface soil as the drill blades or augers penetrate below the ground surface. The augers shear and mix the soil while the hot air or steam is being injected, causing stripping and thermal desorption of the organic contaminants from the soil particles and volatilization of the contaminants (La Mori, 1994). Injected air transports contaminant vapors to the surface for collection in a shroud that operates under a slight vacuum and rests firmly on the ground above the area being treated.

Once collected in the shroud, the contaminant vapors are sent to an off-gas treatment system for destruction or collection. Condensation, activated carbon adsorption, or thermal destruction can treat contaminants, depending on the type and concentration removed. The treated air and steam can be reinjected for further soil treatment (U.S. DOE, 1998). During this demonstration, vapors were treated with a portable catalytic oxidation (CATOX) unit, neutralized with an acid-gas scrubber, and then released to the atmosphere.

Several companies have developed mobile treatment technologies based on this process. In-Situ Fixation (ISF) of Chandler, Arizona, developed and operated the dual-auger rotary steam-stripping system subcontracted for this remediation demonstration. Figure 1 illustrates the ISF System operation at Pinellas.

A patent on certain aspects of the steam-stripping technology exists. According to ISF, no patent infringements occurred during the Pinellas Project. Anyone considering placing contracts for the use of this technology should be aware of the potential for patent-related issues (U.S. DOE, 1998).

FIGURE 1. The ISF dual-auger rotary steam stripping system at Pinellas.

Rotary Steam-Stripping Performance Goals. The Pinellas remediation demonstration was to accomplish the following goals:
- Determine optimum system operating parameters for efficient treatment during an initial treatment efficiency characterization (TEC) phase.
- Treat approximately 10,000-yd^3 (7,600-m^3) area to reduce VOCs in the soil and groundwater from levels in excess of 1,000,000 µg/L in the groundwater and 1,000,000 µg/kg or less in the soil to a level of 100,000 to 200,000 µg/L in the groundwater and 100,000 and 200,000 µg/kg in the soil.
- Measure overall contaminant removal from the treatment area.
- Determine system operational effects on the environment (e.g., aquifer, surface water, air) outside the treatment area.
- Determine the cost effectiveness of the Pinellas demonstration.

Rotary Steam-Stripping Performance Assessment. The following steps were used to assess system performance:
- Collect pre- and post-treatment soil and groundwater samples to verify contaminant removal and the absence of contaminant migration.
- Perform an initial TEC to gain further insight into the technology and to establish optimum treatment parameters.
- Collect air samples to verify off-gas treatment and destruction.
- Monitor system operations to assess fugitive emissions.
- Collect site-specific treatment costs to quantify the cost effectiveness of steam stripping.

RESULTS AND DISCUSSION

Project Operations. This schedule was used for project operations:

Mobilization:	September 25 – December 24, 1996
Assembly/Start-up:	January 6 – January 20, 1997
TEC:	January 21 – February 11, 1997
Treatment:	February 12 – April 2, 1997
Demobilization:	April 3 – April 16, 1997

During mobilization, delays were experienced because the appropriate model of trackhoe was not available and the acid-gas scrubber had to be constructed. After assembly and testing of the system, seven holes were treated during the TEC in predetermined areas of varying contamination. Several operational problems with field gas chromatograph operation, off-gas destruction and neutralization, and fugitive emissions outside the shroud created significant delays and 6 days of downtime during the TEC.

During the TEC and the treatment operations (January 21 through April 2, 1997), 48 holes were treated to a depth of approximately 32 ft (10 m) below ground surface, or approximately 2,000 yd^3 (1,529 m^3) of the 10,000-yd^3 (7,600-m^3) treatment area. Problems experienced with operational control and mechanical breakdown resulted in 50% downtime for the overall project duration (6 days downtime during the TEC; 17 days downtime during project operations). However, downtime was reduced to approximately 25% when good operational control was established after initial startup.

Treatment Efficiency. Table 1 presents contaminant reduction data and associated operational details.

Project Costs. The subcontract financial terms with ISF were

Mobilization	$ 95,000
Demobilization	51,000
<u>Time and Materials</u>	<u>773,651</u> (not to exceed)
Subcontract Total	$919,651

On the basis of on-line percentages, the operational costs of the ISF system at this site ranged from $50/yd^3 ($65/m^3) to $400/yd^3 ($523/m^3) of treated soil and groundwater, or about $300/lb ($661/kg) to $500/lb ($1,102/kg) of contaminant removed, depending on contaminant levels (U.S. DOE, 1998).

The inability of the off-gas treatment system to process all the contaminant vapors removed by the dual-auger system was a shortfall in the design process. This deficiency severely affected the subsurface VOC removal rate, the cost per cubic yard of soil and groundwater treated, and the overall cost effectiveness of the system (U.S. DOE, 1998). Most notably, the presence of toluene (having a high heat of combustion) severely affected the ability to control the CATOX operations.

TABLE 1. Pre-treatment and post-treatment soil and groundwater concentrations.

Hole	Pre-treatment concentrations (max.)		Post-treatment concentrations (max.)		Reduction (%)	Removal Based on FID[a] Data (%)	Treatment Method
	Soil (μg/g)	Ground-water (mg/L)	Soil (μg/g)	Ground-water (mg/L)			
1A	<1	5,170	120	1,484	69	93	Air only
1C	1,860	2,480	106	724	81	55	Air, then steam
3A	20	1,426	325	1,019	7	95	Air only
3B	7	6,952	11	1,135	84	30	Air, then steam
3C	82	1,860	29	341	81	95	Air and steam
4A	28	NA[b]	143	NA	–	90	Air and steam
4B	900	NA	26	NA	97	95	Air and steam
4C	204	NA	158	NA	23	95	Air and steam
MP14	19	251	<1	2	99	No data	Air and steam
MP18	<1	1,290	2	198	85	75	Air and steam
MP19	<1	1,364	6	198	85	45	Air and steam

[a]FID = flame ionization detector.   [b]NA = not analyzed.

Performance. An assessment of the remediation demonstration at Pinellas provides these performance observations:

- The ISF dual-auger system demonstrated the ability to remove large masses of contaminants rapidly from the soil and groundwater.
- On the basis of pre- and post-treatment sampling analyses, the ISF dual-auger system was able to remove an average of 77% of the contamination in the treated areas.
- The rotary steam-stripping system was not able to treat an approximate 10,000-yd^3 (7,600-m^3) treatment volume to remove VOCs in the soil and groundwater to a level of 100,000 to 200,000 μg/kg and 100,000 to 200,000 μg/L, respectively, within the not-to-exceed amount of the time and materials phase of the subcontract.
- The off-gas treatment system efficiently destructed volatilized contaminants recovered from the treatment area only after system modifications and limitations of the air/steam injection rates were implemented.
- Fugitive emissions in the treatment area necessitated engineering controls.

CONCLUSIONS

Lessons Learned. Important lessons learned during the Pinellas remediation demonstration of dual-auger rotary steam stripping include

- Pre-project discussions and a cooperative relationship with regulatory agency personnel are essential to conduct innovative technology demonstrations.

- Regulatory permits necessary for project operations (e.g. air emissions, fuel storage) should be of long enough duration to allow for unforeseen delays.
- Identification of contaminants, concentrations, and distribution is critical for the design of an off-gas treatment system.
- Pre-project research of rotary steam-stripping vendors should include average downtime during past projects; the effect of fugitive emissions and potential controls needed; sufficient capacity of off-gas treatment systems and the ability to increase capacity quickly; and potential patent issues.
- Delays in post-treatment sampling of 2 to 3 weeks may be experienced because of soft, loosened soil over boreholes.
- Subsurface conditions will be significantly altered after treatment (e.g. soil conditions, hydraulic conductivity); the long-term effects are unknown.

Summary. During the operating period, 48 treatment holes were drilled to a depth of approximately 32 ft (10 m), resulting in the treatment of more than 2,000 yd^3 (1,529 m^3) of soil and groundwater and the removal of approximately 1,200 lbs (544 kg) of VOCs (U.S. DOE, 1998). This removal operation demonstrated that the ISF dual-auger rotary steam-stripping system is capable of in situ remediation of VOC-contaminated soil and groundwater. More specifically, the ISF dual-auger system demonstrated the ability to liberate large masses of contaminants from the Pinellas soil and groundwater. However, operational problems, limitations because of off-gas treatment system capacity, and mechanical breakdowns during the Pinellas remediation demonstration prevented many of the expected benefits from being fully realized.

ACKNOWLEDGMENTS

The rotary steam-stripping demonstration was made possible through the assistance of David Ingle (now with the DOE Grand Junction Office), the DOE Pinellas Area Office, and the ITRD Program.

REFERENCES

La Mori, P. N. 1994. "Using In-Situ Hot Air/Steam Stripping (HASS) of Hydrocarbons in Soils." HazMat International '94, Philadelphia, PA.

Milgard Environmental Corporation. 1993. *Advanced Techniques for In Situ Remediation,* MEC–195. Livonia, MI.

U.S. Department of Energy. 1998. *Cost and Performance Report, Dual Auger Rotary Steam Stripping, Pinellas Plant Northeast Site.* Innovative Treatment Remediation Demonstration Program, Sandia National Laboratories/New Mexico, Albuquerque, NM.

U.S. Environmental Protection Agency. 1991. *Toxic Treatments, In Situ Steam/Hot-Air Stripping Technology, Applications Analysis Report,* EPA/540/A5–90/008. Risk Reduction Engineering Laboratory, Cincinnati, OH.

IN SITU HYDROTHERMAL OXIDATIVE DESTRUCTION OF DNAPLS IN A CREOSOTE CONTAMINATED SITE

Roald N. Leif, Marina Chiarappa, Roger D. Aines, Robin L. Newmark and Kevin G. Knauss, Lawrence Livermore National Laboratory, Livermore, CA, USA and Craig Eaker, Southern California Edison Company, Rosemead, CA, USA

ABSTRACT: Hydrous Pyrolysis / Oxidation (HPO) is an *in situ* thermal remediation technology that uses hot, oxygenated groundwater to completely mineralize a wide range of organic pollutants. A field demonstration of HPO was performed at a creosote contaminated site during the summer of 1997. The groundwater was heated by steam injections and oxygen was added by coinjection of compressed air. The remediation was monitored from multiple groundwater monitoring wells. Dissolved organic carbon levels increased in response to steam injections as a result of the enhanced dissolution and mobilization of the creosote into the heated groundwater. Elevated concentrations of partially oxidized organic compounds (i.e. phenols, benzoic acid, fluorenone, anthrone and 9,10-anthracenedione), decreased levels of dissolved oxygen and isotopic shifts in the dissolved inorganic pool were indicators of partial to complete oxidative destruction of the creosote in the heated aquifer as a result of the HPO process.

INTRODUCTION

The 4.3 acre Southern California Edison Pole Yard located in Visalia, California was in operation for 80 years as a wood preservation treatment facility. As a result of this operation, this site has become contaminated with a DNAPL mixture composed of pole-treating creosote and an oil-based carrier fluid containing pentachlorophenol. Placed on the EPA Superfund list in 1977, pump and treat technology was deployed to reduce and contain the contaminant plume. Over a period of nearly 20 years an estimated 10,000 lbs. of contaminant were removed from the soil and groundwater.

In the summer of 1997 Southern California Edison began the application of two thermally enhanced remediation technologies to accelerate the clean-up. The first method, Dynamic Underground Stripping (DUS), involves steam injection coupled with vacuum extraction to enhance the mobilization and removal of free product (Newmark and Aines, 1995). The second method, Hydrous Pyrolysis / Oxidation (HPO), is a novel *in situ* thermal remediation technology that uses hot, oxygenated groundwater to destroy organic contaminants by completely oxidizing the organic pollutants to carbon dioxide. The supplemental oxygen is delivered in the form of injected air. HPO is needed to destroy the residual DNAPL components not readily removed by the DUS process.

Initial laboratory-based feasibility experiments were conducted to investigate the HPO of actual DNAPL material with excess dissolved O_2 under conditions similar to those achievable during thermal remediation (Knauss et al., 1998; Leif et al., 1998). These experiments demonstrated that dissolved O_2 readily reacts with the compounds making up the DNAPL creosote mixture to form products ranging from partially oxidized intermediates, such as phenols and benzoic acid (Figure 1), to the fully oxidized product CO_2 (Figure 2). Field implementation of HPO remediation at the Southern California Edison Pole Yard site was initiated in May, 1997 using 11 steam injection wells encircling the creosote DNAPL pool. An aquifer situated 75-102 ft. below ground surface was targeted for the HPO field demonstration.

Oxidative Destruction of Aqueous Creosote Components

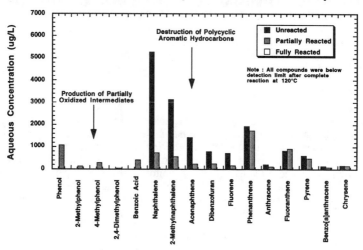

FIGURE 1. Histogram showing aqueous concentrations of organic compounds during a laboratory scale HPO experiment. Starting concentrations are shown in black and partially reacted are shown in gray. Complete destruction of creosote was achieved at 120°C.

O_2 Consumption and CO_2 Generation During HPO

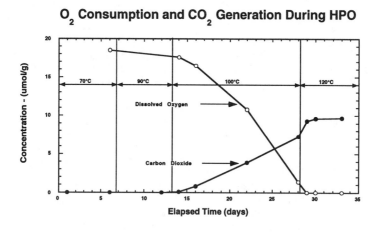

FIGURE 2. Oxygen consumption and carbon dioxide generation during a laboratory scale HPO experiment where the reaction temperature was increased from 70°C to 120°C over a duration of 33 days. Complete oxidation of creosote components is shown by the production of CO_2.

ANALYTICAL METHODS

Priority pollutants were extracted and concentrated by solid phase extraction (SPE) prior to analysis by gas chromatography - mass spectrometry (GC-MS). Typically a water sample ranging in volume from 1 to 4 liters was flowed by positive pressure through an SPE cartridge packed with 200 mg of ENV+ (International Sorbent Technology), a highly crosslinked styrene-divinylbenzene resin suitable for extraction of nonpolar and polar compounds from water. After sample extraction, the SPE tubes were dried and eluted with 4.5 mL of a dichloromethane / isopropanol eluent (1:1). The extracts were spiked with a six component internal standard mix and volumes adjusted to 5 mL. Bottles were extracted with a dichloromethane / isopropanol solvent mix (1:1) to extract organic compounds adsorbed to the glass. The extracts were spiked with a six component internal standard mix and the volumes adjusted to 5 mL. GC-MS analyses of the SPE extracts were performed on a Hewlett-Packard 6890 gas chromatograph equipped with a 30 m x 0.25 mm i.d. HP-5ms (5% phenyl methylsiloxane) capillary column (0.25 μm film thickness) coupled to a Hewlett-Packard 6890 Series Mass Selective Detector operated in electron impact mode (70eV) over the mass range 35-450 dalton with a cycle time of 1.1 s. The GC oven temperature was programmed at isothermal for 2 min. at 50°C, ramped at 8°C/min. to 300°C, and isothermal for 6.75 min., with the injector at 250°C and helium as the carrier gas. The MS data were processed using Hewlett-Packard Chemstation software. Internal standard method, using a relative response factors, was used to quantitate the target compounds.

RESULTS AND DISCUSSION

The creosote-derived groundwater contaminants present in the intermediate aquifer of the Southern California Edison Poleyard exhibited large variations in both compound distributions and contaminant amounts depending on when and where the water samples were taken. Observations consistent with the process of DUS were the increased concentrations of dissolved organic compounds following increases in groundwater temperature as a result of the steam injections. This is represented by the elevated levels in the aqueous concentrations polycyclic aromatic hydrocarbons (PAH) following the injections of steam (Figure 3). The relative abundances of the higher molecular weight PAH (i.e. fluoranthene, pyrene and chrysene) were also observed to increase as a result of the steam injections.

One result from the HPO process was the rise in the groundwater concentrations of partially oxidized organic compounds. These oxygenated compounds (i.e. low molecular weight phenols, benzoic acid, fluorenone, anthrone and 9,10-anthracenedione) represent the partially oxidized intermediates formed during the HPO of a complex creosote mixture. Fig. 3 shows how the concentrations of these oxygenates changed in response to the steam injections. The levels of total oxygenates maximized following both steaming events and their presence is consistent with the aqueous phase oxidations expected under these conditions.

The measurement of dissolved oxygen also aided in the evaluation of the HPO process. A knowledge of the dissolved oxygen level in the groundwater was critical during the application of HPO because the fundamental principle of HPO is the ability of hot, oxygenated water to completely mineralize organic compounds to carbon dioxide. The aqueous phase oxidation will occur as long as sufficient dissolved oxygen is present. Figure 4 is a plot of the dissolved oxygen measurements in the aquifer as a function of time during the field test. A steady decrease in the level of dissolved oxygen was observed during the field test and is consistent with the HPO chemistry where the dissolved oxygen is the oxidant during the chemical oxidation of the aqueous organic species.

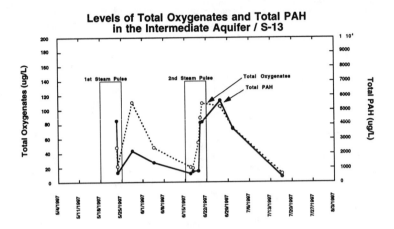

FIGURE 3. Concentrations of total oxygenates and total PAH
(polycyclic aromatic hydrocarbons, EPA Method 8270C) in the
aquifer during the HPO test period. Elevated PAH concentrations
reflect enhanced mobilization due to DUS. Oxygenate increases are
consistent with partial hydrocarbon oxidation by HPO.

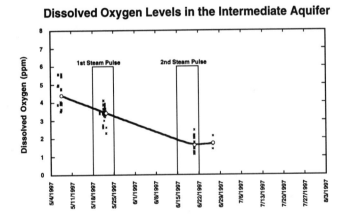

FIGURE 4. Concentration of dissolved oxygen as a function of time
during the HPO demonstration. Average dissolved oxygen
concentrations (open circles) were calculated using the combined
values from three different analytical techniques. Dissolved oxygen
levels dropped from 4.4 ppm to 1.7 ppm during the test period.

Another analytical tool used for evaluating the progress of the HPO remediation process was the measurement of carbon isotope abundances (^{12}C, ^{13}C and ^{14}C) of the dissolved inorganic carbon. Because both $^{13}C/^{12}C$ and $^{14}C/^{12}C$ values of the creosote are distinct relative to the groundwater, these measurements were used to trace carbon derived from the oxidation of the creosote compounds. Figure 5 shows the variations in ^{14}C versus $\delta^{13}C$ values of dissolved inorganic carbon in the groundwater. The groundwater end-member value was the isotopic signature prior to steaming. The dissolved inorganic carbon became "older" after steaming, consistent with the production of dissolved inorganic carbon by the oxidation of "dead" creosote carbon.

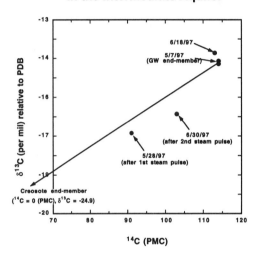

Variations in ^{14}C versus $\delta^{13}C$ values in the Intermediate Aquifer

FIGURE 5. **Variations in ^{14}C versus $\delta^{13}C$ values of dissolved inorganic carbon (DIC) in groundwater. The groundwater end-member value was the isotopic signature prior to steaming. The DIC became "older" after steaming, consistent with the production of DIC by the oxidation of "dead" creosote carbon.**

CONCLUSIONS

It is clear from the Visalia Field Test at the Southern California Edison Pole Yard that the combined applications of two *in situ* thermal remediation technologies, Dynamic Underground Stripping and Hydrous Pyrolysis / Oxidation, have greatly accelerated the remediation of this creosote-contaminated site. The application of DUS to the site accelerated the mobilization and removal of creosote. The application of HPO to the Southern California Edison Pole Yard has accelerated the site remediation by oxidizing creosote components. Observations consistent with the process of HPO were increases in groundwater oxygenate concentrations, decreases in dissolved oxygen levels and shifts in carbon isotope abundances in the inorganic carbon pool.

ACKNOWLEDGMENTS
    The authors thank Allen Elsholz and Ben Johnson for field. We also thank the employees of Visalia Southern California Edison Co. and the employees of SteamTech Environmental Services for expert assistance at the site. This work was performed under the auspices of the U.S. Department of Energy by the Lawrence Livermore National Laboratory under Contract W-7405-Eng-48. Partial support provided by the Southern California Edison Company was greatly appreciated.

REFERENCES
Knauss K. G., M. J. Dibley, R. N. Leif, D. A. Mew and R. D. Aines. 1998. "Aqueous oxidation of trichloroethylene (TCE) : A kinetic analysis." *Geochim. Cosmochim. Acta* (submitted).

Leif R. N., R. D. Aines and K. G. Knauss. 1998. *Hydrous Pyrolysis of Pole Treating Chemicals: A) Initial Measurement of Hydrous Pyrolysis Rates for Naphthalene and Pentachlorophenol; B) Solubility of Flourene at Temperatures up to 150°C.* Lawrence Livermore National Laboratory, Report.

Newmark R. L. and R. D. Aines. 1995. *Summary of the LLNL gasoline spill demonstration - Dynamic Underground Stripping Project.* Lawrence Livermore National Laboratory, UCRL-ID-120416.

SOIL VAPOUR EXTRACTION AS A CONTAINMENT TECHNIQUE

Almar M. Otten (Tauw Environmental Consultants, Deventer, Netherlands)
Frank Spuij, Renate G.M. Lubbers and Johannes C.M. de Wit

ABSTRACT: Soil vapour extraction (SVE) as a containment technique is a sensible and (cost-) effective alternative for more conventional techniques, like sealing the topsoil. Soil vapour extraction actively interrupts the diffusive gas flow from the subsurface to the topsoil. This makes it an effective zero risk technique. Theoretical considerations and results from pilot plant studies illustrate that intermittent pumping, with relatively short SVE phases and long standstill phases, suffice to contain the pollution and to minimize risks.

INTRODUCTION

Soil vapour extraction (SVE) is probably the most popular in situ soil remediation technique. The principle of the technique is stripping the contaminant from the vadose soil zone by forced air flow in the gas-filled pores. As a bonus, SVE stimulates aerobic biodegradation, due to enhanced aeration. SVE is either used as the single remediation technique, or, in combination with other (in situ) techniques like pump and treat, air sparging or partial excavation. An alternative application of SVE is to use it as a containment technique.

Volatilization of contaminants is a pathway that highly determines the potential toxicological risk of soil pollution. To reduce exposure via this pathway, impermeable foil, or other measures that seal the top soil, are used. In this paper, we will illustrate that SVE can be an interesting and cost-effective alternative. In the next section, we give a description of the concept of SVE as a containment technique. This is followed by the presentation of two case studies at which the SVE containment technique was successfully tested. In the discussion section we will discuss the pro and contra's of the SVE technique compared with more conventional isolation techniques.

DESCRIPTION OF THE SVE CONTAINMENT TECHNIQUE

The SVE technique involves the installation of air extraction, injection and monitoring filters in the subsurface, pumping devices, and, (if necessary) off-gas treatment. Mostly, the site is (at least partly) a build up area, which strongly reduce the degrees of freedom in the location and type of extraction filters that can be installed. The principles of the design of a SVE extraction (as an in situ remediation technique) are rather well established[1,2]. In the optimal design, the radius of influence of the extraction devices is such that the soil system is fully ventilated, that stagnant phases are not present, and, that the time to flush the system once, is short. Geohydrological calculations based on site characteristics, like (spatial variable) permeability, help to optimize the design.

For SVE as a containment technique, the principles of the installation are similar, but, the optimal conditions may be somewhat different. Now, our goal is the interruption of the natural diffusive gas flow. Preferably, the removal of the contamination should be small, since a large removal will increase the costs of the off-gas treatment.

Under natural conditions, diffusive gas flow from the contaminant source (in the subsurface or at the groundwater table) to the top soil is a slow process. Apparently, the simplest and most secure way to control the system is to maintain a (very) small continuous flow-rate throughout the system. In practice, however, this is hardly feasible. The radius of influence of the extraction filters is proportional to the flow rate. Slow flow rates involve a dense network of extraction filters. Mostly, the installation of such a network will be too expensive or not possible due to site-specific factors, like for instance buildings and infrastructure.

A sensible alternative is intermittent pumping at a higher pumping rate and a large radius of influence. Pumping at higher rates successfully interrupts the diffusive flow, and, removes the readily available part of contamination that is present in the volume where convective gas flow is induced. The readily available fraction is, for instance, the amount present in the macropores, either in the gas phase, dissolved in the water phase, or, reversibly sorbed to the soil solid matrix. When this fraction is removed, and, the concentrations in the monitoring filters are below target levels, the extraction can be stopped for a while.

During the standstill the natural flow patterns are restored. Slow diffusion controlled mass transfer, from the areas of the soil that are still contaminated, will refill the readily available fraction removed during the SVE. The contaminated areas can be the unsaturated zone outside the radius of influence of the extraction, the saturated zone, or, local areas within the radius of influence, where convective flow was absent, for instance in aggregates.

Owing the slow subsequent delivery, the concentration in the monitoring filters increases. When the concentrations in the monitoring filters become close to the target values, extraction is started again and continues until the good available fraction is removed. The process of extraction followed by standstill is repeated during the duration of the containment.

Clearly, the cost-effectivity of the SVE containment technique will depend highly on the length of the pumping phase and the standstill phase. The duration of the pumping phase depends on the size of the readily available fraction; Is the contaminant only present in the gas phase, or, is it mainly sorbed?

When the unsaturated soil zone is strongly contaminated, for instance due to a spill of pure product, we expect an (almost) exponential concentration decrease during the first one or two years of the remediation. Such long pumping phases are unfavourable for SVE as a containment technique, but, one may consider that such a site can be efficiently treated by SVE as a remediation technique.

A more favourable situation exists when the top soil itself is not strongly contaminated, and, the source of the concentration is located deeper in the subsurface, for instance at the groundwater table. For this situation we expect a rapid concentration decrease within the first weeks of the extraction.

The duration of the standstill phase depends on the quickness of the restoration of the natural concentration profile due to the subsequent delivery from the "sources", the areas that are still contaminated. The time to restore the natural profile depends on (1) the kinetics of contaminant release in the source, (2) the distance from source to the critical monitoring filter in the subsoil, (3) the storage capacity in the readily available fraction (reversible sorption and dissolution in the water phase) (Schoen, 1994).

CASE STUDY ARNHEM

Problem. In some 30 monumental houses around a former dry cleaner's, indoor air concentrations of perchloroethylene (PCE) above the maximum acceptable level of 250 $\mu g/m^3$ were found. Soil investigation showed that very high concentrations were present in the soil, the soil gas (up to 25 gr/m^3) and the groundwater (120,000 $\mu g/l$). The maximum soil gas concentrations were found near the groundwater level at 20 m-bsl. Because of serious health problems, remediation measures were necessary on a short term.

Sealing of the floor was not effective in these old houses. Despite the sealing, concentrations in the crawl space remained high. This was probably caused by leakage along or through the old brick walls. As an additional measure, SVE was chosen, and, a pilot plant study was carried out to investigate the possibilities and the effectiveness of this technique. The pilot plant study was carried out at the house with a PCE concentration of about 5 mg/m^3 in the crawl space. The air concentration in the living room was about 2.5 mg/m^3.

Pilot plant. The pilot plant consisted of two air extraction drains, that were placed at a depth of 1.5 m below the lowest part of the house at a mutual distance of 10 m. At a depth of 1.0 m below the house, three drains were installed that could be used for additional air infiltration. This appeared not to be necessary. In and around the house 17 monitoring filters were placed at the following depths: 3 at 1 m above the drains, 6 at 0.5 m below the drain, 2 at 2 m below the drain and 6 at 2.5 m below the drain. Monitoring consisted of measuring the underpressure and the PCE concentrations in the monitoring filters, and the indoor air concentration. Figure 2 shows the location of the drains and the monitoring filters.

The execution of the pilot plant was divided into three phases. In the first 19 days soil vapour was extracted at total discharge-rate of 150 m^3/hour. Phase two, from day 20 to 30, was the standstill phase and in phase 3, from day 31 to 40, soil vapour was again extracted at a total discharge rate of 150 m^3/hour.

The soil gas concentrations in the monitoring filters are given in Table 1. During SVE the soil gas concentrations in the filters above the drain decrease rapidly. The concentrations in the deeper filters, especially those south of the house, remained relatively high. Most likely, these high concentrations are caused by rather, small and local sources of PCE at an unknown depth between surface level and the groundwater level. We expect that these sources are less-permeable zones in the unsaturated zone in which the PCE has stayed behind. During the standstill phase the concentrations in the 3 filters above drain level increased significantly, but did not reach the initial values.

The indoor air concentrations were measured before phase 1 and after the flushing phases. After phase 1, 15 days of SVE, in the crawl space the air concentration had decreased from 5 to 0.05 mg/m^3. In the living room at the first floor of the house the concentration had decreased from 2.5 mg/m^3 to 0.008 mg/m^3, which is far below the acceptable level for indoor air quality. After phase 3 the indoor air concentrations were comparable to the concentrations after phase 1.

The indoor air concentrations were about 100 times lower than the soil gas concentrations. This can be explained by assuming that the gas phase is diluted by convective ventilation of the crawl space and the house. Tracer experiments support this assumption.

Despite the dilution, the indoor air concentration seems to be directly related to the soil gas concentrations in the upper monitoring filters; The observed concentration decrease, after phase 1, in the indoor air and the filters is of the same order.

Table 1. Case Study: Measured soil gas concentrations

filter	depth (m. below drain)	concentration (mg/m^3)				
		day 0	day 6	day 13	day 31	day 41
101·	0.5	165		5		1
101	2.5	466		6		6
102	0.5	428		3		2
102	2.5	1991		13		8
103	-1.0	2351	3	3	100	2
103	2.0	3761		86		25
104	-1.0	3783	3	25	276	2
104	0.5	4269		8		16
104	2.5	4890	241	160	908	41
105	-1.0	4075	3	1	11	1
105	2.0	4269		5600		2003
106	0.5	5821		80		38
106	2.5	7761		150		290
107	0.5	8052		350		85
107	2.5	11947		200		350
108	0.5	5675		980		571
108	2.5	8915		2100		1013

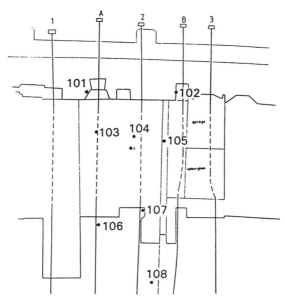

Figure 1: Case Study: Overview of the installation

Full scale system. The pilot plant study proved SVE to be an effective containment technique. Therefore, it was decided to carry out the SVE for all 30 houses located around the former dry cleaner's with indoor PCE concentrations above the acceptable level of 250 $\mu g/m^3$.

Beneath each house an horizontal drain was installed by means of remote control so that inconvenience for residents could minimized. Alongside of each house a soil gas monitoringfilter was placed.

The drains are divided into three clusters. Cluster 1 where the highest soil gas and indoor air concentrations were measured consists of 6 drains. The other two clusters consist of 12 drains. Drains of one cluster are connected to the same vacuumpump. All three vacuumpumps have a maximum capacity of 250 m^3/hr. Each drain was provided with a valve that could be operated automatically. Within one cluster drains are opened sequentially and there is only one drain opened at the time. The required opening time of one drain depends on the rate of decrease of the soil gas concentration and the rate of recontamination of soil gas due to diffusion from the source of PCE. A first estimation of an optimum sequence was based on the results of the pilot-plant.

For treatment of the contaminated air a choice had to be made between catalytic oxidation and regenerable active carbon. Because of an uncertainty with regard to the total amount of PCE to be treated, regenerable active carbon was selected as being the less risky and thus the most cost effective system.

On line monitoring of soil gas concentrations and indoor air measurements are used to optimize the extraction system and to minimize the operational costs. This has been achieved by reduction of the discharge rate as well as extension of the period of standstill of the system.

Results of the full scale system. First results showed that soil gas concentrations decreased within a period of days. Indoor measurements were carried out one month after starting vapour extraction and showed that all indoor air concentrations had been reduced to below the acceptable level.

The system has been operated since January 1995. Before starting the system the average indoor air concentration was 460 $\mu g/m^3$. In 1995, 1996 and 1997 the average indoor air concentrations were less than 5 $\mu g/m^3$. From large studies in different Dutch towns it was concluded that a concentration of 2 to 3 $\mu g/m^3$ can be regarded as urban background level.

DISCUSSION

Conventional isolation techniques are based on the construction of an impermeable barrier in the subsurface or at the topsoil. Clearly, the seal is only effective at the area where it is present and when the leakage through or along the impermeable layer remains negligible in time. By the construction of new buildings it is possible to meet the requirement of an everlasting negligible leakage by choosing the right material and proper installation. At contaminated sites, where old houses are present (e.g. Case study), our experience is that installation of a sealing layer can be a problem.

SVE proved to be very effective in reducing soil gas concentrations and indoor air concentrations. An advantage of SVE is that it is a dynamic and flexible system, based on the idea of actively interrupting the contaminated, diffusive gas flow. SVE is a containment measure with few risks of failure. Possible failures, such as the breakdown of a pump or the purification plant, are easily traceable and repairable on a short term. Since the increase of soil gas concentrations occurs relatively slow, the breakdown of one part of the system will not immediately lead to health risks.

Compared with sealing, SVE has the disadvantage of current costs. The current costs of SVE are determined by the required discharge rate and the duration of the flushing phases compared with the standstill phases. To estimate the costs of SVE, it is important to determine site characteristics like: type, degree and sources of the contamination, depth of the groundwater table, the sorption capacity and the spatial distribution or heterogeneity of these properties. The current costs can be reduced by optimizing the intermittent pumping regime. This requires frequent monitoring during the first phase of the operation of the system.

The decision whether to choose sealing or SVE as containment measure depends on costs and on the risks of failure. If sealing fails, the origin of failure is hard to detect and probably expensive to repair. Therefore, in old house sealing is less suitable than SVE.

Construction of new building gives the opportunity to install a proper sealing layer. Nevertheless, we advise to install some air extraction drains below the sealing. The installation costs are small and, in case of failure of the sealing, drains can be easily connected to a vacuum pump.

SVE can be applied to site types where volatilization of contamination causes health problems or bad smell: dry cleaner's (e.g. case study), gas stations, former gasworks, industrial sites and tank terminals. Former gasworks are situated close to old town centres, and, frequently, became domestic areas after dismantling. At these sites, volatilization of contaminants, like benzene and naphthalene, may cause high indoor concentrations and health problems. At industrial sites and tank terminals, volatilization of spilled contaminants may lead to unacceptable toxicological risks. Since a full and rapid remediation of former gasworks and industrial sites is often not feasible, SVE can be used to reduce the actual exposure and to prevent further spreading of contaminants.

At present, we are optimizing the SVE containment technique and are working on the automated process operation that integrates on line monitoring with descriptive and predictive modelling and automated process control. With minor modifications this automated SVE system can be used as a combined early warning/soil remediation system for gas stations, depots for organic solvents and for industrial sites.

CONCLUSIONS
- Since, SVE actively interrupts the natural diffusive gas flow from the contaminant source (in the subsurface or at the groundwater table) to the top soil, it is an effective technique to achieve a zero risk situation.
- An intermittent pumping regime, with short active SVE phase and long standstill phases, suffices to reduce the volatilization.
- SVE can be used as a separate containment technique or, in combination with more conventional sealing techniques. With slight modifications the SVE system can also be used as a combined early warning/soil remediation system for gas stations, depots for organic solvents and for industrial sites. This modified system is currently in development.

REFERENCES

Johnson, P.C., Stanley, C.C., Kemblowski, M.W., D.L. Byers and Colthart, J.D. (1990), *Ground Water Monitoring Review*, Spring Issue, 159-177.

Johnson, P.C., Kemblowski, M.W. and Colthart, J.D. (1990), *Ground Water*, Spring Issue, **28**:413-429.

Schoen, J. (1994) *Diffusion controled mass transfer and soil vapour extraction*, MsC thesis, Department of Environmental Technology, Wageningen Agricultural University, The Netherlands (In Dutch).

ACCELERATED REMEDIATION OF CHLORINATED SOLVENTS VIA CYCLIC MULTI-PHASE EXTRACTION

Mark A. Culbreth, Donald R. Ehlenbeck, and Robert R. Colberg (Environmental Consulting & Technology, Inc., Tampa, Florida, USA)
Andrew C. Bailey (First Union Corporation, Charlotte, North Carolina, USA)

ABSTRACT: A remedial strategy consisting of a three-component cycle (dewatering, soil vacuum extraction, and resaturation) was developed to mitigate solvent contamination in the surficial aquifer from a dry cleaning facility. The cycle is initiated by dewatering the surficial aquifer using a well point dewatering system and a liquid-ring pump. Dissolved contaminants are extracted through advective transport. As the aquifer is dewatered, an induced vacuum mobilizes soil vapors in the previously saturated zone volatilizing residual solvents. The final element of the cycle is a resaturation and equilibration period during which the multi-phase extraction system is off.

Active remediation at the dry cleaning facility began on May 1, 1996. Solvent concentrations were significantly reduced in all but two wells within 3 months of startup. After 1 year, solvent concentrations in samples from all but one well had decreased to near or below the drinking water standards. The increase in solvent concentrations in samples collected from well MW-9 revealed the presence of a residual solvent source beneath the dry cleaning facility. To accelerate site remediation, three angled wells were installed into the residual solvent target by employing blind bore directional drilling. After 6 months of cyclic dual-phase extraction, 70% to 97% reductions in tetrachloroethene have been observed in samples collected from the angled wells. Exponential decay analysis of the data from MW-9 showed that without active remediation in the source area, the tetrachloroethene concentration would approach an asymptote at 122 µg/L. Since multi-phase extraction has been implemented in the residual source area, the concentration of tetrachloroethene has decreased to the site rehabilitation level.

INTRODUCTION

Construction of the Manatee Towne Centre, a retail shopping center, was completed in late 1989. The site is located in western Manatee County, Florida. Prior to development, the site was characterized by pine and palmetto scrub with organic rich soils and a shallow water table, less than 0.7 meters below land surface. During development, the site was cleared and approximately 1.2 meters of fill was brought in to elevate the site to current grade.

At the rear of each retail unit is a gravel-filled sump, approximately 1 meter deep, intended to facilitate infiltration of storm waters from the roof of the shopping center. A dry cleaning facility, one of the original tenants, began operation in February 1990 using tetrachloroethene as the dry cleaning solvent. In

September 1990, chlorinated solvents were detected in ground water samples collected during a phase II property transaction assessment. Although no spills were reported, there was some evidence of staining on the carpet near the back door. A ground water sample collected from the sump just outside the back door contained 5,400 µg/L of tetrachloroethene.

CONTAMINATION ASSESSMENT SUMMARY

The surficial aquifer consists of undifferentiated sands, approximately 2.4 meters thick. With the approximate 1.2 meters of fill material present beneath the site, the total thickness of the unconsolidated sands is approximately 3.6 meters. The surficial aquifer at the site is approximately 1.8 meters thick. The soils throughout the surficial aquifer are fine to very fine grain with disseminated organic material, silts, and clays. The direction of ground water flow at the site is toward the northeast at approximately 0.01 meters/day under a hydraulic gradient of approximately 0.017 meter/meter with a hydraulic conductivity of approximately $1.7x10^{-4}$ centimeters per second (cm/sec).

In the study area, the Hawthorn Group is approximately 120 meters thick and separates the surficial and Floridan aquifers (Scott, 1988). The Hawthorn Group lies unconformably beneath the surficial aquifer, and is represented by thick, dense, phosphatic dolosilts of the Arcadia Formation. Permeability analyses on a Shelby tube sample of the dolosilt yielded a permeability of $2x10^{-6}$ cm/sec.

Soil samples containing detectable concentrations of chlorinated solvents were collected from a boring adjacent to the sump. Tetrachloroethene was detected at 16 µg/kg at 1.2 meters below land surface, above the water table, but wet due to infiltration through the sump. 1,2-dichloroethene was detected in the soil sample collected from the water table at 150 µg/kg.

Ground water in the surficial aquifer beneath the site is anaerobic, with less than 1 mg/L of dissolved oxygen. Ground water in the undeveloped area east of the paved property is aerobic with dissolved oxygen as high as 4 mg/L. Ground water samples collected from the deep wells document the extent of dissolved solvents as limited to the surficial aquifer. Figure 1 shows the

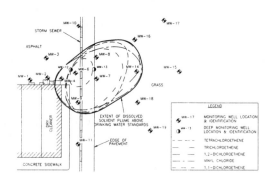

FIGURE 1. Site layout and extent of the solvent plume.

site layout and the extent of dissolved chlorinated solvents at concentrations greater than the drinking water standards. Table 1 is a summary of the ground water quality data from selected wells.

TABLE 1. Peak solvent concentrations in ground water samples from selected wells during the assessment.

Well	PCE	TCE	1,2-DCE	VC
MW-4	4,200	2,100	10,000	180
MW-6	180	800	6,200	390
MW-7	96	170	1,110	80
MW-8	179	461	1,400	37
MW-9	28	59	920	115

Notes: All values reported in µg/L.
PCE = Perchloroethene and tetrachoroethene.
TCE = Trichloroethene
1,2-DCE = Sum of cis and trans isomers of 1,2-dichloroethene.
VC = Vinyl Chloride

REMEDIAL ACTION SUMMARY

In January 1995, a dual-phase extraction system was designed to reduce the concentrations of chlorinated solvents in the surficial aquifer to predefined site rehabilitation levels. The major components of the remedial system include a liquid-ring pump, a network of seven well points, a diffuser tank for water treatment, and carbon canisters for vapor phase treatment.

Each well point was constructed from 5.08-cm-diameter PVC well with 1.8 meters of 0.025-cm slotted screen. The spacing of the wells were based on ground water modeling to estimate the zone of influence for each well. Five well points were placed around the edge of the plume and two placed within the plume. The location of the well points are shown in Figure 2.

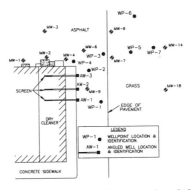

FIGURE 2. Location of wellpoints and angled wells.

Each well point contained a drop tube lowered to the bottom of the well. Both the drop tube and the well screen were used for extraction. The drop tube was used for extracting ground water while both soil vapors and ground water could be extracted from the well screen. Valves and gauges were used to control the vacuum distribution in the well and drop tube. Both the well and the drop tube were connected to a common header pipe, which led back to a treatment compound.

Operation of the extraction system occured in three phases. First, the surficial aquifer was dewatered, thus removing nearly all of the mobile dissolved solvents via advective transport. Ground water held at residual saturation in the soils contains dissolved solvents; therefore, the second phase was soil vapor extraction through the well screens from the previously saturated zone. The concentrations of dissolved and adsorbed solvents in the dewatered surficial aquifer are reduced via volatilization. The extraction phase operates for approximately 3 weeks. During the third phase, the extraction system was stopped, allowing the surficial aquifer to resaturate and the residual solvents to diffuse into the ground water. Ground water samples collected during the monitoring of the remediation system are collected after the remediation system has been shutdown for at least a week.

Operation of the remediation system began on May 1, 1996. Within 1 year of operation, the concentrations of dissolved solvents had decreased to near the drinking water standards in all wells except MW-9. Table 2 is a comparison of the water quality data from selected wells prior to remediation and after 1 year of active remediation.

TABLE 2. Ground water quality data from selected wells during startup and the first year of remediation.

Well	Period	PCE	TCE	1,2-DCE	VC
MW-4	Pre-remediation (April 1996)	1.8	< 1.0	< 1.0	< 1.0
	After 6 weeks (beginning May 1, 1996)	50	4	< 1.0	< 1.0
	After 1 year of remediation	7	< 1.0	< 1.0	< 1.0
MW-7	Pre-remediation (April 1996)	32	76	283	80
	After 6 weeks (beginning May 1, 1996)	9	44	176	20
	After 1 year of remediation	< 1.0	6	59	7
MW-8	Pre-remediation (April 1996)	47	110	267	20
	After 6 weeks (beginning May 1, 1996)	2	8	28	< 1.0
	After 1 year of remediation	< 1.0	< 1.0	< 1.0	< 1.0
MW-9	Pre-remediation (April 1996)	17	21	26	32
	After 6 weeks (beginning May 1, 1996)	5,400	1,000	1322	36
	After 1 year of remediation	180	13	34	< 5

See Table 1 for notes.

Ground water samples from wells MW-2, MW-4, and MW-6 exhibited high concentrations of solvents during the early phases of the assessment. During the latter part of the assessment, the concentrations of solvents in wells MW-2 and MW-6 decreased to below the site rehabilitation level. Wells MW-7 and MW-8 are property line wells located hydraulically downgradient of the dry cleaner and have continued to show a decrease in solvent concentrations over time. Samples collected from well MW-9 have never had significant concentrations of tetrachloroethene or trichloroethene until just prior to startup of the remediation system. The ground water sample from this well contained elevated concentrations of 1,2-dichloroethene, but subsequent analyses showed a consistent decrease in concentration. After startup of the remediation system, the solvent concentrations showed significant increases in samples collected from MW-4 and MW-9.

The increase in solvent concentrations in MW-9 led to the suspicion that a residual source may be present beneath the dry cleaning facility. Three angled wells were designed and installed in July 1997 to assess the ground water quality characteristics beneath the facility. Each well is approximately 7.6 meters long and enters the subsurface at an angle of 30° from horizontal. The wells contain 2.3 meters of stainless steel channel packed well screen with 0.025 cm slotted screen and 20/45 silica sand filter pack, threaded to a PVC riser. These angled wells reach a total vertical depth of 3.6 meters. The center well (AW-2) is screened directly beneath the dry cleaning machine. The other two wells are located approximately 4.6 meters north and south of the center well (Figure 2). Analysis of ground water samples collected from these wells shows that angled well AW-2 contains the highest concentrations of solvents while the samples from the other two wells contain significantly lower solvent concentrations. This suggests that the residual solvent source beneath the building is limited in areal extent.

In August 1997, each of the angled wells were fitted with a well head assembly similar to that of the other extraction wells and connected to an existing header pipe. Figure 3 is a schematic illustration of the construction details of the angled wells. The three angled wells and well points WP-1, WP-2, and WP-4 were allowed to operate in the cyclic fashion as previously described. Table 3 summarizes the water quality data from the three angled wells and additional water quality data for MW-9.

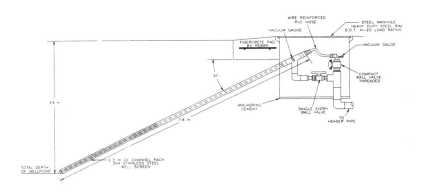

FIGURE 3. Schematic illustration of angled well construction details.

TABLE 3. Ground water quality summary from selected wells.

Well	Period	PCE	TCE	1,2-DCE	VC
AW-1	Initial sample (July 1997)	45	< 10	< 10	< 10
	After 3 months (beginning August 1997)	5	< 1.0	< 1.0	< 1.0
	After 6 months	1	< 1.0	< 1.0	< 1.0
AW-2	Initial sample (July 1997)	3,700	420	64	< 10
	After 3 months (beginning August 1997)	380	130	890	< 25
	After 6 months	1,100	100	56	< 20
AW-3	Initial sample (July 1997)	60	< 10	< 10	< 10
	After 3 months (beginning August 1997)	8	< 1.0	< 1.0	< 1.0
	After 6 months	10	2	4.9	< 1.0
MW-9	After 3 months	16	4	< 1.0	< 1.0
	After 6 months	3	< 1.0	< 1.0	< 1.0

See Table 1 for notes.

These data show that after 6 months of operation, reductions in solvent concentrations from 70% to over 97% were achieved in the angled wells. A regression analysis was conducted using an exponential decay model to evaluate the reduction in tetrachloroethene detected in ground water samples collected from MW-9 during the first year of operation (Figure 4). The regression analysis showed that without active remediation in the residual source area the

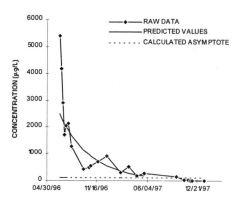

FIGURE 4. Regression analysis of tetrachloroethene in samples from MW-9.

tetrachloroethene concentration would approach an asymptote of 122 μg/L. Dual-phase extraction from the angled wells was initiated in month 16 of active site remediation. The subsequent monitoring data from well MW-9 shows the tetrachloroethene concentration has decreased to the site rehabilitation level.

REFERENCES

Scott, T.M. 1988. *The Lithostratigraphy of the Hawthorn Group (Miocene) of Florida.* Bulletin No. 59. Florida Geological Survey, Tallahassee, Florida.

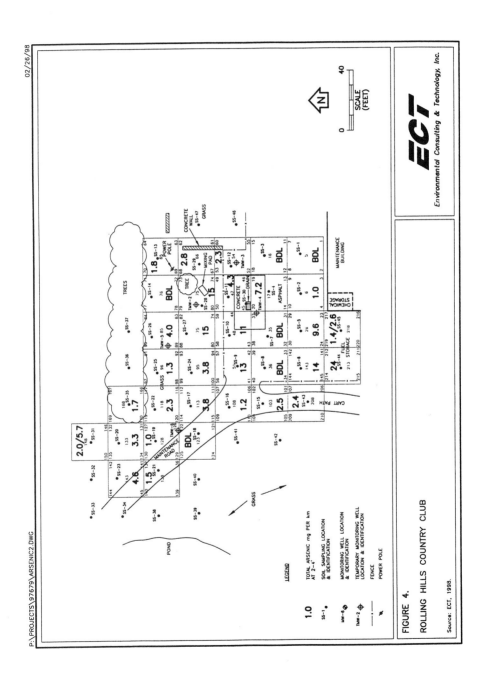

FIGURE 4.

ROLLING HILLS COUNTRY CLUB

Source: ECT, 1998.

USE OF DIFFUSION MODELING TO AID ASSESSMENT OF RATE-LIMITED VAPOR TRANSPORT FOR SVE CLOSURE

Dominic C. DiGiulio (U.S. EPA, Ada, Oklahoma)
Mark L. Brusseau (University of Arizona, Tucson, Arizona)
Varadhan Ravi (Dynamac Corporation, Ada, Oklahoma)

ABSTRACT: In this paper, we provide an example of a partial assessment of field-scale, rate-limited vapor transport modeling to support SVE closure decision making. Specifically, we simulate concentration reduction in a clay lense surrounded by more permeable soils in which mass removal during SVE is completely controlled by diffusion. We estimate the time to reach 10 ug/kg trichloroethylene in soil at four moisture saturation levels. At the saturation level observed in the field (95%), it may take in excess of 80 years to reach 10 ug/kg TCE while for a significantly reduced saturation level (86%) this concentration can be reached in just over 15 years. In the former case, for a well characterized, designed, and monitored SVE system where future contaminant mass flux to ground-water is expected to be insignificant, it would appear that this simulation supports closure. Decision making in the latter case is less clear and will likely depend heavily on the adequacy of site characterization, design, monitoring, and mass flux modeling.

INTRODUCTION

SVE is usually quite effective during the early stages of operation. In the later stages however, removal is characterized by low vapor concentrations and extensive effluent tailing. Upon periodic cessation and restart of vacuum extraction, rebound of vapor concentration is often observed. Confirmatory soil sampling often reveals that soil-based concentration standards, which are typically in the low parts per billion (ppb) have not been met. These effects are largely due to poor site characterization, design, monitoring, and rate-limited mass transfer (e.g., diffusion from "immobile" domains, sorption kinetics, and immiscible phase dissolution). It is apparent that alternative methods of SVE closure are needed.

The U.S. EPA is currently devising a comprehensive strategy for SVE closure involving assessment of site characterization, design, monitoring, rate-limited vapor transport, and mass flux to and from ground water. In this paper, we provide a site-specific example of a partial assessment of rate-limited vapor transport. The assessment is partial in that there are a variety of tests and data interpretation techniques which could potentially be used to support evaluation of field-scale rate-limited vapor transport. The simulations conducted in this paper do not necessarily constitute a sufficient condition to meet assessment of rate-limited vapor transport for SVE closure. Specifically, we simulate concentration reduction in a clay lense surrounded by sandy soils in which mass removal during SVE is completely controlled by diffusion. We estimate the time it takes to reach an average concentration of 10 ug/kg trichloroethylene in soil and illustrate the sensitivity of moisture content in diffusion modeling.

SITE DESCRIPTION

The following information on site description was extracted from a report prepared by Earth Tech (1997). The area where diffusion modeling was conducted (Building 763) was used for aircraft maintenance at Norton AFB for over 50 years. Soils under and around building 763 are characterized by extreme heterogeneity. The most common soil type near the surface is silty sand containing discontinuous lenses of silt and sand. The silty sand can be absent in places or be as thick as 20 feet. This unit is underlain by 15 to 35 feet of sand containing discontinuous lenses of silty sand and silt. Gravel and cobbles, ranging in size from 1/8 to 2 inches in diameter, were frequently encountered at depths beginning at about 10 feet below ground surface (bgs). This sandy unit is underlain by silt and silty sands containing discontinuous lenses of clay. Sands are very fine to coarse grained and moderate to poorly sorted while the clays and silts range from loose to very stiff. The average depth of these finer-grained soils is about 35 feet bgs. Information for diffusion modeling was extracted from borehole CB-1 where a silty clay unit was found between 38.0 and 40.5 feet bgs. Gravelly, silty sandy soils above and clayey, sandy, silty soils below this unit contained TCE concentrations near the detection limit (2 ug/kg) while as illustrated in Figure 1, the silty clay unit contained concentrations ranging from 1300 to 6100 ug/kg TCE. This suggested that vapor transport was limited by diffusion from the silty clay lense to surrounding soils of higher permeability. The current cleanup standard for TCE contaminated soil prescribed in the Record of Decision (ROD) is 5 ug/l TCE concentration in leachate as determined using the Toxic Characteristics Leaching Procedure (TCLP) by EPA Method 1311/8240. Using site-specific calculated partitioned coefficients, this is equivalent to approximately 10 ug/kg which will be the goal set for modeling.

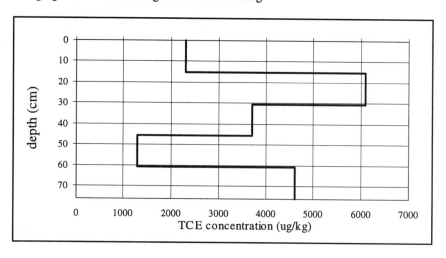

FIGURE 1. Concentration (dry wt.) profile in clay lense.

MODELING

The governing equation describing one-dimensional diffusion is given as:

$$\frac{\partial C_T}{\partial t} = k \frac{\partial^2 C_T}{\partial x^2} \qquad (1)$$

and initial as boundary conditions:

$$C_T(0,t) = g(t) \ (1a), \qquad C_T(L,t) = g(t) \ (1b), \qquad C_T(x,0) = f(x) \ (1c)$$

where:

t	=	time (T)
L	=	width of diffusive (L)
f(x)	=	initial total soil concentration distribution of chemical i (M/L^3),
g(t)	=	soil concentration of chemical i at boundaries as a function of time(M/L^3),
C_T	=	total soil concentration of chemical i [M/L^3], and
k	=	effective diffusion coefficient of chemical i [L^2/T].

Boundary conditions (1a) and (1b) require knowledge of vapor concentration at the top and bottom of the clay lense as a function of time. As illustrated in Figure 2, the decrease in vapor concentration at the extraction well during the period of operation can be adequately described by a simple analytical expression.

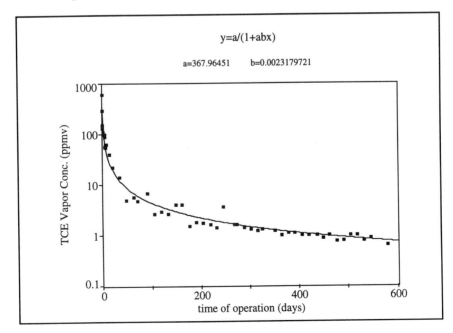

Figure 2. TCE vapor concentration as a function of time from extraction well.

This expression can then be used to estimate future effluent vapor concentrations at the well if SVE were to continue indefinitely. We use vapor concentrations at the well to estimate vapor concentrations at the top and bottom of the clay lense although it is preferable to have actual vapor concentration data as a function of time near the soils being simulated. Unfortunately, this information was not available.

The effective diffusion coefficient k is defined as:

$$k = \frac{\xi_a D_a H + \xi_w D_w}{H\theta_a + \theta_w + K_d \rho_b + \dfrac{M_i \rho_o \theta_o}{S_w \displaystyle\sum_{j=1}^{N} X_j M_j}} \qquad (2)$$

Table one summarizes site-specific input data used for simulation. Since identification and quantification of all constituents in a NAPL mixture is impractical at most hazardous waste site, an average molecular weight for the NAPL can be used in place of mole fractions. For soil-air and soil-water tortuosity estimates, we use a theoretically based model derived by Millington and Quirk (1961).

Table 1. Input for diffusion modeling

	Description	Value	Units
θ_a	volumetric air content	varies	cm^3_{air}/cm^3_{soil}
θ_w	volumetric water content	varies	cm^3_{water}/cm^3_{soil}
θ_o	volumetric NAPL content	0.00	cm^3_{NAPL}/cm^3_{soil}
ϕ	porosity	0.38	cm^3_{pore}/cm^3_{soil}
ρ_b	bulk density	1.68	g_{solids}/cm^3_{soil}
ρ_o	density of NAPL	1.462	g_{NAPL}/cm^3_{NAPL}
S_w	water solubility for TCE	0.0011	g/cm^3
H	Henry's Law Constant for TCE	0.38	$(ug/cm^3_{air})/(ug/cm^3_{water})$
K_d	soil-water partition coeff.	0.857*	$(ug/g_{solids})/(ug/cm^3_{water})$
MW_o	avg. molecular weight of NAPL	131	g/mole
MW_i	molecular weight of TCE	131	g/mole
D_a	free air diffusion coeff. for TCE	6366.8	cm^2/d
D_w	free water diffusion coeff. for TCE	0.804	cm^2/d

*Estimated by Kd=Koc*foc where foc=0.0068, Koc=126 cm3/g

A solution to equation (1) with initial and boundary conditions (1a) through (1c) is found in Carslaw and Jaeger (1959), p. 104

$$C_T(x,t) = \frac{2}{L}\sum_{n=1}^{\infty} e^{-kn^2\pi^2t/L^2}\sin\left(\frac{n\pi x}{L}\right)$$

$$\left[\int_0^L f(x')\sin\left(\frac{n\pi x'}{L}\right)dx' + n\left(1-(-1)^n\right)\frac{\pi k}{L}\int_0^t e^{kn^2\pi^2t/L^2}g(t')d(t')\right] \tag{3}$$

The time dependent function f(t') is given in Figure two. Boundary condition (1c) requires an initial soil distribution. This information is typically provided as a discrete total soil concentration values with depth as illustrated in Figure one. The expression for f(x') can be integrated in j partitions where:

$$\int_0^L f(x')\sin\left(\frac{n\pi x'}{L}\right)dx' = \frac{L}{n\pi}\left\{f_1 + \sum_{i=1}^{j-1}\left(f_{i+1}-f_i\right)\cos\left(\frac{x_i n\pi}{L}\right) + (-1)^{n+1}f_j\right\} \tag{4}$$

An average matrix concentration is useful for comparison as a function of time. Integration of (3) with respect to x from 0 to L, followed by dividing by L leads to:

$$C(t)_{T_{avg}} = \frac{2}{L\pi}\sum_{n=1}^{\infty}\left(\frac{1-(-1)^n}{n}\right)$$

$$\left[e^{-kn^2\pi^2t/L^2}\int_0^L f(x')\sin\left(\frac{n\pi x'}{L}\right)dx' + \left(1-(-1)^n\right)\frac{n\pi k}{L}\int_0^t e^{-kn^2\pi^2/L^2(t-t')}g(t')d(t')\right] \tag{5}$$

RESULTS AND DISCUSSION

Figure 3 illustrates the average concentration in the clay lense as a function of time and moisture saturation. At the moisture saturation at the time of sampling (95%), in excess of 80 years would be required to reach a concentration of 10 ug/kg because the high moisture content results in a very low diffusion coefficient. This level is reached in just 15 years when saturation is reduced to 86% because of the nonlinear effect of the diffusion coefficient on soil concentration. For a well characterized, designed, and monitored SVE system where contaminant mass flux to ground-water will be insignificant, it would be difficult to justify an additional 80 years of SVE operation to meet a 10 ug/kg TCE concentration level. Decision making in the latter case is less clear and will likely depend heavily on the adequacy site characterization, design, monitoring, and mass flux modeling..

SUMMARY

Vapor diffusion modeling has been used here to aid assessment of rate-limited vapor transport. We emphasize that diffusion modeling alone does not necessarily constitute a sufficient condition to meet this portion of the SVE closure strategy. We also stress that assessment of rate-limited vapor transport is only one of five assessment factors for SVE closure. SVE closure also requires assessment of site-characterization, design, monitoring, and mass flux to and from ground water.

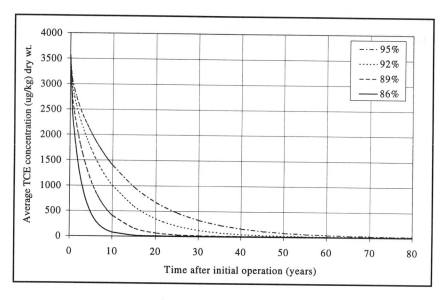

Figure 3. Average TCE concentration (dry wt) in clay lense as a function of time and moisture saturation.

REFERENCES

Earth Tech. 1997. Installation Restoration Program (IRP), Soil Vapor Extraction Closure Report, Buildings 673 and 763, TCE Source Area Remedial Action, Central Base Area Operable Unit, Norton AFB, San Bernardino, CA, Prepared for AFBCA/DD Norton, Norton AFB, San Bernardino, CA, USAF

Carslaw, H.S. and J.C. Jaeger. 1959. *Conduction of Heat in Solids.* Oxford University Press Inc., New York

Millington, R.J. and J.P. Quirk. 1961. Permeability of porous solids, *Trans. Faraday Soc.*, 57: 1200-1207.

ACKNOWLEDGMENTS

We are grateful to the EPA Superfund remedial project manager, Ms. Kathleen Sayler for providing the information necessary to conduct this modeling and Mr. James Cummings of the EPA Technology Innovation Office for support in the SVE closure initiative.

DISCLAIMER

PASSIVE SOIL VAPOR EXTRACTION FOR INTERIM REMEDIATION AT THE SAVANNAH RIVER SITE

Joseph Rossabi and Brian D. Riha (Savannah River Technology Center, Aiken, SC)
Robert S. Vanpelt and Tom Kmetz (Bechtel Savannah River Inc., Aiken, SC)
Bradley E. Pemberton (Gregg In Situ Co., Aiken, SC)

ABSTRACT: Passive soil vapor extraction (PSVE) or barometric pumping has been used as a rapid, interim remediation strategy at a waste site contaminated with chlorinated solvents. Cone penetrometer-installed wells have removed more than 50 lbs. (23 kg) of solvent in the first year of operation. The cone penetrometer was used to both characterize the site and install 2" (0.051 m)-diameter, vadose zone remediation wells fitted with Baroball check valves to enhance the passive removal rate of vapor phase contaminants. The complete vadose zone characterization and remediation system installation at the 3-acre waste site was completed in three weeks.

INTRODUCTION

The Department of Energy's Savannah River Site has operated to manufacture nuclear materials in support of federal programs since the early 1950s. During the course of metal fabrication operations, process chemicals and solvents were released to the subsurface at several areas on site. The Miscellaneous Chemical Basin (MCB) waste unit, located in the northwest portion of SRS, received a small portion of the liquid chemical wastes consisting of waste solvent and used oil over an 18 year period. It is also believed that partially full drums were emptied at this site. The basin is located approximately 1.5 miles (2.4 km) south of the 3/700 Area operations and 3 miles (4.8 km) east of the SRS boundary. The exact boundaries of the basin have not been determined; however, the location has been estimated based on site photographs. Photographs indicate that the 20 by 20 by 1-ft (6.1 x 6.1 x 0.31-m) deep basin was in use from about 1956 to 1974. In 1974 the basin was re-graded and the original near-surface basin sediments were distributed in a field at the site with approximate dimensions of 350 by 350-ft (107 x 107 m). The surface of the waste unit exhibits a slight slope of approximately 3 percent to the east-southeast. Weeds, grasses and small pine trees currently cover the site.

TECHNOLOGY DESCRIPTION - BAROMETRIC PUMPING

Natural atmospheric pressure fluctuations are transmitted through the unsaturated subsurface, however, the pressure waves are damped and delayed in phase to degrees dependent on the effective permeability of the formation. As a result of the attenuation and delay of the transmitted pressure wave, at a given time the atmospheric pressure at the surface and the soil gas pressure in the subsurface will be different. If these two zones are directly connected (by a vadose zone well), the pressure differential will result in a flow either into or out of the well. If the subsurface contains VOCs in the gas phase, flow out of the well will result in removal of the contaminants from the subsurface without mechanical pumping (Rossabi et al., 1994). This phenomena has been observed for more than a century, (the first reference in the literature is an article in *Science*, 1896 by Fairbanks) but has only recently been proposed for environmental remediation. The technology is also known as barometric pumping. The duration of flow events (either flow into the well or flow out of the well) can be as long as two days continuously but as expected by mass balance considerations, the average time of flow in is equal to the average time of flow out. The flow rates for a typical well are generally low (28 to 280 lpm) but if the concentration of the contaminant in the gas phase is high, mass removal by this process can be significant (as high as 1-2 kg/day/well). Mass removal can be enhanced using low cost or natural techniques. Enhancements include the addition of wind powered turbine vacuum pumps, installation of one way mechanical valves to prevent dilution of the contaminated soil gas by clean air from the surface, and solar heat injection to raise vapor pressures and the partitioning of the VOCs from the liquid or aqueous phases to the gas phase.

The Baroball® is a device developed by the authors specifically for barometric pumping applications. It uses a lightweight ball in a conical seat to permit gas flow in one direction with a minimal pressure requirement (approximately 1 millibar) but effectively prevents gas flow in the reverse direction. Depending on the configuration, the Baroball can be used to allow contaminated soil gas out of a well and prevent clean air from diluting the soil gas in the subsurface (standard mode of operation), to inject air and/or nutrients into the subsurface to enhance bioremediation, to control or confine the movement of a subsurface gas phase plume in the vadose zone, or to passively transfer heated, water-saturated air into the subsurface to enhance volatilization in the subsurface. The Baroball device is attached directly to the top of the well casing at the surface.

SITE CHARACTERIZATION/CPT WELL INSTALLATION

Geology. The Savannah River Site is underlain by a thick wedge (approximately 300-m) of unconsolidated Tertiary and Cretaceous sediments consisting primarily of sand, clayey sand, and sandy clay. Two significant clayey layers in the vadose zone at the Miscellaneous Chemical Basin are located at approximately 0-15 ft (0-4.6 m) and 75-90 ft (22.9-27.4 m) below ground surface. The water table at the site is at approximately 120-ft (36.6 m) below the ground surface. The ground surface elevation is approximately 340-ft msl (103.6 m) (Riha et al., 1996).

Cone penetrometer (CPT) sleeve friction/tip pressure ratio logs confirm the presence of a fine grain sediment layer 10 to 15 ft (3-4.5 m) thick near the surface. Below this layer is a course grained (sandy) layer approximately 50 ft (15 m) thick with a few narrow silty layers. Below the sandy layer is an interbedded clayey/silty/sandy zone ranging in thickness from 5-ft (1.5 m) on the northwest to

15-ft (4.5 m) on the southeast of the investigated area. The sediment layering dips to the southeast and tends to follow the surface contour.

Soil Gas. Previous screening activities conducted at the MCB include soil gas surveys, soil confirmation sampling, and groundwater monitoring. Two phases of shallow soil sampling were conducted in 1986; sampling results indicate the presence of trichloroethylene (TCE) and perchloroethylene (PCE) in concentrations up to 5,520 µg/kg and 44,800 µg/kg, respectively at a depth of 1.5 to 2 feet below surface. The highest concentrations were detected north of the basin and contaminant migration appeared to be in a north-northeasterly direction. In 1987 and 1989, soil samples collected from soil borings, monitoring wells, and test pits were analyzed and found to contain volatile organic contaminants (VOCs) (e.g. TCE, PCE, carbon tetrachloride) from just below surface (0-2 feet) to depths of approximately 225 feet. However, the highest concentrations appear to occur within the upper 120 feet (vadose zone). Inorganics (e.g. metals such as aluminum, lithium and lead) have also been detected at levels in excess of drinking water standards. Analysis of the groundwater since 1985 show TCE and PCE levels in excess of the EPA Maximum Contaminant Levels (MCLs) for drinking water standards in 9 of the 13 monitoring wells which surround the basin.

During September and October of 1996, a total of 128 soil gas samples were collected at varying depths at 25 locations using a CPT soil gas method (Riha et al., 1996). Sampling depths were chosen during the CPT push by analyzing the friction ratio logs and determining permeable locations near the clayey zones. The contaminants were expected to be located in the fined grain sediments but it is difficult to obtain sufficient soil gas flows for sampling because of lower effective permeability in these zones. A soil gas sample was also taken near the center of the 50-ft thick sandy zone to determine the vertical distribution of soil gas concentrations.

Concentrations ranged from 140 ppmv TCE, 99.5 ppmv PCE, and 28.7 ppmv CCL4, in the vicinity of CPT-MCB-4 to non detect levels at the outer perimeter of the basin (Riha et al., 1996). Concentrations are highest for TCE and the center of the plume is located near CPT-MCB-4, 13, and 22. Soil gas concentrations generally decrease with depth and distance from the upper fine grain zone. The extent of the soil gas plume increases laterally with depth in a bell shape. The lateral soil gas migration tends toward the southeast away from CPT-MCB-4.

Sediment Concentrations. Five locations at the MCB were sampled for sediment analyses in 1996. Two-ft (0.61 m) long core samples were collected with the CPT to determine the vertical extent of the sediment contamination. The core was also inspected visually and used to correlate the CPT logs with lithology.

As expected, higher concentrations of contaminants appear primarily in the two zones of fine grained sediments. The results from the sediment samples indicate a liquid contaminant source was released in the area near CPT-MCB-S2 and CPT-MCB-4. Sediment VOC contamination is consistent with releases of

PCE and TCE to the subsurface over an 18 year period. The contaminants have migrated down through the upper clay layer and to the southeast into the lower clay layer.

PSVE Well Installation. Twenty-five vadose zone wells were installed at the MCB in the same push hole made during the CPT geophysical and soil gas pushes. The wells were installed by threading a steel push tip onto the PVC well pipe and pushing the tip with the steel CPT rods in the center of the PVC, effectively pulling the PVC well down with the push tip. The steel push tip remained in the ground.

The wells were installed so the formation was compressed against the outer surface of the casing and screen, leaving no annular space. The potential for vertical migration is minimal when wells are installed this way. All vadose zone wells were installed with screen zones set at depths to maximize removal of contaminant. To this end, a balance was struck between proximity to contaminants (fine grain or clay strata) and maximum flow zones (coarse grain or sandy strata). Generally, wells were screened near fine grain zones occurring at depth between 70 and 90 ft or just below the shallow fine grain zone at a nominal depth of 15-30 feet. Concrete pads were installed around the wells at the surface to prevent contamination by spills at the surface. The entire characterization and installation of the PSVE system was completed in three weeks for less than $60K. Figure 1 shows the location of the CPT-installed wells and standard wells at the MCB.

RESULTS AND DISCUSSION

Concentration measurements were made on the vadose zone wells approximately monthly during barometric pumping outflow events. Well vapor was sampled and analyzed using a Bruel and Kjaer Model 1302 photo acoustic infrared gas analyzer. The instrument was calibrated to measure TCE, PCE, CCl_4, and CO_2 (carbon dioxide) and periodic samples were obtained in Tedlar sample bags and analyzed on an HP 5890 series gas chromatograph for instrument verification. Pressure and flow measurements were also recorded.

For the MCB, BaroBalls were used to enhance removal of contaminated gas. An idealized conceptual model of the remediation can be described in the following way. Source contamination, either in the aqueous phase or as NAPL, resides primarily within fine grain materials in the vadose zone, held by capillary forces (Looney, 1992). The contaminant moves slowly into the more coarse grain, and generally more permeable zones by a variety of mechanisms including interphase mass transfer, and liquid and gas diffusion. Once in the coarse grain material, gas diffusion rapidly distributes the contaminant throughout. Mass will be continually transferred to the coarse zone in an attempt to reach an equilibrium distribution determined by the Henry's Law constant of the contaminant or its vapor pressure but will be limited by the diffusion rate of movement through water or gas in the tortuous and often occluded path through the fine grain material. The contaminant in the coarse grain material (gas phase) will be most readily removed during barometric outflow events and the transfer from fine grain to coarse grain will begin anew. If the removal is approximately periodic and consistent, contaminant removal will appear to follow a first order rate and decline exponentially.

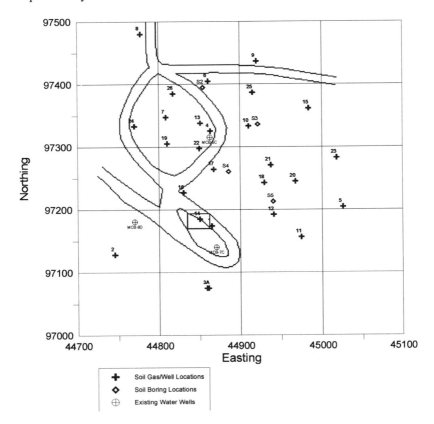

FIGURE 1. Plan view of MCB Site with CPT wells.

In figure 2, concentration data from representative wells collected over a year is shown along with fitted exponential curves of decline. Based on these fits, the time for cleanup in the zone of influence of each of the wells can be estimated. For all of the wells except three near a shallow source zone, cleanup to below concentrations of 1 ppmv will occur before the year 2000. For the three recalcitrant wells, cleanup will occur before the year 2020. The amount of mass removal can be determined by integrating the area under the concentration decline curve and multiplying by the average flow rate from the well. Flow rates ranged from 6 lpm to 180 lpm averaging 72 lpm. From the data, more than 23 kg of contaminant has been removed by the PSVE system. Figure 3 shows a contour map of the contaminant plume (based on well concentrations) immediately following well installation and after one year of passive soil vapor extraction.

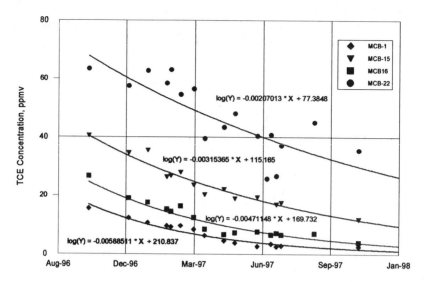

FIGURE 2. Representative well contaminant gas concentration decline rate and first order fits.

Application of this technology is directed to any site where volatile substances (chlorinated solvents, petroleum products, etc.) have contaminated the vadose zone. Natural pressure fluctuations and their damped and delayed transmission through the subsurface occur in all environments but passive soil vapor extraction is particularly well-suited to sites with large vadose zones or substantial low permeability layers that increase the damping and delay of the atmospheric pressure signal.

REFERENCES

Fairbanks H. A. (1896) Notes on a Breathing Gas Well. Science, v 3., 693.

Looney, B. B. (1992) Assessing DNAPL Contamination, A/M Area, Savannah River Site: Phase I Results. Westinghouse Savannah River Company, WSRC-RP-92-1302.

Riha, B.D., B.E. Pemberton, and J. Rossabi. 1996. *Miscellaneous Chemical Basin Expedited Site Characterization Report.* WSRC-TR-96-0407.

Rossabi, J., B. B. Looney, C. A. Eddy-Dilek, B. D. Riha, and V. J. Rohay. (1994) Passive Remediation of Chlorinated Volatile Organic Compounds Using Barometric Pumping." Proceedings of the Water Environment Federation: Innovative Solutions for Contaminated Site Management, Miami.

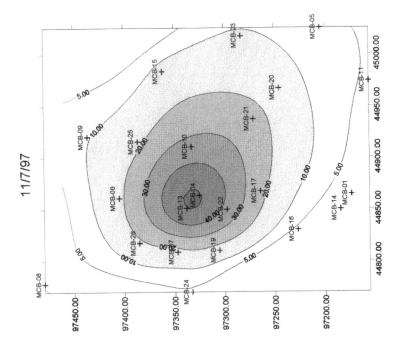

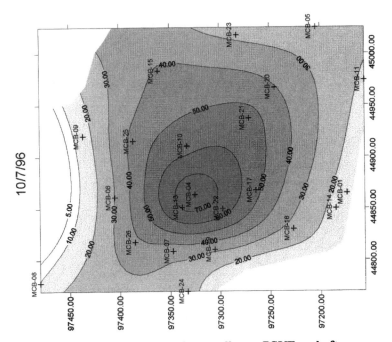

FIGURE 3. Soil gas concentrations from wells pre-PSVE and after one year.

DESIGN AND OPERATIONAL CONTINGENCY FOR REMEDIATION OF LOW PERMEABILITY STRATUM

Christopher G. A. Ross, Leo S. Leonhart, Jim D. Schwall, and Denise M. Bresette (Hargis + Associates, Inc. La Jolla, California)
Tim J. Allen (Raytheon Missile Systems Company, Tucson, Arizona)

ABSTRACT: A full-scale multi-remedial technology groundwater remediation system was designed to maximize mass removal of trichloroethylene (TCE) and 1,1-dichloroethylene (1,1-DCE) from a variable saturated, low permeability stratum. The remedial technologies were selected based on cost, effectiveness, implementability, compatibility, and ability to allow for uncertainties in the subsurface conditions over the 200-acre area of interest. The flexibility-designed remediation system incorporated dual-phase extraction (DPE), in situ bioremediation (ISB), and conventional extraction (CE). The groundwater remediation system also incorporated use of existing remediation systems to reduce capital costs and utilized a decision tree for operation of the groundwater remediation system to capitalize on the most effective technology based on the variable saturated conditions of the low permeability stratum. The design incorporated 41 wells, associated reversible piping network, the existing groundwater treatment system, existing mobile vapor phase treatment systems, an amendment injection system for ISB, and a control system. Design was completed in 1996, with construction completed in March 1997. The system has operated for nine months as of January 1998. The early focus of the remediation system was use of DPE in areas where groundwater elevations were relatively low to capitalize on the relatively high efficiency of mass removal in the unsaturated zone.

INTRODUCTION

TCE and 1,1-DCE are halogenated volatile organic compounds (HVOCs) with Maximum Contaminant Levels (MCLs) of 5 micrograms per liter (ug/l) and 7 ug/l, respectively. These compounds tend to be readily transported with groundwater in the dissolved phase and are usually persistent in most groundwater systems.

Low permeability strata tend to transmit relatively small quantities of groundwater under most horizontal hydraulic gradients. This explains the low groundwater production rates of wells screened in these types of strata. However, saturated, low permeability strata over- or underlying relatively higher permeability strata (aquifers) can transmit significant quantities of groundwater over relatively large areas where a downward or upward vertical hydraulic gradient exists. If the low permeability strata contain elevated concentrations of a contaminant, the underlying or overlying aquifers can be impacted by contaminated groundwater transmitted through the low permeability strata.

The success of a remedial strategy focused on reduction of mass of HVOCs in low permeability stratum is usually tied to the capability of the remedial technology to transmit fluids (liquid or vapor). Removal efficiencies for TCE and 1,1-DCE from low permeability stratum are generally greatest using unsaturated zone technologies to move vapors, primarily due to the properties of the fluid, the nature of the contaminants, and the relatively high diffusion rates for vapors as compared to water. Unsaturated conditions can be induced by dewatering the saturated zone. This can be enhanced by simultaneously extracting groundwater and vapor, also referred to as DPE. This can also be realized if the groundwater levels decrease due to other circumstances, such as seasonal fluctuations or changes in recharge or discharge from the groundwater basin.

The project duration is expected to be approximately 5 years. The project includes use of the following remedial technologies: DPE, ISB, and CE. This paper focuses on the application of DPE. Future topics may include an evaluation of the effectiveness of the ISB as the operational and monitoring data is obtained and evaluated over the next several years.

Objective.   The objective of the project is to reduce the mass of TCE and 1,1-DCE in a low permeability stratum referred to as the shallow groundwater zone (SGZ). The goals of the SGZ removal action are to protect human health and the environment, to minimize the migration of contaminants, and to reduce contaminant levels in groundwater in the SGZ.

The concentrations of TCE and 1,1-DCE within the SGZ have ranged from non-detect to 3,200 ug/l, and non-detect to 1,700 ug/l, respectively, based on monitoring data collected from 1981 to June 1995 (Hargis + Associates, 1996).

Site Description. The Installation Restoration Program at AFP 44 in Tucson, Arizona, managed by Raytheon Missile Systems Company (formerly Hughes Missile Systems Company), has targeted multiple sites for remedial action. The SGZ, also referred to as Site 14, comprises the uppermost water-bearing unit at the site. The SGZ is contaminated primarily with TCE and 1,1-DCE.

The SGZ is a fine-grained geologic unit which underlies portions of AFP 44. This fine-grained unit is found between a generally coarser-grained unit that extends downward from the ground surface and the regional aquifer. There is recharge to the SGZ from either channel infiltration or, in some cases, recharge wells within the regional aquifer. SGZ monitor wells have exhibited slowly declining water levels between recharge events. This indicates that the water is slowly, but continuously, migrating downward to the regional aquifer.

The footprint of the shallow zone is presently limited to the northwestern portion of AFP 44 over an area of approximately 200 acres. Preliminary efforts to extract groundwater from the SGZ suggest that the fine-grained unit has extremely low well yields (mean = 0.17 gallons per minute [gpm] [0.6 liters per minute {lpm}]). As a result of these low yields and the variability in grain size within the unit, removal actions involving conventional pump-and-treat technologies would be difficult to implement. Additionally, natural drainage of the SGZ would be extremely slow and probably could not be accomplished without eliminating recharge.

MATERIALS AND METHODS

A CERCLA removal action was initiated in 1996 beginning with an Engineering Evaluation/Cost Analysis (EE/CA). Given the saturated and unsaturated zone conditions in the vicinity of the SGZ, the EE/CA study produced a flexibility-designed remediation system incorporating DPE, ISB, and CE. The design incorporated 41 wells, associated reversible piping network, the existing groundwater treatment system, existing mobile vapor phase treatment systems, an amendment injection system for ISB, and a control system.

Well Field Construction. Forty-one wells comprise the extraction well field (35 new and 6 existing). The new wells were constructed using air rotary techniques. Boreholes for remediation wells are approximately 10 inches in diameter, up to a depth of approximately 115 feet and have a 50-foot screened interval. Dedicated electrical submersible pumps were set near the bottom of the wells after well development was completed. Well construction activities started in November 1996 and were completed in January 1997.

Groundwater Sampling. Baseline groundwater samples were collected from all but three of the new wells, which were either dry or had an insufficient amount of water to sample. The groundwater samples were analyzed for TCE and 1,1-DCE using EPA Methods 502.2 and 524.2.

Soil Vapor Testing. Vapor extraction tests were performed at all new wells. Soil vapor tests were performed for a period of 2 hours. Vapor extraction rates and vacuum readings were noted, and soil vapor samples were collected. The soil vapor samples were analyzed for TCE and 1,1-DCE using a capillary gas chromatogram/electron capture device.

Initial Operation. The start-up configuration included 16 DPE, 7 ISB, and 14 CE wells (4 wells not currently used). This configuration was focused toward the use of DPE in areas where water level elevations were relatively low, allowing the added benefit of increased rates of mass removal within the current unsaturated zone conditions. The operation of the SGZ remediation system is based on a decision tree developed in the EE/CA. The decision tree logic incorporated threshold vapor concentrations and vapor flow rates for DPE operation and groundwater quality and hydraulic considerations for CE and ISB operation.

Operational Monitoring. Water level data was collected and extraction/injection rates were monitored to evaluate the effects of injection and/or extraction on the SGZ and the associated aquifer system. Groundwater samples were collected and analyzed for TCE and 1,1-DCE after the remediation activities were initiated. Vapor extraction rates and concentrations were monitored at the DPE extraction wells and the vapor treatment unit.

RESULTS AND DISCUSSION

Baseline Conditions. Figures 1(a) and 1(b) shows the concentration of TCE and 1,1-DCE, respectively, in soil vapor and groundwater samples collected from 32 of the new wells. In general, as the concentration of TCE and 1,1-DCE increase in the water phase so does the concentration in the vapor phase. Although the correlation coefficient of the linear regression is relatively poor, the linear regression for TCE indicates that the concentration of TCE is generally greater in water than in vapor (see Figure 1(a)). 1,1-DCE has the opposite concentration relationship (see Figure 1(b)). These concentration relationships generally tend to relate to the Henry's Law Constant for these two compounds.

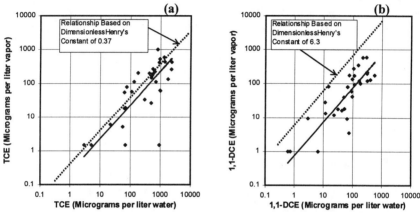

FIGURE 1. Baseline groundwater and soil vapor concentrations at 32 new wells. (a) Trichloroethylene. (b) 1,1-Dichloroethylene.

Vapor extraction tests were conducted at the 35 new wells. The vapor extraction rates varied from 18 to 189 standard cubic feet per minute (scfm [510 to 5,300 lpm]). Vacuum was measured at 34 of the new wells during the test and ranged from approximately 1.5 to 11 inches of mercury (in Hg [3.8 to 28 cm Hg]). Figure 2 shows that the lower vacuums were measured at the wells with generally higher flow rates. The flow rates at the wells with high flow rates could be increased by increasing the equipment size (not formation limited). The relatively coarse grained unsaturated zone overlying the SGZ probably provides a substantial portion of the vapor flow to the higher flow rate wells, as the well screens extend into this coarser zone.

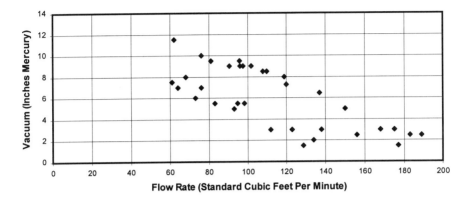

FIGURE 2. Baseline flow rates and vacuums at 34 new wells.

Operational Progress Monitoring. The extracted vapor from 16 DPE wells is treated using two mobile vapor treatment units, one has been operational since April 1997 and the other since May 1997. The extracted groundwater is conveyed to the existing groundwater treatment.

The mass of TCE and 1,1-DCE removed during the first eight months of operation was approximated for the vapor and water phases.

Figures 3a and 3b show the operational flow rates and operating time, respectively, for the vapor potion of the DPE system. Figures 3c and 3d show the average mass removal rate and cumulative mass removal for the vapor extraction portion of the DPE system. The average mass removal rates were based on the TCE and 1,1-DCE vapor concentrations and extraction rates. The cumulative mass removal was estimated using the operating time and average mass removal rates.

Approximately 520,000 gallons (2,000,000 liters) of groundwater was extracted from 15 of the 16 DPE wells (one well was dry) during the first eight months of DPE operation. The average cumulative groundwater flow rate from the 15 wells during this time period was less than approximately 1.5 gpm (5.5 lpm). The mass of TCE and 1,1-DCE in the water phase of the DPE system was estimated based on the initial groundwater concentrations (prior to DPE start up) and total groundwater extracted from the respective wells over the eight-month period. Based on this, less than 4 pounds (less than 2 kgs) of TCE and 1,1-DCE were removed in the water phase of the DPE system during the first eight months of operation.

The comparison of estimated mass of TCE and 1,1-DCE removal in the vapor and water phases indicates that over 99 percent of the mass of TCE and 1,1-DCE were recovered from the vapor phase operations of the DPE system. This percentage may be influenced by VOCs contained in the unsaturated zone above the SGZ. In other words, if all of the VOCs removed were contained beneath the water table (prior to initiating DPE), the percentage of recovered in the vapor phase would likely be lower than that observed at this site. However, it is worth noting that the mass of VOCs removed using DPE is substantially greater than the mass removed using groundwater extraction alone.

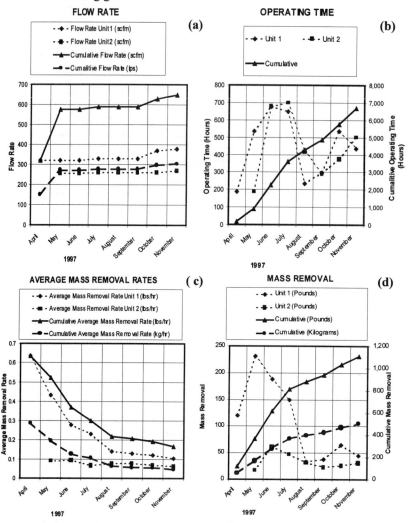

FIGURE 3. Vapor phase flow rate (a). Vapor treatment operating time (b).
Average mass removal rates (TCE and 1,1-DCE) (c). Mass removal
(Total TCE and 1,1-DCE) (d).

REFERENCES

Hargis + Associates, Inc. 1996. *Shallow Groundwater Zone Removal Action Engineering Evaluation/Cost Analysis For Air Force Plant 44 Tucson, Arizona.* Final Draft, June 18, 1996.

DOES MULTI-PHASE EXTRACTION REQUIRE SOIL DESATURATION TO REMEDIATE CHLORINATED SITES?

Ralph S. Baker and Daniel M. Groher (ENSR, Acton, Massachusetts, USA)

ABSTRACT: Multi-phase extraction (MPE) is a rapidly emerging technology that applies to sites contaminated with chlorinated hydrocarbons present in soil and/or groundwater. By simultaneously extracting soil vapor, residual dense non-aqueous phase liquid (DNAPL), if present, and groundwater from the same well, MPE offers a potentially more rapid cleanup solution for moderate- and low-permeability soils. Conventional soil vapor extraction (SVE) addresses unsaturated soils. MPE is generally viewed as an enhancement of SVE capable of dewatering saturated soils, both within the capillary fringe and to some extent beneath the pre-existing water table. It is reasoned that MPE can remediate chlorinated volatile organic compound (CVOC)-contaminated sites by pulling residual DNAPL toward extraction wells during dewatering, and through more effective advective transport due to the resulting increases in air permeability.

Neutron profiling of the subsurface has been performed via access tubes located near MPE wells during five recent pilot studies at chlorinated solvent sites. The results have demonstrated that despite application of high vacuums [16-25 in. Hg (54-85 kPa)], medium- and fine-textured soils tend to remain at or near saturation during MPE. Monitoring point pressure data, when compared to capillary pressure-saturation curves, suggest that these soils are remaining nearly saturated because their air-entry values are not being achieved during MPE.

INTRODUCTION

MPE is an in situ remediation technology that simultaneously extracts multiple fluid phases from a well or wells: air, water (i.e., including dissolved contaminants), and non-aqueous phase liquid (NAPL) (USEPA, 1997). The primary purposes for using MPE are to enhance SVE and/or NAPL recovery.

Let us define water saturation, S_w, and oil saturation, S_o, as the volume fractions of the soil pores occupied by water and NAPL, respectively. Total liquid saturation, $S_t = S_w + S_o$. Any soil pores not occupied by water or NAPL are air-filled, such that air saturation, $S_a = 1 - S_t$. Air permeability, k_a is a function of S_a. Subsurface airflow is restricted to unsaturated portions of the soil, i.e., above the capillary fringe, at locations where $S_t < 1$, $k_a > 0$.

Application of a vacuum during SVE causes the water table to upwell, and the capillary fringe to translate upward commensurately. It is commonly thought that when applied for the purpose of enhancing SVE, MPE can counteract the effects of upwelling by desaturating soils adjacent to and above the liquid intake point of the well, thereby exposing to airflow soils that were previously within or below the capillary fringe. The extent to which soil undergoes desaturation during MPE, however, has not been documented.

Most MPE reports focus on data collected aboveground, typically showing that application of increasing vacuums to an MPE well results in enhancement of air extraction rate and a corresponding increase in CVOC mass removal. Neither of these lines of evidence, however, indicate the degree to which the subsurface is being dewatered, or whether airflow is impinging upon the entire target zone. Increase in the overall radius of influence is often viewed as constituting additional proof of SVE enhancement. One needs to closely scrutinize such data to determine how much of the target zone is actually undergoing treatment. These issues were examined during five separate MPE pilot tests, each of which utilized neutron logging to examine subsurface water/air saturation during extraction.

MPE PILOT TEST SITES AND PROCEDURES

The effectiveness of MPE was evaluated at five pilot test sites at which chlorinated hydrocarbons are present in soil and groundwater (Table 1). At each of these sites MPE had been identified as a potentially feasible remediation tech-

TABLE 1. MPE pilot test site conditions.

SITE	Primary Contaminants	Soil Type	Depth to Water Table ft(m)bgs	Extraction Well Screen Interval ft(m)bgs	Hydraulic Conductivity (cm/sec)
Squibb Mfg. Co. Site, Humacao, PR	Dichloromethane (MeCl$_2$), MIBK, xylenes	fill: clay	0.5 (0.15)	3 to 20 (0.9 to 6.1)	1 x 10^{-6} (a) 5 x 10^{-4} (b)
Confidential Site, S. CA	1,2-DCA, TCE, VC	silty sand, silty clay	20 (6.1)	20 to 30 (6.1 to 9.1)	3 x 10^{-7} (c) 4 x 10^{-5} (d)
LCAAP OU18, Lake City, MO	TCE, PCE, MIBK, toluene	alluvium: silty clay	5 (1.5)	8 to 18 (2.4 to 5.5)	9 x 10^{-6} (e) 2 x 10^{-4} (f)
LCAAP NECOU, Lake City, MO	TCE, PCE, toluene	residual colluvium: silty clay	7 (2.1)	5 to 26 (1.5 to 7.9)	2 x 10^{-7} (g) 3 x 10^{-5} (f)
Silresim Superfund Site, Lowell, MA	1,1,1-TCA, TCE, 1,1-DCE, Freon 113, MeCl$_2$, ethylbenzene, benzene, styrene	lacustrine: silts and sandy silts	5 (1.5)	11 to 32 (3.4 to 9.8)	4 x 10^{-5} to 1 x 10^{-3}

Laboratory determinations on: (a) 1; (c) undetermined number; (e) 8; and (g) 5 intact soil cores (mean is reported where applicable). Field determinations based on: (b) Mean of slug tests; (d) Numeric flow model calibrated to MPE test; (f) Modified pumping test conducted during MPE.

nology. The first four sites are manufacturing facilities where solvents were utilized, stored and/or disposed. The Silresim site is a former chemical waste reclamation facility. All testing occurred during 1996. With the exception of the two shallow well pilot tests at the Lake City Army Ammunition Plant (LCAAP), which were located at sites approximately 850 m apart in geomorphologically dissimilar settings, these pilot tests were conducted independently of one another; therefore, test conditions and methods varied.

All tests made use of a slurp tube positioned within the screened interval of the extraction well. Vapor and liquid were suctioned via the slurp tube to an aboveground equipment skid/treatment system. The applied vacuum given in Table 2 represents the range of gage vacuums measured within the extraction well during the steady state portions of the tests. The extraction rates of vapor and liquid were measured downstream of the vapor-liquid separator.

TABLE 2. MPE pilot test operating conditions and results.

SITE	Test Length (hr)	Applied Vacuum in. Hg (kPa)	SVE Rate scfm (std. m³ per min)	GWE Rate gpm (L/min)	VOC Mass Extracted as vapor / as liquid	Test Designer/ Operator
Squibb Mfg. Co. Site, Humacao, PR	128 [1]	6-19 (20-64)	18 (0.5)	0.38 (1.4)	5 kg / < 1 kg	ENSR Corp.
Confidential Site, S. CA	160	4-8 (14-28)	25 (0.7)	0.07 (0.3)	1,360 kg / 900 kg	ENSR Corp.
LCAAP OU18, Lake City, MO	162	9-16 (31-54)	35 (1.0)	0.85 (3.2)	379 kg / 17 kg	Radian Int. LLC
LCAAP NECOU, Lake City, MO	162	16-24 (54-81)	2.4 (0.07)	0.15 (0.6)	70 kg / 0.5 kg	Radian Int. LLC
Silresim Superfund Site, Lowell, MA	64 [2]	7-25 (24-85)	2 (0.06)	0.8 (3.0)	12 kg / U	Foster Wheeler Env. Corp.

(1)Data are representative of MPE with drawdown phase of test (128 hr); bioslurping (i.e., MPE without drawdown) had first been conducted for 102 hr. (2)Data are representative of MPE with drawdown portion of test, conducted for 64 hr. High vacuum SVE had first been conducted for 72 hr. Following MPE, SVE with dewatering using submersible pumps was conducted for 456 hr. (U) indicates undetermined.

Three to five neutron probe access tubes were installed at each site in the vicinity (1 to 14 m) of the MPE extraction wells. The tubes consisted of 2-in. (5.1 cm) diameter steel pipe driven directly into the ground. In the case of the two LCAAP sites, pilot holes the same diameter as the access tubes were first created by driving and retrieving core samplers, after which the access tubes were driven into the pilot holes. Neutron moisture meters were used repeatedly and non-destructively to profile changes in the liquid content in the formation surrounding the access tubes. The depth of resolution varies as a function of saturation, from about a 16-cm radius at $S_w = 1$, to 70 cm at near zero water content (Gardner, 1986). As water and NAPL are both hydrogen-rich, neutron thermalization measurements cannot distinguish between them; reported soil moisture values are therefore inclusive of any NAPL that might be present, and reflect changes in S_a.

Analyses of intact soil cores collected from the MPE target zone at each site included moisture characteristics (i.e., capillary pressure-saturation curves) by ASTM D 2325-68(81) and D 3152-72. These data can be interpreted to indicate the degree to which an initially saturated soil sample will be dewatered, at equilibrium with a given level of applied vacuum.

RESULTS AND DISCUSSION

Table 2 presents the approximate steady-state vapor extraction rate and the groundwater extraction (GWE) rate during each of the tests. A significant contaminant mass was removed during several of the tests. For example, at the LCAAP Operable Unit Area 18 (OU18), 834 lb (379 kg) of total VOC were removed as vapor and 37 lb (17 kg) as liquid during the test. At the LCAAP Northeast Corner Operable Unit (NECOU), the corresponding values were 155 lb (70 kg) as vapor and 1 lb (0.45 kg) as liquid. In addition, the zone of influence for each of the tests, which was characterized by vacuum measurements made at soil gas monitoring points, varied substantially. It is beyond the scope of this paper to present the vacuum influence data in detail, but it was for the most part spatially variable, with some monitoring points that showed little or even no influence closer to the extraction well than other points that showed greater influence. We attribute this behavior to the occurrence of preferential flow pathways, such as structural cracks in the clayey matrix, that are intercepted by some but not all of the monitoring point screens.

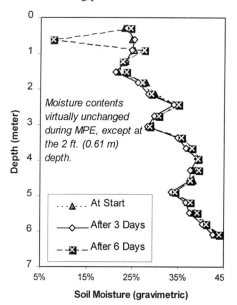

FIGURE 1. Moisture profiles 3 ft. (0.9m) from the MPE well at the Squibb Mfg. Co. site.

For each of these tests, the neutron moisture profiles exhibit small differences in liquid content pre-MPE versus during MPE (Figures 1 and 2). Silresim was an exception, with relative reductions in soil moisture ranging from 0 to as much as 20%. Thus, only a limited amount of dewatering was evident during MPE, and that which did occur was limited to only portions of the subsurface.

The capillary pressure-saturation data, although obtained for core samples that are much smaller in scale than the scale of the MPE tests, do help to explain these findings. For example, at LCAAP OU18, the capillary pressure-saturation data indicate that the air-entry pressure head is 40 in. H_2O (10 kPa), or as much as 275 in. H_2O (27 kPa) when determined to be at the inflection pressure, by the method of White et al. (1972). These values are large compared to those observed in vacuum probes during MPE. The largest vacuum observed during the test was 95 in. H_2O (24 kPa), at a monitoring point 3 ft. (0.91 m) from the extraction well and screened 8-9 ft. (2.4-2.7 m) bgs.

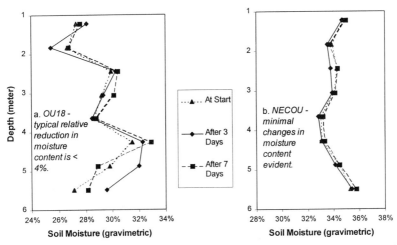

FIGURE 2. Moisture profiles at LCAAP: (a) 4 ft (1.2 m) from the OU18 MPE well, and (b) 5 ft (1.5 m) from the NECOU MPE well.

Much smaller vacuums were measured at all other monitoring points during the test than would need to be applied to the soil to initiate air entry. Thus the soil for the most part remained saturated despite the large vacuums applied. Following this analysis, the same conclusions are reached for all test sites considered.

If vacuum probes are screened across long depth intervals in low permeability soils, the chance that the screen will intercept a preferential flow pathway is greater. Vacuum measurements made at such probes are more apt to represent the elevated vacuum being propagated within the permeable (dewatered) pathway than the substantially lower vacuum within the adjacent, still saturated soil matrix. Such measurements overoptimistically portray what is occurring within the soil as a whole.

IMPLICATIONS FOR REMEDIAL EFFECTIVENESS

There are two notable, and seemingly contradictory, conclusions from these data: (1) MPE did not significantly dewater the bulk soils around MPE wells screened in low permeability soils; and (2) MPE can remove large amounts of contaminant mass under some conditions in otherwise low permeability soils. We conceptualize that there are three primary scenarios that can prevail during MPE. At moderately permeable locations where air-entry pressures are low, MPE should be able to significantly enhance airflow and do so cost-effectively (Figure 3a). At sites having lower permeability soils with high air-entry pressures, MPE will not be capable of dewatering the soil. As a result, airflow will be limited to a few preferred pathways, while the bulk of the soil will remain saturated (Figure 3b). If a significant portion of the contaminant mass lies within matrix blocks that remain water-saturated, mass transfer from such blocks will be diffusion-limited and very slow (McWhorter, 1995). To overcome this limitation, extraction wells may need to be spaced very close together and/or positioned so that multiple, narrow screen intervals encompass the target zone, neither of which may be practical. However, in lower permeability soils (such as exist at the S. CA site and at LCAAP OU18), preferential flow paths may contain a substantial portion of the contaminants of concern. If MPE wells intersect such preferential pathways, then mass removal may be possible even under otherwise discouraging conditions. MPE has, in fact, been selected for full-scale implementation at these two sites. Finally, at sites with excessively high permeability, MPE wells will tend to be flooded with water, very little airflow will be induced, and the efficiency of the process will be compromised (Figure 3c). Given these implications, the challenge for design practitioners is to determine which scenario will prevail at a given site.

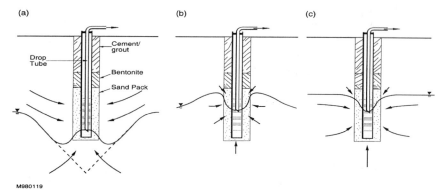

Figure 3. Hypothetical scenarios that can prevail during MPE. Length of arrows indicates fluid velocity.

REFERENCES

Gardner, W.H. 1986. "Water Content." In A. Klute (Ed.), Methods of Soil Analysis, Physical and Mineralogical Methods. 2nd ed. *Agronomy* 9(1):493-544.

McWhorter, D.B. 1995. "Relevant Processes Concerning Hydrocarbon Contamination in Low Permeability Soils." In T. Walden (Ed.), *Petroleum Contaminated Low Permeability Soil,* pp. A1-34. Pub. No. 4631. American Petroleum Institute, Washington, DC.

USEPA. 1997. *Presumptive Remedy: Supplemental Bulletin Multi-Phase Extraction (MPE) Technology for VOCs in Soil and Groundwater.* EPA 540-F-97-004. Office of Solid Waste and Emergency Response, Washington, DC.

White, N.F., D.K. Sunada, H.R. Duke, and A.T. Corey. 1972. "Boundary Effects in Desaturation of Porous Media." *Soil Sci.* 113(1):7-12.

PUMP AND TREAT *MAY BE* THE SOLUTION

Adam H. Hoffman, P.E. (Shepherd Miller, Inc., Fort Collins, Colorado)

ABSTRACT: Shepherd Miller, Inc. (SMI) is operating a system in northern Illinois designed to remediate chlorinated solvents and petroleum hydrocarbons in soil and ground water. This project demonstrates that sometimes, pump-and-treat, despite its reputation, is the most appropriate remediation alternative, especially when utilizing innovative technologies such as horizontal wells, enhanced by soil venting.
Results of a site characterization included:

- Contamination is present in a thin, saturated zone between approximately 5 and 10 to 13 feet (1.5 and 3 to 4 meters) below land surface. The perched, saturated zone is underlain by a highly impermeable clay/silt.

- Contaminants of concern include solvents and petroleum hydrocarbons.

- Ground water contamination may have migrated beneath the manufacturing building and potentially off site.

This project demonstrates that, under certain conditions, pump-and-treat may be a viable and cost-effective remediation alternative, as well as the applicability of horizontal wells, and innovative ways to use soil venting.

INTRODUCTION
Shepherd Miller, Inc. (SMI) is operating a treatment system in northern Illinois designed to remediate soil and ground water impacted by chlorinated solvents and petroleum hydrocarbons resulting from historic leaking of six underground storage tanks (USTs). The project began in February 1993 with the removal of the USTs, four 5,000-gallon (19,000-liter) and two 15,000-gallon (57,000-liter) USTs used to store solvents and fuel at the site. Impacted soil and water were identified in the excavations at the time of UST removal. Following site characterization, an innovative remediation design was implemented. Remediation included ground water extraction from horizontal wells and soil venting in the source area to remediate soil and ground water.
Pumping and treating ground water has been much-maligned recently. Recent papers (National Research Council, 1994, National Academy of Sciences, 1994, and Haley, 1991) have indicated that pump-and-treat will not work to remediate impacted ground water in a timely or cost-effective manner. The work at this site demonstrates that sometimes. Ground water pump-and-treat may be an important part of a cost-effective program to reach cleanup objectives in a timely manner.

SITE HISTORY

The facility was constructed in the early 1950's, was purchased by the client in the 1970's, and was used to manufacture electrical and telecommunications equipment until it was sold in 1995. The facility is located in a mixed land use area of Chicago that includes manufacturing, office space, warehouses, retail, and residential. When the client owned the property the facility included warehouse space, offices, and manufacturing lines. During peak production, operations included light machining, cleaning, painting, and assembly operations. Two boilers were fed by three USTs that contained heating oil and diesel fuel until the building heat was converted to natural gas in the mid-1980's. Three additional USTs contained solvents used in the manufacturing process.

SITE CHARACTERIZATION

A site characterization was initiated based on contamination identified during removal of the USTs. The site characterization included a Geoprobe™ soil and ground water investigation both inside and outside the manufacturing building. A mobile gas chromatograph was used to provide real-time data (with some laboratory confirmation), which were used primarily to direct installation of monitoring wells and identify the extent of impacted soil in the source area.

Eight monitoring wells were installed to characterize the extent and degree of impacts to ground water. Conclusions from the site characterization included the following.

1. Contamination is present in a thin (5- to 8-foot [1.5- to 2.5-meter] thick), perched, saturated zone between approximately 5 to 12 feet (1.5 and 3 to 4 meters) below land surface (bls). The saturated materials consist of silty sand grading to sand and gravel at depth. The perched zone is underlain by a highly impermeable clay/silt.

2. Two ground water plumes exist beneath the site (Figure 1). A 10-acre (4-hectare) area impacted primarily by solvents (volatile organic compounds [VOCs], including trichloroethylene (TCE) and 1,2-dichloroethene (DCE)) and to a lesser degree by benzene, toluene, ethylbenzene, xylenes, and semi-volatile organics (SVOCs) from the fuel tanks. A second, smaller plume (less than 1 acre [one-half hectare]) was minimally impacted by diesel fuel components.

3. No free product, either as light non-aqueous phase liquid (LNAPL) (fuel) or dense non-aqueous phase liquid (DNAPL) (solvent), has been detected in monitoring wells.

4. Ground water at the site flows to the north-northeast and is locally influenced by the stormwater recharge pond located on the western portion of the site.

5. Ground water contamination has migrated beneath the manufacturing building and potentially off site to the northeast.

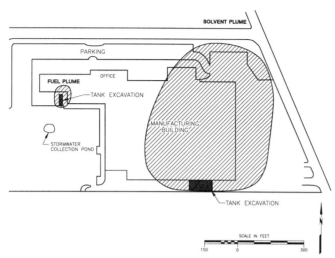

Figure 1. Approximate Extent of Impacted Ground Water.

General ground water concentration data are summarized in Table 1.

Table 1. Ground Water Concentration Data

	TCE*	1,2-DCE*
Source Area Concentration Range	100 - 40,000	300 - 17,000
Mid-plume Concentration Range	ND - 1,200	ND - 230
Toe of Plume Concentration Range	ND - 18	ND - 31
IEPA Class II Ground Water Cleanup Objective	235	200

* - concentrations in micrograms per liter

REGULATORY AUTHORITY

The project is administered through IEPA's Pre-Notice Site Cleanup Program. This program, which was new at the start of this project, was designed for facilities at which an expedited, minimal-oversight approach would be acceptable. The program was designed to work in a manner similar to a Brownfield program, facilitating remediation of industrial sites to enhance their marketability and to make sale and future productive use of the facility viable.

The Pre-notice program has been a success for this project. The client is interested in pursuing remediation, and is obligated by its sale agreement with the new owner to demonstrate progress and closure.

REMEDIATION CONCEPTUALIZATION

Durfing conceptualization, concerns included: the thin saturated sequence, requiring numerous vertical extraction wells; the proximity of the building and the foundation, complicating ground water flow paths; and the potential for off-site impacts, both upgradient and downgradient, requiring some type of containment component to the design.

The primary objectives of the ground water extraction system were and are to dewater the perched zone creating an unsaturated zone that can then be efficiently remediated by soil venting. Secondly, the ground water extraction system is designed to contain impacted ground water on site. Horizontal wells were selected over a network of vertical wells because of the very thin saturated sequence which would result in a very small radius of influence of the wells (estimated to be less than 50 feet [15 meters]), requiring as many as 40 vertical wells. The high number of vertical wells, pumps, and piping were cost-prohibitive; however, two horizontal wells could be installed quickly and achieve superior results. Horizontal wells would require the use of only four aboveground pumps, substantially reducing both capital and O&M costs compared to the vertical wells.

One 500-foot (150-meter) long horizontal well (Figure 2, Horizontal Well A) was installed parallel to the direction of flow along the axis of the plume. The 800-foot (250-meter) long well (Horizontal Well B) is located through the source area, approximately perpendicular to the direction of flow.

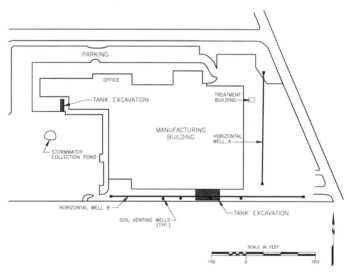

Figure 2. Horizontal Wells, Proposed Soil Venting Wells, and Treatment Building

Horizontal well A (500 feet [150 meters] long) is screened with stainless steel well screen. The formation material was allowed to cave in around the well creating a natural sand pack. The 800-foot (250-meter) long well (Horizontal Well B) was constructed of pre-packed stainless steel well screen. Both wells were installed with directional drilling technique and were located on top of the aquiclude.

Multiple treatment technologies were evaluated based on cost and effectiveness. The diffused bubble aeration system (DBAS) system was selected because of the anticipated low flow rate (less than 10 gallons [38 liters] per minute total flow), high efficiency, relative simplicity of the system, and the low cost to purchase, install, and operate the system. The effluent is discharged to the storm sewer under a National Pollutant Discharge Elimination System (NPDES) permit. Effluent air from the DBAS is discharged under a State of Illinois air permit.

The entire system is connected to a telemetry system that can be monitored via modem. Valves in the system are electronically-actuated ball valves and can be opened and closed remotely allowing flow rate adjustment. Flow rates can also be monitored remotely making NPDES reporting efficient.

The remediation system was constructed in the winter of 1996 and started in April 1996. The remediation system includes the two horizontal wells, 1,500 feet (460 meters) of trench, more than 7,500 feet (2,300 meters) of piping and conduit, four aboveground ground water pumps, three pump houses, motorized valves, the DBAS, secondary activated carbon treatment, a new 3-phase power connection, a new telephone/modem line, a free-standing treatment building, and the telemetry system. The system was installed with piping to accommodate four vertical soil venting wells to be installed in the summer of 1998 in the source area.

IEPA personnel were on site for installation of one horizontal well and were impressed by the speed, lack of spoils, and limited exposure of personnel to impacted soil and water.

INNOVATION

Several key components of this system expected to allow it to overcome the typical problems associated with many pump-and-treat systems. A major reason pump-and-treat systems have not been effective, as has been widely reported, is because when ground water elevations fluctuate due to pumping irregularities or seasonal fluctuations, contaminated ground water (and LNAPL if present) can become entrained in the pore spaces, effectively locking contamination in the media.

Pump-and-treat systems are also seen as ineffective at sites where DNAPL is present. This constant source is not easily controlled by any method, and most pumping systems have little effect. DNAPL was not detected at the site.

This pumping system's primary objectives are to dewater the source and contain the plume. The objective in the remainder of the plume will be to draw impacted ground water toward the middle of the site and treat as much water in this area as possible. As concentrations in the source area are rapidly reduced by the SV system, dissolved concentrations downgradient of the source will decline and will be kept on site by the containment pumping.

OPERATION AND CLOSURE

In the first year of operation, ground water concentrations in the solvent source area dropped 90 percent. Ground water concentrations in the monitoring well furthest downgradient from the source have been reduced to near cleanup objectives for the last two quarterly monitoring events in 1997. Concentrations in the smaller, petroleum hydrocarbon plume have attenuated naturally to below cleanup standards since the fuel tank was removed. Plans to install a small, separate remediation system in the petroleum hydrocarbon plume have been put on hold indefinitely, pending additional confirmatory ground water monitoring data.

Impacted soil in the solvent source area was covered with asphalt, concrete, and a large trash compactor as part of site redevelopment. We will attempt to demonstrate (through methods specified by IEPA's Tiered Approach to Corrective Action Objectives [TACO] guidance) that impacts to ground water from residual contamination in soil in both the solvent and petroleum source areas are negligible, and that the existing ground surface covers (vegetated soil in the hydrocarbon source, and asphalt and concrete in the solvent source) are sufficient to prevent mobilization of residual contaminants.

We anticipate that the system may run between three and ten years. Utilizing IEPA's TACO process, Shepherd Miller expects to achieve cleanup standards, and ultimately a "No Further Remediation" letter from IEPA, which, when issued, will fulfill the client's obligation to the current property owner.

By agreeing to sell the property but maintain the environmental liability associated with this contamination, the client facilitated the rejuvenation of a major manufacturing facility. At the time the client sold the property, the facility was no longer operating, employing less than 50 people. The teamwork demonstrated between the client, the IEPA, and the current property owner has resulted in a redeveloped property at which manufacturing operations operate three shifts, 24 hours per day, employing more than 1,300 people. It is hard to imagine a more positive result for the community, the former owner, and the current owner.

REFERENCES

Haley, J.L., et. al., *Ground Water Monitoring Review*. 1991, Winter, 119-124.

National Academy of Sciences, 1994. Alternatives for Ground Water Cleanup. Report of the National Academy of Science Committee on Ground Water Cleanup Alternatives. Washington, D.C.: National Academy Press.

National Research Council, *Environmental Science and Technology*. 1994, 28, 362-368.

PETROCHEMICAL REMOVAL FROM CONTAMINATED WELLS

Galoust M. Elgal (McClellan AFB, CA)

INTRODUCTION

McClellan Air Force Base is located northeast of Sacramento, California and is surrounded by residential and agricultural districts that can be affected by migration of pollution. The vadose (unsaturated) zone of the soil is composed of layers of sand, silt and clay. For several decades land disposal of petrochemicals (PCHEM), composed of solvents and fuels, has resulted in contamination to a depth of 150 feet (50 m). The clays and sand impede but do not halt the migration of spilled liquids (Anderson, et. al.1992; Ground Water Handbook, 1990). To prevent migration of the ground water plume into the surrounding community and to accomplish the clean-up, two water processing methods have been used, (1) air stripping with natural gas combustor, and (2) activated charcoal filter. Water pump and treat was started in 1987. In 1987 the concentration of petrochemicals was 60 mg/L and now it is 0.5 mg/L. If the pump and treat operations are continued at the present costs, the cost per pound of removal will increase exponentially. It is shown that not to rely on natural attenuation and biological degradation to complete the clean-up, if pump and treat is continued until non-detect, the cost of remediation can reach the range of $200,000 per pound of petrochemical removal.

GROUND WATER TREATMENT PLANT NO.1 WITH AIR STRIPPING

The main Ground Water Treatment Plant (GWTP No.1) comprises of air stripping and combustor (Appendix I). The water is pumped from wells located at Areas B, C, and D and piped to GWTP centrally located in Area C. The process involves pumping the water from the wells through preheater heat exchangers to the air stripper where it is separated into two streams, vapors containing petrochemicals and water. The petrochemical vapors go through a preheater heat exchanger to the combustor to be burned using natural gas. The combustion products composed of carbon dioxide, halides and water vapor flow to the scrubber, which uses sodium hydroxide to absorb the hydrogen chloride, before exhausting into the atmosphere. The water from the bottom of the air stripper passes through the granulated activated charcoal (GAC) filters where it receives final polish. The treated water from the plant is used to maintain a man-made recreational pond with natural habitats and the excess is fed to Magpie Creek. From the start 1987 to the present a total of 1,250,000,000 gallons of water (4,700,000,000 liters) have been pumped through the plant and total of 6200 gallons (23,000 liters) of petrochemicals have been extracted. This quantity is equivalent to 52,000 pounds or 23,000 kg. Currently, preparations are being made for an upgrade.

The concentration of the petrochemicals in the water initially fluctuated due to pockets of high concentration being swept clean by the water flow. The initial average influent concentration was 40 mg/L and has decreased asymptotically to approximately 0.5 mg/L (Figure 1).

Petrochemicals Detected at the GWTP No.1. The chemical analysis profile of the water at the inlet and outlet of the plant is listed in Table 1. The predominant petrochemicals are trichloroethylene and dichloroethylene. Compounds such as vinyl chloride, a biodegradation product, also appear in the individual wells but when diluted with the water of the other wells its concentration approaches non-detect.

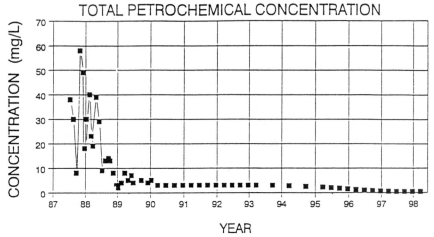

FIGURE 1. Plant influent concentration versus time. Additional wells were installed in 1995 and flow rate was increased from 300 to 700 gpm.

TABLE 1. Current Chemical Analysis at the Inlet of the G W T P No. 1.

PETROCHEMICALS IN WATER	CONCENTRATION ug/L	
	INFLOW	OUTFLOW
1,1-Dichloroethane	1.2	ND
1,1-Dichloroethylene	110.0	ND
1,1,1-Trichloroethane	1.9	ND
Trichloroethylene	310.0	ND
cis-1,2-Dichloroethylene	17.2	ND
Tetrachloroethylene	0.9	ND

The total of the petrochemicals is 441 ug/L (0.44 mg/L)

Petrochemicals Removed at the GWTP No.1. The components and the percent of the petrochemicals have varied through the years. For example, xylene, toluene and benzene have reached non-detect. For calculations the average density of the mixture was determined to be close to the density of water, therefore for the purpose of these calculations the average was selected to be 1.0 g/ml or 62.4 lb / cu ft or 8.34 lb/gal. (The Installation Restoration Guide, 1989).

The data source is from the flow meters located at each well which read gallons per minute. Since the plant performance is reported weekly, the calculations have been made on weekly basis. The conversion factor for time is 10,080 min/wk. A sample calculation for the period Mar 1987 to Dec 1987 is as follows. The water flow was 100 gpm and average concentration 40 mg/L:

gal/wk = 100gal/min x 40ppm x 1/mil x 10,080min/wk = 40.3 gal/wk
gal = 42 wks x 40.3 gal/wk = 1692.6 gal
pounds = 1693 gal x 8.34 lb/gal = 14,120 lb petrochemicals

The pounds and gallon can be converted to international units with,
Conversions: pounds x 0.4536 = kilograms
                      gallons x 3.7854 = liters

   The high initial concentration resulted in a high rate of petrochemical removal
and in time gradually decreased with lower rate of pounds removed (Figure 2). As
additional wells were installed the flow rate also increased.

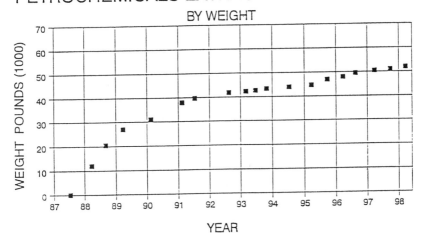

PETROCHEMICALS EXTRACTED AT GWTP NO. 1
BY WEIGHT

FIGURE 2. Cumulative petrochemicals removed. Additional wells were
installed in 1995.

Cost Evaluation of the GWTP No.1. The original cost of the plant design and
construction in 1986 was $ 4,000,000. The major portion of the cost of operation and
maintenance is contractor operation. The reimbursables involve plant system
maintenance. The over and above costs are outside of plant maintenance, such as
water well pump replacement and repairs to conveyor pipes. The breakdown of costs
is tabulated:

Contractor Operation		Utilities	
Labor	$ 300,000	Electricity Wells	$ 20,000
Tech Support	350,000	Electricity Plant	40,000
Reimbursable	100,000	Natural Gas	40,000
Over and Above	100,000	Chemical Analysis	40,000
		In-House USAF	80,000

   Total Cost of Operation and Maintenance: $ 1,070,000

Summary Average Cost per Pound of Petrochemicals Extracted:
   Cost per pound not including plant construction: $136
   Cost per pound including initial plant construction: $228

GROUND WATER TREATMENT PLANT NO.2 WITH CHARCOAL

The GWTP No. 2 utilized charcoal. This system installation initially cost
$300,000 and the yearly cost of operation and maintenance was $ 60,000. The water
was pumped from two wells and purified passing it through activated charcoal and
then to the aeration ponds. The total flow rate was low, approximately 5 gpm. The
petrochemicals comprised of trichloroethylene, tetrachloroethylene and freon. The
total concentration was 4 mg/L. Hence, the charcoal loading was low. From the start
of operation May 1991 to the end in June 1994 the total petrochemicals extracted was
33.6 gallons (376.4 pounds or 127.2 liters or 186 kg). During the operation, the two
series installed 20,000 lb (9070 kg) charcoal bins did not have to be replaced. This
plant operation was discontinued in June 1994 to install an experimental technology
two phase extraction system.
   Cost analysis of petrochemicals extracted:
   Cost per pound not including plant construction: $ 425
   Cost per pound including system installation: $ 1222

NATURAL ATTENUATION AND BIODEGRADATION

Relying on the use of natural attenuation and biological degradation of
petrochemicals in soil and ground water has always been known, but recently has
received increased importance. (Symposium on Natural Attenuation of Chlorinated
Organics, 1996). The cost of remediation has become excessive and it can continue
for 50 to 150 years in trying to achieve maximum contaminant level (MCL), which is
0.005 mg/L for TCE. Once the concentration versus time stabilizes and becomes
asymptotic to the time coordinate, it is time to start considering a switch to natural
attenuation. The cost of operation and maintenance is not proportional to the decrease
in concentration as the cleanup progresses. The yearly cost of the ground water
treatment systems was $1,070,000 in 1987. The petrochemical concentration in water
was 60 mg/L. Now it is about 1 % of what it was initially. If the cost of operation
was proportional then it should now be costing about $10,000 per year. It is not. In
1997 it was $500,000. The cost per pound of removal has been calculated as the
concentration approaches non-detect. The basis of calculation uses $500,000 per year
cost of operation and maintenance and 500 gallons per minute water processed.
Sample calculation is as follows:
 Water Processed: 500 gal/min x 525,600 min/yr x 8.34 lb/gal = 2,191,752,000 lb/yr
 Petrochemicals Extracted: 2,191,752,000 lb/yr x 0.005 parts/million = 11 lb/yr
 Cost per Pound when Concentration is 0.005 ppm: $ 500,000 / yr x (1 / 11 lb/yr) =
   = $ 45,450 / lb

The data for 60 to 0.001 mg/L has been plotted (Figure 3). As 0.001 mg/L (1 ug/L)
is approached asymptotically, the cost will be rising toward $200,000 per pound of
PCHEM removal.

CONCLUSION:

Two basic water processing systems have been evaluated, first a ground water
treatment plant using air stripping and combustor , and second, a simple carbon filter
system. The first is a powerful process that destroys and processes practically all
petrochemicals and is economical for high flow rates and high concentrations. The
second system with carbon only is economical for low flow rates, such as 5 gpm, and

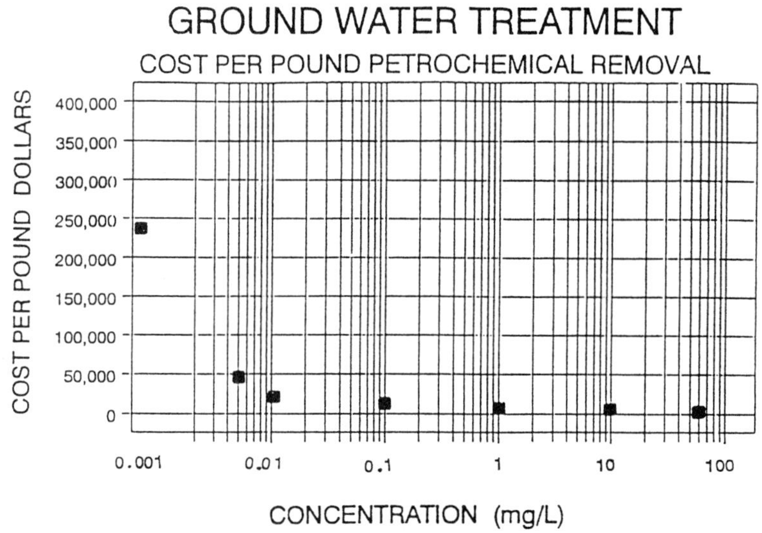

GROUND WATER TREATMENT
COST PER POUND PETROCHEMICAL REMOVAL

CONCENTRATION (mg/L)

FIGURE 3. As concentration decreases the cost per pound of PCHEM removed increases. Currently the concentration is in the range of 0.1 to 1 mg/L.

low concentrations. For the latter the cost per pound removed is higher than the air stripping system, but the initial capital investment and the operation and maintenance is lower. Thus the over all cost is lower. One of the compounds which necessitates the use an air stripping and combustor system is vinyl chloride. Carbon does not filter vinyl chloride well. The ground water treatment plant No. 1 has encountered this compound and has successfully destroyed it. With comprehensive sampling, it has been verified that not only clean-up is occurring but also the containment of the contaminated plume. It is shown that the cost of pump and treat becomes excessive as the concentration approaches non-detect. The decision to convert to natural attenuation and biological degradation can not be defined here, but is dependent on the available funding, the decision of the community members and the regulatory agencies concerning extent of clean-up requirements.

REFERENCES

Anderson, M. R., Johnson, R. L., Parkow, J. F.,"Dissolution of Dense Chlorinated Solvents into Ground Water: Dissolution from a Well-Defined Residual Source", Ground Water, pp. 250-256, March-April 1992, Vol 30, No. 2.

Ground Water Handbook, Vol 1: *Ground Water and Contamination*, U.S. Environmental Protection Agency, EPA/625/6-90/016a, Sept 90

Symposium on Natural Attenuation of Chlorinated Organics in Ground Water, Office of Research and Development, U.S. Environmental Protection Agency, Washington D.C., EPA/540/R-96/509, September 11-13, 1996.

The Installation Restoration Program Toxicity Guide, Harry G. Armstrong Aerospace Medical Research Laboratory, Wright Patterson AFB, Ohio 45433-6573, Biomedical & Environmental Information Analysis, Oak Ridge, TN 37831-6050, Vol 1- , 1989.

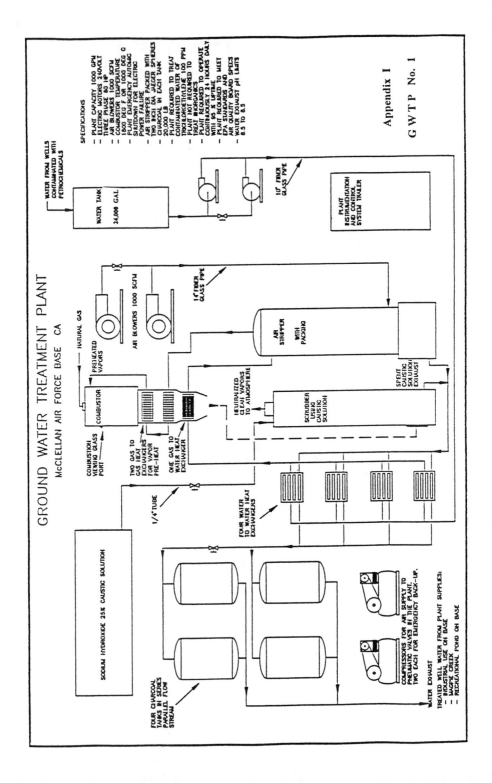

GROUND WATER TREATMENT PLANT
McCLELLAN AIR FORCE BASE CA

Appendix I

GWTP No. 1

MODIFICATION OF SORPTION CHARACTERISTICS IN AQUIFERS TO IMPROVE GROUNDWATER REMEDIATION

Gregory Smith (ENSR Consulting and Engineering, Westmont, Illinois)
Bruce Dumdei (ENSR Consulting and Engineering, Westmont, Illinois)
Valdis Jurka (Lucent Technologies, Inc. Morristown, New Jersey)

ABSTRACT: The authors adapted oil well stimulation technologies[1] to groundwater remediation. Chlorine dioxide has been used in the oil industry as a bactericide, to remove sludge, improve formation permeability and improve oil production. We utilized chlorine dioxide in the field over a period ranging from May 1996 to March 1997 to temporarily alter Eh characteristics of the aquifer matrix and its response to pH, thereby modifying its sorptive characteristics.

The applications were performed in the field at a site undergoing full-scale soil and groundwater remediation for trichloroethene and 1,1,1-trichloroethane that has been underway since 1991, using steam flooding and simultaneous enhanced biotransformation[2] The site is a former electronics manufacturing plant located in the northern suburbs of Chicago, Illinois. Dense non-aqueous phase liquids (DNAPL) had been removed from the subsurface with the exception of four areas. Two of these areas appeared to have significant oils (cutting oils, hydraulic oils) which were suspected to be limiting the DNAPL removal. It was decided to apply chlorine dioxide for removing oily materials from subsurface formations to aid remediation.

After injection of chlorine dioxide, we observed concentrations of 1,1,1-trichloroethane in groundwater increase approximately two-fold in response because of desorbtion and dissolution of the compounds. Net concentrations after treatment, decrease by one order of magnitude.

INTRODUCTION

The release of organic substances into the subsurface and groundwater changes the oxidation/reduction (redox) conditions in the subsurface through the oxidation of the organic matter. When oxygen in the groundwater is consumed, oxidation of organic matter occurs, but the oxidizing agents are NO_3^-, MnO_2, $Fe(OH)_3$, and SO_4^{2-} (Freeze and Cherry, 1979). As these oxidizing agents are consumed, the groundwater environment becomes more reduced to the point where methanogenic conditions may result if there are: 1) sufficient oxidizeable organics; 2) sufficient nutrients; and 3) temperature conditions conducive to bacterial growth.

---

[1] U.S. Patent No. 4,871,022
[2] U.S. Patent No. 5,279,740

The oxidation of organic matter lowers the pH in aqueous systems. Reduction in pH also affects the sorptive capabilities of the minerals in the aquifer matrix (Stumm and Morgan, 1981). Disturbing the redox equilibrium conditions further disturbs the electrostatic charges on soil grains, and hence the sorptive characteristics of the soil. However, localized available ligands or oxidation states may be more important in the sorptive characteristics of the soil.

As pollutants are transported with groundwater they are adsorbed and desorbed on the aquifer matrix. This process has been likened to a chromatographic dispersion in the aquifer (Jackson and Patterson, 1987). Most compounds elute with the groundwater in a pattern consistent with their respective octanol-carbon and octanol-water partition coefficients. Because of the adsorption desorbtion phenomena, we have reasoned, that until the pollutants are removed or reduced to an insignificant quantity on the aquifer matrix, it may be very difficult to achieve regulatory based clean-up criteria in remediating groundwater through conventional pump and treat methods.

CHLORINE DIOXIDE

Chlorine dioxide has been used in the oil production industry for years to rehabilitate oil wells. In the oil industry, it has been observed that it acts as a bactericide, facilitates the removal of sludge and improves formation permeability. It is believed that the improvement of formation permeability is at least due in large part to modification of the mineral surfaces within the formation. Chlorine dioxide is a relatively unstable gas (boiling point 11°C), and compression of the gas has the potential of explosion. Therefore, it must be generated at the point of use.

Chlorine dioxide breaks down in water according to the following two reactions:

$$ClO_2 + 1e \rightarrow ClO_2^- \qquad\qquad E_0^{25°C} = 1.15V \text{ as gas}$$
$$\qquad\qquad\qquad\qquad\qquad\qquad\qquad = 0.95V \text{ dissolved as liquid}$$
$$ClO_2^- + 4e + 2H_2O \rightarrow Cl^- + 4OH^- \qquad E_0^{25°C} = 0.78V.$$

Chlorine dioxide does not react with aliphatic compounds, but the properties of interest in this study are:

- Oxidation of inorganic compounds

- High diffusion into hydrophobic lipid layers (Masschelein, 1991).

- Using the Rio Linda process for generating ClO_2 allows adjustment of pH, approaching the point of zero charge in soil water systems, desorbing the adsorbed compounds.

The oxidation of the inorganic compounds allows redox processes from the release of oxidizeable organics into aquifer systems to be temporarily reversed (see discussion on the field data).

The ability of chlorine dioxide to diffuse into hydrophobic materials (Masschelein, 1991) is particularly significant in remediating non-aqueous phase liquids (NAPLs) in groundwater flow systems. Chlorine dioxide diffuses into NAPL zones comprised of trichloroethene and 1,1,1-trichloroethane, improving the ability for removal through vapor extraction.

BENCH SCALE TESTING

Soil samples were obtained from a site currently under a voluntary soil and groundwater remediation in the suburban Chicago area. A high concentration of chlorinated solvent solution in water, containing trichloroethene and 1,1,1-trichloroethane with groundwater, and biotransformation daughter compounds (cis & trans-1,2-dichloroethene, 1,1-dichloroethene, 1,1-dichloroethane, chloroethane and vinyl chloride) was obtained from two remediation wells to spike the soil samples. Six test cells were filled with soil and then spiked with the chlorinated solvent mixture, infiltrating the soil sample from the bottom up to provide for complete saturation of the soil and reduce the opportunity for trapped air in the samples.

The test cells were allowed to sit for 48 hours. After sitting, samples were obtained to establish initial concentrations of chlorinated solvents in the soils. Two test cells were treated by flushing distilled water, and two cells were treated by flushing with two different concentrations of chlorine dioxide. Each cell was treated by passing a one pore volume flush from the bottom of the cell upwards to reduce the potential for preferential migration of the fluids through the cells. The chlorine dioxide flushes were followed with distilled water flushes.

Samples of the distilled water and soils were then analyzed for concentrations of chlorinated solvents and compared to the original concentrations. The distilled water flushes showed no significant change in concentrations of trichloroethene. The soils in the cells that were treated with chlorine dioxide showed approximately 80% concentration reduction, one pore volume flush. Interestingly, there was no significant difference in the reduction observed in the two different strengths of chlorine dioxide. This indicated to the authors that the chlorine dioxide was reacting with the soil rather than the chlorinated solvents, suggesting there was no significant oxidation of the solvents.

FIELD SCALE TESTING

We have been conducting a soil and groundwater remediation in the suburban Chicago area since January 1991 for chlorinated solvents in the form of trichloroethene and 1,1,1-trichloroethane. The solvents were present in the groundwater in both a dissolved or aqueous phase liquid (APL), and in an undissolved or dense non-aqueous phase liquid (DNAPL).

After a period of four to five years using a steam flood and simultaneous enhanced biotransformation process, it was observed that although concentrations had been significantly reduced, and DNAPL had been removed, there were still some areas where concentrations were persistently high. It was suspected that the presence of co-released oils in the soils may be responsible for several persistently high areas of concentration.

For field testing, we selected an isolated extraction well identified as well F13) that had shown persistent concentrations of 1,1,1-trichloroethane ranging between 10,000 and 40,000 µg/l for, while surrounding areas showed decreasing concentrations and achieved the cleanup standards. This well had periodically shown the presence of oil.

The soils at this site have hydraulic conductivities ranging from 1.05×10^{-5} to 1.23×10^{-4} cm/sec. The treatment consisted of injecting approximately 9500 litres (2500 gallons), corresponding to a calculated one pore volume in the area defined by the well and the surrounding wells in the grid. The solution was left in the subsurface for approximately two days before being pumped out. As the solution was pumped out, the Eh and pH conditions were monitored, as well as the concentrations of chlorinated solvents.

Figure 1 presents the Eh and pH conditions measured to track the break-down of ClO_2 in the subsurface, and the resultant changes in the saturated zone induced by the introduction of the ClO_2. From Figure 1, it can be seen that the pH initially decreases after the acidic solution is introduced and then shows a generally increasing trend. Correspondingly, the Eh increased in response to the introduction of the ClO_2 and decreases as the ClO_2 breaks down. Interestingly, the pH after treatment is less acidic that before and the Eh is more negative, with both of these parameters tending towards what would be expected for natural conditions.

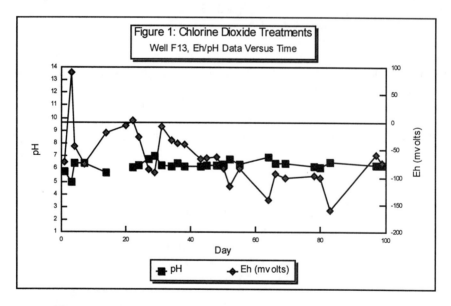

Figure 2 presents the concentrations of volatile organic compounds over time for the initial three months of treatment at well F13. It can be seen that

1,1,1-trichloroethane concentrations increase in response to ClO_2. After peaking at Day 24, the concentrations of 1,1,1-trichloroethane decrease to Day 55, whereupon an increase is observed. Temperatures in the groundwater increase at this time in response to the steam flushing after the ClO_2 injection. Concentrations decrease to Day 99. Therefore, we observe three phases of mass removal: 1) the initial release apparently associated with the decrease in pH; 2) mass removal as the ClO_2 reacts with the soil matrix; and 3) an initial peak and drop off as the groundwater or injected steam is flushed through the system and the soils return to equilibrium.

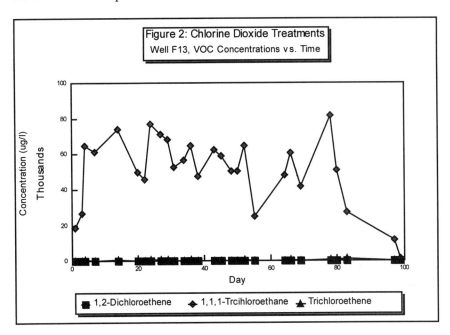

Figure 2: Chlorine Dioxide Treatments
Well F13, VOC Concentrations vs. Time

CONCLUSIONS

Chlorine dioxide temporarily oxidizes soil minerals and modifies pH and redox conditions in the water, which changes the soil minerals and their sorbtive properties, allowing for improved mass removal via conventional groundwater extraction techniques.

In the bench scale and field study, chlorine dioxide treatment exhibited the following characteristics:

- Barring extremely tight geologic conditions, ClO_2 appears to reach the highly contaminated zones;

- It modifies the soil matrix rather than oxidizing the organics;

- Releases compounds of concern from the soil matrix;

- Increases removal of tightly bound compounds; and

- Decreases flush time associated with aquifer remediation, which in turn reduces operational time.

The specific reactions that occur in the subsurface have not yet been determined, but field studies demonstrate effectiveness. More data will result in generalizing techniques and improving predictability.

REFERENCES

Freeze, R.A. and J.A. Cherry (1979). *Groundwater.* Prentice-Hall Englewood Cliffs, N.Y.

Jackson, R.A. and R.J. Patterson (1989). "A Remedial Investigation of an Organically Polluted Outwash Aquifer". *Groundwater Monitoring Review,* Vol. 3 No. 9.

Masschelein, W.J. (1991). "Chlorine Dioxide". In: *Chemical Oxidation, Treatment Technologies for the 90s.* Proceedings of the First International Symposium, Vanderbilt University. Eckenfelder, W.W., A.R. Bowers and J.A. Roth, Eds. Technomic Publishing Co., Inc. Lancaster, PA 17609.

Stumm, W. and J.A. Morgan (1981). *Aquatic Chemistry,* An Introduction Emphasizing Chemical Equilibria in Natural Waters. John Wiley & Sons, N.Y., N.Y.

HYDRAULIC IN-SITU REMEDIATION TECHNIQUES WITH SURFACTANTS: OPTIMIZATION OF HYDRAULIC SYSTEMS

Reinhold Josef, Baldur Barczewski, Hans-Peter Koschitzky, Jürgen Braun
University of Stuttgart, Germany

ABSTRACT: Two different artificial heterogeneous large scale aquifers with the dimensions of 9 x 6 x 4.5 m (L x W x H) and two contaminants (Trichloroethylene as a DNAPL and Xylene as a LNAPL) are used for 4 hydraulic remediation experiments. The goal is to optimize contaminant extraction from the source zone by adapting groundwater flow regime with packer-systems and increasing solubilization/mobilization with surfactants. Tracer test results were combined with numerical modeling to estimate in detail the "real" permeability of the emplaced sands. During remediation mass balances were done which found over 85 % of the emplaced 29 kg TCE. These results will be compared with those of 2 other projects using different hydraulic remediation techniques under identical conditions.

INTRODUCTION

One of the five main research topics of the German groundwater research facility *VEGAS* is "Optimization of existing and development of new hydraulic remediation techniques". *VEGAS* is a facility of the University of Stuttgart linked to the Institut für Wasserbau, Lehrstuhl für Hydraulik und Grundwasser, but also available to off-campus research institutions, industrial companies and cooperation partners abroad. It is conceived as a connecting link between conventional small-scale laboratory experiments and field investigations. In contrast with field investigations on contaminated sites, experiments in *VEGAS* are conducted under controlled and reproducible conditions, and therefore mass balances and remediation efficiencies can be specified [Barczewski, Koschitzky].

In applying hydraulic remediation techniques, problems are caused by the high variability of the hydraulic conductivity, the difficult accessibility of zones with low permeability, the low solubility of some contaminants in water and low mass-transfer rates. Several methods for increasing the extraction efficiency in saturated zones will be tested. This includes arrangements with five vertical circulation wells combined with packer systems. Furthermore, the combination of hydraulic schemes with the application of biodegradable surfactants will be investigated.

Objectives. The main objective of this project is to enhance contaminant removal by manipulation of ground water flow fields through heterogeneous aquifers using specially designed wells (optimization of hydraulics). The optimized flow field results in low energy consumption. Additionally, the enhancement of technological scale extraction of Xylene from the source zone using surfactants will be investigated Studies were carried out to quantify the reduction of hydraulic permeability in the presence of surfactants.

MATERIALS AND METHODS

For the large scale remediation experiments three different sands of a non-uniform grain size distribution (the maximum permeability ratio was 50) were used in the construction of two artificial heterogeneous aquifers with the dimensions 9 x 6 x 4.5 m (L x W x H): a block model with 60 discrete blocks and a layered model (cf. Figure 1) with 4 inclined layers [Josef]. The remediation flow was superposed over a simulated groundwater flow during the remediation experiment.

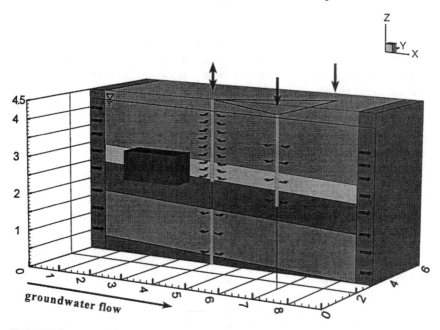

FIGURE 1: cross section of the layer model with screened well sections and location of the TCE contamination

While using the **Dense Non-Aqueous Phase Liquid (DNAPL)** groundwater levels in both aquifers were up to 4.1 m above the bottom. The emplacement of the contamination source was achieved by excavating the original sand and refilling it mixed with 29.3 kg TCE in residual saturation to avoid DNAPL migration towards the aquifer bottom (cf. Figure 1). The groundwater level was lowered during the emplacement of the source zone. Precautions were taken to minimize TCE loss by evaporation.

For the experiments with the **Light Non-Aqueous Phase Liquid (LNAPL)** the groundwater level was lowered to 3.7 m to create a vadose zone. 31 kg of Xylene were injected into four probes at different heights while raising the groundwater level. The method for emplacing was designed to achieve a contaminated zone in residual saturation below groundwater level.

Based on tracer experiments the numerical model "SICK100" (GKW Ingenieurgesellschaft mbH, Bochum, Germany) was used to optimize the flow rates and installation dimensions of the remediation plant which consists of the well system including the packers, an extraction pump and active carbon filters. The piezometric head was measured at 378 sampling points, this provided an another mean of verifying the numerical model.

Every effort was made to monitor the progress of the remediation and to achieve good mass balance. The change of concentrations during the experiments were monitored by sampling 378 points distributed regularly in the entire aquifers. For the mass balances the concentrations of contaminant and fluxes in and out of the aquifers were determined.

RESULTS AND DISCUSSION

Hydraulics. In an ideal remediation flow field every contaminated streamline is captured by the extraction wells and it is not diluted by uncontaminated streamlines. In reality, extracted water flux always exceeds the injected water flux because of unknown heterogeneities and diffusion.

For the contaminant TCE the optimization of flow is done in the two large scale aquifers. As an example results of the experiment in the layer model are shown.

TCE in layer model. Figure 2 and 3 show two TCE distribution of the remediation experiment in the layered model. Figure 2 shows the distribution before starting hydraulic remediation without any groundwater or remediation flow. Higher than expected concentrations in the lower levels are mainly due to preceding experiments. Despite the emplacement of DNAPL at residual saturation migration is presumed to have occurred. During remediation water was extracted in the upper central well and injected after treatment in the other three screened sections of the wells (figure 1) and the extraction flux was varied. Figure 3 shows the TCE distribution after 54 days of hydraulic remediation. Figure 4 and 5 shows the flux, the response in concentration and the calculated cumulative mass in the two outflows: extracted water for remediation (figure 4) and aquifer effluent (figure 5). Due to the high volatility of TCE high concentrations were found far from the contaminated zone before starting the remediation experiment (figure 2). Pump problems in the first few days resulted in a concentration peak downstream in the aquifer outflow. Interruptions of the hydraulic remediation (at 54 and 112 days) resulted in a concentration peak when restarting and a small peak downstream in the aquifer outflow. In general the concentration in the extraction well was decreasing exponentially and with temporary peaks after interruptions in flow. Each decrease in the extracted water flux resulted in a higher extracted concentration.

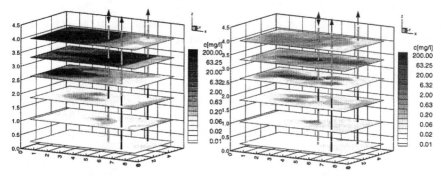

FIGURE 2: TCE-Distribution after initial rising of the ground water table.

FIGURE 3: TCE-Distribution after 54 days of hydraulic remediation

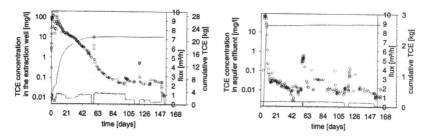

FIGURE 4: TCE concentration, cumulative TCE removal and flux in the extraction well

FIGURE 5: TCE concentration, cumulative TCE removal and flux in the aquifer effluent

Xylene in layer model: In the first 40 days of hydraulic remediation of Xylene in the layer model (phase 1) the concentration of the extracted water has remained at approximately 10 % of Xylene solubility (figure 6). No contaminant was found downstream (figure 7). Every decrease of the extracted water flux resulted in a higher extracted concentration. The calculated cumulative xylene mass was nearly 39% of the emplaced Xylene. In phase 2 it is expected that the application of surfactants will increase the concentration of xylene in the extracted water (concentration peak). As a result the mass removal rate is expected to increase temporarily. In phase 3 (after 2 days of surfactant injection) hydraulic remediation will be continued with the same extraction and injection points and the same pumping rates i.e. the same flow field. The mass removal rate is expected to be lower after surfactant use.

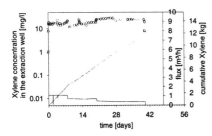
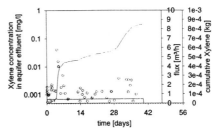

FIGURE 6: Xylene concentration, cumulative xylene removal and flux in the extraction well

FIGURE 7: Xylene concentration, cumulative xylene removal and flux in the aquifer effluent

Investigations for the application of surfactants: For the investigation of the reduction in hydraulic permeability while using surfactants, 0.8 m long columns were used with constant head at inflow and outflow. In each column the piezometric head could be measured at 10 points along the length of the column. This experimental setup allowed to investigate and to locate the loss in hydraulic permeability. The measurements of the decrease of hydraulic permeability in columns while applying surfactants showed large differences between the surfactant mixtures. In addition to higher viscosities several other parameters were also found to contribute in a direct or indirect way to permeability loss. As an example the influence of water hardness and concentration is shown in the figure 8. The surfactant mixture contains non-ionic and anionic surfactants. Certain ratios of anionic surfactants to water hardening components causes precipitation of liquid crystal-like particles. With higher water hardness the permeability decrease is lower as the figure 8 shows. Surfactant screening is done and several surfactant mixtures were selected for different contaminants in cooperation with several partner projects throughout Germany. Another partner project designed a surfactant reuse and groundwater treatment plant.

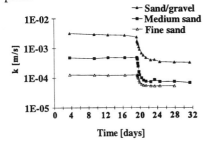
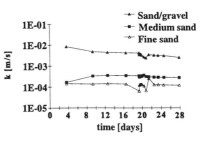

FIGURE 8: the progress of hydraulic permeability starting with pure water, from 19 days on applying a solution of a mixture non-ionic and anionic

A: concentration 0.5%, hardness 23°dH   **B:** concentration 1%, hardness 23°dH

One mixture will be applied in the large scale experimental remediation of Xylene. During the application of surfactants additional plant components will be integrated. For the large scale application of surfactants one basic requirement is to know the location of the contamination source and to adapt the flow field for remediation so that surfactants reaching uncontaminated regions is minimized. Another requirement for economical application is the reuse of the surfactants by selectively removing only the contaminant from the contaminant-surfactant-water mixture. In this project packer systems were used for the adaptation of the flow field. A partner project is handling surfactant reuse and water treatment.

OUTLOOK

At the end of the 3 years project (May 1998) the results of each large scale experiment will be compared with other hydraulic remediation techniques (ground water circulation well, air injection well) which were used by other projects in the same aquifers with the same initial conditions [Luber, Brauns] [Scholz et al.].

ACKNOWLEDGMENTS

This project is financed by the German Federal Ministry of Education, Science, Research and Technology BMBF (registration number WT9527/3).

REFERENCES

Barczewski, B., Koschitzky, H.-P., 1996. "The *VEGAS*-Research Facility: Technical Equipment and Research Projects". In: *Groundwater and Subsurface Remediation.* (Eds.: H. Kobus, B. Barczewski, H.-P. Koschitzky), Springer, Berlin Heidelberg.

Josef, R., Barczewski, B., Koschitzky, H.-P., Braun, J., 1997. "Untersuchungen zur Optimierung hydraulischer Verfahrenstechnologien zur Schadstoffeliminierung aus Grundwasser mittels Tensiden. In: *Möglichkeiten und Grenzen der Reinigung kontaminierter Grundwässer. Resümee und Beiträge des 12. DECHEMA-Fachgesprächs Umweltschutz.* Frankfurt a.M.: DECHEMA.

Luber, M., Brauns, J., 1997. Air Injection Wells for Groundwater Remediation - Large Scale Experiment. Proceedings of the international conference GREEN II - Contaminated and Derelict Land, 8.-11 Sept. 1997, Krakau, Polen.

Scholz, M., Weber, O., Stamm, J., Eldho, T.I., Jirka, G.H., 1997. "Large-Scale Laboratory Investigation of in situ Groundwater Remediation Using Vertical Circulation Flow Systems". *Proc. XXVII Congress Int. Assoc. for Hydraulic Research, San Francisco, Theme C,* 120-125.

GROUNDWATER TREATMENT USING AIR STRIPPING: A SUCCESSFUL CASE STUDY

Lois J. Weik (Dames & Moore, Bethesda, Maryland, USA)

ABSTRACT: At a former industrial research and development (R&D) facility in Carroll County, Maryland, an air stripping system installed into fractured bedrock successfully remediated an aquifer contaminated with 1,1,1-trichloroethane (TCA) and 1,2-dichloroethene (DCE) at levels as high as 39,000 micrograms per liter (μg/L) and 1,300 μg/L, respectively. A recovery/recharge system consisting of 10 wells was designed and installed in 1985. The groundwater was treated using a 1.5-foot-diameter, 20-foot tall vertical air stripping tower. The treated water was discharged upgradient to prevent offsite point source discharge to surface water and to expedite cleanup. Contaminant levels decreased rapidly until approximately 1992, when the removal rates began to reach an asymptote. In June 1995, the system was cyclically operated to flush the cones of influence of the recovery wells. Monitoring indicated that groundwater elevations returned to prepumping conditions; and that--though contaminant concentrations increased between cycles, as expected--they remained below treatment goals. In April 1997, the system was shut down. Contaminants remain below treatment goals. Post-closure monitoring will continue for 2 years.

ENVIRONMENTAL SETTING

The 20-acre site is a former industrial R&D facility located in a rural area of Carroll County, Maryland. The site is predominantly bounded by woods and farmland. Beaver Run and an unnamed tributary are located near the site. The topography of the area is very hilly, with elevations ranging from 720 to 840 feet mean sea level (msl; Figure 1).

Surface soil on the hills consists of saprolite or residual soil derived from the weathering of underlying bedrock. The soil is typically about 40 feet thick on the hilltops; it thins until bedrock is exposed on the slopes near Beaver Run. The soil texture is typically silty to clayey. The saprolite grades downward into relatively unweathered bedrock, which consists of the greenish-to-grayish chlorite Prettyboy Schist. Joints following its irregular foliation are typically spaced a few inches to 1 foot apart in outcrops along Poole Road. The strike of the foliation near the site varies from north to about 30 degrees east of north. Dips are generally within 10 degrees of vertical. The total depth of the schist near the site is probably in the range of thousands of feet.

Groundwater occurs in fractures of various sizes within the schist, and to a lesser extent in the overlying soil weathered from it. Fractures are not uniform and are often concentrated in fracture zones, which can be identified as linear features on

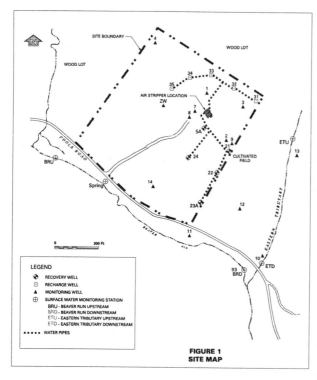

LEGEND

◆ RECOVERY WELL
▫ RECHARGE WELL
▲ MONITORING WELL
⊕ SURFACE WATER MONITORING STATION
  BRU - BEAVER RUN UPSTREAM
  BRD - BEAVER RUN DOWNSTREAM
  ETU - EASTERN TRIBUTARY UPSTREAM
  ETD - EASTERN TRIBUTARY DOWNSTREAM
▪▪▪▪▪ WATER PIPES

FIGURE 1
SITE MAP

aerial photographs or through their association with stream valleys. The schist is not an extremely productive aquifer. Well yields are typically from a few gallons to a few tens of gallons per minute (gpm). Groundwater flow from the site is generally southward, but is affected by rock structure, which appears to hinder groundwater movement toward the stream east of the site. In general, groundwater is recharged by precipitation on the upland and discharges in stream valleys; these valleys are not necessarily barriers to lateral groundwater movement. Groundwater in the saprolite has a downward component of movement into the bedrock. Groundwater deeper in the bedrock flows more nearly horizontally, or even somewhat upward, as it moves toward stream valley discharge areas.

NATURE AND EXTENT OF CONTAMINATION

R&D activities at the site ceased in 1969; there are currently no facilities onsite. Groundwater is primarily located within fractures and was found to be contaminated with elevated levels of TCA and DCE, at maximum concentrations of approximately 39,000 µg/L and 1,300 µg/L, respectively. No source was identified, and no dense nonaqueous phase liquids (DNAPLs) were present. Figure 2 shows TCA concentrations prior to installation of the treatment system.

TREATMENT SYSTEM DESIGN AND OPERATION

Based on an extensive groundwater investigation--including fracture analysis, groundwater flow direction evaluation, and groundwater sampling and analysis for TCA/DCE--Dames & Moore recommended installation of an air stripping system with groundwater recovery wells and an upgradient recharge system. Recovery wells were established within or slightly downgradient of the

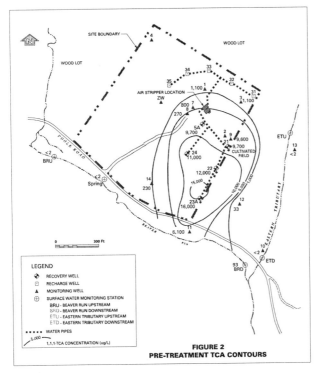

FIGURE 2
PRE-TREATMENT TCA CONTOURS

area containing the highest levels of TCA and DCE to inhibit plume advancement and to reverse natural gradients by forming a cone of depression near the plume centroid. The recovery well network was intended to contain and capture contaminants on the facility property.

A recovery/ recharge system consisting of 10 wells was designed and installed in 1985. The water quality monitoring network included 13 monitoring wells and five surface water sampling stations (including one spring). Groundwater was treated using a 1.5-foot-diameter vertical air stripping tower containing 20 feet of 1-inch pall ring packing. The air stripper was designed for an average flow of 50 gpm and an air-to-water ratio of 40:1. The treated water was discharged upgradient to prevent offsite point source discharge to surface water and to expedite cleanup. The system does not require air treatment.

Groundwater elevation monitoring indicated that the system effectively established a cone of influence near the plume centroid that prevented further offsite migration of contaminants. Contaminant levels decreased rapidly until approximately 1992, when the removal rates began to reach an asymptote. In June 1995, all treatment goals had been met, but concentrations were elevated in one monitoring well (well 11, see Figure 1) at the edge of the recovery well network. It is likely that groundwater in this well was relatively stagnant, being subject to the opposite influences of the recovery system and stream discharge. Although the system had a positive influence on this well, it was relatively weak compared to the influence on wells closer to the recovery zone, and concentrations decreased in this well at a relatively slower rate.

CYCLIC OPERATION

Cyclic operation was initiated to enhance site cleanup by flushing the cones of influence of the recovery wells. This operation was also intended to address

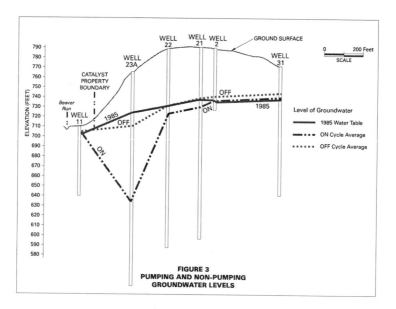

**FIGURE 3
PUMPING AND NON-PUMPING
GROUNDWATER LEVELS**

concerns of the State of Maryland Department of the Environment (MDE) that concentrations would increase when the system was turned off because of a release of contaminants previously held within the cones of influence, as elevations and flow gradient returned to normal. In addition, there was concern that contaminants still remaining in well 11, located at the edge of influence of the recovery system, would be discharged to the stream if the system was turned off, in turn causing elevated levels of TCA and DCE within Beaver Run. Because the system had been operated continuously since 1985, the recovery process had not flushed the dewatered cones of influence.

Prior to beginning cyclic operations, the potential for offsite contaminant migration from the area near well 11 during non-pumping cycles was assessed. Hydrogeologic and watershed evaluations were conducted to assess flow rates into the overall cone of influence and the potential for discharge of affected groundwater into nearby Beaver Run. These modeling efforts indicated that no significant impacts to human health or the environment were predicted as a result of cyclic operation. In June 1995, cyclic operation commenced. The system was operated for nearly 2 years, with an on/off cycle of 2 weeks.

Monitoring during cyclic operation indicated that groundwater elevations (measured bi-weekly) returned to approximately 98 percent of prepumping conditions (see Figure 3), thereby thoroughly saturating the area of the cone of influence of the recovery wells. The seven wells shown in Figure 3 approximate a northeast-southwest cross-section of the site and are aligned approximately parallel to groundwater flow direction during nonpumping conditions. The figure shows water levels prior to the initiation of pumping activities in 1985, and average on- and-off cycle levels during cyclic operation.

Groundwater samples continued to be collected for chemical analysis on a quarterly basis. In addition, every 2 weeks during the first 7 months of cyclic operation, groundwater samples were collected from the five site recovery wells. Although contaminant concentrations in wells increased between cycles, they remained below treatment goals. Figure 4 illustrates the typical fluctuation in observed concentrations. The initial TCA measurement in early June 1995 represents the groundwater concentration prior to cyclic operation (i.e., after operating the recovery system steadily since 1985). Well 11--located at the edge of the recovery system--exhibited a decrease in contaminant levels, probably as a result of the natural attenuation that occurred primarily during the off cycle. Contaminants were not detected in Beaver Run or the spring.

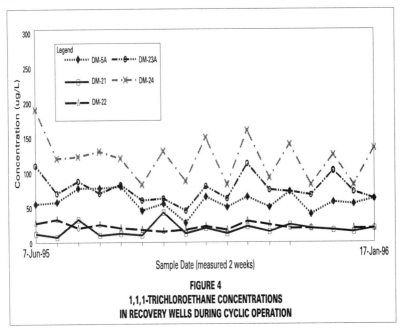

FIGURE 4
1,1,1-TRICHLOROETHANE CONCENTRATIONS
IN RECOVERY WELLS DURING CYCLIC OPERATION

SYSTEM CLOSURE

In April 1997, the system was shut down. Overall, contaminant concentrations in all wells have, as expected, remained the same or slightly increased, but are still well below treatment goals. There have been no measurable discharges to Beaver Run. Table 1 summarizes the TCA concentrations for key wells during the first three sampling events after system shutdown. Quarterly post-closure monitoring will continue for 2 years.

Table 1. TCA Concentrations After System Shutdown

Date						

Recovery Well Number

Date	5A	11	21	22	23A	24
Mar (1)	66	580	11	14	55	89
May	47	730	12	11	46	40
Aug	120	1100	33	18	47	46
Nov	33	140	32	23	47	30

(1) Last sampling event during system operation.

CONCLUSIONS

Although pump-and-treat is generally not very effective for groundwater contamination in fractured bedrock, it was successful at this site because of:

- The localized, defined plume.
- The addition of upgradient recharge to aggressively flush contaminants from within fractures.

Cyclic operation of the system provided an excellent interim step to system closure, providing verification of minimal contamination in the cone of influence. In addition, monitoring during cyclic operation provided reassurance that the discharge of contaminated groundwater at the edge of the treated plume would not negatively affect the stream. The successful application of pump-and-treat at this site indicates that proper site conditions and a flexible system design and operation combined with predictive modeling, can remediate groundwater contamination.

In retrospect, cyclic operation may have been implemented several years earlier, when contaminant recovery rates first began to level off. This may have led to an earlier system shutdown. Because of the innovative nature of cyclic operation, Dames & Moore was required to conduct modeling prior to the initiation of cyclic operation to provide a level of confidence that there would be no negative impacts to human health or the environment.

CHLORINATED SOLVENT GROUNDWATER REMEDIATION AT THE FAIRBANKS DISPOSAL PIT

Wm. Gordon Dean, P.E. (WRS Infrastructure & Environment, Tallahassee, FL)

ABSTRACT: The Fairbanks Disposal Pit in Gainesville, Florida is an abandoned 10-acre borrow pit used to dispose over 2,000 drums of asphalt and spent solvents, construction debris, and other wastes. The groundwater in four separate aquifers underlying the site became contaminated (primarily with trichloroethene and 1,1-dichloroethene) as a result. The groundwater contamination plume covered approximately 70 acres (28,000 square meters, m^2) in the surficial aquifer and extended over two miles (3.2 kilometers, km) downgradient in the underlying aquifers. WRS Infrastructure & Environment, Inc. designed and implemented Corrective Action Plans for each of the four affected aquifers. Technologies used to address the groundwater contamination include deep horizontal recovery trenches, vertical recovery wells, conventional air stripping towers, Floridan aquifer injection wells, and discharge to a retention pond. The various treatment systems are integrated via state of the art computer based telemetry. The treatment systems were installed and started between December 1995 and July 1996. To date, the area of surficial aquifer contamination has been reduced from 70 acres (28,000 m^2) to two monitor wells, the contaminant plume in the H1 aquifer has been reduced by over 75%, and the contaminant plumes in the lower two aquifers have been reduced to one isolated area. Monitoring data from the lower aquifers indicate that no rebound effect is being observed in the remediated areas. The estimated time to complete groundwater remediation at the site is two to four years. The following activities are being evaluated to expedite the remediation effort: soil vacuum extraction in the surficial soils exposed by the groundwater recovery system and in-situ chemical oxidation in the surficial aquifer.

INTRODUCTION

The Fairbanks Disposal Pit in Gainesville, Florida is an abandoned 10 acre (28,000 m^2) borrow pit used to dispose over 2,000 drums of asphalt and spent solvents, construction debris, and other wastes. Disposal activities occurred from 1956 to 1982. The groundwater in four separate aquifers underlying the site became contaminated (primarily with trichloroethene and 1,1-dichloroethene) as a result. Over 300 private potable supply wells existed within a one-mile (1.6 km) radius of the site. Over 100 of these wells were contaminated by chlorinated solvents and a nearby public wellfield was also threatened. The groundwater contamination plume covered approximately 70 acres (28,000 m^2) in the surficial aquifer and extended over two miles (3.2 km) downgradient in the underlying aquifers. Site activities were conducted under both a Resource Conservation and

Recovery Act (RCRA) Closure/Post-Closure Permit and an Administrative Order under Sections 104, 106(a), 107, and 122 of the Comprehensive Environmental Response, Compensation, and Liability Act (CERCLA).

Hydrogeologic Setting. Four separate aquifers (the surficial, H1, H2, and Floridan) are present beneath the site. Each exhibited different groundwater flow directions and aquifer properties, and numerous karstic features were present. The surficial aquifer is the first aquifer encountered. It is an unconfined, water table aquifer consisting of approximately 60 feet (18.2 m) of sedimentary deposits of sand, silty sand, and clayey sand. A phosphatic gravel/sand/clay layer is located at approximately 35 to 45 feet (10.7 to 13.7 m) below land surface (bls) and was used as a division between the upper and lower zones of the surficial aquifer. The silty sands underlying this layer become progressively less permeable with depth. The surficial aquifer thins to the north with the topographic slope. The thinnest portion [approximately 25 feet (7.6 m) thick] is located beneath a small creek north of the site. Numerous karstic features were encountered close to this creek.

The Hawthorn Group is a regional aquitard separating the surficial aquifer and the Floridan aquifer. It consists of tight, plastic, marine clays interbedded with water producing zones of sand, limestone, and shell. Two such zones (the H1 and H2 aquifers) have been identified at this site. The H1 aquifer is located between approximately 90 and 120 feet (27.4 to 36.6 m) bls in the Hawthorn Group. The maximum thickness is approximately 30 feet (9.1 m) north of the site and the minimum thickness is approximately 15 feet (4.6 m) far southwest of the site. The H1 aquifer exists under confined conditions as demonstrated by analysis of data from two pump tests.   However, a number of residential wells interconnected the H1 aquifer to the surficial, H2, and/or Floridan aquifers and may have provided conduits for contaminant migration through the aquitards. Additionally, the upper confining unit is breached by karstic features in the area of the creek. The lithology found within the H1 aquifer characteristically consists of upper and lower sharp contacts between the Hawthorn clays and a dolomitic limestone.   The characteristics of the limestone vary from well cemented to porous and well indurated possessing angular phosphatic fragments, shell casts, and fossil molds. Increased permeability due to the existence of fractures or solution cavities was encountered in a number of wells. The silty, clayey sands in the lower 20 feet (6 m) of the surficial aquifer and approximately 35 feet (10.7 m) of Hawthorn clays combine to act as an upper aquitard. Approximately 50 feet (15.2 m) of Hawthorn clays act as a lower aquitard and separate the H1 and H2 aquifers.

The H2 aquifer is located in the lower portion of the Hawthorn Group. The H2 is encountered at depths ranging between 165 feet to 170 feet (50.3 to 51.8 m) bls. The H2 aquifer varies in thickness and depth across the study area with a maximum thickness of approximately 40 feet (12.2 m) and a minimum thickness of approximately 15 feet (4.6 m). The lithology found within the H2 aquifer characteristically consists of a phosphatic, sandy limestone which ranges in porosity from low to high. Areas of high porosity are generally related to that portion of the rock where fossil molds are present or where the rock is relatively soft and fossiliferous. The confining clay layer between the H2 and Floridan aquifers is a maximum of 10 feet (3 m) thick and is not continuous across the site.

The depth to the Floridan aquifer ranges from approximately 190 to 225 feet (57.9 to 68.6 m) bls and it is estimated to be over 400 feet (122 m) thick in the study area. The Floridan aquifer was encountered throughout the study area and exists under confined conditions. The lithology found within the Floridan aquifer characteristically consists of a very dense milky white limestone with an abundance of shell fragments. Increased permeability due to the existence of fractures or solution cavities was encountered in a number of wells. The Floridan aquifer is the primary source of potable water for the Gainesville, Florida area.

Source Removal. The RCRA closure activities began in 1983 with the removal of 1,046 drums containing a variety of wastes, primarily asphalt and chlorinated solvents. An additional 345 drums were removed in 1990. No soils were removed during these activities. As part of the CERCLA order, a complete clean closure of the soils was implemented in 1995. A total of 111,000 cubic yards of soil was removed and processed, with approximately 100,000 cubic yards determined to be clean following the on-site processing and 11,000 cubic yards disposed as contaminated media. Another 35 drums were also discovered and removed. The physical closure was completed in 1996 with a modified RCRA cap over the entire 10 acre (4,000 m^2) site.

All residential wells within a 1 mile (1.6 km) radius of the site were geophysically logged. All improperly constructed wells which could act as a conduit for contaminant migration were properly abandoned. Properly constructed Floridan wells were provided if requested by the home owner. These activities removed all known sources of contamination and all known preferential pathways for migration.

SURFICIAL AQUIFER REMEDIATION

The surficial aquifer contained the highest concentrations of contaminants and was acting as a source of the dissolved phase contamination to the underlying aquifers. The highest contaminant concentrations encountered in surficial monitor wells at the site are summarized in Table 1. The cleanup standards were determined by the State of Florida and are the Maximum Contaminant Levels (MCLs) and Practical Quantification Limits (PQLs) listed in the table. The concentrations in the monitor wells proved to be deceptively low. A maximum concentration of 2,100 µg/L of 1,1-dichloroethene was encountered in the influent to one of the surficial aquifer treatment systems.

A variety of technologies were evaluated to remediate the groundwater contamination in the surficial aquifer. Vertical recovery wells and air sparging were eliminated due to the size of the contaminated area and the relatively low permeability of the soils. Another concern was the sedimentary nature of the deposits, which could create a number of perched layers of contamination.

TABLE 1. Surficial Aquifer Contaminants

Contaminant	Highest Concentration, $\mu g/L$
Benzene MCL = 1 $\mu g/L$	10.00
Chloroethane PQL = 2.12 $\mu g/L$	52.60
1,1-Dichloroethane PQL = 1.12 $\mu g/L$	32.60
1,1-Dichloroethene (1,1-DCE) MCL = 7 $\mu g/L$	261.00
1,1,1-Trichloroethane MCL = 200 $\mu g/L$	264.00
Trichloroethene (TCE) MCL = 3 $\mu g/L$	241.00
Vinyl Chloride MCL = 1 $\mu g/L$	1.28

Following a series of pilot tests, a design consisting of 14 horizontal recovery trenches [each approximately 300 feet (91.4 m) long] was selected. Installation of the recovery trenches began by removing the soil overburden in the trench areas to within 2 feet (0.6 m) of the water table with conventional excavation equipment. The recovery trenches were installed with a special trenching machine which cuts the trench, places the perforated piping, and fills the trench with clean backfill in one operation. The trenching machine can cut a trench 28 feet (8.5 m) deep, so the trenches extend 26 feet (7.9 m) beneath the water table. The native soil removed to create the trenches was left in the bottom of the overburden excavation and covered with the previously excavated native overburden. The ends of the trenches were fitted with vertical standpipes so conventional submersible pumps and level controls could be used for groundwater recovery. The design flow rate for each of the 14 trenches was 81 gallons per minute (gpm) (306.6 liters per minute, Lpm). Actual production rates at startup ranged from 35 to 105 gpm (132.5 to 397.4 Lpm).

Packed tower air strippers are used to treat the recovered groundwater. The treated effluent is discharged to a man-made retention pond created during the site backfilling operation. The surficial aquifer system was started in December 1995 and has successfully treated 190 million gallons (719 million liters) as of February 16, 1998. Water levels in the area have been drawn down 10 to 12 feet (3 to 3.6 m), and the flow rates have correspondingly reduced. The area of contamination has been reduced to two monitor wells, with a maximum concentration of 27 $\mu g/L$ 1,1-DCE. Pilot tests for soil vacuum extraction to address the unsaturated zone created by the pumping operation and in-site chemical oxidation by hydrogen peroxide are planned to complete remediation of the surficial aquifer.

H1 AQUIFER REMEDIATION

Groundwater contamination in the H1 aquifer extended from 0.5 miles (0.8 km) north of the site to 1.5 miles (2.4 km) south of the site. The highest contaminant concentrations encountered in the H1 monitor wells are summarized in Table 2. As previously noted, over 100 residential wells were also contaminated. The open hole interval for these wells typically began in the H1 aquifer and extended to the depth necessary to produce a usable amount of water (usually into the H2 and Floridan aquifers).

TABLE 2. H1 Aquifer Contamination

Contaminant	Highest Concentration, μg/L
1,1-Dichloroethane PQL = 1.24 μg/L	36.40
1,1-Dichloroethene MCL = 7 μg/L	96.50
Methylene Chloride PQL = 4.0 μg/L	15.50
Trichloroethene MCL = 3 μg/L	13.30

Due to the depth and areal extent of contamination, vertical recovery wells were the only feasible recovery option. Two on-site H1 aquifer recovery wells and five H1 aquifer injection wells north of the site were installed. The low permeability limestone encountered in four of the five injection wells precluded their use for injection. The fifth well was installed in a cavernous zone that was subsequently determined to be contaminated. That well was converted to a recovery well. Pumping rates from each of these three wells range from 60 to 90 gpm (227 to 341 Lpm). Pumping the three wells resulted in a reduction in the potentiometric surface beneath the site of approximately 20 feet (6 m).

These three recovery wells satisfactorily adduced the on-site contamination. Previous assessment work indicated very low permeability in the H1 aquifer to the southwest of the site, so recovery wells were not initially installed in this area. After startup of the on-site system, a properly constructed H1 aquifer residential well was located in the area of contamination southwest of the site. This well was capable of producing 25 gpm (94.6 Lpm) and was used to recover groundwater from the off-site areas of contamination.

All recovered groundwater is treated by packed tower air strippers. The effluent from the on-site wells is discharged to the man-made retention pond and the effluent from the off-site residential well is discharged to a Floridan injection well. The H1 aquifer system was started in February 1996 and has successfully treated 150 million gallons (568 million liters) as of February 16, 1998. The area of contamination has been reduced by over 75%. The highest remaining concentration of 1,1-DCE is 34 µg/L.

H2 AND FLORIDAN AQUIFER REMEDIATION

The contamination in the H2 aquifer extended from the creek north of the site to 1,000 feet (305 m) south of the site, and was also found in an isolated area east of the site. The contamination in the Floridan aquifer was found in three isolated areas south and east of the site. The contaminant concentrations in both aquifers were generally low (maximum 44 µg/L 1,1-DCE and 8 µg/L TCE) and all contamination is believed to be a result of point sources from improperly constructed residential wells. One off-site H2 recovery well and three on-site H2 aquifer recovery wells were installed. Two of the three on-site wells were remediated to below MCLs by 24-hour pump tests. The third on-site well was remediated to below MCLs after 9 months of operation. No contaminant rebound has been observed in any of these wells. The off-site H2 aquifer recovery well is in the same location as the easternmost Floridan aquifer recovery well. This well is still in operation and exhibits a 1,1-DCE concentration of approximately 20 µg/L.

Three off-site Floridan aquifer recovery wells and six Floridan aquifer injection wells were installed to address the three isolated areas of contamination in the Floridan aquifer. One well was remediated to below MCLs after 30 days of operation and a second after 90 days of operation. No contaminant rebound has been observed in either of these wells. The third recovery well is still in operation and exhibits a 1,1-DCE concentration of approximately 9 µg/L. The H2 and Floridan aquifer systems were started in June 1996. As of February 16, 1998, a total of 27.5 million gallons (104 million liters) had been recovered from the H2 aquifer and 210 million gallons (795 million liters) from the Floridan aquifer.

SUMMARY

Treatment systems to address the four contaminated aquifers were installed and started between December 1995 and June 1996. A total of 577.5 million gallons (2.2 billion liters) has been successfully removed and treated as of February 16, 1998. The area of surficial aquifer contamination has been reduced from 70 acres (28,000 m^2) to two monitor wells, the contaminant plume in the H1 aquifer has been reduced by over 75%, and the contaminant plumes in the lower two aquifers have been reduced to one isolated area. Monitoring data from the lower aquifers indicate that no rebound effect is being observed in the remediated areas. The estimated time to complete groundwater remediation at the site is two to four years. As a result of these activities, the contamination threat to the public wellfield has been removed.

PLUME STABILIZATION USING PUMP AND TREAT TECHNOLOGY

Dudley Patrick (U. S. Navy, Charleston, South Carolina)
Roy Hoekstra (Bechtel Environmental, Inc., Oak Ridge, Tennessee)

ABSTRACT: Pump and treat technology was used to stabilize a chlorinated solvent plume at a U. S. Navy site in Key West, Florida. Solvents were routinely released for many years to the soil surrounding the site, a former jet engine test facility. Concentrations of trichloroethylene (TCE), cis 1,2-, and trans 1,2- dichloroethylene (DCE) were found in the range of 3-41 microgram/liter (μ/L), 74-1560 μg/L, and 170-2800 μg/L, respectively. The Navy provided evidence that the plume was naturally biodegrading and therefore proposed natural attenuation as a final remedy. The Environmental Protection Agency (EPA) Region IV and Florida's Department of Environmental Protection (FDEP) rejected the proposal, directing instead that the Navy begin immediately to address the solvent plume, which was migrating towards a large saltwater inlet, connected to the Atlantic Ocean. The Navy designed and installed a simple pump and treat system. Calculations showed that it would take one to two years to stabilize the plume. Pumping began in July 1996 and was terminated in June 1997 after effluent results demonstrated that objectives had been met. Despite a growing trend towards other technologies, this much-maligned technology continues to offer Navy engineers and managers an alternate means to address groundwater contaminant problems.

INTRODUCTION

Site Description. The Jet Engine Test Cell at Naval Air Station Key West is located on Boca Chica Key, in the lower Florida Keys. Boca Chica Key is a hazardous waste treatment, disposal, and storage site under Resource Conservation and Recovery Act (RCRA) permit signed into effect in 1990. Since the mid-1980's, jet engines were brought to the Test Cell, mounted on concrete pads, where they were repaired and tested. Chlorinated solvents and degreasers were routinely used during maintenance and testing, and some of it leaked into the surrounding soil. Reportedly, these types of substance were not used after 1995. However, the Test Cell itself was not shut down until 1997.

The Test Cell is situated (Figure 1) near a runway in a remote area off the Boca Chica airfield. It is bordered to the south by an asphalt road that parallels a runway and to the east and west by grassy areas. To the north, a large saltwater inlet empties into the Atlantic Ocean. Despite its remoteness, the site is in an environmentally sensitive location with its close proximity to the inlet. The saltwater inlet is a spawning ground for many threatened and endangered species which habitat the Florida Keys National Marine Sanctuary, recently established in this area of the Atlantic Ocean.

The geology and hydrogeology of the site were determined from soil borings and monitoring wells. Oolitic limestone and medium to fine grain sand were interspersed with abundant shell fragments down to the termination of the

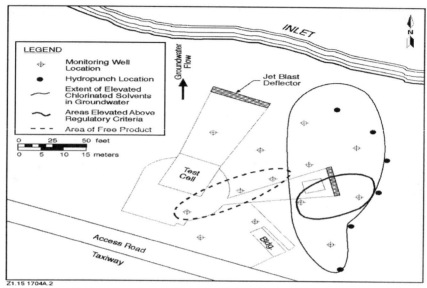

FIGURE 1. Jet Engine Test Cell at Naval Air Station Key West

soil borings at 20 feet (6.0 m) below land surface (BLS). The hydrogeologic unit associated with the oolitic limestone is the surficial aquifer. Recharge to the aquifer is directly through rainfall. Groundwater flows predominately to the north towards the inlet, with some tidal influence. Groundwater elevations were measured to be 1 to 3 feet (0.3 to 0.9 m) BLS. (U.S. Navy, 1997a)

Problem. The problem at the Jet Engine Test Cell was groundwater contamination due to accidental releases of chlorinated solvents over many years. In addition, there was a minor jet fuel spill in the late 1980's. A further exacerbating condition was that the plume was determined to be migrating towards the large saltwater inlet to the north (Figure 1). The Navy determined that the solvents were biodegrading and proposed (Bechtel Environmental, Inc., 1995) to the State and EPA regulatory agencies that natural attenuation would likely finish off any remaining contaminants before they were able to enter the saltwater inlet. The regulatory agencies rejected the proposal and directed the Navy to address the potential migration of the solvent plume towards the inlet.

MATERIALS AND METHODS

Selection of Remedial Technology. Several different technologies were considered for the interim remedial action at the Jet Engine Test Cell site. These included air sparging, bioslurping, natural attenuation, and pump and treat. Air sparging and bioslurping were rejected as not providing hydraulic control for the chlorinated solvent plume. Additionally, the low hydraulic conductivity of the soils at the site and high water table levels were also reasons that these two technologies were not considered further. Natural attenuation was rejected by the state regulators because of concerns of the free product at the site and the migration of the chlorinated solvent plume toward the inlet. Pump and treat was selected for the interim remedial action for the site.

Groundwater Modeling. The groundwater-modeling program *Visual Modflow* was used to model the aquifer and to estimate the effect the pump and treat system would have on the contaminants at the site. The modeling was used to determine the following: recovery well quantities and placement, flow rates required to provide hydraulic control of the site, recharge gallery location and sizing, duration of pumping to achieve remedial goals, and estimated contaminant loading with time.

The model used the hydrogeologic data collected from slug tests and a pumping test conducted at the site. The model used the following aquifer characteristics: horizontal hydraulic conductivity of 13.46 ft/day (4.1 m/day), vertical hydraulic conductivity of 1.346 ft/day (0.41 m/day), specific storage of 0.0001 ft^{-1} (0.00003 m^{-1}) a porosity of 0.35 and an effective porosity of 0.25. (Bechtel, 1995)

The groundwater model developed for the Jet Engine Test Cell site was based on existing site conditions. A model domain of 260 by 260-foot (79.2 by 79.2-meter) area was used with the north boundary being the inlet. The north and south boundaries were modeled as constant head boundaries and no-flow boundaries were used to model the east and west boundaries. The surficial aquifer was represented in the model as a 15-foot (4.6-meter) thick upper system and a 20-foot (6.1-meter) lower system. The vertical extent to the surficial aquifer was divided into six layers. The model was calibrated to match the existing groundwater gradients.

The modeling effort proved to effectively match the existing conditions. The drawdowns and hydraulic control that the remedial system provided were very close to those predicted by the modeling. The recharge gallery was of particular concern at this site. The site had a very high water table of 1 to 3 feet (0.3 to 0.9 meters) below grade and a low recharge rate. One of the regulatory requirements for the recharge gallery was that the treated water could not breach the ground surface. The model was used to optimize the configuration and location of the recharge gallery and to ensure that excessive groundwater mounding did not occur during system operation. Several different schemes were modeled and the recharge gallery was designed to provide the best recharge for the least construction costs.

Using the model, three four-inch diameter recovery wells were located as shown on Figure 2. Two recovery wells (RW-1 and RW-2) were required in the area of the floating product with a flow rate of 1.25 gallons (4.73 liters) per minute. One recovery well (RW-3) was required in the solvent plume with a flow rate of 2.5 gallons (9.5 liters) per minute.

Groundwater Recovery System Design. The groundwater recovery and treatment system consisted of the following elements:

- A groundwater extraction system with three total fluid, top loading, pneumatically operated recovery pumps, associated tubing and instrumentation and controls.
- An oil/water separator to remove any recovered free product from the recovered groundwater.
- A low profile air stripper to remove the volatiles from the groundwater. The discharge limits for groundwater were the drinking water maximum contaminant levels (MCL). The estimated maximum air emissions from the system were 3-lbs (1.36-kilograms) per day, since this was below 15-lbs (6.8-kilograms) per day threshold, air emissions treatment was not required.
- A pumping and distribution system to pump the treated water to the recharge gallery.
- A recharge gallery (Figure 2) consisting of 630 feet (192 meters) of perforated pipe in gravel filled trenches. The recharge gallery covered an area of 180 feet by 50 feet (54.9 meters by 15.2 meters).

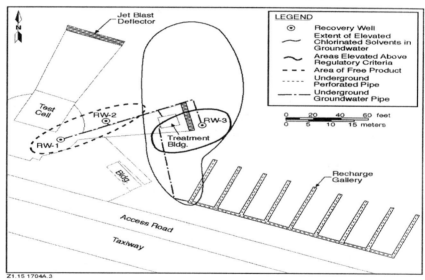

Z1.15 1704A.3

FIGURE 2. Piping schematic showing recharge gallery

RESULTS AND DISCUSSION

System Operation. The system was designed with automatic controls to allow safe operation and to provide automatic shutdown should a problem occur. The pneumatic pumps were able to deliver the required pumping rates. The air stripper was able to reduce the contamination levels to below the cleanup goals. This system was installed very close to the inlet and the water had high salinity. Even though the system was specified for this type of environment, corrosion was a problem throughout the operation of the system.

Results. There were four contaminants of concern at this site that were in concentrations above MCL: TCE; cis 1,2-DCE; trans 1,2-DCE and benzene. Additionally there was a very thin layer of free product in two of the monitoring wells at the site. TCE was considered to be the indicator contaminant, because of the low clean up limit and high concentrations. TCE showed the highest percentage of reduction. (Table 1 and Figure 3)

TABLE 1. Solvent concentration reductions over the last two months

Contaminant	Highest detection	Average detection last two months of operation	% reduction
TCE	342 µg/L	37 µg/L	89%
cis 1,2-DCE	763 µg/L	331 µg/L	56%
trans 1,2-DCE	1410 µg/L	739 µg/L	47%
benzene	22.9 µg/L	5 µg/L	78%
Petroleum free product (thickness)	0.52 ft (0.16 m)	0.1 ft (0.03 m)	81%

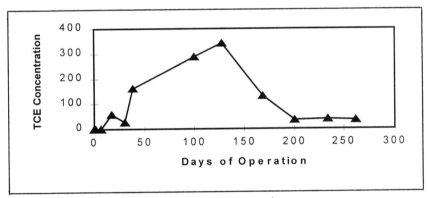

FIGURE 3. Changes in TCE concentration over time

Unexpected Occurrences. Several unexpected events threatened to impact the Navy's plans for meeting objectives within a 12-month period. First, groundwater chemistry results revealed that calcium and magnesium were present in concentrations high enough to cause scaling. A dissolving block sequesterant system was added. Second, a pneumatic transfer pump failed due to inadequate lubrication of its internal parts. The lubrication system was upgraded. Third, severe electrical storms in the area caused power surges that damaged motors and controllers. The motors and controllers were replaced, and a surge suppressor was installed. Lastly, and most significantly, a black slime or biomass began growing in the wells and began covering internal surfaces of the pumps and other equipment. The concern was that the biomass would eventually clog the system and cause a shutdown. The solution was to add hydrogen peroxide to the wells once a month to control the growth. (U.S. Navy, 1997b)

Partnering. A concept that has gained national attention, partnering played a key role in the success of this technology and the decision to use it at the Jet Engine Test Cell. The Key West partnering team, as it is called, consists of the Navy Remedial Project Manager (RPM), the FDEP RPM, the EPA RPM, NAS Key West personnel, and representatives from the study consultants and cleanup contractors. The Key West partnering team is empowered to make tough decisions and to take calculated risks if there looks to be a payoff. The decision to use the more conventional approach to the groundwater contamination problem at the Test Cell went against the grain of technical advice being offered from those outside the partnering team. Later, when it was clear that the decision to implement pump and treat technology was being supported by laboratory data, and that objectives had been achieved, the partnering team acted quickly to shutdown pump and treat operations.

Conclusions. The pump and treat system that was installed at NAS Key West's Jet engine test cell was a relative inexpensive interim action that was able to reduce the concentrations of contaminants, especially TCE, over time. The reductions of TCE and free product thickness have reduced the regulators concerns about migration of the plume into the adjacent inlet. The long-term remedy for this site is planned to be natural attenuation. The Corrective Measures Study will identify the next steps to be taken at the site.

REFERENCES

U.S. Navy. 1997a. *Supplemental RCRA Facility Investigation and Remedial Investigation Report for Naval Air Station Key West High-Priority Sites.* Prepared for Southern Division, Naval Facilities Engineering Command, by Brown & Root Environmental, Inc.

Bechtel Environmental, Inc. 1995. *Technical Memorandum: Groundwater Evaluation, Jet Engine Test Cell, NAS Key West, Florida.*

U.S. Navy. 1997b. *Project Completion Report for Delivery Order No. 004, Naval Air Station Key West, Florida.* Prepared for Southern Division, Naval Facilities Engineering Command, by Bechtel Environmental, Inc.

REMOVAL OF VOCS FROM GROUNDWATER USING AIR-SPARGED HYDROCYCLONE STRIPPING TECHNOLOGY

Laurie L. LaPlante (University of Utah, Salt Lake City, Utah)
J.D. Miller and Dariusz Lelinski (University of Utah, Salt Lake City, Utah)

ABSTRACT: The air-sparged hydrocyclone (ASH) technology offers the unique opportunity to achieve efficient removal of volatile organic compounds (VOCs) from contaminated water at a high specific capacity, many times that of conventional air stripping technologies. Originally developed for the fast and efficient flotation of fine particles from suspensions, VOC stripping is the first application of the ASH as a contacting reactor rather than a flotation device. After initial research revealed that the ASH stripping system performed well in a laboratory setting, field tests were conducted at the Tooele Army Depot to determine the effects of typical groundwater chemistry on VOC removal capabilities. The Tooele site is currently treating VOC contaminated groundwater via conventional aeration techniques. Initial results indicate that the ASH technology has comparable efficiency while operating in a space many times less than that required for conventional air stripping equipment of similar capacity. Effects of fouling due to build-up of scale forming salts were also examined.

INTRODUCTION

Chlorinated hydrocarbons are among the most toxic of the volatile organic compounds found in industrial wastewaters and other contaminated water sources. Groundwater can become polluted with VOCs from waste disposal sites, surface chemical spills, leaking underground tanks, and above ground storage tank failures. Groundwater contamination is a critical environmental issue, since groundwater is the source of drinking water for 50% of the citizens in the US and 97% of the rural population. In addition, most organic chemicals have characteristics that make them a genuine health risk and difficult and costly to clean up in groundwater. A cursory review of the National Priority List (NPL) indicates that approximately 68% of all Superfund sites in the continental United States have VOC contaminated groundwater. Clean up efforts on some of these sites require treatment at a rate of more that 10 million gallons per day for up to 30 years. With such extensive VOC contamination nationwide, the need for improved high efficiency, high capacity treatment is evident.

Objective. Previous fundamental research into the air-sparged hydrocyclone (ASH) stripping performed at the University of Utah pilot plant has demonstrated success of the ASH technology in air-stripping VOCs from contaminated waste water streams.[1] The intent of this current research is to (1) determine the feasibility of using the ASH technology for removal of VOCs from contaminated groundwater under actual field conditions, and (2) compare the effectiveness of the ASH technology to other pump-and-treat systems in use. The first stage of this testing took place at the Tooele Army Depot in Tooele, Utah from November to December 1997. This site, which has approximately 38 billion gal (136 billion L) of trichloroethylene (TCE) contaminated groundwater, is currently being treated via two packed column air strippers at a rate of 7500 gpm (26,395 L/min). This treatment, which began in 1993, is scheduled to continue until the year 2023.

Site Description. The Tooele Army Depot (TEAD) north site is located approximately 35 miles southwest of Salt Lake City in the southern portion of the Tooele Valley. From 1968-1988, approximately 125,000 gal (473,125 L) of industrial waste and storm water was discharged daily to four unlined ditches which flowed to an unlined Industrial Waste Lagoon. These disposal practices created a groundwater plume measuring approximately 2.5 miles (4.0 km) long by 1.5 miles (2.4 km) wide, contaminated primarily with TCE. Sixteen extraction wells are currently being used to collect the water, which is treated by an air stripping treatment facility which utilizes two 50 ft (15.2 m) high randomly packed towers. The average combined influent TCE concentration to the stripping towers is approximately 26 µg/L and the stripper effluent concentration is consistently non-detect, i.e., below 1 µg/L. The maximum contaminant level (MCL) for TCE as set forth in the National Primary Drinking Water Regulations is 5 µg/L. The treated groundwater is recharged back into the aquifer via 13 injection wells.

MATERIALS AND METHODS

The ASH unit consists of two concentric right-vertical tubes and a conventional cyclone header at the top (Figure 1). The inner tube is a porous tube through which air is injected (or sparged); the outer nonporous tube simply serves to create an air jacket. Contaminated water is fed tangentially at the top through the header to develop a swirl flow on the porous tube wall, leaving an air core centered on the axis of the ASH. The high velocity swirl flow shears the injected air to produce a high concentration of small bubbles. As a result of the intimate interaction between these numerous fine bubbles and the contaminated water, chlorinated hydrocarbons in the water are stripped and transferred by the air flow, which exits through the vortex finder of the cyclone header. Water stripped of its contaminants is discharged as an underflow product into a receiving tank.

Mobile Unit Description. In order to conduct the field testing, a self-contained

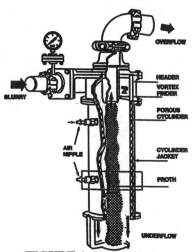

FIGURE 1. Schematic drawing of air sparged hydrocyclone.

mobile ASH stripping unit was designed and built (Figure 2). Contaminated water from the TEAD site was obtained from a surge tank, which held the influent water for the stripping towers. Elevation head fed water through flexible tubing from the surge tank to the mobile unit holding tank (T-1), where it was pumped through the first 2-in (5.08 cm) dia. ASH (ASH-A). Stripped water leaving the

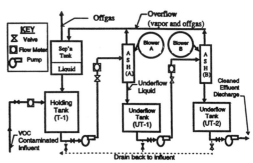

FIGURE 2. Mobile ASH system flow diagram.

underflow of the first ASH was collected in an underflow tank (UT-1). For 2-stage operation, the water was then pumped to the second 2-in (5.08 cm) dia. ASH (ASH-B), where it was stripped again and collected in an identical underflow tank (UT-2). The cleaned water was then discharged back to the stripping tower for additional treatment. The ASH overflow (vapor and gas) was separated in a separation tank and the liquid was returned to the holding tank. The offgas was discharged to the atmosphere in accordance with local air pollution control regulations. The water flowrate to each ASH was controlled by manually operated globe valves and was measured using a pitot tube flowmeter located on the discharge side of each pump.

Air was supplied to each ASH via identical regenerative blowers (EG&G Rotron DR/CP 606). When necessary, air flowrates were measured using a single rotometer, which was positioned to allow for the airflow measurement from either blower to the corresponding ASH. When maximum airflow was required, air from a single blower was routed directly to the corresponding ASH, by-passing the rotometer and subsequently reducing unnecessary resistance in the line. Airflow at the blower outlets was adjusted using manually operated globe valves. Air pressure in the air plenum of each ASH was measured with a pressure gauge. Water samples for VOC analysis were collected from the influent holding tank and from each underflow tank.

Sampling and Analysis. Water samples were collected in 40-ml glass sample bottles with Teflon-lined rubber septa and screw caps. Care was taken to insure that no vapor space remained in the bottles and samples were kept at 4°C until analysis, which was never longer than seven days after collection. The frequency of sample collection varied depending on the phase of the experimental program (described below). Samples were analyzed for TCE by EPA Method 524.2, which has a lower detection limit of 0.5 µg/L and an error of approximately 10%. All results were validated according to EPA Contract Laboratory Program procedures.

Experimental Phases. The experimental program consisted of 2 phases involving both parametric and extended duration testing. The groundwater temperature remained constant at around 18°C and the ambient air temperature ranged from 4 to 16°C. At least one influent sample was taken during each day of the testing. Although the influent TCE concentration to the stripping towers was known to be fairly constant (around 26 μg/L), our analytical values varied from 24 to 30 μg/L. The cause for this deviation was presumed to be a combination of inconsistencies with the sampling method and analytical error. It was therefore decided to average all values taken over the course of the testing, and a value of 26.4 μg/L for influent concentration is used throughout this paper. All phases of testing were performed using an influent water flowrate of 50 L/min.

Phase 1. The objective of this phase of the field tests was to confirm pilot plant test results comparing stripping efficiency to air-to-water ratios, only at influent concentrations 10^3 times lower than used in the laboratory. In order to simulate pilot plant tests, Phase 1 was conducted with the system operating in a single-stage configuration, i.e., the contaminated water passed through only one ASH unit before being discharged. As stated above, the influent water flowrate was kept constant at 50 L/min, while the air flowrate was adjusted from 88, 71, 53, and 35 cfm (2500, 2000, 1500, and 1000 L/min). The corresponding values of air-to-water ratio (Q*) were 50, 40, 30, and 20 respectively. Three samples of the stripped effluent were taken from each underflow tank (UT-1 and UT-2) at each value of Q*, with enough time between sampling events to allow for the cycling of approximately 7 tank volumes of water.

Phase 2. At the conclusion of Phase 1 , new porous tubes were inserted into each ASH unit and the system configuration was modified to allow for 2-stage treatment, i.e., the contaminated water passed through both ASH units before being discharged. The objective of Phase 2 was to simulate relatively long-term operation to determine effects of fouling on removal efficiency and system air pressure. For this phase of testing, the air from each blower to the corresponding ASH unit was unrestricted, i.e., the blowers were operating at maximum output, by-passing the rotometer except for calibration purposes. Pressure readings for each ASH were recorded every 30 min. It was determined that the maximum pressure that could be overcome by each blower (adjusted for elevation and ambient air temperature) was approximately 2.75 psig (18.9 kPa). Phase 2 tests were conducted intermittently for approximately 32 hrs until such time that the pressure in ASH A appeared to reach a plateau at 2.67 psig (18.4 kPa). At this time, the temperature of the internal motor windings of blower A was measured to be approximately 178°F (81.1°C), which was above the manufacturer's recommended value of 70°C. The decision was made to terminate Phase 2 testing at that time. The increase in air pressure was presumed to be a result of the build-up of scale-forming salts (specifically iron and calcium) on the inner surface of the porous tubes. Water samples were collected from each underflow tank every hour for the duration of the testing to determine the TCE removal efficiency following each stage of ASH stripping.

RESULTS AND DISCUSSION

Phase 1. As discussed above, three effluent water samples were taken at each value of Q* and the results averaged. Figure 3 shows the field test TCE removal efficiencies plotted against those obtained during pilot plant testing. The pilot plant results were presented in a previous publication and were explained from fundamental considerations of air-stripping technology.[1] The field test results are particularly significant because they demonstrate that the first order stripping characteristics of the ASH have been extended from pilot plant feed solutions of 500 mg/L down to the field test feed solution of 25 µg/l. As can be seen, field test removal efficiencies for Q* 40 and 30 met or exceeded expected results, whereas those for Q* 50 and 20 were less than expected by 10 and 20%, respectively. Although the exact reason for this is not clear, it should be noted that tests at Q* 40 and 30 were conducted with a porous tube that had formerly been used to treat approximately 27,000 L of groundwater. A new tube was used for testing at Q* 50 and 20, and the results were a decrease in removal efficiency. One possible cause could be manufacturing defects in the porous tube which allowed for an uneven air flow distribution through the pores. As a result, an uneven bubble distribution would reduce removal efficiency. With time, the tube may have become "conditioned" as fouling of the larger pores resulted in a more even air distribution through other portions of the tube. This might explain the higher removal efficiencies for the "conditioned" porous tube.

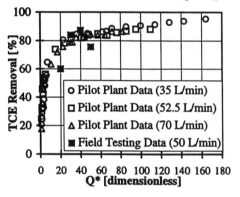

FIGURE 3. Results of Phase 1 testing. Pilot plant TCE removal efficiencies compared to those achieved in the field.

Phase 2. Figure 4(a) shows the increase in air pressure due to fouling of the porous tubes as measured in the air plenum of each ASH over the duration of Phase 2 testing. Air pressure increased from 1.35 psig (9.3 kPa) in both ASH units at the beginning of testing to 2.67 psig (18.4 kPa) and 2.60 psig (17.9 kPa) in A and B, respectively, at the termination of the test. Except for the first 3 hrs, the air pressure in ASH A was consistently higher than that in ASH B by an average of 7%, an indication that fouling was more predominant in the first stage of treatment. The discontinuities in pressure at hours 6, 13, 20, and 23 denote shut-down of the unit for at least 16 hrs. Upon resuming operation, the pressures in both ASH A and B were lower by an average of 17.5% and 20.7%, respectively. This is true except for the shut-down at hour 6 where the pressure in ASH B is

228

shown to be higher at start-up, a discrepancy thought to be a result of gauge reading errors. The largest pressure drop in both ASH units (approximately 25%) occurred at hour 20 following a 2 day shut-down period.

While at this time there is no conclusive evidence as to why the drop in air pressure during system shut-down occurred, several explanations are being investigated. The first is the possible re-dissolving of calcium and iron salts which may have precipitated on the porous tube during system operation. After ceasing operation, the porous tube remains in a saturated state for a period of time, allowing precipitated calcium and iron salts which are blocking the pores to be re-dissolved. When high pressure air is again forced through the porous tube at start-up, it is plausible that even more fouling could be expelled. The combined effect could be a significant reduction in pressure at start-up. A second possible explanation could be the build-up of water in the air plenum during operation due to an assembly defect such as a leaking gasket. Over time, the water level in the plenum would reach a height such that the air flow through a major portion of the porous tube would be restricted, thus causing an increase in air pressure. During an extended period of shut-down, the water would be allowed to drain out of the air plenum and the pressure would subsequently be lower at start-up.

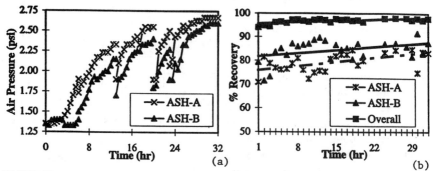

FIGURE 4. Results of Phase 2 testing. (a) Air pressure and (b) TCE removal efficiency versus time.

As discussed earlier, the air pressures in both ASH units seemed to reach a plateau at approximately hour 30, which was taken to be an indication that fouling was no longer continuing to plug the porous tubes at the rate observed at the beginning of the test. Because the pressure increased during operation, it was expected that the air flow had decreased, the effect being a reduction in removal efficiency. Surprisingly, this was not observed. Instead, there was a slight increase in removal efficiency over the duration of the test [Figure 4(b)]. (Note that in the interest of cost control, no samples were analyzed during hours 20 through 24 of the testing.) The average percent recoveries for ASH A and B were 79.5% and 84.4%, respectively, with the average overall percent recovery for 2-stage treatment being 96.8%. A 4.0% rate of increase in the overall recovery was observed over the duration of the test, which may be a result of the "conditioning" of the porous tube as discussed above.

CONCLUSIONS

The ASH stripping system represents a promising technology for the removal of VOCs from groundwater. Field testing has shown that treatment using a 2-stage ASH system can provide TCE removal efficiency comparable to that of a packed tower air stripper (> 96%), but requires an operating space many times less. As a result, preliminary estimates show that a significant savings in capital costs are possible. Additionally, site specific services for small volume flows could be provided with a compact mobile system. Although the potential for fouling of the ASH system components over long-term operation remains uncertain at this time, initial observations indicate that control of hard scale build-up has the possibility to be less problematic than with traditional aeration techniques. There is, however, a need for more long-term field testing to confirm this observation.

ACKNOWLEDGMENTS

This research was financially supported by funding from the Great Plains-Rocky Mountain Hazardous Substance Research Center with support from the U.S. EPA, and by the Utah Department of Environmental Quality, who provided assistance with analytical costs. Appreciation is also extended to ZPM, Inc. for funding of the design and construction of the mobile unit.

REFERENCES

1. Lelinski, D., and J.D. Miller. 1996. "Removal of Volatile Organic Compounds From Contaminated Water Using Air-Sparged Hydrocyclone Stripping Technology." In *Proceedings of the International Conference on Analysis and Utilization of Oily Wastes AUZO '96*, (Gdansk, Poland), supplement pp. 53-62.

COMPARATIVE ANALYSIS OF TRICHLOROETHENE REMOVAL FROM THE SANTA MARGARITA AQUIFER

John Aveggio, P.E., SHN Consulting Engineers & Geologists, Eureka, CA
Brian Bandy, Watkins-Johnson Company, Scotts Valley, CA

ABSTRACT: Long term operational data from a federal Superfund site are evaluated to determine the cost efficiency of each specific remedial process in removing trichloroethene (TCE) from the Santa Margarita aquifer. Five innovative and traditional remedial processes have been used at the site. These are an activated carbon pump and treat system, an aeration pump and treat system, a subsurface drain system, an infiltration and extraction system, and a soil vapor extraction system. Removal costs range from a low of $6,800/kilogram (kg) of TCE removed for the aeration pump and treat system to a high of $110,000/kg TCE removed for the infiltration and extraction system.

INTRODUCTION

A chlorinated solvent release into a wastewater treatment system resulted in the distribution of chemicals, primarily TCE, throughout soil and groundwater at the Watkins-Johnson Company (WJC) facility in Scotts Valley, California. The WJC facility is situated above the Santa Margarita Aquifer, a federally designated sole source aquifer that provides drinking water to the city of Scotts Valley. In response to the release, WJC initiated an investigation to determine the extent of the source area and groundwater degradation. We found that the source area consisted of several distinct septic tanks and their related leachfields, and dry wells which distributed the release across a wide area both horizontally and vertically. Dry wells were found to extend to 45 feet below ground surface (BGS). Infiltration of TCE laden wastewater through the dry wells was a rapid and efficient delivery mechanism to accelerate the transport of TCE to the Santa Margarita aquifer.

The Santa Margarita aquifer consists of medium to course sands with gravel interbeds and is highly transmissive. The vadose zone of the Santa Margarita aquifer is approximately 80 feet thick at the WJC facility. The sand and gravel lithology has a low affinity for adsorption of organic material and resulted in low concentrations of TCE being detected in soil samples. However, the partitioning of TCE into soil gas was found to be a significant source of TCE.

A shallow perched zone exists in the vicinity of the WJC facility at approximately 90 feet BGS. The aquitard that supports the perched zone consists of a moderately cemented conglomerate and is known to be semi-permeable. Its saturated thickness is on the order of a few feet throughout the perched zone. The perched zone hydrology has been locally altered due to the long term disposal of wastewater through the drywell and leachfield systems. A perched zone TCE groundwater plume was found to exist directly below the center of the facility.

The maximum concentration of TCE detected in the perched zone was approximately 13,000 micrograms per liter (ug/l).

The regional zone of the Santa Margarita aquifer is at approximately 150 feet BGS and contains an approximate 40 foot thick saturated zone. A regional zone TCE groundwater plume extends from the facility to approximately one mile downgradient. The plume configuration was a classic cigar shape, which ultimately discharged into the surface water of Bean Creek. The maximum concentration of TCE detected in the regional zone was approximately 5,600 ug/l.

In response to our findings, we initiated a remedial action program to remove TCE from soil, soil gas, surface water, and groundwater to below regulatory levels. A variety of innovative and traditional response actions have been completed at the site:

- A regional zone pump and treat system has removed the greatest mass of TCE and has been operating since 1987.
- A passive subsurface vertical drain system was functional for 6 years and transported TCE from the perched zone to the regional zone for extraction.
- The perched zone drain system was replaced by a perched zone infiltration and extraction system, which has been in operation since 1994.
- A soil vapor extraction (SVE) system has been operational since 1994 and has extracted approximately 25% of the total TCE removed from the site.

The site investigation and remedial action program were overseen by the United States Environmental Protection Agency (EPA) Superfund program.

REGIONAL ZONE ON-SITE PUMP AND TREAT SYSTEM

Regional zone on-site groundwater extraction consists of four pumping wells, which produced up to 350 gallons per minute (GPM). On-site groundwater is processed through a granular activated carbon (GAC) treatment system. The on-site system has been operating since October 1987. Components of the system include four, twelve inch diameter extraction wells, which were completed to 150 feet BGS. The capture zone of the on-site system extends to approximately 40% of the overall plume length. Individual submersible pumps transferred groundwater to a holding tank through an underground plumbing network. Piping runs ranged from 100 to 400 feet and extraction piping diameters were either 2 or 3 inches. Contaminated groundwater exits the holding tank and is pumped through a Calgon Model 10 system, which contains two downflow GAC adsorption vessels, each filled with 20,000 pounds of virgin Calgon Filtrasorb 300 carbon.

Following treatment, water is directed into three disposal options. The first priority is to meet the water supply needs of the facility. Treated water has been shown to meet all state and federal water quality objectives and is used throughout the manufacturing facility. The next use is for infiltration to maintain the groundwater mounds used for hydraulic control of the perched zone. Any remaining water is then discharged to Bean Creek.

The on-site treatment system has been effective in reducing the concentration of TCE in the regional zone. The plume has been significantly reduced in both size and concentration. Approximately 95% of the regional zone wells are now below negotiated cleanup levels and the site may soon be entering the compliance monitoring period. The on-site extraction and treatment system is an example of the successful application of a pump and treat system.

The on-site system has been operational for over 10 years. The capital cost associated with the installation of the on-site system was approximately $750,000. Operational costs for the system average $65,000 annually, and therefore, the total amount spent to date is approximately $1,300,000. The on-site extraction and treatment system has processed approximately 1,011,000,000 gallons of groundwater and the total estimated mass of TCE removed is approximately 225 kg. This yields an average cost per kg of TCE removed as $7,300/kg.

OFF-SITE PUMP AND TREAT SYSTEM

The off-site regional extraction system was in operation from 1987 to 1988. The extraction wells of the off-site system were placed near the leading edge of the plume. Extraction from these wells stopped the discharge of contaminated groundwater to Bean Creek and captured the migrating groundwater which was beyond the range of the on-site system. The off-site system consisted of two, twelve inch diameter extraction wells which extended approximately 50 BGS and produced up to 350 gpm. Individual submersible pumps transferred groundwater through an aboveground six inch diameter piping system, which discharged to a passive aeration system. The passive aeration system consisted of a series of fine mesh screens in which the discharge flowed across by gravity. Treated water was then discharged to Bean Creek.

The off-site system was successful in eliminating the discharge to Bean Creek and in "cutting off the head" of the plume. None of the off-site monitoring wells or the surface water sampling points exhibit detectable concentrations of TCE. The off-site treatment system is an example of a "low tech" solution to a complex problem. This approach was only made possible by cooperation between the various regulatory agencies involved in the project.

The off-site system was operational for approximately 14 months. The capital cost associated with the installation of the off-site system was approximately $50,000 and the operational costs for the system were $15,000. The total amount spent was approximately $65,000. The off-site extraction and treatment system processed approximately 184,000,000 gallons of groundwater and the total estimated mass of TCE removed is approximately 9.6 kg. This yields an average cost per kg of TCE removed as $6,800/kg.

PERCHED ZONE DRAIN SYSTEM

While the aerial extent of perched zone contamination is small compared to the regional zone, a significant amount of TCE is present in the perched zone. The saturated thickness of the perched zone is only a few feet and the hydraulic conductivity is low. Therefore the removal rate achievable by extraction wells in the perched zone is small. The inefficiencies of groundwater extraction in the perched zone led to the development of the perched zone drains.

The perched zone drains enhanced the ongoing natural leakage of the perched zone aquitard by creating an efficient pathway for TCE to move from the perched zone into a regional zone extraction well. The design of the drains consists of a screened area across all of the perched zone and a screened area below the regional zone water table. This allowed contaminated groundwater to flow by gravity from the perched zone directly into the regional zone below, bypassing the longer process associated with vertical flow through the aquitard and into unsaturated portion of the regional zone. The perched zone groundwater would then mix with the regional zone groundwater and would be contained within the capture zone of a regional zone extraction well.

The perched zone drain system was considered an innovative passive remediation strategy, which had minimal operational expense. While the traditional wisdom is to not cross-connect groundwater zones, we concluded that the installed drains only improved upon a natural and inefficient pre-existing drainage system. During the life of the perched zone drain system the average TCE concentration in the vicinity of the drains decreased by approximately 75%.

The drains operated over a six year period. The capital cost associated with the installation of the perched zone drain system was approximately $190,000. The average annual operational cost for the system was approximately $5,000. Therefore the total amount spent was approximately $220,000. The perched zone drain system processed approximately 16,000,000 gallons of groundwater and the total estimated mass of TCE removed is approximately 22 kg. This yields an average cost per kg of TCE removed as approximately $10,000/kg.

PERCHED ZONE INFILTRATION AND EXTRACTION SYSTEM

We found that with an aggressive program of infiltration of treated water, the migration of the perched zone contaminant plume could be hydraulically controlled and that the saturated thickness could be increased to the point where groundwater extraction was feasible. We constructed a U-shaped configuration of perched zone infiltration wells to build groundwater mounds, which funneled the plume toward extraction wells. Thirty five (35) gpm are infiltrated by gravity, through a series of dry wells located at what was once the edge of the perched zone contamination. Six perched zone extraction wells are located within the center of the previous plume area. These wells extract groundwater with a combined flow rate of approximately 10 gpm. Extracted groundwater is piped to the holding tank of the on-site treatment system for processing. The perched zone infiltration and extraction system has reduced the average TCE concentration in the perched zone by approximately 98%.

The perched zone infiltration and extraction system has operated over three years. The capital cost associated with the installation of the perched zone infiltration and extraction system was approximately $185,000. The operational costs for the system average $20,000 annually, and the total amount spent is approximately $245,000. The perched zone infiltration and extraction system has processed approximately 15,000,000 gallons of groundwater and the total estimated mass of TCE removed is approximately 2.2 kg. This yields an average cost per kg of TCE removed as $110,000/kg.

SOIL VAPOR EXTRACTION SYSTEM

For remediation purposes we divided the vadose zone at the site into three layers. Each layer contains a series of monitoring and extraction wells. The extraction wells were manifolded to a common positive displacement vacuum pump and carbon adsorption treatment system. The combined system removes approximately 700 standard cubic feet per minute (CFM) of TCE laden soil gas. The most contaminated zone is the "A" zone which is just above the saturated portion of the perched zone. The middle or "B" zone is the next most contaminated area followed by the shallow or "C" zone, which is lightly contaminated.

The TCE soil gas plume was approximately 5 acres in size, with a maximum soil gas concentration of 140 mg/m^3. Soil gas cleanup levels are based on the threat to groundwater as calculated by the vadose zone transport model VLEACH. VLEACH output has indicated that the C zone has achieved cleanup goals and extraction from this zone has been terminated. Limited soil gas contamination remains in the central portion of the A and B zones. The area is approximately ½ acre in size and the concentrations have decreased by 90%. The SVE system has been successful in removing the vadose zone source of contamination and the system will soon be entering into the compliance monitoring period.

The soil vapor extraction system has operated over three years. The capital cost associated with the installation of the soil vapor extraction system was approximately $360,000. Operational costs for the system average $60,000 annually, and the total amount spent has been approximately $600,000. The soil vapor extraction system has processed approximately 1,052,000,000 cubic feet of soil gas and the total estimated mass of TCE removed is approximately 80 kg. This yields an average cost per kg of TCE removed as $7,500/kg.

SUMMARY AND CONCLUSIONS

The WJC site has successfully used a variety of remedial techniques to mitigate the TCE release. Mitigation costs for these techniques are summarized in Table 1. Four of the methods fell within a narrow cost range of $6,800 to $10,000 per kg of TCE removed. The difference in cost between the on-site regional and off-site regional pump and treat systems can be attributed to the water treatment method. The off-site aeration system was "homemade" and had virtually no operational costs associated with its use, while the GAC system requires periodic carbon changes and a significant capital purchase. Both the regional extraction systems have been successful in achieving aquifer restoration.

TABLE 1. Cost per kilogram of trichloroethene removed for various remedial options.

REMEDIAL TECHNIQUE	TOTAL COST	KG TCE REMOVED	$/KG TCE REMOVED
REGIONAL ZONE ON-SITE PUMP AND TREAT	$1,650,000	226 KG	$7,300
OFF-SITE PUMP AND TREAT	$65,000	9.6 KG	$6,800
PERCHED ZONE DRAINS	$220,000	22 KG	$10,000
PERCHED ZONE INFILTRATION AND EXTRACTION	$245,000	2.2 KG	$110,000
SVE	$600,000	80 KG	$7,500

The SVE system is responsible for approximately 25% of the overall TCE removed to date and has contributed to the overall decrease in the projected length of cleanup. TCE concentrations in the source area have been efficiently removed by the SVE system at essentially the same cost as would be incurred if the TCE was allowed to migrate to the regional zone.

The greatest disparity in cost is associated with removal of TCE from the perched zone, for which there is an order of magnitude difference in cost. The greatly increased cost is due to the infrastructure required for injection and extraction and the increased maintenance required to operate the system. From a strictly economic standpoint, it would seem appropriate to return to the drain system for TCE removal. However, we believe that the overall length of cleanup would be increased. This is due to the longer time associated with moving the TCE plume to the drains and then removing the TCE from the regional aquifer. The infiltration of clean water into the perched zone also flushes out any TCE globules that would have remained out of the capture zone of the drains.

A total of approximately 340 kg of TCE has been removed from the site. This is equivalent to one 55 gallon drum of product and believed to be the amount originally released. At the time of the release, TCE cost approximately $2.00 per kg. It is now being removed at roughly 3,500 times its purchase price.

BENCH-SCALE EVALUATION OF DISSOLVED CHLORINATED SOLVENT CAPTURE USING CATIONIC SURFACTANTS

Bruce M. Sass, Lili Wang, Arun R. Gavaskar, Eric Drescher, Sarah McCall, Bruce Monzyk, and Daniel Giammar (Battelle, Columbus, Ohio, USA)

Abstract: Cationic surfactants have the capability to compete with metal cations for exchange sites on fine-grain mineral surfaces. Sorbed cationic surfactants render these mineral surfaces hydrophobic, thus enhancing their sorption properties for nonpolar and weakly polar organic compounds. The objective of this bench-scale study was to investigate sorption of various cationic surfactants onto natural aquifer materials and to determine their effectiveness for causing trichloroethene (TCE) to partition from groundwater mixtures onto treated mineral surfaces. This study demonstrates the potential of cationic surfactants to immobilize hydrophobic organic substances in contaminated areas and thus prevent contaminant plumes from spreading. It is significant that one highly studied surfactant, hexadecyl trimethyl ammonium chloride, had poorer affinity for a soil than other surfactants tested. This suggests that higher expectations for cationic surfactant treatment may be possible. The most promising surfactants appear to form thick hydrophobic bilayers consisting of long, hydrocarbon chains that provide a large volume within which TCE sorption occurs. This research work suggests that promising new leads should be exploited in more detailed studies and thereafter utilized in actual site remediation field tests.

INTRODUCTION

Cationic surfactants have been reported to significantly enhance the sorptive capacity of soils for hydrophobic organic contaminants (HOCs) and have potential to immobilize HOCs in contaminated areas and thus prevent contaminant plumes from spreading. Previous studies have indicated that quaternary ammonium cations can compete effectively with metals for clay mineral exchange sites, and that sorption of organic cations can cause partioning of HOCs from aqueous solution onto the treated soil (Mortland et al., 1986; Burris and Antworth, 1992). The conglomeration of alkyl chains of the surfactants creates a hydrophobic organic medium at the mineral surface, favoring sorption of nonionic organic contaminants. The effectiveness of quaternary ammonium surfactants on the sorption of nonionic organic contaminants may be determined by the sorption affinities of surfactants to both the soil materials and the organic contaminants. The purpose of this work was to examine the sorption behavior of different cationic surfactants on soils and their enhancement of trichloroethene (TCE) sorption on treated soils. Five quaternary ammonium surfactants were selected for this study to provide a range of surface-sorbed configurations of alkyl and aromatic groups.

MATERIAL AND METHODS

All surfactants and chemical reagents used in this study were of analytical grade or higher. All aqueous solutions were prepared in deionized (DI) Milli-Q grade water.

Surfactant Solutions. Table 1 lists chemical information on five quaternary ammonium surfactants used in this study. Table 1 includes one well-studied surfactant, hexadecyl trimethyl ammonium (HDTMA) chloride, and four less common surfactants whose structures are illustrated in Figure 1. The surfactant solutions were prepared by diluting concentrated surfactants in DI water. The

TABLE 1. CHEMICAL CHARACTERISTICS OF CATIONIC SURFACTANTS

Trade Name	Active Ingredient	Cation MW	Manufacturer
Accosoft 501	Methyl *bis*(tallowamido ethyl)-2-hydroxyethyl ammonium methyl sulfate	800	Stepan Company
Aliquat 336	Tricaprylyl methyl ammonium chloride	425	Henkel Corporation
HDTMA-Cl	Hexadecyl trimethyl ammonium chloride	285	Aldrich Chemical Company
Maquat MC1416	*N*-alkyl (60% C_{14}, 30% C_{16}, 5% C_{12}, 5% C_{18}) dimethyl benzyl ammonium chloride	340	Mason Chemical Company
Maquat SC18	Stearyl dimethyl benzyl ammonium chloride	390	Mason Chemical Company

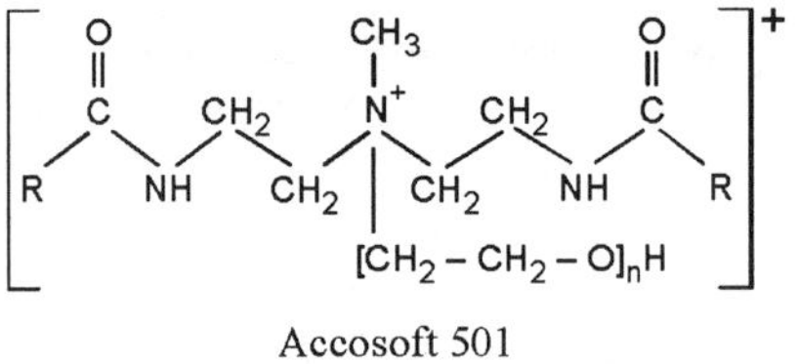

Accosoft 501
R = tallow

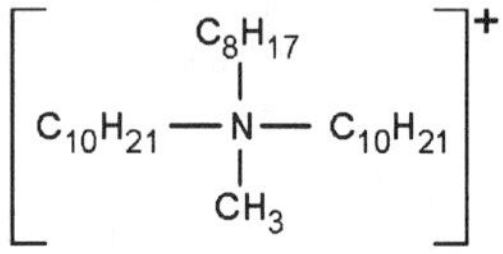

Aliquat 336

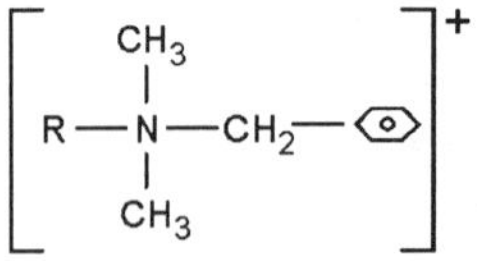

Maquat MC1416
R = *N*-Alkyl (C_{14-16})

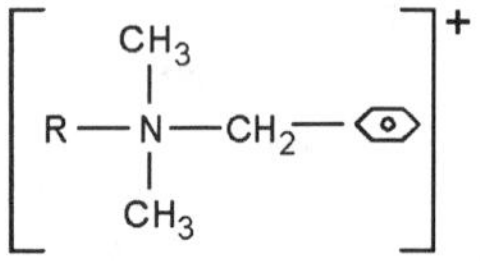

Maquat SC18
R = *N*-Alkyl (C_{18})

Figure 1. Structures of Four Quaternary Ammonium Cations

concentrations of the surfactant solutions were chosen to be below the critical micelle concentration (CMC) to avoid micellization, which might hamper sorption onto mineral surfaces.

Soil Selection. The soils used for this study were collected from pristine sites at Tyndall Air Force Base, Florida (Tyndall soil), and Battelle's West Jefferson Research Center, Ohio (WJ soil). The Tyndall soil was characterized as sandy and the WJ soil was more fine-grained with high clay content. After collection, the soils were air-dried and passed through a 2-mm sieve. The sieved soils were dried in an oven at 105°C prior to use in the experiments. Analyses for residual moisture content, total organic carbon (TOC), and cation exchange capacity (CEC) are summarized in Table 2. These data provide some insight into the soils' native sorptive capacities for HOCs and the potential for enhancement by cationic surfactant treatment. Partitioning of HOCs onto soil generally increases with organic carbon content. Because the TOC of WJ soil is about 9 times lower than that of Tyndall soil, the sorptive capacity of WJ soil for HOCs should be less than that of Tyndall soil. On the other hand, WJ soil should be able to sorb a higher concentration of cationic surfactant molecules because its CEC is about twice that of Tyndall soil. Both of these factors indicate that WJ soil would benefit more greatly from treatment with cationic surfactants than Tyndall soil. Initial tests showed that these assumptions were correct. The remainder of this paper will focus on experiments that were conducted with WJ soil.

TABLE 2. CHARACTERISTICS OF SOILS

Soil Type	Moisture	TOC	CEC
	(%)	(%)	(meq/100 g)
Tyndall Soil	0.43	0.9	3.6
WJ Soil	0.85	0.1	6.3

Sorption of Surfactant on Soil. Surfactant sorption experiments were performed using batch equilibration methods. Measured amounts of soil were added to 100 mL of individual surfactant solutions. The suspensions were shaken for 48 hours at room temperature (23±1°C) in a rotary tumbler. After centrifugation, the surfactant concentration in the supernatant was determined by a modified colorimetric disulfine blue test (Waters and Kupfers, 1976). The amount of surfactant sorbed onto the soils was calculated from the difference between initial and final solution concentrations.

Disulfine blue solution was prepared by dissolving 0.5 g disulfine blue in 25 mL of water. A 250 μL aliquot of disulfine blue solution was added to 50-mL aqueous samples and standards. The mixture was poured into a 250 mL separatory funnel containing 25 mL of chloroform. After shaking for 1-2 minutes and allowing phase separation to occur, the organic layer was withdrawn and its absorbance at 635 nm was recorded to calculate the concentration of the surfactant. This method was capable of detecting cationic surfactants at concentrations as low as 0.1 mg/L. Figure 2 shows the absorbance of a 1 mg/L solution of HDTMA ion in the extract. Absorbance of the chloroform blank (maximum at approximately 655 nm) was subtracted from the background. Absorbance was linear up to approximately 5 mg/L, depending on the surfactant used.

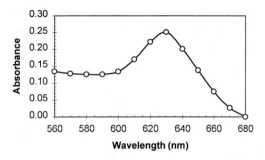

Sorption of TCE on Surfactant-Treated Soils. Surfactant-treated soils were prepared by adding 110 g of dry soil to a ½-gal plastic bottle containing 2 L of a solution with sufficient surfactant to saturate exchange sites. The surfactant concentration was below the CMC. After agitating the bottle in a rotary tumbler for 48 hours, the suspension was rinsed with DI water five or six times, using centrifugation to remove the excess surfactant. Rinsing was stopped when the surfactant concentration in the rinsate was less than 1 mg/L. The rinsed, surfactant-treated soil was then dried in an oven at 45°C and stored in a plastic bottle at room temperature.

Figure 2. Absorbance of a 1 mg/L Standard of HDTMA (Chloroform Blank Subtracted)

Experiments to measure the sorption of TCE by the untreated and surfactant-treated soils were conducted in 10-mL glass test tubes. TCE solutions were prepared with concentrations of 1.0, 2.5, 5.0, 7.5, and 10.0 mg/L by serially diluting a 500 mg/L stock solution. Six (6) mL of each TCE solution was added to 10-mL test tubes containing 10 g each of surfactant-treated soil so that headspace was minimal. The test tubes were capped tightly with Teflon® liners and were placed on an orbital shaker for 48 hours at 2,500 rpm. After shaking, the test tubes were centrifuged to sediment the soils. A 2.0-mL aliquot of the supernatant was immediately transferred into a 2-mL amber glass vial for TCE analysis. Control samples were prepared similarly without soil. In addition, some experiments were conducted with soils that were not treated with surfactant to compare native sorption capability with that of the surfactant enhancement.

TCE in the solution was analyzed by gas chromatography, using a flame ionization detector, GC-FID (HP5890), and purge and trap. The amount of TCE sorbed was determined from the difference between the initial and final concentrations of TCE.

RESULTS AND DISCUSSION

Sorption isotherms for surfactants on WJ soil are shown in Figure 3. These isotherms have a characteristic Langmuir shape, where the amount of surfactant sorbed increases rapidly at low concentrations to plateau values at higher concentrations. Therefore, the data were modeled using the Langmuir equation to describe the sorption of surfactant onto the soil:

$$q = q_m K_q C / (1 + K_q C)$$

In this equation, q is the sorption density (mg surfactant/g soil), q_m is the maximum sorption density, C is the solution surfactant concentration (mg/L), and K_q is the sorption constant (L/mg). The data were fitted to the Langmuir equation using a nonlinear least squares regression routine. Results are shown in Table 3.

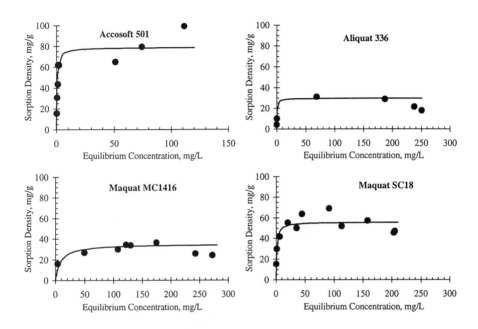

Figure 3. Sorption Isotherms for Cationic Surfactants on WJ Soil

The K_q term in the Langmuir equation is a measure of the affinity of the surfactant for the soil. In Figure 3 it can be seen that surfactants with higher K_q terms have characteristically steeper initial slopes than those with lower K_q terms. Based on the fitted values for K_q (Table 3), the affinity of the surfactants for WJ soil increases in the following order: HDTMA ~ Maquat MC1416 < Maquat SC18 < Aliquat 336 < Accosoft 501. Numerically, the affinities of Accosoft 501 and Aliquat 336 for WJ soil are about 20 times greater than that of HDTMA; the affinity of Maquat SC18 for WJ soil is about half that of Aliquat 336. Maquat MC1416 and HDTMA-Cl had similar affinities for soil.

Selection of a particular surfactant that has an optimum affinity for a soil would be important in a field application, where the surfactant would have to compete for exchange sites with inorganic cations. In an aquifer setting, surfactants with higher K_q would be distributed over shorter distances, making the treatment zone more manageable in size and also reducing cost of materials and monitoring expenses. It is significant that the highly studied surfactant, HDTMA, performs poorest of the set of test compounds with regard to soil affinity, and suggests that higher expectations for cationic surfactant treatment may be possible.

TABLE 3. LANGMUIR FITTING PARAMETERS FOR SORPTION OF SURFACTANTS ON WJ SOIL

Surfactant	K_q (L/mg)	q_m (mg/g soil)	Surfactant Molecules per Exchange Site
Accosoft 501	2.0	79	1.6
Aliquat 336	1.7	30	1.1
HDTMA-Cl	0.1	50	2.8
Maquat MC1416	0.12	35	1.6
Maquat SC18	0.80	56	2.3

The q_m term in the Langmuir equation describes the average density of sorbed surfactant molecules on soil particle surfaces. Table 3 shows that Accosoft 501 has the highest q_m and Aliquat 336 has the lowest. The value of q_m is one indication of a surfactant's ability to enhance partitioning of HOCs. However, the effectiveness of the sorbed surfactant to collect HOCs depends on surface organization of hydrophobic groups as well. To better understand how the surfactants are distributed on soil surfaces, it is useful to compare sorption density to the number of cation exchange sites.

The ratio of surfactant molecules to the number of cation exchange sites was calculated using the fitted values for q_m (Table 3), cation molecular weights (Table 1), and CEC data (Table 2). To a first approximation a unit ratio indicates monolayer coverage, a ratio of 2 indicates bilayer coverage, and so on up to formation of "foam" film layering. However, due to surface heterogeneity, actual surfaces may not contain discrete layers, but may have regions where different types of layering are favored. According to the calculations shown in Table 3, only Aliquat 336 seems to exhibit near monolayer coverage, which may stem from its tipodal branching and short hydrocarbon chain lengths, which could inhibit formation of multi-layer structures. Accosoft 501 and Maquat MC1416 indicate a mixture of monolayer and bilayer coverage, and HDTMA-Cl and Maquat SC18 exceed bilayer coverage. In general, bilayer films do not favor HOC partitioning because formation of admicelles (tail-to-tail configuration) exposes a charged, hydrophilic surface to the aqueous solution. In addition, the highly organized hydrophobic region of these layers prevents HOCs from penetrating into it.

PARTITIONING OF TCE ON SURFACTANT-TREATED SOIL

Sorption isotherms for partitioning of TCE onto untreated and surfactant-treated WJ soils are shown in Figure 4. Results of experiments using HDTMA are not reported here. In Figure 4, lines with steeper slopes indicated greater sorption of TCE than lines with lesser slopes. Comparison of plots for untreated and surfactant-treated soils shows that partioning increases in the following order: Maquat MC1416 < Aliquat 336 < Accosoft 501 < Maquat SC18.

To compare the partioning results, analytically, the data were modeled by a linear isotherm:

$$K_d = C_S/C_W$$

In this expression K_d is the partition coefficient (mL/g soil); C_S is the concentration of TCE sorbed on the soil (µg/g soil); and C_W is the TCE concentration in solution (mg/L). K_d has important ramifications in contaminant transport, because it determines the surface contribution to the retardation factor. Soils with high K_d have a higher retardation factor than those with low K_d.

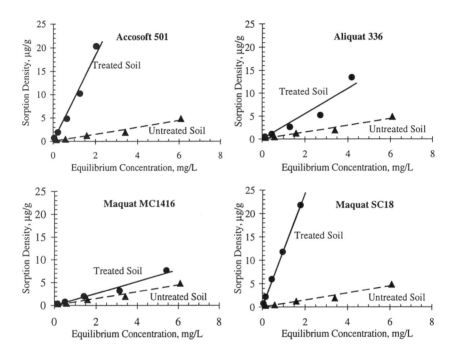

Figure 4. Sorption Isotherms for TCE on Surfactant-Treated WJ Soil

The following relationship is also important:

$$K_{OC} = K_d/f_{OC}$$

where K_{OC} is the partition coefficient normalized for organic carbon (mL/g OC) and f_{OC} is the fraction of organic carbon (g OC/g soil). This expression shows that K_{OC} measures the intrinsic ability of the OC to partition HOCs, whereas K_d measures total partitioning (i.e. depends on mass of OC). In this study f_{OC} was estimated based on q_m (Table 3), rather than measured directly, because of instrument limitations.

Results of fitting the data using linear regression are shown in Table 4. The K_d term shows the combined effects of surfactant density on soil surfaces and the ability of the hydrophobic moieties to stabilize TCE. It can be seen that Maquat SC 18 and Accosoft 501 possess the highest K_d terms, which exceed that of untreated soil by approximately 16 times and 12 times, respectively. Aliquat

336 and Maquat MC1416 were only 4 times and 2 times higher than untreated soil, respectively.

TABLE 4. FITTING PARAMETERS FOR TCE SORPTION ON SURFACTANT-TREATED WJ SOILS

Surfactant	K_d	Relative K_d	f_{OC}[a]	K_{OC}[b]
	(mL/g soil)	(ratio)	(g OC/g soil)	(mL/g OC)
Accosoft 501	9.25	12	0.056	166
Aliquat 336	2.75	4	0.024	114
Maquat MC1416	1.31	2	0.029	45
Maquat SC18	12.2	16	0.047	262
Untreated soil	0.744	1	0.001	744

(a) f_{OC} for surfactant-treated soil was calculated based on q_m (Table 3); f_{OC} for untreated soil was measured.

(b) K_{OC} for surfactant-treated soil was estimated based on measured K_d and calculated f_{OC} values; K_{OC} for untreated soil is not considered to be accurate because f_{OC} was near the detection limit.

An attempt was made to calculate K_{OC} for the purpose of comparing TCE partitioning when each surfactant is normalized by mass. However, as stated above, f_{OC} was not measured directly, and therefore calculations based on these values must be considered approximate. Results are shown in Table 4. For comparison, a typical K_{OC} for TCE sorption by natural humic matter is approximately 100 to 125 mL/g OC (Lyman et al., 1990). The calculations in Table 4 show that K_{OC} for two surfactants, Accosoft 501 and Aliquat 336, is similar to that of natural humic matter. K_{OC} for Maquat MC1416 is approximately half that of natural humic matter and K_{OC} for Maquat SC18 is approximately twice as large. The K_{OC} for untreated WJ soil was based on measured OC content and is not considered to be very accurate because OC was near the detection limit.

CONCLUSIONS

Table 5 summarizes the results of this study by numerically ranking the surfactants according to the parameters that were investigated. It can be seen that Accosoft 501 scores highest in terms of affinity (K_q) for WJ soil, as well as having the highest sorption density (q_m). Furthermore, Accosoft 501 ranks highly in terms of its total (K_d) and intrinsic (K_{OC}) ability to partition TCE. Maquat SC18 ranks highest in TCE partitioning but ranks second in its affinity to the soil. The high affinities of Accosoft 501 and Aliquat 336 for WJ soil is most likely related to the strength of the ionic interaction between the cation group and soil exchange sites. Taken together, it is concluded that enhanced TCE partitioning onto soils treated with Maquat SC18 and Accosoft 501 may be due to particularly advantageous organization of the nonpolar surfactant moieties on soil particle surfaces.

The most promising surfactants appear to form thick hydrophobic bilayers consisting of long, hydrocarbon chains that provide a large volume within which TCE sorption occurs. This model implies that the hydrophilic outer surface of the bilayer must not represent a significant impediment to TCE penetration. The energy barrier to sorption might be overcome by a high amplitude molecular motion and ample equilibration time. As to why thick layers are needed to achieve enhanced TCE sorption, it might be rationalized that the strong dipole of the TCE molecule must be accommodated within the bilayer by allowing it to dimerize with anther TCE molecule, or otherwise interact in a dipole canceling manner to attain a thermodynamically stable state. The dipole may be too large for the shorter surfactants to stabilize, especially if they are only one monolayer thick.

TABLE 5. NUMERICAL RANKING OF SORPTION STUDY PARAMETERS FOR WJ SOIL[a]

Surfactant	K_q	q_m or f_{OC}	K_d	K_{OC}
Accosoft 501	4	4	3	3
Aliquat 336	3	1	2	2
HDTMA-Cl	1	3	NA	NA
Maquat MC1416	1	2	1	1
Maquat SC18	2	3	4	4

(a) 1 = least favorable; 4 = most favorable; NA = not available.

REFERENCES

Burris, D.R. and C.P. Antworth. 1992. "In situ modification of an aquifer material by cationic surfactant to enhance retardation of organic contaminants." *J. Contam. Hydrol. 10:*325-337.

Lyman, W.J., W.F. Reehl, and D.H. Rosenblatt. 1990. *Handbook of Chemical Property Estimation Methods.* American Chemical Society. Washington D.C. Chapter 4.

Mortland, M.M., S. Shaobai, and S.A. Boyd. 1986. "Clay-organic adsorbents for phenol and chlorophenols." *Clays and Clay Miner. 34:*581-585.

Waters, J. and W. Kupfers. 1976. *Analytica Chimica Acta 85:*241.

MULTI-SITE COMPARISON OF CHLORINATED SOLVENT REMEDIATION USING INNOVATIVE TECHNOLOGY

Douglas N. Dreiling
(Burns & McDonnell Waste Consultants, Inc., Wichita, Kansas)
Leo G. Henning and Robert D. Jurgens
(Kansas Department of Health & Environment, Topeka, Kansas)
Deborah L. Ballard
(Burns & McDonnell Waste Consultants, Inc., Kansas City, Missouri)

ABSTRACT: A 90-day pilot test was conducted during 1997 to evaluate three technologies at separate locations in Hutchinson, Kansas to evaluate cost-effective remediation options for removing tetrachloroethylene (PCE) from groundwater. The remedial technologies tested included air sparging and soil vapor extraction (AAS/SVE), ozone and air injection (C-Sparger™), and in-well stripping (NoVOCs™). Each test location consisted of a single or combined remediation well configuration and monitoring wells used to observe PCE removal rates. Actual treatment performance varied between sites, with PCE concentration reduction ranging from approximately 66 to 91 percent during the test period. A cost comparison of the pilot scale technologies indicate that the AAS/SVE system was the least expensive to install and the C-Sparger™ most economical to operate.

INTRODUCTION

In July 1995, the Kansas legislature enacted the Drycleaning Environmental Response Act which established a trust fund for remediation of soil and groundwater affected by chlorinated hydrocarbons originating from former and existing drycleaning facilities. This program is actively managed by Kansas Department of Health & Environment personnel.

In order to compare and evaluate cost-effective remedial technologies that can be applied to various sites across the state, three pilot tests were initiated at three sites in Hutchinson, Kansas in the summer of 1997. The technologies evaluated included air sparging with soil vapor extraction, ozone and air injection using micro-porous sparge points and groundwater circulation, and in-well stripping combined with groundwater circulation. The tests were conducted over a 90-day operation period.

Objectives. The relative effectiveness of each technology was evaluated with respect to contaminant removal, overall cost, ease of operation, and applicability to various locations. Contaminant removal effectiveness of each treatment system was assessed primarily by observing the reduction in dissolved-phase PCE concentrations over time. Cost-effectiveness of each system was determined by evaluating capital hardware and installation costs, on-going operation costs, and labor requirements to operate, maintain, and optimize each system. The final evaluation criteria was the applicability and ease of implementation of a particular remediation system at multiple locations prioritized for remediation.

Site Description.   All pilot test sites were located near former and existing drycleaning facilities within the city limits of Hutchinson, Kansas. The AAS/SVE system was located on private property near 13th and Main Streets, the C-Sparger™ system on state-owned property east of 25th and Main Streets, and the NoVOCs™ system was on private property near Independence Drive and 30th Street (Northgate).

Sediments underlying Hutchinson consist of Pleistocene age unconsolidated stream and terrace deposits comprised of sand, silt, and clay near land surface. These deposits grade to coarser sand and gravel containing thin lenses of silt or clay in the lower part of the formation. Subsurface deposits are laterally persistent across the test sites and consist of clay and silty clay near the surface to approximately 7 feet (2.1 meters; m) below ground surface (bgs). The clay and silty clay grades to very fine-grained sand to coarse-grained sand with gravel at the bottom of the aquifer.

The Equus Beds Aquifer is the main source of usable groundwater in the Hutchinson area. At the test locations, the water table of the unconfined aquifer was encountered from 14 to 16 feet (4.3 to 4.9 m) bgs. The hydraulic conductivity value calculated at one location was estimated to be approximately 500 to 770 feet/day (152 to 235 m/day) with a general hydraulic gradient of 0.001 foot/foot or meter/meter. The resulting groundwater flow velocity was calculated to be approximately 3 to 4 feet/day (0.9 to 1.2 m/day). Dissolved-phase PCE in the vicinity of the source areas appeared to be limited to the top 15 feet (4.6 m) of the aquifer with maximum concentrations at the test sites ranging from 30 to 600 micrograms per liter (μg/L). No free-phase PCE has been encountered in the aquifer.

TEST SET-UP AND OPERATION

Each test configuration consisted of above ground remediation hardware in a temporary enclosure or trailer, a single or combined remediation well configuration, above- and below-grade piping, and groundwater monitoring wells. The placement and construction of monitoring wells varied for each site to accommodate technology-specific data collection requirements.

13th and Main Street Location.   The AAS/SVE remediation system utilized a 4-inch (10 centimeter; cm) diameter polyvinyl chloride (PVC) extraction well completed in the vadose zone to 14 feet (4.3 m) bgs constructed with a 0.020-inch (0.05 cm) slot screen section from 9 to 14 feet (2.7 to 4.3 m) bgs. In addition, a 2-inch (5 cm) diameter PVC air sparging well was completed to a total depth of 42 feet (12.8 m) bgs; the 0.010-inch (0.025 cm) slotted screen section was placed from 39.5 to 42 feet (12 to 12.8 m) bgs. Groundwater information was collected from six monitoring wells installed 25 to 100 lateral feet (7.6 to 30.5 m) down-gradient of the remediation well configuration in three nested locations.

The AAS and SVE wells were connected to hardware within a mobile remediation trailer using flexible hosing with quick-connect fittings. Air was injected into the subsurface through the AAS well at an average rate of 30 standard cubic feet per minute (scfm) or 0.8 standard cubic meters per minute (scmm). Once the off-gas from sparging reached the vadose zone, it was extracted using the SVE well operated at an average rate of 140 scfm (3.9 scmm). Recovered vapor was vented without treatment to the atmosphere within allowable state discharge rates.

25th and Main Street Location. The C-Sparge™ process combines in-situ air stripping with ozone injection. Dissolved-phase chlorinated hydrocarbons are first extracted from aqueous solution and through a gas-to-gas reaction with ozone are converted to hydrochloric acid, carbon dioxide, and water. An ozone/air mixture is injected sequentially into the formation through lower and upper sparge points. Intermittent operation of a submersible pump installed in the well creates vertical circulation of groundwater. The average rate of injection per remediation well during testing was 3 scfm (0.08 scmm).

The C-Sparger™ system included a 4-inch (10 cm) diameter PVC remediation well installed to a total depth of 35 feet (10.7 m) bgs. A micro-porous sparge point was placed in the lower part of the borehole between 33 and 35 feet (10 and 10.7 m) bgs in a 60-mesh filter pack. Four-inch (10 cm) PVC casing was then installed with two 0.010-inch (0.025 cm) screened sections placed from 13 to 18 feet (3.9 to 5.5 m) bgs through the vadose and saturated zone, and in the saturated zone from 23 to 28 (7.0 to 8.5 m) bgs. A self-contained down-hole unit including a second sparge point and fluid pump was then installed in the casing. To better understand the exclusive effect of ozone for PCE removal, a second, identically-configured treatment well was installed to inject only air, instead of ozone, through the sparge points. Both wells were connected to remediation equipment using an above-ground tubing system.

Groundwater information was collected from two clusters of 2-inch (5 cm) diameter monitoring wells situated around the remediation wells. Five monitoring wells were installed as the first cluster from approximately 10 to 40 lateral feet (3.0 to 12.2 m) from the ozone injection well with staggered 0.010-inch (0.025 cm) slot screened intervals. Three monitoring wells comprising the second cluster were installed from 10 to 25 feet (3.0 to 7.6 m) laterally from the air injection well (no ozone). Final well screen intervals were selected to evaluate treatment effectiveness in the groundwater circulation zone.

30th Street and Independence Drive Location (Northgate). The NoVOCs™ process consists of in-well air lifting with volatile organic compound (VOC) stripping. Air lifting and stripping is accomplished by injecting air through a diffuser located inside the well casing. Once the air is introduced, groundwater is "lifted" upward in the well casing. This turbulence converts dissolved-phase VOCs to the vapor phase. A three-dimensional area of groundwater circulation is established using the two screen sections in the well. Groundwater flows into the lower screen installed in the saturated zone and is returned to the aquifer via the vadose zone. This process creates the circular influence in groundwater movement. During operation, air was injected into the remediation well at a rate of approximately 70 to 95 scfm (2.0 to 2.7 scmm) and the water flow rate through the well was approximately 40 to

50 gallons (182 to 227 liters) per minute.

The NoVOCs™ system included one 8-inch (20.3 cm) diameter PVC remediation well installed to a depth of 38 feet (11.6 m) bgs. Two stainless steel 0.010-inch (0.025 cm) screened sections were placed from 8 to 18 feet (2.4 to 5.5 m) bgs across the vadose and saturated zones and in the saturated zone from 28 to 33 feet (8.5 to 10.1 m) bgs. An air diffuser constructed of 1.5-inch (3.8 cm) diameter PVC with 0.060-inch (0.15 cm) screen was placed inside the well casing approximately 4 feet (1.2 m) below static water level. Two, one-half-inch (1.3 cm) diameter PVC casings were installed in the borehole to monitor pH and carbon dioxide levels. An infiltration gallery with a 12-foot (3.7 m) radius and 13 feet (4.0 m) in depth was installed around the well to allow treated groundwater to efficiently re-enter the formation. The remediation well was connected to hardware within a mobile remediation trailer using flexible hosing with quick-connect fittings. Four 2-inch (5 cm) diameter PVC monitoring wells were installed surrounding the remediation well. Each monitoring well was constructed to a total depth of 35 feet (10.7 m) bgs; the wells were located approximately 30 to 80 lateral feet (9.2 to 24.4 m) from the remediation well.

TEST METHODS

Pilot test activities for all sites were conducted over five months and included monitoring well and system installation, pre-test groundwater sampling, a 6-day system start-up period, on-going data collection and operation and maintenance during the 90-day test period, and a post-test groundwater sampling event. During the start-up period, intensive data collection events were performed to record system operation and groundwater parameters. Groundwater information collected included water levels, dissolved oxygen readings, pH, conductivity, and samples for both onsite gas chromatograph analysis and laboratory analysis using EPA Method 8010. In addition, vapor sampling of system influent and effluent was also performed, where applicable. To evaluate ongoing remediation progress, groundwater samples were collected at approximate three-week intervals at each location and submitted for VOC analysis. Routine system operation and maintenance of all systems was performed by Burns & McDonnell personnel with the support of remediation hardware suppliers. A post-test groundwater data collection and sampling event was completed approximately two months following system shut-down to assist evaluation of remediation effectiveness.

TEST RESULTS

Contaminant Removal: Because each test site exhibited varying concentrations of VOCs and each system operated at different flow rates, the comparison of treatment effectiveness was evaluated by comparing the percentage reduction in dissolved-phase PCE concentrations at each monitoring well. Figure 1 shows contaminant reduction rates exhibited at the monitoring well located nearest each remediation well. Contaminant removal rates are presented as a percentage reduction between system start-up (SU), system shut-down (SD), and lowest (Low) PCE concentrations (in µg/L). Results suggest that each treatment system was effective at removing dissolved-phase PCE. Similar, but lesser reductions were observed in remaining monitoring wells at each test site. Treatment systems were located downgradient of source areas and VOCs continually migrated through the treatment area.

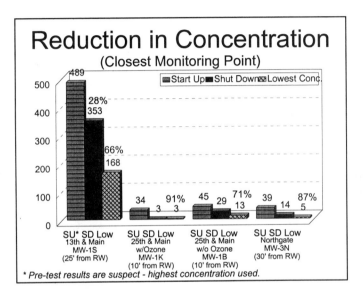

FIGURE 1. Reduction in PCE concentrations (µg/L) observed in the monitoring well located nearest the remediation well for each location.

Installation Cost Comparison. Figure 2 shows installation costs for the pilot test including temporary electrical service, above ground hoses, piping, and fittings, and

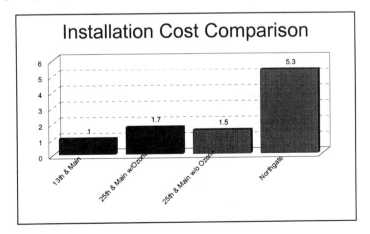

FIGURE 2. "Normalized" installation cost comparison includes electrical service, connections and piping, and remediation well costs.

installation of the remediation wells. The comparison was developed by "normalizing" the actual dollar amount of the cost of each treatment system. For example, the calculated cost for the 25th and Main Street C-Sparger™ ozone system (1.7) is 70 percent higher than the 13th and Main AAS/SVE system assigned a cost comparison factor of one. In general, the simplest installation resulted in the lowest cost; the AAS/SVE system at 13th and Main Street consisted of hardware housed in a mobile remediation trailer and "typical" PVC remediation well construction. The C-Sparger™ wells with and without ozone injection required additional well construction labor and installation of a micro-porous sparge point. The majority of cost for the Northgate location was for installation of the large-diameter NoVOCs™ remediation well. Aside from site-specific underground piping costs, the comparative costs for implementation for a full-scale remediation system should be consistent with those encountered during pilot testing. The costs for monitoring well installation is not included in the comparison.

Monthly Operating Cost Comparison. Figure 3 shows a comparison of monthly operation cost for each system, normalized to an assigned value of 1.0. The monthly operating cost includes electrical consumption, granular activated carbon use, and operation and maintenance labor to effectively operate each system. The comparison shows that the C-Sparger™ system is the most cost-efficient to operate, mainly due to low electrical consumption using 10-volt single phase service. The AAS/SVE and NoVOCs™ systems exhibited similar electrical use; the normalized cost for the NoVOCs™ system also includes the cost for carbon dioxide injected to reduce calcium carbonate precipitation in the well borehole. The comparative costs for the pilot tests should be relatively similar for implementation of a full-scale system.

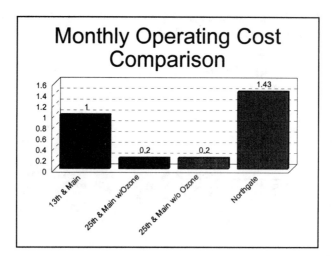

FIGURE 3. This comparison presents monthly operating costs including electrical and granular activated carbon consumption, and O&M labor.

NUMERICAL PERFORMANCE EVALUATION
OF A PILOT SCALE RECIRCULATING WELL SYSTEM (NoVOCs™)

Donald M. Dwight, P.E. (Metcalf & Eddy, Wakefield Massachusetts)
Pier F. Mantovani (Metcalf & Eddy, Wakefield Massachusetts)
Joseph English (Formerly of EG&G Environmental)

ABSTRACT: A pilot test of the NoVOCs™ recirculating well technology was performed at the Massachusetts Military Reservation (MMR) on Cape Cod, Massachusetts. The system consisted of two NoVOCs™ wells located in the center of a TCE plume. The system was monitored for six months using a network of nested monitoring wells. Three-dimensional groundwater flow modeling was performed to evaluate and illustrate recirculation, capture zones and system performance. The model was calibrated to operational drawdowns measured at monitoring wells. Particle tracking was performed to evaluate recirculation patterns, the number of treatment cycles, capture zone, and the size of the recirculation zone. A Dipole Flow Test analytical model was used with measured drawdown data to verify anisotropy values and demonstrate that the drawdown pattern formed is consistent with that of a recirculation pattern. The numerical evaluation results in conjunction with graphical representation of the field data formed the basis of the performance evaluation that was presented to demonstrate the effectiveness of the NoVOCs™ system in treating and recirculating groundwater.

INTRODUCTION

The Massachusetts Military Reservation (MMR, also known as Otis Air National Guard Base) is located on Cape Cod near Falmouth, Massachusetts. Several groundwater contaminant plumes exist on the site, reportedly the result of routine operations and accidental spills on MMR. Pilot programs were undertaken by the Air Force Center for Environmental Excellence (AFCEE) to evaluate the effectiveness of recirculating wells in reducing contaminants to the cleanup goals and in providing plume containment to prevent offsite migration of contaminants. A six-month pilot study conducted from December 1996 to June 1997 by Metcalf and Eddy (M&E) examined the effectiveness of NoVOCs™ in attaining the performance goals for the Site. System operation was performed by M&E and monitoring was performed by AFCEE's onsite contractor, Jacobs Engineering Group (JEG).

NoVOCs™ Technology Background. NoVOCs™ is the trade name for a patented technology for in well stripping of volatile organic compounds (VOCs). It relies on air-lift pumping to extract goundwater from the aquifer in the lower (intake) screen of a dual-screen well, strip VOCs from the groundwater within the well and return the treated water to the aquifer in the upper (recharge) screen creating a toroidal circulation pattern of treated water within the aquifer. Further discussion of the NoVOCs™ technology can be found in Mantovani et. al. (1998).

Objective. The objective of the pilot study was to evaluate the effectiveness of NoVOCs™ at MMR for use in contaminant reduction and plume containment. Field data were collected to document the decrease in groundwater concentrations resulting from NoVOCs™ system operation and to measure the size of the recirculation zone created by the system. Although these data showed impressive performance of the NoVOCs™ system, they were limited in extent due to time and cost constraints. Field data results are presented in Mantovani et. al. (1998). The evaluation of system performance through the use of numerical models is discussed herein. Numerical models were used to further demonstrate NoVOCs™ system performance at MMR by calibrating them with pilot system data and documenting recirculation through illustration of flow paths.

MEANS AND METHODS

Groundwater Flow Modeling. Groundwater flow modeling using MODFLOW and particle tracking using MODPATH were performed to further substantiate that recirculation in the pilot NoVOCs™ system was occurring and to evaluate its effectiveness. Numerical modeling was performed by both M&E and AFCEE's onsite contractor, JEG, to evaluate pilot system performance. M&E's model was taken a step farther than JEG's model by performing extensive model calibration using field operational data. This resulted in a highly predictive model for use in demonstrating system performance. Performance was demonstrated by using the calibrated flow model in conjunction with particle tracking to compute groundwater flow paths and recirculation patterns. As will be discussed below, this exercise demonstrated that under the operational head conditions measured in the field, recirculation of groundwater was occurring.

The model consists of 42 rows, 49 columns and 29 layers with areal dimensions of 1500 feet by 1500 feet and a vertical dimension of 210 feet. The layer properties (i.e. thicknesses, horizontal hydraulic conductivities, K_h, vertical hydraulic conductivities, K_v, and anisotropy ratios, K_h/Kv) were provided by JEG and were based on interpretation of boring logs, soil properties and site information from other studies at MMR. Although differences in layer properties were subtle, several layers were required to attain the resolution necessary to model steep groundwater gradients and recirculation patterns. Based on the boring logs, three zones of varying hydraulic properties were identified (i.e., upper middle and lower zones). The recharge screen was installed within the middle zone and the intake screen was installed within the lower zone. Figure 1 illustrates a cross section of the NoVOCs™ wells, the model layer zones and the seven monitoring depth horizons (A - G) within the pilot area.

The approach to calibration was to compare model computed drawdowns to field measured drawdowns at seven depth intervals within 27 monitoring wells during operational periods of the NoVOCs™ pilot system.

Steady-state model simulations were performed under an iterative process of adjustment of layer parameters in the middle and lower zones, to approach a better match with the field drawdowns. In the end, it was found that small increases in

K_h/K_v from initial conditions were necessary to attain a suitable calibration. In addition, adjustments to K_h and K_v values were necessary between the recharge and intake screens to attain a better match with field observed drawdowns. The initial and final model parameters are shown in Table 1.

Figure 1. Cross section of NoVOCs™ pilot system

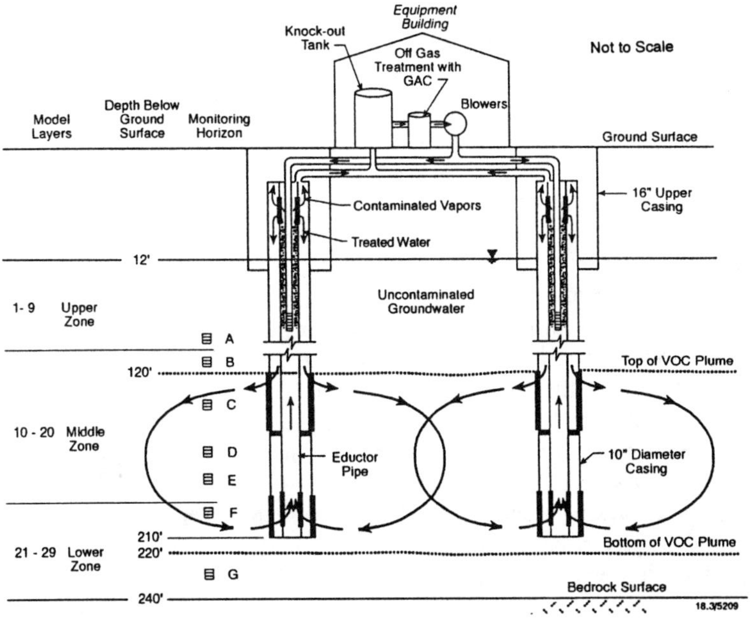

TABLE 1. Calibration Scenarios (Average Values)

Parameters	Average values		
Init. Conditions	Upper	Middle	Lower
K_h (ft/day)	235	162	143
K_v (ft/day)	188	29	32
K_h/K_v	1.25	5.6	4.5
Final Calib.			
K_h (ft/day)	235	159	235
K_v (ft/day)	188	21	48
K_v/K_v	1.25	7.5	4.9

A comparison of the average drawdowns at each depth interval from all 27 monitoring wells is shown in Figure 2. The average absolute value of model

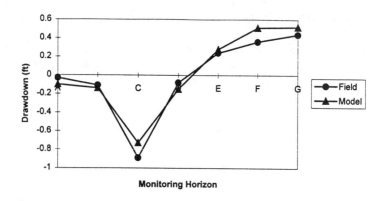

Figure 2. Comparison of field measured and model computed drawdowns

calculated drawdowns at all monitoring points was 0.37 which compares extremely well with the average field drawdown of 0.35 feet.

Particle Tracking Simulations. To illustrate the model flow paths, particle tracking simulations were performed with MODPATH using model-computed heads from the calibrated flow model. Two particle tracking simulations were performed: 1) Particle release from the intake screens tracked backward in time to illustrate the origin of captured particles and the size of the captured zone and 2) Particle release from the recharge screens tracked forward in time to illustrate the recirculation pattern of the NoVOCs™ system and

Particle tacking results for each of these scenarios are shown in Figure 3. Figure 3(a) indicates that the NoVOCs™ system effectively capture all particles within its capture zone (i.e. no particles pass through the system without being captured). Figure 3(a) also illustrates the size of the recirculation zone formed by the NoVOCs™ system and the recirculation radius. This figure indicates that the capture zone of the pilot system has a width of approximately 450 feet and the recirculation radius of each well is approximately 100 feet. Figure 3(b) demonstrates that recirculation of groundwater does occur within the NoVOCs™ system. The figure shows that 34 of the 48 groundwater particle paths are recirculation paths and 14 of the groundwater flow particle paths are down gradient release paths. This translates to a recirculation rate of 3 to 4, meaning that on average, each groundwater flow particle is treated 3 to 4 times before it is released down gradient.

Dipole Flow Analysis. The drawdown data from the pilot test were analyzed using a Dipole Flow Test procedure developed by Kabala (1993). The dipole flow test analysis is based on an exact, closed form solution to the groundwater flow equation for the pressure head distribution in the vicinity of a recirculating well such as a NoVOCs™ well. The objective was to demonstrate that the measured drawdowns match the calibrated dipole model (using calibration parameters within a reasonable range based on other observed/measured site conditions) and thus indicate that the head distribution necessary to create recirculation patterns has been established.

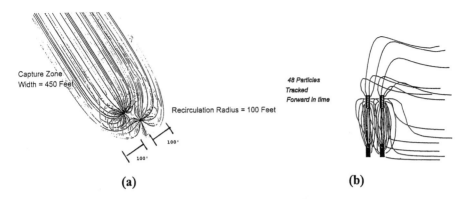

**Figure 3. Particle tracking results. (a) Plan view showing backward
particle tracking from the intake screen. (b) Cross section view showing
forward particle tracking from the recharge screen.**

A secondary purpose was to further confirm the value aquifer anisotropy used in
design calculations and groundwater modeling.

In an initial evaluation of the model sensitivity, an automated parameter
identification (PI) routine was used to calibrate the model by adjusting the two
calibration parameters (K_h and K_h/Kv). The PI routine was set up with the
constraints that the hydraulic conductivity could vary by an order of magnitude and
the anisotropic ratio could vary between 1 and 10.

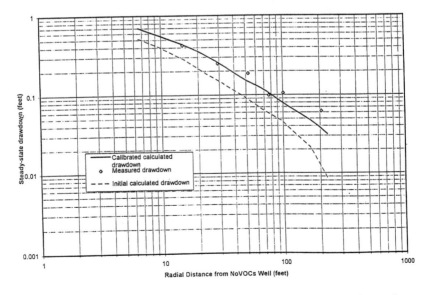

**Figure 4. Comparison of initial results, calibrated results and
observed data for the dipole analysis.**

The best fit solution obtained by the PI routine was with a hydraulic conductivity of 156 ft/day and an anisotropic ratio of 10, which is consistent with site data and the groundwater flow modeling results. A comparison of the initial results, calibrated results, and observed drawdown data are presented in a typical log-log distance-drawdown plot in Figure 4. These calibrated results show a very good match with the observed results. One additional model simulation was performed with an anisotropic ratio of 20. The model results did not fit the observed data as well as the calibrated results using an anisotropic ratio of 10.

RESULTS AND DISCUSSION

The results from the numerical performance evaluation of the NoVOCs™ pilot system at MMR demonstrated that the system was successfully recirculating treated groundwater. This was demonstrated by calibrating a 3-dimensional groundwater flow model to the observed head conditions and illustrating resulting groundwater flow lines. The model demonstrated that plume containment was occurring within the recirculation zone, the recirculation radius was approximately 100 feet and groundwater is effectively recirculated and treated approximately three times before it is released downgradient. Recirculation was also confirmed through calibrating a dipole model. The analysis confirmed that the pressure head distribution developed by the NoVOCs™ system is consistent with a recirculating (dipole) flow regime. The analysis also confirmed values of anisotropy for the aquifer. The numerical results were presented to support an already impressive demonstration of NoVOCs™ effectiveness at MMR through significant reductions of VOCs within the recirculation zone (Metcalf & Eddy, 1997). AFCEE has authorized continued operation of the pilot wells to gain more long term information on the application of NoVOCs™ at this site. The system is still in operation today and addItional performance results are being compiled.

REFERENCES

Kabala, Z. J. 1993. "The Dipole Flow Test: A New Single-Borehole Test for Aquifer Characterization." Water Resources Research, 29, 99-107.

Mantovani, P.F., Donald M. Dwight, Joseph English, 1998. "Pilot Recirculating Well System (NoVOCs™) for Remediation of a Deep TCE Plume." Presented at The First Intenational Conference on Remediation of Chlorinated and Recalcitrant Compounds.

Metcalf & Eddy, Inc., 1997. "CS-10 Pilot Preliminary Operating Data and Result Interpretation." Draft. Presented to Jacobs Environmental and the Air Force Center for Environmental Excellence (AFCEE), Massachusetts Military Reservation.

PILOT RECIRCULATING WELL SYSTEM (NoVOCs™) FOR REMEDIATION OF A DEEP TCE PLUME

Pier F. Mantovani (Metcalf & Eddy, Wakefield Massachusetts)
Donald M. Dwight, P.E. (Metcalf & Eddy, Wakefield Massachusetts)
Joseph English (Formerly of EG&G Environmental)

ABSTRACT: A pilot test of the NoVOCs™ recirculating well technology was performed at the Massachusetts Military Reservation (MMR) on Cape Cod Massachusetts. NoVOCs™ is an innovative technology based on dual-screened recirculating wells. Air-lift pumping and in-well stripping are used to circulate groundwater and remove VOCs. The pilot system consisted of two NoVOCs™ wells located in the center of a deep TCE plume (i.e., CS-10 plume) in a sand and gravel aquifer. The plume is approximately 110 feet thick and exists at the bottom of a 210 feet thick overburden aquifer. The combined recirculation flow rate of the two wells was 300 gpm. Water levels and water quality parameters were monitored for five months using a network of monitoring wells nested at seven depth intervals. After five months of operation, the average TCE concentration within the recirculation zone was reduced by 83%. In-well stripping efficiency averaged 91% over the period and an average of 3 lb/day of TCE was removed by the two wells.

INTRODUCTION

The Massachusetts Military Reservation (MMR), also known as Otis Air National Guard Base is located on Cape Cod near Falmouth, Massachusetts. Several chlorinated solvent plumes exist on the site, reportedly the result of routine operations and accidental spills on MMR. Pilot programs were undertaken by the Air Force Center for Environmental Excellence (AFCEE) to evaluate the effectiveness of recirculating wells in reducing contaminants to the cleanup goals and in providing plume containment to prevent offsite migration of contaminants. A six-month pilot study conducted from December 1996 to June 1997 by Metcalf and Eddy (M&E) examined the effectiveness of NoVOCs™ in attaining the performance goals for the Site. System operation was performed by M&E and monitoring was performed by AFCEE's onsite contractor, Jacobs Engineering Group (JEG).

NoVOCs™ Technology Background. NoVOCs™ is the trade name for a patented technology for in well stripping of volatile organic compounds (VOCs). The NoVOCs™ system remediates VOCs in groundwater using in situ air stripping. The system relies on an airlift system to circulate water through a well and facilitate transfer of contaminants from the liquid phase to the vapor phase. The density gradient produced by diffusing pressurized air into the well causes groundwater to be drawn in through the lower screen of a dual-screened well and lifted upwards inside the well casing. As the water rises, the contact of the air bubbles with the water causes VOCs to be stripped from the groundwater. The vapor phase is drawn off and treated through an above ground air treatment unit and the treated water is returned to the aquifer through the upper screen of the well. The extraction of groundwater through the lower screen and the injection of water through the upper screen causes a groundwater circulation pattern to developed within the aquifer. Within the circulation pattern, groundwater can be recirculated through the well several times based on system design parameters to repeatedly treat groundwater before it is released down gradient. The size of the recirculation pattern and number of recirculation cycles are controlled by aquifer parameter such as hydraulic conductivity and heterogeneity and by design parameters such as pumping rates, submergence depth of the diffuser and vertical screen spacing. The in-well stripping efficiency is controlled by the chemical properties of the contaminant and design parameters such as the air-to-water ratio (AWR).

Site Background.   The pilot system site is located within the CS-10 plume on MMR. The aquifer consists of glacial outwash composed of sands, silty sands and gravels. The average horizontal hydraulic conductivity is approximately 160 ft/day. At the pilot system site, the plume is approximately 220 feet deep. The principle contaminant is trichloroethylene (TCE). The site was originally selected by the client with the intent of evaluating plume containment rather than hot-spot treatment. However, after the system was installed, concentrations as high as 2,700 μg/L of TCE were detected in the vicinity of the pilot wells and as high as 5000 μg/L 200 feet upgradient of the pilot wells, indicating that the site was closer to the hot spot of the plume than originally anticipated.   The groundwater flow direction was also determined by the client to be different from the design basis after pilot system installation, resulting in the system being oriented at a 45 degree angle, rather than perpendicular to the groundwater flow direction.

Objective.   The objective of the pilot study was to evaluate the effectiveness of NoVOCs™ in reducing contaminant concentrations and preventing the offsite migration of contaminants at MMR.   Water quality data measurements were collected to document the decrease in groundwater concentrations resulting from NoVOCs™ system operation, and pressure head measurements were made to determine the size of the recirculation zone created by the system. These data were presented in conjunction with numerical analyses of the data to demonstrate the performance of the system.   The results of numerical analyses are presented in Dwight et. al. (1998).

MEANS AND METHODS

System Layout. The pilot system consists of two NoVOCs™ installed at a horizontal spacing of 100 feet and to a depth of 210 feet. The well design is illustrated in Figure 1. The depth of the plume and the large flow rates required for treatment, necessitated the use of large casing diameter relative to other NoVOCs™ applications. The wells are ten inches in diameter (except for a 16" upper casing) and each have two 15- ft long screens with a vertical separation of approximately 45 feet spanning the thickness of the plume. The wells were constructed with an inner six inch diameter eductor pipe to facilitate recharge in the deep portion of the aquifer. The system was designed to induce recirculation within the plume only and not to impact clean water. The system was operated at an air flowrate of 530 scfm per well to achieve the desired recirculation flow rate of approximately 300 gpm (total) and AWR of 20:1. The AWR of 20:1 was determined to achieve the desired level of contaminant reduction and efficient air-lift pumping. The contaminant-laden air was removed via vacuum blowers, conveyed through moisture separators and treated through granular activated carbon. The system was operated in a closed loop in which the treated air was recycled for reinjecton back down the wells. All blowers and ancillary equipment were located in an aboveground 20' by 20' building .

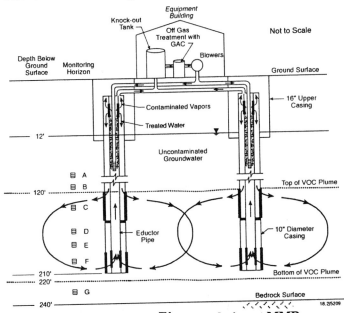

Figure 1. NoVOCs™ system design at MMR

The pilot system monitoring well network, consisting of 27 monitoring wells nested in 16 well clusters is illustrated in Figure 2. The majority of the wells were installed parallel to the alignment of the NoVOCs™ wells at distances of 15 feet and 30 feet downgradient of the NoVOCs™ wells. A cross-gradient and a downgradient well were installed to monitor concentrations near the edge of the recirculation zone. An upgradient well was installed to monitor incoming groundwater concentrations. Monitoring wells were installed at seven different monitoring horizons (see Figure 1). The monitoring wells and NoVOCs™ influent and effluent were sampled on a monthly basis and analyzed for VOCs, pH, iron, manganese and dissolved oxygen.

Data Interpretation. Monthly snapshot vertical profiles of TCE concentrations were created to illustrate decrease in TCE concentrations over time. Vertical contours of piezometric levels measured in the monitoring well network were created and flow lines were drawn to demonstrate the recirculation pattern formed by the pilot wells.   Numerical groundwater flow modeling was performed to investigate system operation and to further demonstrate system performance. The model was calibrated to piezometric heads measured during pilot system operation. The model confirmed that a recirculation pattern was formed and illustrated the extent of the recirculation (treatment) zone through use of particle tracking. Analytical modeling of streamlines and a dipole analysis of recirculating wells were performed to further evaluate the recirculation patterns. The results of these analyses agreed with the numerical modeling results and further confirmed the recirculation patterns of the pilot wells (Dwight et. al., 1998)

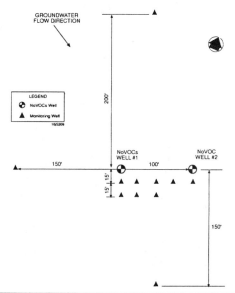

Figure 2. NoVOCs™ pilot system monitoring network.

RESULTS AND DISCUSSION

The average and maximum concentrations of TCE in the treatment zone prior to start up of the system were 1,110 µg/L and 2,700 µg/L, respectively. Figure 3(a) illustrates the average decrease in TCE concentrations within the recirculation zone during the pilot testing period and the associated reduction percentage. Figure 3(b) illustrates average in well stripping performance based on influent and effluent concentrations. Figure 4 illustrates the change in concentration from baseline (prior to system start-up) to the end of the fourth month of operation. As shown in Figure 3(a), after one month of operation, the average TCE concentration within the recirculation zone was reduced by 60% to 440 µg/L. After five months of operation, the average TCE concentration within the recirculation zone was reduced by 83% to 184 µg/L.   Data from individual monitoring horizons indicate that the TCE concentrations in the intervals immediately above and below the plume (i.e., A and

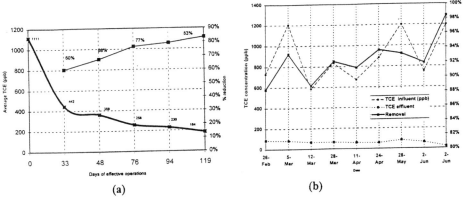

Figure 3. NoVOCs™ pilot system performance. (a) Average TCE reduction and percent reduction in monitored zone. (b) Influent/effluent TCE concentrations and average in-well stripping performance.

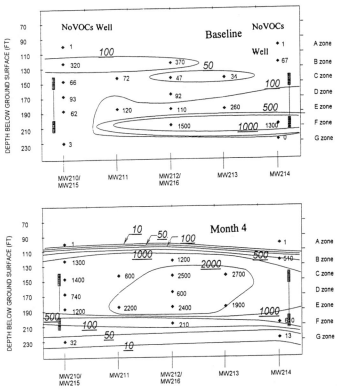

Figure 4. Concentrations before and during NoVOCs™ pilot system operation. (a) Baseline concentrations. (b) Concentrations after month 4.

G monitoring horizons) remained non-detectable, and thus indicating there was no upward or downward migration induced by the circulation of the recharged flows. Concentrations in the intermediate intervals between the extraction and recharge screens decreased 90 to 97%. Although the system was not monitored long enough to directly measure the recirculation radius, numerical analyses performed on the data indicate the fully formed recirculation radius was approximately 150 feet and the capture zone was approximately 450 feet wide.

Interpretation of influent and effluent concentration, shown in Figure 3(b), indicated that the in-well stripping efficiency averaged 91% over the test period and an average of 3 lb/day TCE was removed by the two wells. This translated to a 30 to 1 reduction in concentration. For a short period of time, the wells were operated at a higher air to water ratio to demonstrate the ability to achieve even lower effluent concentrations. Over this period, the wells achieved a stripping efficiency of 98.5%, translating to a 60 to 1 reduction in concentration.

CONCLUSIONS

The NoVOCs™ pilot study performed at MMR demonstrated the effectiveness of the NoVOCs™ technology in remediating a VOC plume. The direct field measurements made during five months of operation conclusively demonstrate that the NoVOCs™ system forms a large capture zone, effectively recirculates groundwater and in the process effectively removes VOCs from the aquifer. The system achieved in-well stripping efficiencies of 91% to 98.5% and removed 300 lbs of TCE from the aquifer over the five month period. Through data interpretation, it was determined that the system was recirculating 70% of reinjected flows, which translates into an average of three stripping cycles for each particle of groundwater.

The radius of the recirculation zone formed by the pilot system was estimated to be 100 feet and the capture zone was estimated to be 450 feet. The system created no measurable disruption of the groundwater levels, and therefore, there was virtually no effect on water supply in the area. AFCEE has authorized continued operation of the pilot wells to gain more long term information on the application of NoVOCs™ at this site. The system is still in operation today and additional performance results are being compiled.

REFERENCES

Metcalf & Eddy, Inc., 1997. "CS-10 NoVOCs™ Pilot Preliminary Operating Data and Result Interpretation." Draft. Presented to Jacobs Environmental and the Air Force Center for Environmental Excellence, Massachusetts Military Reservation.

Dwight, D. M., P. F. Mantovani, J. English. 1998. "Numerical Performance Evaluation of a Pilot Recirculating Well System (NoVOCs™). Presented at The First Intenational Conference on Remediation of Chlorinated and Recalcitrant Compounds.

EFFECTS OF ANISOTROPY AND LOW CONDUCTIVITY ON RECIRCULATION-WELL PERFORMANCE

David Ward and Kent Bostick, Jacobs Engineering, Albuquerque, New Mexico
Jeff Carman, Jacobs Engineering , Otis ANG, Massachusetts

ABSTRACT: Flow-path modeling was used to visualize and interpret the three-dimensional flow and contaminant distribution for three recirculation-well pilot tests. The sites were located in transmissive alluvium up to 300 ft thick, with PCE and TCE present at concentrations up to 1000 μg/L at depths from 60 to 150 ft. Air sparging reduced contaminants in reinjected water to as low as 0–5 μg/L. An array of nested monitoring wells provided a cross-sectional view of groundwater quality and heads between the recirculation wells. Aquifer parameters were estimated from lithologic logs, pumping tests, and regional groundwater models, and were refined by comparing simulated to observed water levels. Model results explained contaminant distributions observed in monitoring wells, showing contaminants moving toward extraction screens while a zone of clean groundwater expanded from the reinjection screens.

Development of recirculating flow is strongly influenced by vertical hydraulic conductivity (K_v) and by anisotropy (ratio of K_v to K_h). In fine-to-coarse sands, K_h was up to 235 ft/day with an anisotropy of 1.2, yielding up to 75% recirculation. Where silty sands separate extraction and reinjection screens, high anisotropy decreased recirculation but increased the width of the capture zone. Where located above or below the well screens, silty sands behave as confining beds, increasing both recirculation and width of capture. Modeling showed recirculation developing where the influent and effluent screens are separated by silty zones with a K_v of greater than 1.5 ft/day and an overall anisotropy of 10 or less.

BACKGROUND

Three recirculation-well pilot tests were conducted in a 300 ft thick section of alluvium in the Mashpee Pitted Plain at the Massachusetts Military Reservation on western Cape Cod. General hydrologic conditions are summarized in AFCEE (1998). In the areas of the pilot tests, the regional hydraulic gradient is approximately 0.002, giving rise to a groundwater flow velocity of ~1.5 ft/day in highly conductive sands (K_h = ~250 ft/day) The pilot tests were conducted to determine the suitability of recirculation wells to remove PCE and TCE in groundwater under a range of lithologic conditions. Measures of recirculation-well performance include the degree of recirculation between injection and extraction screens, and horizontal width and depth of capture zones. Lithologic conditions encompassed by the pilot tests included (1) relatively isotropic lithology with extraction and injection screens completed in a zone of fine to coarse sands, (2) anisotropic lithology with extraction and injection screens straddling poorly conductive silts, and (3) partially confining lithology with the extraction well screen underlain by poorly conductive silts.

Each pilot test consisted of two recirculation wells arranged as a fence approximately perpendicular to groundwater flow (oblique for the anisotropic site), with above-ground structures to contain blowers, controls, and granular activated carbon (GAC) units for air filtration after sparging. Extraction screens were located at the bottom of the wells, and injection screens were placed 28 to 68 ft higher in the section. An array of groundwater monitoring wells measured changes in contaminant concentrations and the influence of the recirculation wells on hydraulic head. Steady-state hydraulic heads were established within hours of start up of the recirculation wells. Significant contaminant reductions occurred within the zone of groundwater circulation, as illustrated in cross sections in Figure 1 showing TCE redistribution before and during operation of the recirculation wells. After 4 months of operation at the isotropic pilot test site, TCE in the cross section has migrated towards the extraction screen, indicating that TCE is moving vertically through the aquifer. At the anisotropic site, pre-operational hydraulic heads indicated a vertical hydraulic gradient across the silt layer, suggesting that the silt may significantly impede vertical flow. During operation of the recirculation wells at this site, groundwater mounding occurred in the vicinity of the injection well screen, confirming that vertical movement was impeded by the silt layer. Under proper design and operation of a recirculation well there should be minimal net change in head.

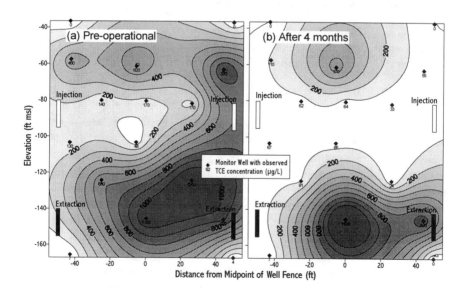

FIGURE 1. Cross sections of downgradient TCE concentration (a) before and (b) during operation of recirculation wells.

Model simulations were used to infer the steady-state groundwater flow distribution and contaminant reduction, which are not easily visualized from the distribution of monitoring points. Groundwater flow paths and potential development of a recirculation cell cannot be determined from hydraulic head measurements alone because of the effects of low-conductivity zones and the degree of vertical anisotropy. Three-dimensional groundwater modeling accounts for these effects, allowing visualization of the zone of recirculation.

MODFLOW (McDonald and Harbaugh, 1984), an industry-standard finite-element numerical model, was used to calculate the head distribution and resultant flow field. To aid in visualizing the flow field, particles were tracked in two directions using MODPATH (Pollock, 1994). One set of particles was tracked backward from the extraction (influent) screens upgradient to their point of entry into the model, thereby delineating the capture zones of the recirculation wells. A second set was tracked forward from the injection (effluent) screens downgradient to their point of exit, thereby delineating the release zones. These codes are integrated under the Groundwater Modeling System (GMS) graphical user interface (BYU, 1996).

Individual groundwater models were created for the test sites. The unconfined groundwater flow system present beneath the MMR was simulated for each site by a rectangular area 2000 ft × 2000 ft. The lower boundary of the models was set to a depth of 230 ft below the top of the water table. The grid cells increase in size from 1.5 ft, which is the diameter of the recirculation wells, to 150 ft × 150 ft at the corners of the model. The models are oriented orthogonal to the lines of recirculation wells. Constant head boundary conditions were used on all sides of the model.

Modeling runs covered a wide range of possible variability in hydraulic conductivity and vertical anisotropy. A sensitivity analysis was performed at each of the pilot test locations by progressively increasing vertical anisotropy.

RESULTS

Under isotropic conditions, approximately 50 to 75 percent of modeled flow lines recirculated between the injection and extraction screens. At this site, the screens were separated by 28 feet of highly conductive sand (K_h = up to 215 ft/day). Various values of anisotropy were modeled to determine its influence on capture and recirculation. As anisotropy increases, recirculation decreases because of reduced vertical conductivity. Pathlines showing recirculation with the most probable value of anisotropy are shown in Figure 2.

Results of modeling at the anisotropic site with a layer of low hydraulic conductivity between the injection and extraction screen show that recirculation is drastically curtailed. At this site, the screens were separated by 68 feet of highly conductive sand (up to 235 ft/day) with a thin layer of silt (0.12–2.2 ft/day) located at approximately the midpoint between the screens. The pumping rate was 30 gpm. Reliable determinations of hydraulic properties of the silt were unavail-

Particles backtracked from extraction screen

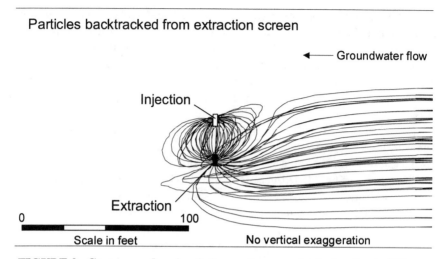

FIGURE 2. Capture and recirculation patterns under isotropic conditions.

able, so a sensitivity analysis was conducted for a range of reasonable values for K_h and K_v. Upgradient and downgradient stagnation points and the radius of recirculation expand with increasing anisotropy, in concert with declining recirculation. Pathlines showing the lack of recirculation in the model of the site with an intervening poorly conductive silt layer are shown in Figure 3.

Modeling of partially confined conditions was performed for the same pilot test site as was used for isotropic conditions. In this scenario, a zone of low hydraulic conductivity beneath the lower extraction screen increased the degree of recirculation. The underlying poorly conductive zone was assumed to have K_h of 32 to 36 ft/day. A range of reasonable parameters was modeled for the anisotropy

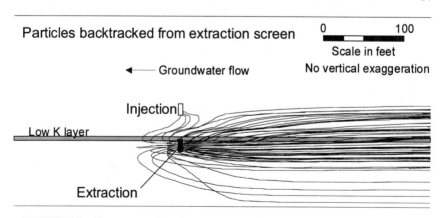

FIGURE 3. Capture and recirculation patterns with an intervening low-conductivity zone (anisotropic case).

of this zone. Increasing the anisotropy limited the depth of the capture zone but increased the width of the recirculation cell and the capture zone. Pathlines showing the effect of the underlying silt layer are illustrated in Figure 4.

Table 1, a summary of properties used in the pilot-test models, shows that the degree of recirculation decreases when the recirculation well screens straddle a layer of low conductivity, and increases when partially confined by an underlying zone of low conductivity.

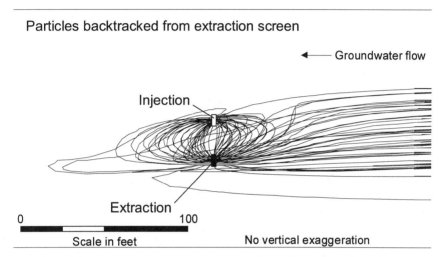

Particles backtracked from extraction screen

◄——— Groundwater flow

Injection

Extraction

0   100

Scale in feet   No vertical exaggeration

FIGURE 4. Capture and recirculation patterns with an underlying low-conductivity zone (partially confined case).

TABLE 1. Recirculation and hydraulic conductivities for pilot test models.

Scenario	Pumping (gpm)	Screen Separation (ft)	Range of K_h (ft/day)	Range of K_v (ft/day)	Averaged Anisotropy	Recircu-lation (%)
Isotropic						
Case #1	50	28	149 to 163	74 to 136	1.2 to 2.0	67
Case #2			203 to 215	169 to 179	1.2	59
Anisotropic						
Case #1	30	68	42 to 235	2.2 to 188	9.9	13
Case #2			27 to 235	0.12 to 118	108	0
Partially Confined						
Case #1	50	28	150 to 163[1]	125 to 136[1]	1.2	75
			32 to 36[2]	2.1 to 2.4[2]	15	
Case #2			204 to 215[1]	170 to 179[1]	1.2	71
			32 to 36[2]	2.1 to 2.4[2]	15	

[1]Range of hydraulic properties between the extraction and injection screen.
[2]Range of hydraulic properties below the extraction screen.

CONCLUSIONS

Successful implementation of recirculation wells for removal of VOCs from groundwater depends on adequate definition of lithologic layers and proper placement of extraction and injection well screens. To maximize recirculation and contaminant capture, design of extraction and injection well screens should include:

- Accurate delineation of lithologic zones.
- Completion in the same lithologic zone.
- No intervening lithologic layer with a contrast in hydraulic conductivity of more than two orders of magnitude.
- An overall anisotropy between the extraction and injection well screens less than ~10 for significant recirculation and effective capture.
- Low-conductivity zones above and below the extraction and injection screens to confine flow lines and expand laterally the zone of circulation.
- Lack of pre-operational vertical hydraulic gradients within the zone of completion.
- Numerical modeling of lithologic conditions, proposed pumping rates, and well geometry.

The regional hydraulic gradient also needs to be considered in the design process because of its influence on orientation of a well fence, size of capture zone and recirculation cell, and degree of recirculation. After a recirculation well has been install, evidence of a successful implementation includes:

- Lack of groundwater mounding during operation.
- Migration of contaminants to the extraction well screen during operation.
- Calculated pathlines from a representative hydrologic model showing significant recirculation (usually > 50%).

REFERENCES

AFCEE (Air Force Center for Environmental Excellence), 1998. *Plume Response Groundwater Modeling Report.* Prepared by Jacobs Engineering Group, Inc. for AFCEE/MMR Installation Restoration Program, Otis Air National Guard Base, Massachusetts.

BYU (Brigham Young University), 1996. *The Department of Defense Groundwater Modeling System GMS Ver. 2.0 Reference Manual.* Brigham Young University — Engineering Computer Graphics Laboratory, Provo, Utah.

McDonald, M.G., and A.W. Harbaugh, 1984. *A Modular Three-Dimensional Finite-Difference Groundwater Flow Model.* U.S. Geological Survey Open File Report 83.

Pollock, D.W., 1994. *User's Guide for MODPATH/MODPATH-PLOT, Version 3.* U.S. Geological Survey, Reston, Virginia.

KINETIC ANALYSIS OF PILOT TEST RESULTS OF THE C-SPARGE™ PROCESS*

William B. Kerfoot (K-V Associates, Inc., Mashpee, Ma)
C.J.J. M. Schouten (Mateboer Milieutechniek B.V., Almere, The Netherlands)
V.C.M. van Engen-Beukeboom (Bureau Bodemsanering Dienst Water en Milieu, Utrecht, The Netherlands)

ABSTRACT: Plume regions often contain mixtures of dissolved chlorinated solvents and petroleum products, presenting a need for co-treatment. The following report describes a pilot test and kinetic analysis of the C-Sparge™ process for remediation of a deep plume of dissolved chlorinated solvents in groundwater. An 800 ft (246 m) long, 60 ft (19 m) wide plume of dissolved trichloroethene was overlain by a separate plume (BTEX, HVOCs) originated from an underground storage tank leak of dissolved petroleum volatile organic compounds. Both sets of compounds were treated during a test conducted by Mateboer Milieutechniek B.V. for the Provincial Government of Utrecht in the vicinity of Rembrandt Street in Bilthoven, The Netherlands, from March 27th through April 4, 1997.

INTRODUCTION

A C-Sparge™ pilot test was conducted by Mateboer Milieutechniek B.V. for the Provincial Government of Utrecht at Rembrandt Street in Bilthoven, The Netherlands, from March 27, 1997 through April 4, 1997. The test involved installation of a C-Sparge™ well (TW), some additional monitoring wells (four 2-inch ID), the use of previously existing miniwells and a fire well across the site.

The C-Sparge™ Process. The C-Sparge™ process consists of a combination if *in-situ* air stripping, where dissolved chlorinated solvents (like PCE, TCE, DCE or vinyl chloride) are extracted from aqueous solution into small bubbles (Kerfoot, 1997). Then the extracted PCE reacts with encapsulated ozone in gas/gas reaction described by the Criegee mechanism as a primary ozonide. The ozonide is very unstable, composing to form a carboxyl oxide which reacts with water (hydrolyzes) as it exits the bubble to yield reaction end products Cl^-, H_2O, and CO_2 (Masten, 1986).

The process focuses ozone reaction selectively to air strippable compounds which invade the bubbles. As a result, if the encapsulated ozone concentration is maintained at a low multiplier of the strippable VOCs, then <u>no</u> additional ozone is available for side reactions with other dissolved organic compounds which have low Henry's numbers.

---

*Patent pending

Site Description. The field test took place in a small park midway on a long plume of predominantly trichloroethene (TCE) originating at a commercial building and traveling over 800 ft (246 m) across a predominantly commercial and residential area (Figure 1). The plume region lies in a thick fine sand deposit which contains gravel (streambed) deposits. About one-half of the area of groundwater overlying the TCE plume was contaminated with dissolved hydrocarbons (BTEX) from a nearby commercial fuel spill.

Soil borings in the vicinity of the plume showed a shallow surface loam extending to 6 ft (2 m) deep. Groundwater occurred at 9 ft (2.5 m). Fine sand occurred in many wells to over 19 ft (6 m) deep. Often gravel layers were intercepted at 13 ft (4 m) to 19 ft (6 m) deep. A thick clay layer, which probably serves as a bottom confining layer, was found at 124 to 130 ft (38 to 40 m) depth. A hydraulic conductivity (K) of 7.5×10^{-2} cm/sec has been estimated for the sand deposits.

Figure 2 shows the site layout, test spargewell (TW), and monitoring wells. The location of the monitoring wells were varied in distance and depth from the test spargewell (TW) to be able to give a 3-dimensional picture of the test results. The larger-diameter (2-in ID) wells allowed groundwater flow measurements as well as pressure changes to be monitored during treatment. A variety of physical and chemical measurements were performed during the test (Table 1).

TABLE 1. Groundwater monitoring during pilot test

Physical Measurement	Chemical Monitoring
Temperature, turbidity, static water elevation, groundwater flow, head pressure change	pH, Fe, redox potential, dissolved oxygen (DO), HVOCs including PCE, TCE, DCE, VC, TCA DCA, VOCs including benzene, toluene, ethylbenzene, xylenes, ozone concentration

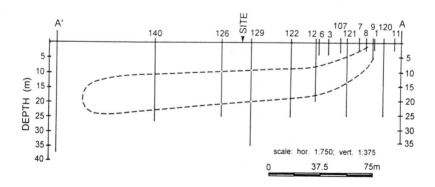

FIGURE 1. Test location, shown crossection, of PCE/TCE plume.

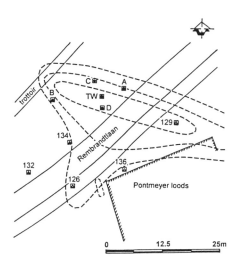

FIGURE 2. Site layout showing lateral spread of D.O. during 10-day test period, test well (TW), and monitoring wells.

C-SPARGE™ WELL INSTALLATION

The C-Sparge™ double-screen well with lower Spargepoint [R] was installed with a recirculating water system and casing. The lower Spargepoint [R] was set at a depth of 84 ft (26 m). A one-half inch tubing extended to the surface from a compression fitting on the Spargepoint [R]. A 4-inch ID triple-screened well extended from 75 ft (23 m) to one ft (.3 m) above grade. Six-foot long screens were placed with bottom edges at 75 ft (23 m), 42 ft (13 m), and 8 ft (2.5 m). The middle casing between the two lower screens received 3 ft (1 m) of bentonite grout, 3 ft (1 m) of cement/bentonite, and 3 ft (1 m) of bentonite to seal the annular space to prevent "short-circuiting" of water. Water and fine bubbles were injected into the formation from the lowest screen, and return water enters the middle screen. The uppermost screen collects gases from just above the water table (8 ft, 2.5 m) to assure vapor control.

RADIUS OF INFLUENCE

The pressure of the gas injection, use of microporous bubblers, and a recirculating well system all function to distribute fine bubbles, containing air/ozone gas through the fine sands under Darcian flow approximating fluid flow. The injection of the air/ozone approximates the injection of water, exhibiting mounding and outward movement until equilibrium is reached.

The presence of microbubbles, gas release, and dissolved oxygen changes normally demark the expansion of the treatment zone. On April 1, operation of the C-Sparge™ unit began at about 2:00 P.M. By the afternoon of April 2, gas bubbles were found discharging at minipoint 129, over 51 ft (17.1 m) from the spargewell installation (TW) (Table 2). Well D, only about 7 ft (2.2 m) from the injection area, showed almost immediate oxygen changes and water which effervesced with fine bubbles.

Table 2 gives the results of field dissolved oxygen (D.O.) measurements taken during sampling.

Figure 2 shows the lateral spread during the ten-day test. Figure 3 depicts the vertical spread of D.O. observed during the test period. The bubble zone was still expanding during the ten-day test. Based upon the time sequence, a long axis extending outwards about 97 ft (30 m) in a westerly and easterly direction would be reached, with a minor axis (at right angles) of about 58 ft (18 m).

TABLE 2. Dissolved oxygen (D.O.) changes observed in monitoring wells near the spargewell.

Well	Day										
	1	2	3	4	5	6	7	8	9	10	11
TW	2.8	11.5	---	---	---	---	---	---	---	---	---
D	1.2	14.2	12.2	11.8	8.9	---	12.0	12.3	13.0	9.2	15.8
129	0.0	1.6	9.3	9.2	7.4	---	7.9	8.6	9.1	8.5	9.4
C	0.0	0.7	1.4	0.9	1.1	---	3.5	3.4	5.7	5.2	8.2
A	0.0	0.0	0.1	3.1	8.6	---	8.6	8.8	11.6	11.7	9.3
B*	0.0	0.0	0.0	0.2	0.0	---	0.0	0.0	0.0	0.0	0.0
126											
(14-15)	0.0	0.0	0.0	3.5	3.7	---	5.9	6.3	6.8	6.7	6.0

*Well B showed a continual increase in redox potential despite exhibiting no oxygen increase. A hydrocarbon plume with high oxygen demand existed in the region.

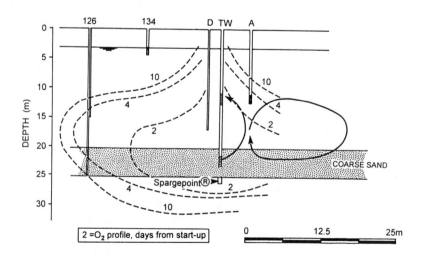

FIGURE 3. Vertical spread of D.O. during 10-day test.

GROUNDWATER FLOW MEASUREMENTS - CIRCULATION PATTERN DEFINITION

Direct groundwater flow measurements were performed with a KVA Model 40 GeoFlo [R] Meter prior to the startup and during the operation to determine background velocity and changes (Kerfoot, 1988). Initial measurements indicated a flow near the spargewell (TW) in a northwesterly direction at a velocity of between .6 and .8 ft/day, coinciding with the direction of movement of the plume.

Additional measurements were taken after beginning injection to determine the velocity of groundwater eddies created by the double-screen well system and rising bubbles which expand the treatment zone both vertically and horizontally across the site. The observed change in direction and rate coincided with a slow vertical mixing rate set up at right angles to the main axis of mixing. At well A (which is 17 ft [5.4 m] east of TW) the change was measured at about 3 ft/d in an easterly direction. At well B (which is 36 ft [11 m] west of TW) a velocity change of .5 ft/d occurred towards the TW well. The A well is shallow (37 ft [11.4 m] below grade) and the B well significantly lower (49 [15 m] ft below grade). Groundwater moving in towards the spargewell would reach a maximum at about 68 ft (21 m) below grade. The vertical eddy for mixing appeared to reach a velocity with a diameter of about 65 ft (20 m) by day 10 of the test, with an estimated velocity of about 11 ft/day (3.3 m/d).

CHEMICAL RESULTS: VOC REMOVAL - CHLORINATED AND PETROLEUM COMPOUNDS

The following differential equation 1 expresses the direct relationship between the rate of elimination and the concentration of the dissolved volatile compound:

$$dc/dt = -bC \tag{1}$$

Where dc = change in concentration of dissolved volatile

dt = change in time

b = fraction of volatile substance that leaves groundwater in one unit of time (day)

C = concentration of volatile compound

The following equation 2 offers the simple solution:

$$C = C_o e^{-bt} \tag{2}$$

Where e = an exponent

C_o = initial concentration of volatile organic compound in groundwater ($\mu g/L$)

The equation shows the behavior of the phenomenon of "exponential decay", since the exponential term e^{-bt} appears in it. The curve starts from a known groundwater concentration and decreases in proportion to the remaining concentration, thus equation 3:

$$\log C/C_o = bt/2.303 \tag{3}$$

$$\text{or } \log C = \log C_o - bt/2.303$$

$$\text{for } C = C_o/2, \, b = 0.693/t_{1/2}$$

The constant $t_{1/2}$ corresponds to the length of time needed for the concentration of Co to decrease by 50%; *i.e.*, $t_{1/2}$ is the half-life of the substance in the groundwater.

In Figures 4a and 4b, the mean -log C/Co is plotted versus time to derive the approximate decay rate observed for the site conditions. The decay constants for both HVOCs and BTEX compounds were computed. In some cases, a linear mean value was clearly being fitted to a dampened oscillating decay as a first approximation.

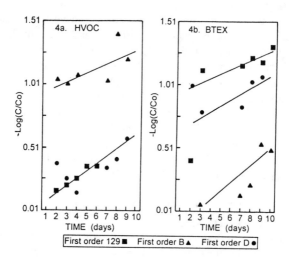

FIGURE 4. Removal rate as a plot of -log C/Co versus time to approximate first order decay rate for observed HVOC (4a) and BTEX (4b) removal.

CONCLUSIONS

The C-Sparge™ system was designed to inject and distribute microbubbles into the fine sandy aquifer. The presence of microbubbles, gas release, dissolved oxygen changes and halogenated volatile organic compounds (HVOC) change defined the enlarging treatment zone over the ten-day period. Based upon time sequence, a long axis extending outwards over 100 ft (30 m) in a westerly and easterly direction was reached with a minor axis (at right angles) of about 56 ft (18 m) from a single treatment spargewell. A gyre generated beside the spargewell had a three-day rotation charctrerized by circular groundwater movement and chemical tracer changes.

A kinetic analysis of the reaction rates was performed. HVOC concentrations of up to 14,500 µg/L were reduced to below 1000 µg/L during the ten-day test.

HVOC removal rates fell between .09 and .14t. BTEX removal rates fell between .07 and .20t. This corresponds to a steady rate of reduction to 1/2 value every 7 to 11 days. In a conservative estimate it would take slightly less than 100 days to reduce the core region (50 ft wide by 30 ft deep by 200 ft long) to below 5 ug/L, assuming no other sources invade the eddies with the treatment volume.

Figure 5 plots the linear decay rates on a logarithmic set of axes to depict the time to bring core region concentrations to 1 µg/L. With HVOCs, the time ranged from 50 to 100 days. For BTEX compounds, the level ranged between 20 to 60 days. Please note that the HVOC removal rate is somewhat slower since the beginning concentration of 2000 µg/L total HVOC is higher than the starting point of the BTEX compounds (50 to 70 µg/L).

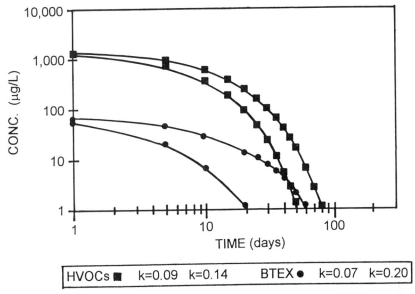

FIGURE 5. Log/log plot of removal (decay) of HVOC and BTEX expected with time based upon observed low and high range.

REFERENCES

Kerfoot, W.B. 1988. "Monitoring Well Construction and Recommended Procedures for Direct Groundwater Flow Measurements Using a Heat-Pulsing Flowmeter." In *Groundwater Contamination: Field Methods*, ASTM, STP 963, Collins and A.J. Johnson, Editors.

Kerfoot, W.B. 1997 "Extremely Rapid PCE Removal from Groundwater with a Dual-Gas Microporous Treatment System." In: *Contaminated Soils, Volume 2*, P.T. Kostecki, E.J. Calabrese, and M. Bonazountas, Eds., ASP Publishers, Amherst, MA, pp. 275-284.

Masten, S.J. 1986. *Mechanisms and Kinetics of Ozone and Hydroxyl Radical Reactions with Model Aliphatic and Olefinic Compounds*, Ph.D. Thesis, Harvard University, Cambridge, MA

AIR SPARGING OF CHLORINATED VOLATILE ORGANIC COMPOUNDS IN A LAYERED GEOLOGIC SETTING

William D. Hughes (Parsons Engineering Science, Cincinnati, Ohio)
Peter I. Dacyk, (Parsons Engineering Science, Inc. Cincinnati, Ohio)

ABSTRACT: Air sparging is considered an inappropriate remedial technology in layered geologic settings. However, with careful design of the air sparging system it can be more effective than groundwater pumping. Chlorinated volatile organic compounds (VOCs) were detected at total concentrations of over 34,000 µg/L in groundwater at an industrial laundry facility. Groundwater pumping, and soil vapor extraction around the suspected source had limited effectiveness. Increases in chlorinated VOC concentrations were detected in groundwater at the down-gradient edge of the property in 1995, in spite of five years of groundwater pumping.

The impacted sand and gravel aquifer contains numerous semi-continuous silty clay strata. Air sparging pilot tests utilizing a helium tracer demonstrated that monitoring and recovery well sand packs provided vertical pathways through the silty clay strata to allow control of the sparging vapors. A 50% reduction in chlorinated VOC concentrations in groundwater in the source area was achieved during the first nine months of air sparging operations.

INTRODUCTION

Layered geologic settings have been regarded as areas where air sparging is of limited value. However, the layered geologic setting also has negative impacts on other remedial technologies such as groundwater pumping. This paper presents a case study of application of air sparging in a layered setting where groundwater pumping has been ineffective in remediation of the site. In this case, air sparging is seen as a lower-cost alternative to discharging groundwater to the publicly-owned wastewater treatment works (POTW), in spite of its limitations given the site geology.

The site of the remedial activities is a commercial laundry facility in southwestern Ohio (Figure 1). The laundry facility sits atop a buried valley aquifer composed primarily of glacial sand and gravel deposits with interbedded clay, silt, and sand stringers. The aquifer is the primary potable water source for the surrounding communities. A groundwater production well field is located approximately 1,070 meters north (downgradient) of the site.

Dry cleaning operations at the facility used tetrachloroethene (PCE) as the primary cleaning solvent through 1986. A release of PCE occurred in 1985 when a 1,000 gallon above ground storage tank ruptured during refilling. The released solvent flowed into a nearby storm sewer which empties into an on-site stormwater retention basin. Emergency response activities included the removal of surface soils from the retention basin. Investigation of the groundwater conditions at the

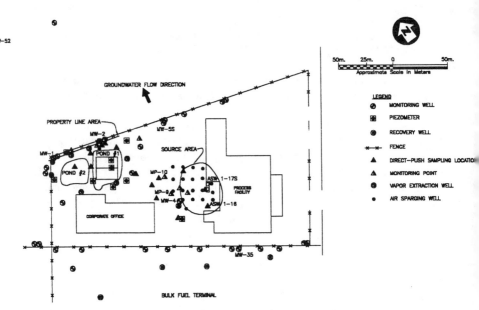

FIGURE 1. Site plan of industrial laundry facility showing source area and property line area.

site revealed dissolved cis-1,2 dichloroethene (1,2-DCE) and trichloroethene (TCE) in addition to PCE. Active remediation of the chlorinated VOCs began in 1989 with the installation of groundwater extraction wells, an air stripper to treat the extracted groundwater, and infiltration galleries within the retention basin to recirculate the contaminated groundwater. Soil vapor extraction (SVE) wells were installed around the retention basin to strip VOCs from the vadose zone soils. These remedial actions were implemented to be in compliance with a consent order that required prevention of offsite migration of water containing more than 5 µg/L PCE and 5 µg/L TCE.

Groundwater treatment and SVE were terminated in the mid 1990's due to poor performance of the infiltration galleries and poor VOC removal rates by the SVE system. Groundwater pumping continued with the intent of providing hydraulic control of the dissolved VOC plume. The untreated water is discharged to the POTW. Groundwater pumping appeared to be controlling the plume and VOC concentrations in groundwater were reduced from a high in 1989 of 973 µg/L in a property line monitoring well (MW-1) to 59 µg/L in April 1994. A sharp increase in VOC concentrations to over 560 µg/L was observed in MW-1 beginning in June 1994. This increase and increases in other wells indicated the groundwater pumping was not effective at aquifer remediation and was not preventing the migration of dissolved VOCs offsite. The rapid increase in VOC concentrations also suggested the possibility of previously unidentified sources of

VOCs. A re-evaluation of the remedial approach was initiated with the primary goal of replacing groundwater pumping with a less costly and more effective remedial technology.

The first step in the re-evaluation was a systematic investigation of groundwater across the entire site. The investigation was conducted using direct-push sampling techniques. The results of the direct-push sampling identified a previously unknown source area with concentrations of up to 42,940 μg/L total ethenes in groundwater near the known surface spill area but over 75 meters upgradient of the stormwater retention basin that had been considered the source area. A review of plant records revealed that underground storage tanks (USTs) adjacent to the tank which spilled the PCE had been used to store detergent that contained chlorinated solvents. The USTs were closed in place several years prior to these activities but had not been recognized as a source of chlorinated VOCs.

AIR SPARGING PILOT TESTING

Recognizing the limitations of the application of air sparging in layered geologic settings, two field pilot tests were performed to evaluate sparging effectiveness at this site. One test was performed in the source area where sparge air was injected below a silty-clay confining layer (Figure 2). The other test was performed in an area along the property line where the low-permeability layers below the water table are discontinuous (Figure 3). Air injection in both tests was at a flow rate of 13 cubic meters per hour (m^3/h) with a wellhead pressure of 0.6 bar. A helium tracer was injected at a concentration of 0.3% during both pilot tests.

The test in the source area was run for three days including three test configurations 1) SVE-only, 2) sparging combined with SVE, and 3) sparging only. Vapor monitoring parameters included monitoring point pressure or vacuum, and oxygen, carbon dioxide, VOC and helium concentrations. Water levels and dissolved oxygen concentrations of groundwater were also monitored. The monitoring results were consistent with site stratigraphy. The low-permeability layers induced greater lateral migration of injected air. Monitoring and recovery well sand packs appeared to allow bypass of vapors from below low-permeability layers into shallower permeable intervals. Helium was detected in monitoring points approximately 6 meters from the injection well. One design element incorporated in the full-scale design was that SVE well screens in the source area were set spanning the low-permeability layers to create pathways for vertical vapor migration.

A 20 hour sparging-only pilot test was performed in the property-line test area. The distribution of air was more uniform in this area due to a lack of continuous low-permeability layers. Influence of air injection was observed 3 to 5.5 meters from the injection well.

SYSTEM DESIGN AND INSTALLATION

The results of the pilot test were used to design a full scale system which was presented to the regulatory agency for approval. The primary concern of the

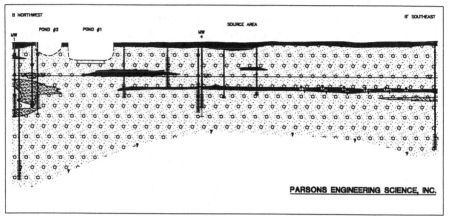

FIGURE 2. Geologic cross - section through source area.

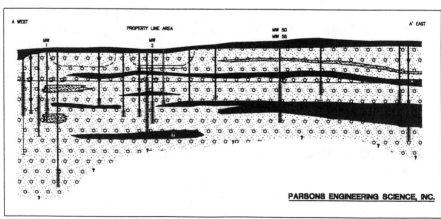

FIGURE 3. East west geologic cross - section along property line.

regulators was the effect of air sparging on the ability of the groundwater recovery wells to maintain hydraulic control of the aquifer to limit the offsite migration of dissolved VOCs. The sparging remedy was approved with agreement that frequent groundwater level monitoring and groundwater sampling would be implemented during start-up. The frequent monitoring would help identify potential problems which would allow timely modification of operations and minimize negative impacts.

The spacing of air injection wells is 9.1 meters. The design air injection capacity was 33 m^3/h per well. Actual air delivery is 8 to 16 m^3/h. The wells are connected by headers in groups of five. Air injection is controlled by a programmable logic controller that allows flexibility in injection cycling.

Soil vapor extraction wells are positioned to achieve capture of sparge-created vapors within a radius of 17 meters from an SVE well. Regenerative blowers are used for soil vapor extraction with a minimum design flow of 116 m^3/h per well. The SVE wells are connected in headers of two to three wells per blower.

Most of the injection wells were drilled to depths of 11 to 13 meters. The wells were completed with 60 cm of continuous slot screen at the bottom of the boring. The wells in the source area were set so the top of the screen was approximately 3 to 4 meters below the highest apparent VOC concentration based on field screening and laboratory analysis of groundwater.

The area of highest VOC concentrations in groundwater is adjacent to the USTs (ASW1-17). In that area the sand and gravel interval is interrupted by two silty-clay lenses. During well installation it was recognized that air injection at multiple depths would be required to effectively treat the area. The well is surrounded by the building, a secondary-containment structure for above ground storage tanks, and the USTs, therefore, multiple drilling locations would not be possible. Nested injection wells were installed in a single boring with one screen set in the shallow interval (11 meters below grade) and one set in the deeper interval (14.6 meters below grade). The two wells were connected to different headers to prevent preferential air flow into the shallower point.

OPERATIONS

The two sparging systems began operation in late March 1997. Table 1 summarizes pre-sparging and January 1998 groundwater VOC concentrations from monitoring wells located within the source area, along the property line and on the edges of the source area. These results indicate a significant decrease in ethene concentrations within the source area (ASW 1-17 & ASW 1-16) and in parts of the fence line treatment area (MW-2). Results in the western portion of the property-line treatment area are less favorable (MW-1). The reduction in concentrations, or lack of a significant change in concentrations, of ethenes in groundwater on the perimeter of the source area (MP-9 & MP-10) suggest that air sparging is not pushing contaminants cross-gradient. Conversely, the increase in ethene concentrations at a property line well (MW-5S) suggests mobilization of additional dissolved ethenes downgradient of the source area.

TABLE 1. Pre-sparging and January 1998 VOC concentrations in groundwater at the commercial laundry facility.

Well	cis-1,2 DCE (µg/L) Pre-sparge/Jan 98	PCE (µg/L) Pre-sparge/Jan 98	TCE (µg/L) Pre-sparge/Jan 98
ASW 1-16	100/8.8	1140/660	170/31
ASW 1-17	26400/24000	6670/1200	9870/<500
MW-1	<25/<25	234/270	25/<5
MW-2	269/<5	502/180	254/130
MW-5S	<25/<25	459/960	<25/<25
MP-9	33/58	83/75	44/18
MP-10	270/25	490/240	240/87

No other VOCs were detected by SW-846 Method 8260

The observed reductions in concentrations within both sparging systems are encouraging. The less apparent effect in the western end of the property-line system and the increases in concentration in monitoring wells on the eastern portion of the property line indicate that additional actions are required.

CONCLUSIONS

The initial results indicate that mechanical removal of chlorinated VOCs by air sparging and SVE can be effective at mass removal in a layered geologic setting. However, the complex stratigraphy within the area of highest VOC concentrations makes it unlikely that the physical removal process will achieve reduction of VOC concentrations to the levels dictated by the regulatory order (drinking water standards). The property owner is currently negotiating with the regulatory agencies to agree on realistic clean-up goals. The commitment to continue source removal by air sparging is being used by the property owner to enhance their negotiation position with the agency.

Modifications to the existing sparging system being considered to increase removal rates include extending the property line system to the east past MW-5S, and increasing air flow on the western end of the property line system. A pilot test of propane injection to induce cometabolism of the ethenes is also being considered.

AN ANALYSIS OF AIR SPARGING FOR CHLORINATED SOLVENT SITES

Richard A. Brown, Fluor Daniel GTI, Trenton, NJ, USA
David H. Bass, Fluor Daniel GTI, Norwood, MA, USA
Wilson Clayton, Fluor Daniel GTI, Golden, CO, USA

ABSTRACT: Air sparging is a frequently applied technology at chlorinated solvent sites. This is in part due to the fact that losses of dense solvents frequently result in contamination of soils below the water table and of groundwater. As a result air sparging is considered a viable technology for solvent sites. However, a fundamental question is how well does air sparging perform at chlorinated solvent sites.

Fluor Daniel GTI has compiled a database of more than 30 completed air sparging projects to address questions of the effectiveness and permanence of air sparging, and to provide predictive indicators of air sparging success. Groundwater concentrations were compared before sparging was initiated, just before sparging was terminated, and in the months following shutdown of the sparging system. The case studies include solvent and hydrocarbon sites, a wide range of soil conditions, and sparge system parameters. Air sparging systems achieved a substantial and permanent decrease in groundwater concentrations at both chlorinated and petroleum sites, both sandy and silty soils, and both continuous and pulsed flow sparging. In some cases, significant rebound of groundwater concentrations was observed.

The success of air sparging is a function of the dynamics of air flow and of the nature of the solvent distribution in the subsurface. This paper will identify the factors that impact the performance of air sparging systems such as air flow, well spacing, well depth, etc. Through an analysis of the data base, this paper will analyze how these factors affect air sparging. The performance is also a function chlorinated solvent distribution in the subsurface. Examining the differences between chlorinated solvent and hydrocarbon sites can help elucidate how performance is impacted by the nature of the solvents as opposed to flow dynamics.

INTRODUCTION

In situ air sparging is a commonly used remediation technology which was developed in the late 1980s as a method for treating dissolved volatile organic compounds (VOCs) in groundwater. Air sparging involves the injection of air under pressure into saturated zone soils. The injected air displaces water and creates air-filled porosity in the saturated soils, volatilizes and removes dissolved and adsorbed phase VOCs, and transfers oxygen into the groundwater. As a result, both physical removal and aerobic biodegradation of contamination in groundwater and saturated zone soil are enhanced. Air sparging has been used to remediate both chlorinated solvents and petroleum hydrocarbons (Bass and Brown, 1991; Billings 1995; Leonard and Brown 1992; Marley et al., 1992; Noonan et al., 1993).

Air sparging offers a means of remediating soils and groundwater without the need for active groundwater pumping, and in some cases, air sparging appears to have produced significant and permanent reductions in groundwater contaminant concentrations. As a result there has been a steady increase in application of air sparging, and hundreds of systems are currently in operation. However, dissolved groundwater concentrations sometimes fall dramatically during sparging, but then "rebound" nearly to original levels once the sparge system is turned off.

The efficacy of air sparging is determined principally by the degree of contact between the injected air and the contaminated soil and groundwater. It is generally agreed that the injected air forms channels through the saturated soil matrix (Noonan et al., 1993; Ji et al., 1993; Johnson et al., 1993). When there is a high density of uniformly distributed air channels and/or significant mixing between channels, then air sparging is expected to be effective. Air sparging is less likely to be effective when the density of air channels is low or non-uniformly distributed and when there is little or no mixing of the water between the air channels.

In this paper, the results of 32 sparging case studies are compiled to shed light on how well and under what conditions air sparging achieves permanent reduction in groundwater contaminant concentrations. Site and system details for the case study sites have been reported earlier (Bass and Brown, 1996).

METHODS

Of the 32 air sparging case study systems, 28 were designed to address source area contamination (6 chlorinated solvent and 22 petroleum hydrocarbon sites). At the other 4 sites, air sparging barriers were installed to treat downgradient contamination (1 chlorinated solvent site and three petroleum hydrocarbon sites). The case studies cover a wide range of geography, soil conditions, and sparge system parameters. Systems were located in 13 states. Soils ranged from silt to coarse sand and gravel, with both undisturbed and backfilled material as the sparged matrix. Sparge well spacings ranged from 12 to 80 feet, and flow rate per sparge well from 3 to 35 scfm. Some systems injected sparge air continuously, others intermittently (pulsed operation). In one case, "in-well" sparging was performed by inserting a sparge pipe to the bottom of existing monitoring wells. Well systems ranged in size from 1 to 16 wells and include both horizontal and vertical wells. Duration of sparging system operation ranged from a few months to more than 4 years.

In each case study, groundwater concentrations have been compared before sparging was initiated, just before sparging was terminated, and in the months following shutdown of the sparging system. Post-shutdown monitoring data are available for only a few months in most cases, but at some sites more than a year of post-shutdown data have been taken. The contaminant reduction, expressed as a percentage, is defined at the point when sparging was turned off :

$$Reduction \ = \ 100\% \bullet (1 \ - \ \frac{C_f}{C_o}) \tag{1}$$

where C_o = dissolved concentration at start of sparging
$\quad\quad\ \ C_f$ = dissolved concentration at termination of sparging

and at the end of the post-shutdown monitoring period:

$$Reduction \ = \ 100\% \bullet (1 \ - \ \frac{C_r}{C_o}) \tag{2}$$

where C_r = dissolved concentration after post-sparging monitoring.

Rebound is defined as

$$Rebound \ = \ \frac{\log(C_r / C_f)}{\log(C_o / C_f)} \tag{3}$$

As defined in Equation (3), rebound is the log of the concentration increase after sparging ends, divided by the log of the concentration decrease during sparging. In other words, rebound is the orders of magnitude of concentration increase after the system is shut down divided by the orders of magnitude of concentration reduction while the system is operating. A rebound less than 0.2 (less than 1 order of magnitude of post-shutdown increase for each 5 orders of magnitude of initial decrease) reflects a permanent reduction in groundwater concentrations, while a value greater than 0.5 (more than 1 order of magnitude increase for each 2 orders of magnitude of initial decrease) reflects substantial rebound.

RESULTS AND ANALYSIS

Overall, of the 28 source area air sparging systems summarized, the following average permanent reductions (i.e., after post-shutdown monitoring, as defined by equation (2)) in dissolved contaminant concentrations were observed:

- 12 systems (43%) achieved an average permanent reduction greater than 95%;
- 17 systems (61%) achieved an average permanent reduction greater than 90%;
- 23 systems (82%) achieved an average permanent reduction greater than 80%.

However, in many cases the average reduction was high, but a single monitoring well either failed to show sufficient contaminant attenuation or else rebounded after the system was shut down, resulting in the need for further action. Another approach to assessment of sparge system performance, then, is the permanent reduction achieved in the least responsive monitoring well. Using this criterion, the following summarizes the minimum permanent reductions in dissolved contaminant concentrations:

- 8 systems (29%) achieved a minimum permanent reduction of 95% or more;
- 12 systems (43%) achieved a minimum permanent reduction of 90% or more;
- 17 systems (61%) achieved a minimum permanent reduction of 80% or more.

In general, systems which produced permanent reductions averaging greater than 90% with all monitoring wells showing a permanent reduction of at least 80% were considered to be successful. The least successful systems produced lower average permanent reductions and often had more than one well displaying significant rebound (the "in-well" sparging system was among the least successful). The grey area in between ("qualified successes") includes sites where permanent reductions somewhat less than 90% were sufficient to effect site closure, or where the substantial reductions achieved were still insufficient to meet stringent remediation goals.

Examination of the characteristics and behavior of the sparging study sites in this limited database leads to the following observations which can be useful in designing and predicting the performance of air sparging systems.

Poor performance was generally characterized by initially reduced concentrations followed by substantial rebound. Sparging usually resulted in an initial reduction in dissolved concentrations, however the maximum rebound at the most successful sites averaged a negligible 0.11, compared with 0.77 (more than 3 orders of magnitude of rebound for every 4 orders of magnitude of initial remediation) at the least successful sites. Furthermore, all of the less successful petroleum sites had a rebound number greater than 0.68 in at least one well, while none of the most successful sites had a maximum rebound as greater as 0.4. Rebound was variable at the qualified successes.

Sparging at chlorinated solvent sites is generally more successful than at petroleum sites. All of the chlorinated sites met the criteria for success or qualified success, as outlined above, but about one-third of the petroleum sites were unsuccessful. On the other hand, when sparging was successful at petroleum sites, the permanent reductions in groundwater concentrations were much greater than at chlorinated sites. For example, the most successful petroleum sites had average permanent reductions ranging from 99.1% to 99.96%. The maximum average permanent reduction observed at a chlorinated site was 96.6%.

More successful systems at petroleum sites had a higher sparge well density covering the entire source area. The successful systems consisted of an average of 6.5 wells spaced an average of 26.6 feet apart (corresponding to an average assumed radius of sparging influence of about 15 feet). In contrast, the less successful systems consisted of only 4.1 wells on average spaced 41.6 feet apart (an assumed radius of influence averaging 24 feet). In addition, several of the less successful systems did not address the entire source area, either because of physical constraints or because the focus was exclusively on the tank pit.

Performance was generally better in systems treating dissolved phase plumes than in systems treating adsorbed contaminants. When released product at petroleum sites did not contact the groundwater (i.e., there was no smear zone of adsorbed product), then remediation by sparging was more effective and permanent, even with less aggressive sparge systems. For example, at two sites where the released product did not appear to extend downward through the entirety of a deep vadose zone, remediation was rapid despite well spacing in excess of 40 feet and a sparging duration of only 4 months. Similarly, the barrier systems which treated low level dissolved plumes were all successful at removing BTEX and chlorinated VOCs, despite large well spacings, in some cases as much as 80 feet. In contrast, among systems addressing smear zones at petroleum sites, those with higher initial dissolved contaminant concentrations (often reflective of a greater prevalence of adsorbed product) did more poorly than those with lower concentrations. The average initial concentration at the most contaminated well at the successful sites was 13,400 μg/l, less than a third of the average initial concentration (44,400 μg/l) at the least successful sites. Likewise, one barrier system which treated a smear zone emanating from an upgradient source produced only modest reductions in BTEX concentrations, which rebounded fully following termination of the system.

These observation are consistent with the generally good performance observed at chlorinated solvent sites. The contamination at the chlorinated sites probably partitioned into groundwater more than at the petroleum sites. The soils at all of the sites in Table 1 were sandy with low organic carbon, and the chlorinated solvents released do not contain a low solubility, high molecular weight fraction (as do most petroleum products). Furthermore, the extent of the releases was more modest at the chlorinated sites (had the releases been much more substantial, mobile DNAPL would have been indicated and sparging would not have been considered). Therefore, the chlorinated solvent sites in the database are essentially dissolved plumes. As discussed above, good contaminant reductions with limited rebound was typically observed after remediation of these sites, despite well spacings as great as 80 feet in some cases.

It is possible that residual LNAPL in the smear zone serves as a reservoir for BTEX, resulting in the observed rebound. Boersma (1993) has described two case histories in which chlorinated solvents were the target contaminants. The first site was a single well 30-day pilot test at a sandy site with dissolved chlorinated concentrations of 2 to 3 ppm. Dissolved concentrations decreased by ½ to 1 order of magnitude, with little rebound six months after system shutdown. The conditions and results are consistent with those at the chlorinated sites in Table 1. At the second site, the source of the chlorinated solvents was a petroleum LNAPL in which the solvents were initially dissolved. Sparging had a smaller and less permanent impact on dissolved concentrations at this site.

When rebound occurred, it sometimes happened many months after sparge system shutdown. Several sites showed only moderate rebound 2 to 4 months following shutdown, but in some source area wells concentrations jumped by another order of magnitude or more within 7.5 to 16 months after shutdown.

Changes in water table levels appeared to be associated with increased rebound. Rising water tables can bring groundwater in contact with fresh sources of contamination. This appeared to be the case at one site where the water table rose by 20 feet over the course of the remediation and post-remediation monitoring. No rebound was observed initially in any monitoring well following termination of sparging, but 7.5 months after sparging was concluded the concentrations increased suddenly by 3 orders of magnitude in one well. Several other sites also experienced substantial rebound in some wells following a post-shutdown water table rise.

CONCLUSIONS

Of the 28 air sparging source area remediation systems, 13 (46%) produced permanent reductions greater than 90% averaged over all monitoring wells, with no monitoring well showing a permanent reduction of less than 80%. An additional 7 sites (25%) had averaged permanent reductions somewhat less than 90%, but this was still sufficient to effect site closure. The performance at the remaining 8 sites (29%) was unsatisfactory.

Sparging appears to clean up dissolved chlorinated solvent and downgradient BTEX plumes more easily than petroleum hydrocarbon source areas. Sparge well depth and well spacing can be greater when treating a chlorinated or downgradient dissolved BTEX plume than when treating a petroleum hydrocarbon source area.

The source area at petroleum sites where a smear zone is present should be addressed using a high density of sparge wells, closely spaced (<20 ft and preferably <15 ft), placed in such a way as to address the entire source area. The higher the initial dissolved concentrations at petroleum sites, the more aggressive the sparge system will have to be to achieve acceptable results. "In-well" sparging, in which a sparge pipe is inserted to the bottom of existing monitoring wells, was not effective.

REFERENCES

Bass, D.H., and R.A. Brown. 1991. "In Situ Air Sparging for Remediation of Contaminated Saturated Zone Soils." *Proceedings of the New Jersey Environmental Expo*, Secaucus, 1991.

Bass, D.H., and R.A. Brown. 1996. "Air Sparging Case Study Database Update." *Proceedings of First International Symposium on In Situ Air Sparging for Site Remediation.* INET, Potomac, MD.

Billings, G. 1995. *Site Summaries Group 1*, Billings and Associates, Inc., Albuquerque, NM.

Boersma, P.M. 1993. "Sparging Effectiveness for Groundwater Restoration." *In Situ Aeration: Air Sparging, Bioventing and Related Remediation Processes*, R. Hinchee, R. Miller, and P. Johnson, Eds. Battelle Press, Columbus, OH, 1993.

Ji, W., A. Dahmani, D.P. Ahlfleld, et al. 1993. "Laboratory Study of Air Sparging: Air Flow Visualization." *Groundwater Monitoring and Remediation.* 115-126.

Johnson, R.L., Johnson, P.C., McWorter, D.B., et al. 1993. "An Overview of In Situ Air Sparging." *Groundwater Monitoring and Remediation.* 127-135.

Leonard, W.C., and R.A. Brown. 1992. "Air Sparging: An Optimal Solution." *Proceedings of Petroleum Hydrocarbons and Organic Chemicals in Ground Water: Prevention, Detection, and Restoration.* NGWA, Dublin, OH.

Marley, M.C, D.J. Hazebrouk, and M.T. Walsh. 1992. "The Application of In Situ Air Sparging as an Innovative Soil and Groundwater Remediation Technology." *Groundwater Monitoring Review.*

Noonan, D.C., W.K. Glynn, and M.E. Miller. 1993, "Enhance Performance of Soil Vapor Extraction," *Chemical Engineering Progress*, (6):89.

METHYL TERT-BUTYL ETHER REMOVAL BY IN SITU AIR SPARGING IN PHYSICAL MODEL STUDIES

Cristin L. Bruce (Arizona State University, Tempe, Arizona)
Craig D. Gilbert (Oregon Graduate Institute, Beaverton, Oregon)
Richard L. Johnson (Oregon Graduate Institute, Beaverton, Oregon)
Paul C. Johnson (Arizona State University, Tempe, Arizona)

ABSTRACT: Two-dimensional laboratory-scale physical aquifer model studies at Arizona State University and the Oregon Graduate Institute have been conducted to simulate MTBE removal by air sparging. Results for saturated media experiments indicate that approximately 85% of immiscible-phase MTBE can be removed from MTBE-containing test mixtures by volatilization under idealized conditions. These experiments also suggest that MTBE removal efficiency by volatilization is affected by the degree of media saturation.

INTRODUCTION

Fuel oxygenates, such as ethanol, methanol, and methyl tert-butyl ether (MTBE), are common additives to gasoline. When added to gasoline, these oxygenates can increase octane and decrease vehicular emissions of carbon monoxide and ozone by promoting more complete combustion. MTBE is the most widely used oxygenate in the United States due to its low cost, ease of production, and favorable blending characteristics.

In a fuel release scenario, the resulting plume of MTBE is larger and more recalcitrant than the resulting plumes of benzene, toluene, ethylbenzene, and xylenes (BTEX). This is because, compared to the other components of gasoline, MTBE is extremely soluble (approximately 40 times more soluble than BTEX compounds at room temperature), resistant to degradation in the aqueous phase, and weakly sorbing to soils.

In 1993, the U.S. EPA classified MTBE as a possible human carcinogen, and issued a draft lifetime health advisory of 20 to 200 µg/L in drinking. In 1996, another draft lifetime health advisory of 70 µg/L in drinking water was released for public comment. Evaluation of current remedial technologies indicate generally low efficiencies and high costs for the removal of MTBE from potable water supplies water (Squillance et al., 1994).

In situ air sparging (IAS) is a technique in which uncontaminated air is injected into the aquifer below the contaminated zone, causing air to rise in the form of channels through the impacted saturated zone. Where these channels come into contact with volatile and semi- volatile contaminants, the chemical partitions into the vapor phase and is transported to the vadose zone, where it can be collected with a vapor extraction system and treated with carbon adsorption.

Trials were conducted in two-dimensional laboratory-scale experimental tanks at Arizona State University (ASU) and the Oregon Graduate Institute (OGI) to determine the possible interactive effects of field variables on the removal efficiency of MTBE.

MATERIALS AND METHODS

Physical Aquifer Models. The two-dimensional experimental physical model at ASU was constructed of Plexiglas, measuring 244 cm (8 ft) long, 122 cm (4 ft) deep, and 5 cm (2 in) wide (Figure 1). The lid was sealed with an O-ring and bolted so that an air-tight seal was maintained. On each end, one inch clear wells extending the depth of the tank were constructed of Plexiglas slots covered with reinforced aluminum screen. The model was filled with 1 mm diameter glass beads to provide a uniform porous medium. A metal plate was placed on top of the beads and compressed by bolts screwed down from the lid to prevent variations in bead consolidation during experimentation. Deformation of the tank was minimized by an external Unistrut frame. Groundwater flow was simulated using a variable speed peristaltic pump (Spectra/Chrom Macroflow Pump, Spectrum, Phoenix) with flowrate controlled by a rotameter (RMA Ratemaster, Dwyer, New Jersey). A compressed air supply line was connected through a port in the bottom of the tank. Offgas concentrations were determined every 13 minutes by an SRI gas chromatograph (Series 9300B, Scientific Research Instruments, Torrance, CA) fitted with a 30 m MXT column (MXT-1, Supelco, Bellefonte, PA). PeakSimple, a software-driven data acquisition system created by SRI, was used to automate sample analysis and data storage.

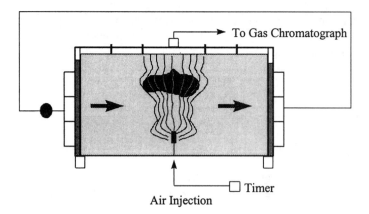

Figure 1. Schematic of the ASU Two-Dimensional Experimental Aquifer Model.

The aquifer model used in the experiments performed at OGI was 5.9 m long, 2.4 m deep, and 0.53 m wide (Figure 2). It was constructed with Plexiglas side and end panels; all supported by a steel frame welded to a steel I-beam bottom. Perforated metal screens covered with 100 mesh wire cloth were installed parallel to and 17.6 cm from the ends of the aquifer to form end reservoirs. These screens were supported by a stainless steel frame to provide a rigid support to the end of the aquifer. The model was filled with an alluvial beach sand to replicate conditions found in an unconsolidated porous media aquifer. The top of the aquifer was constrained by screens identical to those forming the reservoir end wells to prevent fluidization and physical displacement of the sand during air sparging. The top of the steel tank was covered with a leak tight plexiglas plenum creating a 14 cm high head space above the aquifer that allowed all the offgas to be collected. To prevent the loss of the offgas to the end reservoir headspace, leak tight sheet aluminum plates were installed. These plates extended from the plexiglas plenum to 7.62 cm deep in the aquifer.

Nitrogen was injected into the model through a 45.72 cm long, 2.54 cm diameter, 0.2 micron pore size, 316 stainless steel porous metal filter element (Model 2316-G16-18-A00-0.2-AB, Mott Metallurgical Corp.). The injection pressure was maintained at 20 to 25 psi. Flowrates were controlled by a Cole-Palmer floating ball flowmeter (model FM 102-05) at 3.05 L/min. Samples were automatically injected into a Hewlett-Packard 5890 Gas Chromatograph fitted with a 100 m long, 0.25 mm diameter Petrocol column (Supelco, Bellefonte, PA) at one half hour intervals via a 10-port Valco sampling valve. The 10-port valve was activated and GC data analysis acquired by a 3000 Series Data acquisition system (Nelson Analytical Inc.).

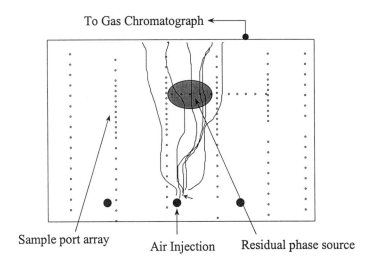

Figure 2. Schematic of the OGI Two-Dimensional Experimental Aquifer Model.

Standard Preparation and Sample Analyses. At ASU, component concentrations were determined by comparing response factors from acquired data with responses from 0.1, 1, and 10 μL-component/ L-air standards created in 10 L Tedlar bags (Model # 232-08, SKC West, Fullerton, CA). Replicates of these standards were measured before and after experimental runs to assess system reproducibility.

At OGI, component concentrations and system reliability were monitored using three types of standards. Dilutions of the source mixture were prepared in precleaned 0.84 L stainless steel canisters pressurized with nitrogen. An independent hydrocarbon standard (Liquid Carbonics) spiked with a known mass of MTBE was also used to verify daily accuracy and reproducibility of the GC analytical system. A hexane internal standard (IS) was analyzed simultaneously with each sample or external standard.

Experimental Approach. At ASU, 40 mL of the immiscible hydrocarbon source mixture (5.9 g MTBE) was injected into the model 10 cm below the bead surface through a ¼" steel tube with a 50 mL Hamilton gas-tight syringe. The mixture was allowed to percolate down to the capillary fringe at 16 cm below the bead surface. The water table was then lowered to 80 cm below bead surface and raised again in order to "smear" the zone of contamination. The compressed air sparge was initiated one hour after the water table had been reinstituted. Groundwater flow was initiated at 10 hours in an attempt to increase product recovery. At 30 hours the water table was lowered to expose all hydrocarbon contaminated media. The experiment was terminated at 40 hours, when offgas concentrations fell below ~0.02 mg-hydrocarbon/L-air.

At OGI, 200 mL of the immiscible hydrocarbon source mixture (MTBE mass 11.76 g) was injected into the model. A 50 mL Hamilton gas-tight syringe was used to inject 50 mL of the chemicals into four source ports centered above the sparge point. The source ports were located 165.1 cm above the model bottom and 264.16 cm, 284.48 cm, 294.64 cm, and 304.8 cm from the south end of the model. In each port the tubing was inserted in the model to a depth of 8.9 cm and 10 mL of the source was ejected from the syringe. The tubing was then advanced to 17.8 cm, 26.7 cm, 35.6 cm, and 44.4 cm into the model and 10 mL of the chemical ejected at each depth.

During injection of the source chemicals the water level in both end reservoirs was 137.2 cm and the capillary fringe in the source area was 165.1 cm. After the source was injected the water level in both end reservoirs was raised to 165.1 cm. This raised the height of the capillary fringe in the source area to 190.5 cm. The nitrogen sparge system was started 48 hours after the source was injected. Sparging was terminated after 12 days.

RESULTS AND DISCUSSION

The results from the data acquisition systems consisted of vapor concentration data over known intervals of time (Figure 3). From these, and the known air/nitrogen-injection rates, instantaneous mass removal rates can be determined. Integration of these mass removal rates over time provides a measure of cumulative mass removal.

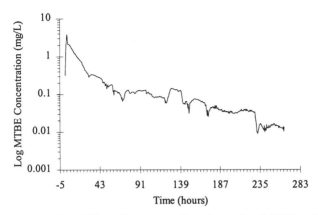

Figure 3. Log Offgas Concentrations from the OGI Physical Model.

Essentially the same qualitative behavior is demonstrated in both models: the observed concentrations (and therefore mass removal rates) are greatest at the beginning of the experiment and then they decline while exhibiting significant variations with time. Peak mass removal rates are comparable for each of the models.

Figures 4 and 5 present the cumulative mass removal expressed as a percentage of the initial mass injected into each aquifer physical model. Long-term removal efficiencies appear to be comparable for the two models, (86.5 and 82.9% recovery). Figure 5 shows that as the immiscible source zone was exposed by draining the model, the recovery is enhanced.

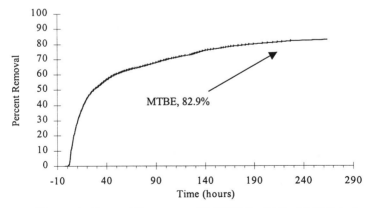

Figure 4. Percent Recovery of MTBE for IAS of 3.05 L/min

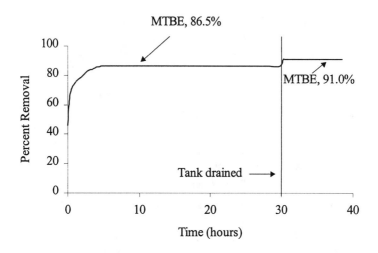

Figure 5. Percent Recovery of MTBE for IAS of 10 L/min

SUMMARY

Results from two-dimensional laboratory-scale physical aquifer models at ASU and OGI indicate that, under idealized conditions, approximately 85% of the MTBE in test-hydrocarbon mixtures containing MTBE can be removed by volatilization during air sparging. Long term removal of MTBE did not seem to be affected by variations in porous media for the range of media tested (1 mm diameter glass beads and alluvial beach sand). It appears, over the short term, that recovery rates are increased with increased air injection rates. The data also suggest that maximum MTBE removal achieved by volatilization is affected by water saturation.

REFERENCES

Rutherford, K. W., 1995. Evaluation of Mass Transfer Rates During a Laboratory Scale Air Sparging Experiment. M.S. Thesis. Arizona State University, Department of Civil and Environmental Engineering.

Squillance, Paul C., James F. Pankow, Nic. E. Korte, and John S. Zogorski. 1994. Environmental Behavior and Fate of Methyl tert-Butyl Ether (MTBE). USGS document

Enhancing Chlorinated Methanes Degradation by Modifying The Fe^0 Reduction System

Shu-fen Cheng and Shian-chee Wu

Graduate Institute of Environmental Engineering, National Taiwan University, Taipei, Taiwan, R.O.C.

Abstract Studies have showed that zero-valent iron, Fe^0, can be used to enhance the degradation of chlorinated organic compounds in groundwater . However, as the progress of dechlorination reaction, the chlorines on the target molecules are removed slower and slower and the degradation of the daughter compounds is greatly reduced or even stoped. If the chemical activity of the surface of iron or the reduction potential of the system can be increased, the reaction will be enhanced. The first part of this study included experiments in the continuous batch reacators to investigate the influence of external electric potention to the degradation rates of CCl_4 and CH_2Cl_2 by Fe^0 and Zn^0 respectively. The second part involved using vials to perform experiments and to compare the reaction rates of CCl_4, $CHCl_3$ and CH_2Cl_2 with Fe, or Zn, or the combination of Fe and Zn.

The results indicated that the external electric potential enhanced the degradation of CCl_4. CCl_4 was entirely depleted in 4 minutes and the half life time for $CHCl_3$ was about 30 hours. Without electric potential, 3 hours was needed to completely degrade CCl_4 and the half life time for $CHCl_3$ was 100 hours. CCl_4 and $CHCl_3$ were degraded rapidly by Zn. In addition, CH_2Cl_2 originated from CCl_4 and $CHCl_3$ was gradually decomposed. Comparatively, the pure CH_2Cl_2 aqueous solution did not show evidence of degradation.

INTRODUCTION

In the last few years, the use of zero-valent iron has been an innovative technology to enhance the degradation of chlorinated organic compounds in groundwater. (Gillham(1994), Matheson(1994), Weber(1996), Siantar(1996)) To date, there are several full-scale and pilot-scale tests in process. However, the laboratory tests showed that the reaction rate decreased significantly as the chlorines being removed from the chlorinated compounds (Matheson, 1994).

In order to promote the reaction rate and make the reductive dechlorination reaction complete, many researches have been dedicated to the improvement of the zero-valent iron remediation technology. Muftikian (1995) employed palladized iron to degrade trichloroethylene(TCE), tetrachloroethylene(PCE), carbon tetrachloride, chloroform, and dichloromethane. Gillham tried to use nickel metal to enhance the degradation of TCE and PCE by zero-valent iron (Appleton, 1996).

For many chlorinated compounds, the degradation by zero-valent iron is pseudo-first-order reaction, and the electron are transfered directly from the iron surface to the target compounds (Gillham, 1994; Matheson, 1994;Weber, 1996, et. al.). The reaction rate depends on the ratio of surface area to volume and surface reactivity of iron. Wang (1997) used synthesized nanoscale iron and palladized nanoscale iron to enhance the degradation reaction. It was also found that the high specific surface area, high surface reactivity, and palladium could prevent the formation of iron oxides on the iron surface(Wang, 1997).

According to Matheson's (1994) research, the probable pathways of the degradation include : (1).direct electron transfer from the metal surface of iron to the adsorbed alkyl halide. (2).catalyzed hydrogenolysis by the H_2 that is formed by the reduction of H_2O. Therefore, if a cathode is introduced into the reaction system, it would provided more electrons on the iron metal surface, regenerate active surfaces on iron particles, and hence increase the available surface area of iron.

The first purpose of this work was to introduce external electrical current into the reactor to regenerate the active surface of zero-valent iron and to enhance the degradation the chlorinated organic compounds such as CCl_4, $CHCl_3$, and CH_2Cl_2. The second purpose of our work was to compare the degradation rates for the CCl_4, $CHCl_3$ and CH_2Cl_2 by Fe, Zn, or bi-metal (Fe and Zn) so that the efficiency of the enhancement made by zinc can be evaluated.

EXPERIMENTAL SECTION

Experimental design   The first part of experiments was to investigate the effect of the use of external electric potential on the enhancement the degradation of CCl_4 and CH_2Cl_2 by iron. Figure 1 is the experimental schematic diagram of the set-up. There are two inter-connected reactors each with capacity about 725mL. The proton-permeable membrane (Nafion 117, Du Pont) was used to separate the two reactors. Fe or Zn foils were used as electrodes, and connected to a DC power supply. Target compounds and metal powder were added in the reactor with cathode. Teflon–coated gas-tight stirrer rods were installed in the two reactors. The rotary rate of the stirrer was controlled at about 500 rpms to make sure the metal particles distributed homogeneously in the reactors. An Oxidation Reduction Potential (ORP) electrode and a pH electrode were installed in reactors to detect the change of ORP, pH, and temperature during the experiments. Samples were taken with a Teflon tube connected to the reactor. Experiments without DC current was conducted for comparison .

In another part of the study, batch experiments were performed to compare the degradation of CCl_4, $CHCl_3$ and CH_2Cl_2 by Fe, Zn, and bi-metal Fe and Zn. 15 ml solution of three chlorinated compounds with concentration of 1.5 mg/L each were mixed with. metal particles of 0, 0.5 or 1.0g in brown glass vials. Vials were sealed with aluminum caps (with PTFE/silicone liner), and rotated with a rotary stirrer at a rate of 30 rpms.

Analysis 5mL of solutions was extracted from the reactors by a gas-tight side-pore syringe and put in a brown vial periodically. The vials were sealed with aluminum caps and put in a 30°C oven for more than one hour to make gas and liquid phases achieve equilibrium. In the batch experiments, 5mL of solution was withdrawn with syringe for analysis.

The gas in the headspace of samples or standard solutions was withdrawn and directly injected to the gas chromatograph (Hewlett-Packard 5890 II) equipped with a DB624 analytical column (30m x 0.53mm x 3.0 um, J&M) and electron capture detector (ECD). The temperature was maintained at 220°C for injection port, 260°C for detector, and 40°C for oven. The column flow velocity was maintained at 3.2 mL/min.

RESULTS AND DISCUSSION

Effects of external potential The external voltage of the reactor cell during the whole reaction was maintained at about 13 Volts. The electrical current was initially 75mA and decreased rapidly to 35mA (in about 1 hour). Then, with the progress of reaction, the current decreased gradually and stayed at 15mA finally. The pH value in the solution reached equilibrium and stayed at about 7.5 since half an hour after reaction started. ORP stayed between –550 and –650mV. In the reactor with external voltage and 30 g of Fe powder, the concentraction of CCl_4 was 14.2 uM initially and was below the detection limit (50 ug/L)within 4 minutes. For the experiments without external voltage, the concentraction of CCl_4 was 10.3 uM initially, and it took 3 hours to completely decompose CCl_4. The concentration of $CHCl_3$ during the reaction was shown in Figure 2. After the reaction started, the concentration of $CHCl_3$ increased rapidly. The maximum concentration of $CHCl_3$ in the reactor was 7.6 uM and was equal to 53.5% of the initial concentration of CCl_4. In the control experiment, the concentration of $CHCl_3$ increased to 8.8 uM, equal to 85.4% of the initial concentration of CCl_4. The half life time of $CHCl_3$ in the reactor was about 30 hours, whereas, it was about 100 hours in the control experiment. With the comparison of the concentrations of $CHCl_3$ in the two sets of experiments, it gives the evidence that external electric current can enhance the degradation reaction for both CCl_4 and its daughter compound, $CHCl_3$. In these experiments, CH_2Cl_2 did not increase with the decomposition of $CHCl_3$. It stayed at relatively low concentration. It was deduced that the decomposition of $CHCl_3$ might not produce CH_2Cl_2 only. More study is needed to clarify the reaction mechanism of $CHCl_3$ decomposition.

Effects of Zn and Fe/Zn The iron powder used in the vial batch reaction was 97% of purity, non-treated commercial iron. The results of experiment showed that the degradation reaction was not obvious. The possible reason was: (1). the reactivity of iron powder was low due to its impurity and non-pretreatment; (2). the stirring was not enough because the rotary stirrer could make iron powder settle and not able to be evenly distributed. Zinc was found able to degrade CCl_4 and $CHCl_3$. The initial concentrations of CCl_4 and $CHCl_3$ in the experiment were 9.7 uM and 20.9 uM, respectively. After 24 hours of reaction, the concentration of CCl_4 and $CHCl_3$ both decreased to below detection limit when only zinc was used (the amount of zinc was 1g/15mL and 0.5g/mL, respectively). The results of experiment were shown in Table 1 and Table 2. For the bi-metal (Fe and Zn at a concentration of 0.5g/15mL each) experiments, the results were obviously different from those when only Fe or Zn was used. CCl_4 in the bi-metal experiment decreased to below detection limit within 24 hours. Almost all CCl_4 was transformed to $CHCl_3$. The mass balance was near 100%. Furthermore, with degradation of $CHCl_3$, CH_2Cl_2 increased but the mass balance cannot be achieved. The total recovery decreased as $CHCl_3$ was degraded. The degradation of CCl_4 by bi-metal was confirmed to be mono-dechlorination, but not for the degradation of $CHCl_3$. In the experiments with addition of Zn to degrade CCl_4 and $CHCl_3$, the concentration of CH_2Cl_2 tended to increase first and decrease later. However, the increasing rate of CH_2Cl_2 did not coincide with the degradation rate of CCl_4 and $CHCl_3$. It might be that some other intermediate products were formed which were adsorbed on the surface of metals, or dissolved in the aqueous phase. The CH_2Cl_2 originated from the degradation of CCl_4 and $CHCl_3$ tended to disappear gradually, whereas, the degradation of pure aqueous solution of

CH_2Cl_2 was not obvious. It was suggested that some compounds produced during the degradation of CCl_4 and $CHCl_3$ could enhance the degradation of CH_2Cl_2.

CONCLUSIONS

The enhancement of the degradation of CCl_4 with external electric potential was confirmed in this study. More study is needed to search for the control strategies to obtain better treatment efficacy. Zn was found to enhance the degradation of CCl_4 and $CHCl_3$ better than the bi-metal, Fe and Zn . Further study to investigate the feasibility of application of Zn in the remediation practices is needed.

REFERENCES

Appleton, E. L. 1996. "A Nickel-Iron Wall Against Contaminated Groundwater." *Environ. Sci. Technol.* 30(12):536A-539A.

Gillham, R. W., and S. F. O'Hannesin. 1994. "Enhanced Degradation of Halogenated Aliphatics by Zero-Valent Iron." *Ground Water.* 32(6):958-967.

Matheson, L. J., and P. G. Tratnyek. 1994. "Reductive Dehalogenation of Chlorinated Methanes by Iron Metal." *Environ. Sci. Technol.* 28(12):2045-2053.

Siantar, D. P. , C. G. Schreier, C. S. Chou, and Reinhard, M. 1996. "Treatment of 1,2-dibromo-3-chloropropane and Nitrate-contaminated Water with Zero-Valent Iron or Hydrogen / Palladium catalysts." *Wat. Res.* 30(10): 2315-2322.

Wang, C. B. and W. X. Zhang. 1997. "Synthesizing Nanoscale Iron Particles for Rapid and Complete Dechlorination of TCE and PCBs." *Environ. Sci. Technol.* 31(7):2154-2156.

TABLE 1. Concentration of CCl_4, $CHCl_3$, and CH_2Cl_2 for the degradation of CCl_4 by Fe and Zn.

Time (hours)	Blank	Zn(1g)	Zn(0.5g)	Zn+Fe(1g)	Fe(1g)	Fe(0.5g)
	CCl_4 conc.(mg/L)					
24	1.266	N.D.	N.D.	N.D.	1.159	1.174
48	1.462	N.D.	N.D.	N.D.	1.194	1.426
72	1.349	N.D.	N.D.	N.D.	1.261	1.392
96	1.394	N.D.	N.D.	N.D.	1.333	1.343
120	1.462	N.D.	N.D.	N.D.	1.496	1.319
	$CHCl_3$ conc.(mg/L)					
24	0.013	0.006	0.008	1.256	0.015	0.008
48	0.010	N.D.	N.D.	0.820	0.016	0.019
72	N.D.	N.D.	N.D.	0.676	0.009	N.D.
96	0.013	N.D.	N.D.	0.570	0.017	0.010
120	0.013	N.D.	0.007	0.447	0.019	0.017
	CH_2Cl_2 conc.(mg/L)					
24	N.D.	0.112	0.123	N.D.	0.113	N.D.
48	0.098	0.179	0.148	0.126	0.076	0.059
72	N.D.	0.718	0.209	0.206	0.040	N.D.
96	N.D.	0.228	0.161	0.160	0.069	0.029
120	N.D.	0.172	0.120	0.171	0.041	N.D.

TABLE 2. Concentration of CHCl₃ and CH₂Cl₂ for the degradation of CHCl₃ by Fe and Zn.

Time (hours)	blank	Zn(1g)	Zn(0.5g)	Zn+Fe(1g)	Fe(1g)	Fe(0.5g)
	CHCl₃ conc.(mg/L)					
24	2.639	N.D.	0.006	2.753	2.538	2.159
48	1.733	N.D.	0.006	1.373	1.823	1.908
72	2.430	N.D.	N.D.	1.962	2.490	2.372
96	2.169	N.D.	0.007	0.566	1.864	1.859
	CH₂Cl₂ conc.(mg/L)					
24	0.171	0.356	0.084	0.175	0.176	0.253
48	0.207	0.044	0.161	0.118	0.060	0.043
72	N.D.	0.091	0.155	0.336	0.050	N.D.
96	N.D.	0.104	0.093	0.167	N.D.	N.D.

N.D. : lower than MDL (CCl₄:50 ug/L; CHCl₃:5.0 ug/L; CH₂Cl₂:15.3 ug/L)

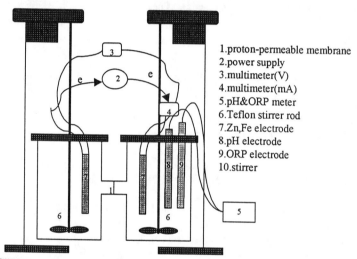

1.proton-permeable membrane
2.power supply
3.multimeter(V)
4.multimeter(mA)
5.pH&ORP meter
6.Teflon stirrer rod
7.Zn,Fe electrode
8.pH electrode
9.ORP electrode
10.stirrer

FIGURE 1. The schematic diagram of the experimental set-up.

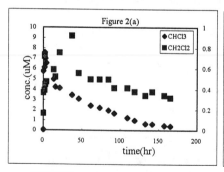
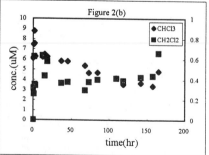

FIGURE 2 . The concentration of CHCl₃ and CH₂Cl₂ corresponded to the time of CCl₄ degradation in the experiments. (a).experiment with external voltage, (b). experiment without external voltage. The main-axis of Y is the conc. of CHCl3(uM), and the by-axis of Y is the conc. of CH₂Cl₂(uM). System : each reactor was added 30 g of iron, unbuffered, mixed at 500 rpms, at room temperature (about 20 °C)

MATRIX EFFECTS ON CATALYST DEACTIVATION IN A Pd/H$_2$ HYDRODEHALOGENATION SYSTEM

N. Munakata, P.V. Roberts, and M. Reinhard (Stanford University, Stanford, CA, USA)

W. McNab (Lawrence Livermore National Laboratory, Livermore, CA, USA)

ABSTRACT: Previous studies indicate that the hydrodehalogenation of halogenated hydrocarbons by supported Pd and H$_2$ may be a promising technology for remediating contaminated waters. Bench-scale laboratory studies have demonstrated the technical feasibility of trichloroethylene (TCE) reduction to ethane by hydrogen in a Pd-catalyzed hydrodehalogenation reaction under ideal conditions. This study investigated catalyst activity in bench-scale continuous-flow columns under a variety of water matrices. Deionized (DI) water caused no observable deactivation, and phosphate caused minimal deactivation when added to DI water. Nitrate, which was amended to DI water, competed with TCE for hydrogen or catalyst sites but caused no deactivation. Carbonate, carbon dioxide and groundwater from Livermore, CA, all caused gradual deactivation of the palladium. Regeneration of the carbonate-deactivated catalyst through evacuation and oxidation improved catalyst activity.

INTRODUCTION

Widespread contamination of groundwaters by halogenated solvents and pesticides has resulted in the need for effective water treatment methods. Conventional technologies such as GAC adsorption simply transfer contaminants from one medium to another, thus requiring additional incineration or regeneration, both of which are expensive. As a result, methods which eliminate, rather than transfer, pollutants are preferable. Previous studies have indicated that reduction by hydrogen in a Pd-catalyzed hydrodehalogenation reaction may be one such promising method. Recent batch experiments showed rapid and fairly complete (>99%) removal of compounds such as TCE and 1,2-dibromo-3-chloropropane (DBCP) from DI water. (Schreier and Reinhard, 1995; Siantar et. al., 1996)

In this work, the performance of the Pd catalyst in column systems with various water matrices (DI water, solutions of known additions, and complex natural groundwaters) is examined, with the objective of defining reaction competitors, catalyst poisons, and potential pretreatment or regeneration needs. The transformation of TCE by a column of Pd/Al$_2$O$_3$ catalyst was tested using various types of water and regeneration potential was assessed. Results are discussed in the following sections.

MATERIALS AND METHODS

Experimental Apparatus. Columns were constructed of 0.7 cm ID glass tubing, filled with Pd/Al$_2$O$_3$ catalyst (UOP) and crushed to a 20-50 mesh (0.3-0.8 mm) size fraction. The catalyst has a measured composition of approximately 1% Pd by weight. The catalyst mass in each column ranges from 0.1-2.0 g.

The flow system (Figure 1) consists of a 3 or 5 gallon reservoir of water, purged with pure H_2 gas for at least 30 minutes. Hydrogen pressure (2-8 psig) pneumatically drives the water from the reservoir to a T-joint, where a saturated solution of TCE is added using a syringe pump. The combined stream then flows through the catalyst column; valves allow the collection of column inlet and outlet samples. Both the water and TCE flow rates can be varied, thus making the feed concentration and the column residence time adjustable. In these experiments, the water flow rate was approximately 1 - 9 mL/min and the inlet TCE concentration was varied between zero and 49 mg/L.

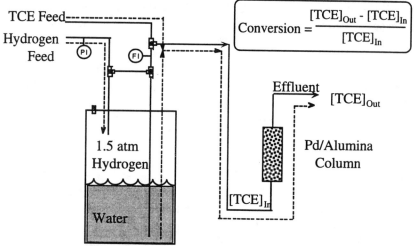

$$Conversion = \frac{[TCE]_{Out} - [TCE]_{In}}{[TCE]_{In}}$$

FIGURE 1. Column experimental set-up.

In each experiment, the reservoir was filled with DI water or groundwater from Livermore, CA (water quality data provided in Table 1). In several runs, the water was amended with nitrate, phosphate, carbonate or CO_2. Carbonate and phosphate were added directly to the reservoir water in the form of sodium carbonate or a 50/50 mixture of mono- and di-basic potassium phosphate. Nitrate was added with a syringe feed to a DI column flow from the reservoir. In two separate tests, CO_2 was added to DI and groundwater by purging with a 90% H_2/10% CO_2 gas mixture (CO_2 partial pressure of approximately 2 psi), rather than pure H_2. The conditions for each experiment are summarized in Table 2.

Analytical Methods. Inlet and outlet TCE concentrations were measured using headspace samples in an ECD detector on the Hewlett Packard 5890 Series II Gas Chromatograph. Inlet and outlet nitrate concentrations were measured using a Hewlett Packard 8451A Diode Array Spectrophotometer at a UV absorbance of 210 nm.

TABLE 1. Livermore water quality data, all values (except pH) in mg/L.

Component	Conc (mg/L)
Carbonate	180
Nitrate	23
Sulfate	8.2
Chloride	33
Ca2+	39
Mg2+	12
K+	1.8
Na+	35
pH	7.4

TABLE 2. Experimental Conditions.

System Parameters	Catalyst Mass (g)	Column Volume (mL)	H_2 Pressure (psig)	Flow Rates (mL/min)	TCE Conc. (mg/L)
DI Water	2.0	3.85	3.5	1.0-9.0	0.4-49
DI Water + Phosphate	0.301	0.69	8	0.90-1.10	0.4-8
DI Water + Carbonate	0.100	0.23	7	1.00-1.20	2-7
DI Water + CO_2	0.100	0.23	8	0.5-5.5	0.5-1.5
W1204 Water	0.300	0.58	8	0.85-1.10	1-6
W1204 Water + CO_2	0.101	0.23	8	1.0	3-19

RESULTS AND DISCUSSION

DI Water/Nitrate Column. In the baseline column test with DI water, inlet TCE concentrations between 1.5 mg/L and 25 mg/L were more than 99% transformed, with no noticeable decrease in catalyst activity over 5 months (Figure 2). TCE concentrations above 30 mg/L exceeded the stoichiometric TCE:H_2 ratio for the complete reduction of TCE to ethane. The degradation rate fell, indicating complete usage of the available hydrogen (approximately 2 mg H_2/L). When nitrate was added to the system, it appeared to compete with the TCE for hydrogen and/or catalyst sites. As the inlet nitrate concentration was increased, the ability of the system to reduce TCE decreased. For example, in the absence of nitrate, TCE was 99.97% reduced, from 25 mg/L to 8.2 µg/L. At a nitrate concentration of 44 mg/L nitrate, TCE was reduced only 52 %, from 6 mg/L to 2.9 mg/L. When nitrate addition was stopped, the column returned to its original TCE removal capacity.

Phosphate Column. To investigate the effects of phosphate on the catalyst life, 100 mg/L of phosphate (50 mg phosphate/L each of mono- and di-basic potassium phosphate) was added to the reservoir water. This addition appeared to cause minor deactivation of the catalyst (Figure 3). The initial temporary drop in conversion may suggest phosphate interactions with the palladium or support;

these interactions stabilized after two days. Further tests are being conducted to verify the phosphate effects.

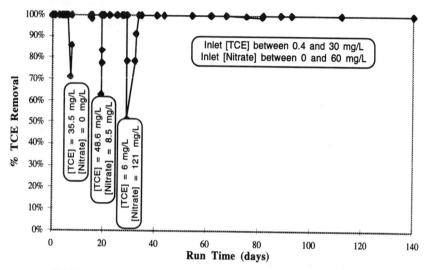

FIGURE 2. TCE Removal in the Column with DI Water.

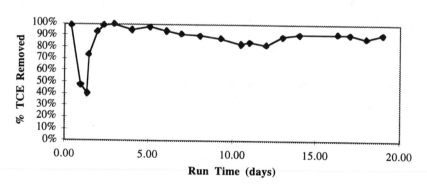

FIGURE 3. TCE Removal in the Column with Phosphate-Amended DI Water.

Carbonate Species Columns. Experiments with Livermore groundwater and the carbon dioxide and sodium carbonate amended waters yielded transformation of TCE but showed deactivation of the catalyst. CO_2 amended DI water completely deactivated the catalyst in 13 days, while the DI water alone showed no deactivation after 5 months. Similarly, in the groundwater test, the CO_2 amended water deactivated the catalyst more rapidly than the water alone (Figure 4).

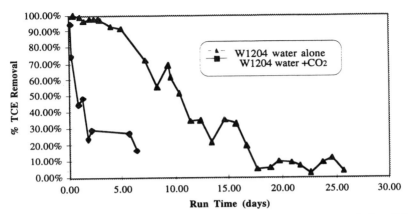

FIGURE 4. Deactivation of the catalyst by groundwater and by CO₂ amended groundwater.

In the Na_2CO_3-amended DI water (500 mg/L Na_2CO_3), TCE conversion decreased to 17.5% after 5.5 days, or 130 hours (Figure 5). An initial attempt was made to regenerate the catalyst by flushing the column with DI water under normal experimental conditions: 1.5 atm of H_2 pressure in the water reservoir, and TCE feed via syringe pump. This resulted in an increase in conversion from 5% to 20%. (Figure 6) Chemical reasoning suggests that a carbonate species (such as carbonate, carbonic acid and carbon dioxide) can sorb to the catalyst surface and be reduced to CO, which can poison Pd. Based on successful regeneration techniques described in the literature, the deactivated catalyst was placed under 0.07 mm Hg vacuum at 300°C for 2.0 minutes, then under oxygen flow at 300°C for 15 minutes (Doering, 1980). During the regeneration process, approximately 25% of the catalyst mass was lost. When the column was restarted on the DI water flow, the catalyst showed virtually complete recovery of activity, given the mass lost. However, the fluctuating performance of the column makes it difficult to fully evaluate catalyst regeneration from this test.

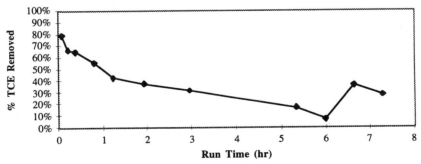

FIGURE 5. Deactivation of the catalyst by sodium carbonate amended DI water.

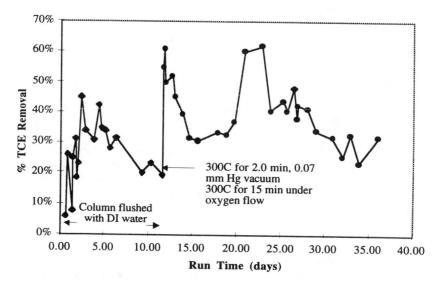

FIGURE 6. Regeneration of the catalyst deactivated by sodium carbonate amended DI water.

SUMMARY

The supported Pd/H_2 system, with its ability to reduce TCE to ethane without significant byproduct formation, has great potential as an alternative remediation technology. The results shown here demonstrate that the use of such a system to remove TCE from relatively clean water is technically feasible. However, it is also clear that some components likely to be present in a groundwater matrix, such as carbonate species, can have a detrimental effect on the catalyst activity and the hydrodehalogenation reaction. Future work will focus on understanding these effects and mitigating them through regeneration, pretreatment and manipulation of the Pd catalyst.

REFERENCES

Doering, D. L., H. Poppa, and J.T. Dickinson, 1980. "Changes Induced on the Surfaces of Small Pd Clusters by the Thermal Desorption of CO." *Journal of Vacuum Science and Technology.* 17(1): 198-200.

Schreier, C.G. and M. Reinhard. 1995. "Catalytic Hydrodehalogenation of Chlorinated Ethylenes Using Palladium and Hydrogen for the Treatment of Contaminated Water." *Chemosphere.* 31(6): 3475-3487.

Siantar, D., C. G. Schreier, C-S. Chou, and M. Reinhard. 1996. " Treatment of 1,2-dibromo-3-chloropropane and nitrate-contaminated water with zero-valent iron or hydrogen/palladium catalysts." *Water Research.* 30(10): 2315-2322.

DEHALOGENATION OF CHLORINATED SOLVENTS USING A PALLADIUM CATALYST AND HYDROGEN GAS

Greg Lowry (Stanford University, Stanford, CA)
Martin Reinhard (Stanford University, Stanford, CA)
Walt McNab, Jr. (Lawrence Livermore National Laboratory, Livermore, CA)

ABSTRACT: A 1% by weight palladium on alumina catalyst and hydrogen was effective for the rapid reductive dehalogenation of several halogenated solvents and pesticides. Three classes of chlorinated solvents were studied: ethylenes (tetrachloroethylene (PCE), trichloroethylene (TCE), dichloroethylene (DCE)), ethanes (dichloroethane (DCA)), and methanes (carbon tetrachloride and chloroform). Chlorinated ethenes transformed most rapidly and had greater than 95% transformation after 17 minutes. The primary reaction products were ethane and hydrochloric acid (HCl). No lesser chlorinated intermediates or products were formed when using a supported palladium on alumina catalyst. The chlorinated methanes had greater than 80% transformation after 153 minutes. The primary reaction products were methane and hydrochloric acid (HCl). Chloroform was produced as a reactive intermediate during the transformation of carbon tetrachloride but never exceeded more than 20% of the initial carbon tetrachloride concentration. The chlorinated ethanes, 1,1-DCA and 1,2-DCA, showed little reactivity and had only 8.9% and 0.4% transformation respectively after 39 hours.

INTRODUCTION

Extensive groundwater contamination by chlorinated solvents and pesticides presents a substantial health hazard throughout the world. For example, TCE has been identified as the major groundwater pollutant at 546 recognized Superfund sites (Cooper, et al., 1993). In response to this widespread contamination, many novel treatment technologies for the transformation of halogenated hydrocarbons are currently under investigation. Specifically, the use of noble metal catalysts for the dehalogenation of halogenated hydrocarbons to innocuous and/or less refractory compounds has recently emerged. In 1992, Kovenkiloglu et al. reported that some chlorinated ethenes, methanes, and aromatics were dehalogenated by palladium catalysts in the presence of hydrogen at room temperature and pressure (Kovenkiloglu et al., 1992). More recently, Siantar and Reinhard reported the rapid transformation of 1,2-dibromo-3-chloropropane (DBCP) and nitrate in water by hydrogen and a palladium catalyst (Siantar and Reinhard, 1996). These findings indicate a potential for using palladium catalysts in the remediation of groundwater contaminated with halogenated hydrocarbons.

MATERIALS AND METHODS

Transformation rates for the chlorinated solvents tested were determined in batch experiments. Two types of reactors were used in this study, a zero headspace reactor and a headspace reactor. Transformations of the chlorinated solvents were achieved with hydrogen and a powdered 1% by weight supported palladium on alumina catalyst. The catalyst, reactors, and experimental procedure are summarized below.

Catalyst. Kinetic experiments were run using a 1% (w/w) palladium on alumina catalyst supplied by Aldrich Chemical Company. The 1% by weight palladium content specified by the manufacturer was assumed to be accurate and not independently verified. The catalyst was supplied and used in a powdered form with particle sizes ranging between 38 and 70 μm (200-400 mesh). The catalyst was supplied in a pre-reduced form and no special precautions were taken to avoid catalyst exposure to air prior to kinetic experiments.

Zero Headspace Reactors. Transformation rate data for each compound were obtained from batch experiments carried out in a 1L glass reactor equipped with two sample ports. Pertinent reactor operating conditions are given in Table 1. The reactor was filled with Milli-Q water, catalyst, and then sparged with pure H_2 gas for 15 minutes. Sparging in this way removes dissolved oxygen and provides an excess of hydrogen well above the required stoichiometric amount such that the reaction rates obtained are independent of the hydrogen concentration. Immediately after sparging, a concentrated aqueous solution of known contaminant concentration was quickly added and the reactor capped, yielding initial contaminant concentrations of 1 to 3 mg/L. One mL aqueous samples were withdrawn from the reactor (displaced by H_2 saturated water to maintain zero headspace) at various times during the reaction and the halogenated hydrocarbon extracted with 0.5 mL of hexane or toluene depending on the compound. The halogenated hydrocarbon concentrations were then determined by GC/ECD.

TABLE 1. Zero Headspace Reactor Operating Conditions.

Reaction temperature, K	293-295
Reaction pressure, atm	1
Reactor volume, L	1.15
Agitator speed, rpm	500
Catalyst concentration, g/L	0.22
Average catalyst particle diameter, μm	54
Initial contaminant concentration, mg/L	1-3

Headspace Reactors. A batch reactor with headspace was used to obtain mass balances during the transformation of TCE and carbon tetrachloride. The headspace reactor was a gas tight 2.0 L glass and stainless steel fermentation reactor equipped with a magnetically driven mixer (Bioengineering Labs). The reactor had a port for gas phase sampling, gas line connections, a small heat exchanger, and a temperature probe. Two 5 cm impellers were attached to the shaft of the magnetic mixer to provide ample mixing within the reactor. The impeller speed and reactor temperature were monitored and controlled by an attached computer. Pertinent reactor operating conditions during runs designed for mass balance determination are given in Table 2.

The reactor was run as a three phase slurry batch reactor. The reactor was filled with 1.4-1.5 L of Milli-Q water. The reactor was then sparged for 15 min. with pure H_2 gas to remove dissolved oxygen and provide an excess of hydrogen well above the required stoichiometric amount such that the reaction rates obtained are not dependent on the hydrogen concentration. Immediately after sparging, a concentrated aqueous solution of the halogenated hydrocarbon was quickly added to the reactor to provide an initial contaminant concentration of 4-8 mg/L. The impeller speed was set to 1000 rpm and the reactor temperature was adjusted to 294 K using the small heat exchanger. The reactor was then pressurized (3-5.7 psig) with pure H_2 gas and allowed to equilibrate for 15-20 min. The powdered 1% Pd on alumina catalyst was then added as a slurry via syringe. At various times during the reaction 200 µL headspace samples were withdrawn from the reactor and analyzed by GC/FID for all carbonaceous gaseous reactants and products.

TABLE 2. Headspace Reactor Operating Conditions.

Parameter	TCE	Carbon Tet.
Reaction temperature, K	294 +/- 0.2	294 +/- 0.2
Reaction pressure, atm	1.2	1.4
Reactor pH		
Initial	6.6-6.8	N/A
Final	4.0-4.2	N/A
Reactor volume, L	2.0	2.0
Liquid Volume, L	1.4	1.5
Headspace Volume, L	0.6	0.5
Agitator speed, rpm	1000	1000
Catalyst concentration, g/L	0.047	0.067
Average catalyst particle diameter, µm	54	54
Initial contaminant concentration, mg/L	4	8.8

RESULTS AND DISCUSSION

Survey of Compounds Amenable to Transformation by a Palladium Catalyst and Hydrogen. Concentration vs. time data were used to determine the reaction order and relative reaction rate constants for priority pollutants in several different compound classes including halogenated olefins and alkyl halides. The pseudo-first order rate constants, (normalized by the aqueous catalyst concentration and palladium content), half life, and half life relative to TCE transformation are given in Table 3 for each of the compounds tested.

TABLE 3. Palladium Catalyzed Transformation Rates of Select Halogenated Compounds.

Compound	Degradation Rate Constant in MQ Water (L min^{-1} g Pd^{-1})	Half Life (min)	Half Life Relative to TCE (min)
Chlorinated Ethenes			
Tetrachloroethene	54.0	5.9	1.3
Trichloroethylene	70.8	4.5	1.0
1,1-Dichloroethylene	70.4	4.5	1.0
c-1,2-Dichloroethylene	83.1	3.8	0.8
t-1,2-Dichloroethylene	77.6	4.1	0.9
Chlorinated Methanes			
Carbon tetrachloride	53.1	6.0	1.3
Chloroform	4.5	70.6	15.7
Chlorinated Ethane/Propane			
1,1,2-trichlorotrifluoroethane	15.5	20.5	4.6
1,2-dibromo-3-chloropropane	60.5	5.3	1.2
1,1-dichloroethane	to slow to measure		
1,2-dichloroethane	to slow to measure		

All compounds tested displayed pseudo first order reaction kinetics. Table 3 shows that many priority pollutants are readily transformed in the H$_2$/Pd system. The reaction rates for most of the compounds tested are extremely rapid, with half life times between 3.8 and 6 minutes even at extremely low catalyst concentrations (0.22g/L). Exceptions are chloroform (t$_{1/2}$=70 min), Freon-113 (t$_{1/2}$=20.5 min), and the dichloroethanes which showed very little reactivity even at elevated catalyst concentrations.

The very fast reaction rates associated with the H$_2$/Pd process imply that both external mass transfer resistance and pore diffusion resistance must be considered. The reaction rates obtained in these preliminary screening tests are therefore specific to the catalyst, reactor design, stirring rates, and quite possibly affected to some extent by external mass transfer resistance. However, the aqueous phase diffusivities of each compound tested are similar and the reactor conditions were essentially constant for each experiment, so significant differences in reaction rates can be attributed to differences in reactivity among the species tested, and not solely due to differences in the liquid-solid mass transfer rates.

TCE Transformation Using Supported Palladium on Alumina Catalyst. The zero headspace reactors showed no indication that lesser chlorinated reaction products/intermediates were formed during TCE transformation, but experiments yielding good carbon mass balances were required to verify those results. The gas tight headspace batch reactor was used to successfully obtain good carbon mass balances for the system. The transformation of TCE in the headspace reactor using the 1% supported palladium on alumina catalyst is shown in Figure 1.

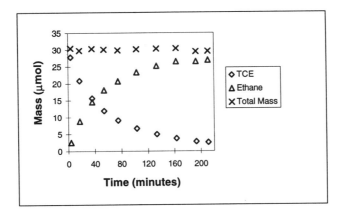

FIGURE 1. Transformation of TCE and subsequent production of ethane. A 99% carbon mass balance was achieved. Reactor conditions: 1.4L water, 0.6L headspace, 1000 rpm mixing rate, 21.5 °C, 66.5 mg of 1% Palladium on alumina powder. Initial [TCE] = 4 mg/L. H_2 pressure=3.0.

Ethane is the final transformation product of TCE when a supported palladium on alumina catalyst is used. Trace amounts of ethylene are formed as a reactive intermediate during the transformation of TCE, but are ultimately hydrogenated to ethane. No indication of chlorinated intermediates/products were seen during GC analysis. A carbon mass balance of 99% provides more evidence that no lesser chlorinated reaction intermediates or by products were produced during the transformation of TCE on the supported palladium on alumina catalyst.

Carbon Tetrachloride Transformation Using Supported Pure Palladium Powder. The primary product in the transformation of carbon tetrachloride by the supported palladium on alumina catalyst is methane (59.7%). Ethane is also produced and accounts for 13.8% of the carbon mass in the system. Chloroform is a reactive intermediate during the transformation of carbon tetrachloride and accounts for as much as 17.9% of the initial carbon mass added to the system. Chloroform dehalogenates at a slower rate to the final non-halogenated products (data not shown). No evidence of methylene chloride or chloromethane was detected in the reactor (data not shown). Minor amounts of ethylene, propane, and propylene were also detected in the reactor. The formation of C2 and C3 gaseous products during the dehalogenation of carbon tetrachloride by a palladium on alumina catalyst implies that a free radical mechanism may be involved. A total carbon mass balance of 93.1% was achieved.

CONCLUSIONS

The use of a palladium catalyst and hydrogen yields very rapid dehalogenation rates for many priority pollutants including TCE, all three DCE isomers, and carbon tetrachloride. The transformation of TCE to ethane occurs without accumulation of any lesser chlorinated intermediates or products. The

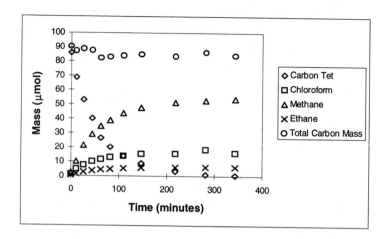

FIGURE 2. Transformation of Carbon Tetrachloride and subsequent production of methane, ethane, and chloroform. A 93% carbon mass balance was achieved. Reactor conditions: 1.5L water, 0.5L headspace, 1000 rpm mixing rate, 22.0 °C, 100 mg of 1% Pd on alumina powder. Initial [Carbon Tet.] = 8.8 mg/L. H_2 pressure=5.5 psig.

transformation of carbon tetrachloride to methane produces about 20% chloroform as a reactive intermediate but no accumulation of methylene chloride or chloromethane is observed. The rapid transformation rates and absence of persistent chlorinated reaction intermediates or products may make palladium technology an attractive option for remediation of chlorinated solvents and pesticides.

REFERENCES

Cooper, W., et al., "Removal of tri- (TCE) and tetrachloroethylene (PCE) from aqueous solution using high energy electrons". *Journal of the Air & Waste Management Association.* 43(10): 1358-1366.

Kovenkiloglu, S., Zhihua, C., Shah, D., Farrauto, R., Balko, E., "Direct catalytic hydrodechlorination of toxic organics in wastewater". *AICHE Journal.* 38(7): 1003-1012.

Siantar, D.P., Schreier, C.G., Reinhard, M., "Treatment of 1,2-dibromo-3-chloropropane and nitrate-contaminated water with zero-valent iron or hydrogen/palladium catalysts", Water Research. 30(10): 2315-2322.

TEMPERATURE EFFECT ON REDUCTIVE DECHLORINATION OF TRICHLOROETHENE BY ZERO-VALENT METALS

Chunming Su (U.S. EPA, NRMRL, Ada, Oklahoma)
Robert W. Puls (U.S. EPA, NRMRL, Ada, Oklahoma)

ABSTRACT: Laboratory batch tests were conducted to investigate the influence of temperature (10, 25, 40, and 55°C) on degradation of trichloroethene (TCE) by zero-valent metals and HCl-treated metals. The metals tested include Fisher electrolytic iron, Aldrich iron, Peerless iron, Master Builders iron, Fisher tin powder, Aldrich tin powder, Aldrich tin granules, and HCl-washed metals of iron and tin. The TCE dechlorination data can be fit satisfactorily with pseudo-first order kinetic rate equations with the rate constants ranging from 0.01 to 0.1 h^{-1} for Fisher electrolytic iron, 0.005 to 0.06 h^{-1} for Peerless iron and Master Builders iron, and 0.0003 to 0.05 h^{-1} for tin metals. The Aldrich iron showed the lowest reactivity. The rate constant increased with increasing temperature and can be described by the Arrhenius equation. The activation energy ranged from 32.2 to 39.4 kJ mol^{-1} for the virgin iron metals, indicative of a diffusion-limited reduction process, and 40.5 to 76.8 kJ mol^{-1} for the virgin tin metals. Temperature effect was more significant for the tin metals. The HCl-washing enhanced the dechlorination rate and decreased the activation energy for all tested metals with exception of Aldrich iron.

INTRODUCTION

Chlorinated hydrocarbons are ubiquitous environmental contaminants of worldwide concern. Recent investigations on reductive dechlorination of chlorinated solvents by zero-valent iron have shown promising potential of applying the technology to the in situ remediation of contaminated groundwater (Gillham and O'Hannesin, 1994, Campbell et al.,1997). Numerous feasibility studies, pilot tests, and field scale demonstration projects have been initiated to use granular iron in the form of in situ permeable reactive barriers (e.g., O'Hannesin and Gillham, 1997). Besides iron, tin has been studied as well (Boronina and Klabunde, 1995). The important factors influencing the degradation reaction include the surface area of metals, impurities of metals, temperature, pH, dissolved O_2, and rate of precipitate formation. In order to explain, predict, and improve the performance of iron barriers, it is necessary to further understand the kinetics of aqueous contaminant/metal systems as affected by environmental variables such as temperature. Little information on the temperature effect on metal-enhanced dechlorination is available and increasing temperature is expected to enhance the reaction.

Objective. The objective of this study was to investigate the effect of temperature on reductive dechlorination of trichloroethene (TCE) by zero-valent iron and tin metals in a laboratory batch experiment. The effect of acid treatment of metals on the kinetics of TCE reductive dechlorination will also be examined.

MATERIALS AND METHODS

The chemicals used were TCE (99+%, Aldrich, Milwaukee, WI), Fisher electrolytic iron (100 mesh, Fisher Scientific, Fair lawn, NJ), Aldrich iron metal filings (325 mesh, 97%, hydrogen reduced, Aldrich, Milwaukee, WI), Peerless iron (Peerless Metal Powders & Abrasive, Detroit, MI), Master Builders iron, Fisher tin powder, Aldrich tin powder (325 mesh, 99.8%), Aldrich tin granules (30 mesh), and HCl-washed metals of iron and tin. Selected metal filings were pretreated by washing 750 g of metals in 400 mL of Ar-sparged 1 M HCl with periodic shaking for 30 min, then rinsed 10 times with vigorous shaking in Ar-sparged deionized water and dried at 100 °C under N_2 atmosphere. The dried metals were stored in Ar-sparged jars to inhibit oxidation. Surface area of the filings (Table 1) was determined by BET N_2 analysis on a Coulter SA3100 surface area analyzer (Coulter Co., Hialeach, FL).

The TCE kinetics experiments were performed in 50-mL (with measured internal volumes of 60 ± 1 mL) clear borosilicate serum bottles (Weaton, Millville, NJ) containing 15 g of metals in contact with approximately 2 mg TCE L^{-1} solution with no headspace. The metals were added to serum bottles that were then filled with the TCE solution prepared in deionized water. The bottles were sealed immediately with aluminum crimp caps with Teflon-lined septa. Two bottles containing TCE and metals and two controls containing TCE only were prepared for each sampling time preset at 2, 4, 8, 12, 24, 48, 72, 96 and 120 h. The tests were conducted at four temperatures (10, 25, 40, 55°C) on a shaker set at 50 rpm in a water-bath . The bottles were removed from the shaker and subsamples were transferred to vials for volatile organic compounds analyses. The concentrations of TCE and its daughter products DCE isomers, vinyl chloride were determined by an automated purge and trap gas chromatography with a flame ionization detector. Dissolved methane, ethane and acetylene were determined using a gas chromatography headspace equilibration technique. The remaining solution was analyzed for Eh, pH, then filtered through 0.1 μm membrane for Fe^{3+}, Fe^{2+} and Cl^- determination.

Table 1. Specific surface area of the metal particles showing the means of duplicate determination with standard deviation.

Material	Specific Surface Area $m^2 g^{-1}$
Fisher Iron	0.091±0.005
Master Builders Fe	1.164±0.047
Peerless Iron	0.699±0.055
Aldrich Iron	0.192±0.001
Fisher Tin Powder	0.139±0.003
Aldrich Tin Powder	0.184±0.006
Aldrich Tin Granules	0.068±0.014
HCl-washed Master Builders Iron	2.136±0.088
HCl-washed Peerless Iron	1.269±0.066
HCl-washed Aldrich Iron	1.930±0.069
HCl-washed Aldrich Tin Granules	0.031±0.005

RESULTS AND DISCUSSION

Pseudo-first order reaction equation was able to describe the degradation of TCE for all the metals as shown in equation (1).

$$\ln C = \ln C_o - k\, t \tag{1}$$

where C is the TCE concentration at time t, C_o the initial TCE concentration, and k the rate constant. Zero-and second-order models gave poorer fitting of the data. To provide a convenient basis for comparison, the half-lives, $t_{1/2}$, were normalized to an area-to-volume ratio of 1 m^2 mL $^{-1}$.

From the rate constants, obtained at different temperatures (10, 25, 40, and 55°C), the energies of activation for the degradation of TCE by metal particles were determined using a logarithmic expression of the Arrhenius equation:

$$\ln k = \ln A - E_a/RT \tag{2}$$

where k is the rate constant, E_a the activation energy, R the molar gas constant, T the absolute temperature, and A is a preexponential factor. The activation energy for the reaction was determined from the slope of a plot of ln k vs. 1/T using linear least-square analysis (Table 2).

Table 2. Rate constants and activation energies (mean ± standard deviation) for the degradation of TCE by metals and HCl-washed metallic materials. Fifteen grams of metals were reacted with about 2 mg TCE L^{-1} in head space-free 60-mL serum bottles for up to 120 h.

Material	Rate constant x 1000				Activation Energy
	10°C	25°C	40°C	55°C	
			h^{-1}		kJ mol^{-1}
Fisher Fe	10.4±0.8	25.8±1.5	58.4±5.4	101±10	39.4±3.6
MSB Fe	8.67±0.23	32.6±1.2	46.9±1.0	59.5±1.7	32.2±0.1
Peerless Fe	5.48±0.01	18.6±0.0	35.7±0.9	48.8±3.4	37.4±0.3
Aldrich Fe	0.08±0.00	0.16±0.03	0.31±0.16	0.55±0.07	32.3±1.9
Fis Sn Powder	0.59±0.14	4.40±0.06	5.84±0.53	47.0±1.6	69.3±3.9
Ald Sn Powder	0.51±0.01	2.29±0.12	8.62±0.02	48.0±2.0	76.8±1.1
Ald Sn Granules	0.31±0.02	0.74±0.03	1.28±0.08	3.62±0.4	40.5±0.2
Fe + Sn†	1.70±0.15	9.15±0.52	30.2±0.90	63.0±2.5	62.5±2.1
MSB (HCl)		41.5±7.0		107±3	25.8±4.0
Peer Fe (HCl)		22.5±0.7		70.6±5.6	31.0±1.3
Ald Fe (HCl)		0.57±0.01		11.5±1.1	81.8±2.6
Ald Sn G (HCl)		0.98±0.06		2.85±0.11	29.1±0.5

†50% Fisher iron + 50% Fisher tin powder by mass.

At each temperature level, Fisher iron showed the greatest rate constants, whereas Aldrich iron the least. The sluggishness of Aldrich iron towards TCE may be due to the fact that it is made by hydrogen reduction and it is less pure (97%) than Fisher iron (99+%). Also the crystal microstructure of different iron samples may affect the interaction of TCE with iron surface. Calculated pseudo-first order reaction constants for the TCE degradation by both virgin and HCl-washed metals increased with increasing temperature. The activation energy values were less than 42 kJ mol[-1] for untreated Fisher, MSB, Peerless and Aldrich iron metals, indicative of a diffusion-limited reduction process. Temperature effect was more significant for the Fisher tin powder, Aldrich tin powder, and an equal mass mixture of Fisher iron and Fisher tin powder compared with untreated iron metals as indicated by the much larger activation energy for the tin metal powders and the mixture. The largest activation energy for the HCl-washed Aldrich iron may be related to the ten-fold increase in surface area (Table 1) and possibly an increase of reactive sites by HCl wash.

The calculated normalized half-lives of TCE transformation by untreated iron metals differed by more than two orders of magnitude between Fisher iron and Aldrich iron with comparable results for MSB and Peerless iron metals (Figure 1). A headspace developed in the Fisher iron serum bottles after 48 h of reaction, whereas no headspace was observed in the other untreated iron metal bottles even after 120 h. The normalized half-lives decreased with increasing temperature with the greatest temperature effect shown between 10 and 25°C.

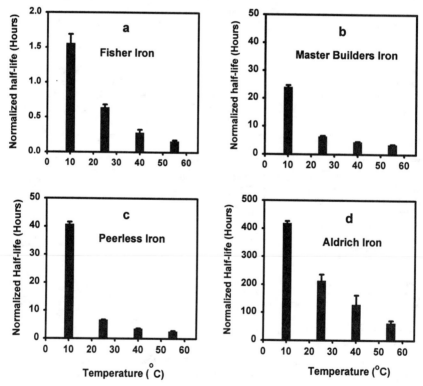

FIGURE 1. Calculated normalized half-lives ($t_{1/2-N}$) determined from fitting the first-order rate equation to the TCE degradation data as a function of temperature for Fisher iron (a), MSB iron (b), Peerless iron (c), and Aldrich iron (d).

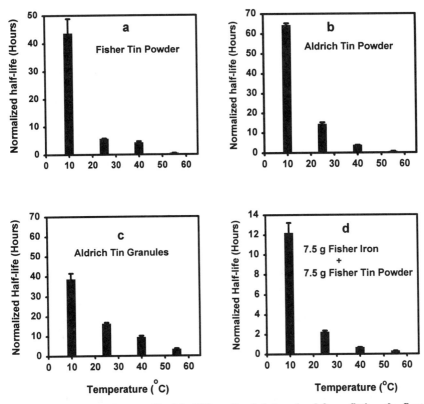

FIGURE 2. Calculated normalized half-lives (t$_{1/2-N}$) determined from fitting the first-order rate equation to the TCE degradation data as a function of temperature for Fisher tin powder (a), Aldrich tin powder (b), Aldrich tin granules (c), and an equal mass mixture of Fisher iron and tin powder (d).

Smaller variation in the normalized half-lives of TCE degradation were observed for Fisher tin powder, Aldrich tin powder and Aldrich tin granules (Figure 2a, b, and c) relative to the iron metals. A mixture of 50/50 by mass of Fisher iron and tin powder (Figure 2d) resulted in half-lives in between the values for the 100% iron or tin systems. At temperatures below 55°C, iron metals are superior over tin metals in TCE transformation. As temperature increased to 55°C, the difference in normalized half-lives diminished between the iron and tin metals with exception of Aldrich iron. The effect of HCl-washing on normalized half-lives were not conclusive with small differences between the HCl-treated and untreated metals.

From 90 to 110% of the chlorine atom present in the starting TCE solution was accounted for by free chloride ions and DCE isomers. *Cis*-DCE was the only detected Cl-containing byproduct of TCE transformation by Fisher and MSB iron, whereas 1,1-DCE was also found in the Peerless iron system. Both *trans*-DCE and vinyl chloride were also detected in the Fisher tin powder system at 55°C. Ethene and ethane accounted for less than 20% of the carbon mass balance for TCE. In the iron system, pH increased with time whereas Eh decreased with time, both reached a steady state within 120 h. The pH was above 9 and Eh near -400 mV for Fisher iron after 24 h. In the tin system, pH decreased initially below 5 then increased slightly over time to 6 and the Eh was never negative. Concentrations of Fe^{3+} and Fe^{2+} were below the detection limit of 0.03 and 0.09 µm.

It is well-established that reductive dechlorination occurs at the surface of metals. The HCl-washing of metals generally resulted in an increase in specific surface area, which in turn led to faster degradation of TCE. Under scanning electron microscope, the HCl-washed Aldrich iron displayed a rougher surface which was due to aggregates of small spheroidal particles on the surface. Some type of coatings scattered across the surface of Peerless iron and MSB iron appear to have been largely removed by the acid wash.

Our findings have practical implication for in situ remediation of chlorinated hydrocarbons by permeable reactive barriers of zero-valent metals in the field. To achieve the same degradation rate, a permeable reactive iron wall in a cold region would require a greater thickness than a wall in a warmer region. Also, since the temperature of most ground water ranges from a few to less than 20°C, an increase of temperature near the metal barriers to >25°C by electrical or radio frequency heating would greatly increase the contaminant degradation rate and thus decrease the thickness requirement of the metal wall and the time for remediation. The feasibility and economics of thermally-enhanced barrier technology may be an area of fruitful research.

ACKNOWLEDGMENTS

C.S. acknowledges the National Research Council Resident Research Associateship program. Although the research described in this article has been funded wholly or in part by the USEPA, it has not been subjected to the Agency's peer and administrative review and therefore may not necessarily reflect the views of the Agency and no official endorsement may be inferred.

REFERENCES

Boronina, T., and K.J. Klabunde. 1995. "Destruction of Organohalides in Water Using Metal Particles: Carbon Tetrachloride/Water Reactions with Magnesium, Tin, and Zinc." *Environ. Sci. Technol. 29*(6): 1511-1517.

Campbell, T.J., D.R. Burris, A.L. Roberts, and J.R. Wells. 1997. "Trichloroethylene and Tetrachloroethylene Reduction in a Metallic Iron-Water-Vapor Batch System." *Enriron. Toxicol. Chem. 16*(4): 625-630.

Gillham, R.W., and S.F. O'Hannesin. 1994. "Enhanced Degradation of Halogenated Aliphatics by Zero-valent Iron." *Ground Water 32*(6): 958-967.

O'Hannesin, S.F., and R.W. Gillham. 1997. "Long-term Performance of an *In Situ* 'Iron Wall' for Remediation of VOCs." *Ground Water 36*(1): 164-170.

ABIOTIC TREATMENT OF TRICHLOROETHYLENE: LAB-SCALE RESULTS AND KINETIC MODEL

Jayant K. Gotpagar, Eric A. Grulke, and *Dibakar Bhattacharyya*
Department of Chemical and Materials Engineering
University of Kentucky
Lexington, KY 40506-0046

ABSTRACT: Fundamental aspects of the reductive dehalogenation of Trichloroethylene (TCE) using zero-valent iron are investigated in the laboratory using zero-headspace system. The rate of degradation is found to be independent of the initial concentration of TCE and the initial pH in the parameter range studied. The reaction is controlled by the available surface area on the iron surface. A kinetic model that includes vapor and liquid phase partitioning is developed for the reaction system. The competitive sorption of TCE to the iron surface is included in the model using a new parameter called '*fractional active sites concentration*'. The kinetic model is also verified using data reported in the literature that uses vitamin B_{12} to reduce TCE.

INTRODUCTION

Remediation of Trichloroethylene (TCE) using zero-valent metals is a promising technology. Several researchers have identified different metals for the process, but little information is reported about the kinetic analysis and fundamental aspects of the reaction. Such process knowledge is important for the successful design of a remediation process employing this technology. The objective of present study is to two fold – 1) to study the effect of various reaction parameters such as initial TCE concentrations, feed pH, surface area of iron etc.- on the rate of reaction and, 2) to develop and apply the kinetic model to suitable system for design of such a remediation system. The model results are verified using the zero-valent iron system and vitamin B_{12} system.

EXPERIMENTAL

Batch runs were carried out in the laboratory for the degradation of TCE using zero-valent iron. Known amount of electrolytic iron powder (obtained from Fisher Scientific, 40-mesh and 100-mesh) was added to 40-ml glass hypovials. The hypovials were then filled with TCE solution of known concentration in deionized water, leaving no headspace and were sealed immediately with caps containing Teflon-lined septa. In order to simulate deoxygenated conditions, sodium bisulfite was added to the hypovials before sealing. The hypovials were then placed in an incubator shaker (150-rpm) allowing for complete mixing. At different times, a set of two hypovials was removed from the shaker and subsamples were transferred to vials for analysis. The detailed description of the experimental procedure and the analytical methods used has been reported in our previous paper [Gotpagar et al., 1997].

SUMMARY OF EXPERIMENTAL RESULTS

The typical profile of the concentration of TCE during the course of the experiment is shown in Figure 1.

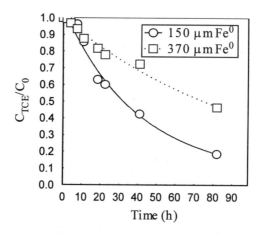

FIGURE 1. TCE concentration profiles during batch experiments

The concentration of the TCE in the control hypovials (no iron) showed insignificant loss throughout the course of the experiment. The concentration of TCE in the reactive hypovials however, showed exponential decline, with the degradation being rapid at the beginning but slowing down considerably as the reaction proceeds.   This particular tailing effect has been attributed to the formation of precipitates on the iron surface. This indeed is due to the changes on the surface of the iron. The data fits the pseudo-first order rate expression. The value of the rate constant obtained (0.021 h^{-1}) matches with literature data [Gillham et al., 1994].

The rate constant of TCE degradation is not affected much by the variation in feed pH over the pH range of 4.6-6.2 studied. Moreover, the rate was also independent of the initial TCE concentrations used, which is a typical characteristic of first order reactions. Iron surface was found to be important for degradation, resulting in two-fold increase in the apparent rate constant when the particle size was decreased from 370 μm (40 mesh) to 150 μm (100 mesh) iron. Similar effect was observed on the chloride formation.

For a successful design of remediation system with this technology, knowledge of intrinsic reaction rate constants is very important. Unfortunately, due to the highly volatile compounds under consideration (TCE and its intermediates – Dichloroethylene (DCE) isomers, Vinyl Chloride, Acetylene, Ethene), it is almost impossible to get such information because of the analytical difficulties involved. Following sections illustrate how mathematical modeling can be used for evaluating the performance of the system under such environments.

KINETIC MODELING

A generalized kinetic model was developed to describe a system with reactions in one phase and mass transfer between two phases. The physical processes used in the development of the model are shown in Figure 2 for two different systems. The model was verified using the experimental data obtained in our laboratory, and an independent study reported in literature [Burris et al., 1996], which used vitamin B_{12} to reductively dehalogenate TCE.

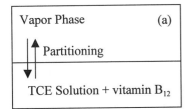 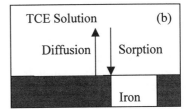

FIGURE 2. Depiction of physical phenomena used for modeling in a) headspace system with vitamin B_{12}, and b) zero-headspace system with iron

The generalized model developed has to be altered to suit the requirements of the system under consideration. For example, in the case of a headspace system (homogeneous vitamin B_{12} system), the partition effects can be incorporated into the model using Henry's law constant and mass transfer coefficients. Furthermore, as the reactions under consideration are very slow, we can assume pseudo-equilibrium between gas and liquid phases and further simplify the analysis. In the case of zero-headspace system using zero-valent iron, there is no gas/liquid mass transfer involved. Due to the presence of solids (iron) however, surface adsorption/sorption phenomenon needs to be considered. Non-reactive sorption of TCE on iron particles has been reported in literature [Burris et al., 1995]. TCE sorption occurs at two sites - reactive and non-reactive sites, and the majority of this sorption is on the non-reactive sites. These researchers have shown that the adsorption of TCE on iron surface (for all sites) follows the modified Langmuir type isotherm given by:

$$C^s_{TCE} = \frac{kb\left(C^w_{TCE}\right)^M}{1+k\left(C^w_{TCE}\right)^M} \tag{1}$$

where C^s_{TCE} and C^w_{TCE} are the sorbed concentration (nmol/g) and aqueous concentration (nmol/ml) respectively. M, k, and b are generalized Langmuir coefficients. The values of generalized Langmuir coefficients (with 40-mesh Fe system) are reported in the literature [Burris et al., 1995]. The fundamental mass balances are then written for all the species involved in each phase. The coupled equations can be solved using numerical methods to get the time profile of concentration of each compound. The vitamin B_{12} results (not reported) showed that the model predicted the concentrations of all the intermediates quite accurately in the entire time range studied. In the analysis of zero-headspace system, some key features of the sorption need to be considered. Equation (1) gives the total sorption to the iron surface. However, reaction is occurring only at the reactive sites. Hence for including the reaction term in the model, following term, called 'fractional active site concentration', and defined as,

$$A_s = \frac{C^{s*}_{TCE}}{C^s_{TCE}} \qquad (\text{* indicates concentration on reactive sites}) \tag{2}$$

was introduced. Figure 3 shows the comparison between experimental and predicted concentrations. As can be seen from Figure 3, there is good agreement between the two at short time, but at longer times, the model predicted higher dechlorination rates than those observed experimentally. This is because, during the modeling analysis, we assumed A_s to be constant. However, in reality, the surface of iron may change due to formation of precipitates. As a result, A_s will also change. We are currently in the process of incorporating this time variation of A_s in the model.

USE OF MODEL FOR OPTIMIZATION

This section illustrates briefly how such a model can be used to optimize the process variables and also to arrive at future research directions without going into the intricate experimental details. The modeling with vitamin B_{12} case showed that the current selectivity of the dechlorination to DCE isomers is around 54% of *cis*-1,2-DCE. However, as the degradation rate constant of *cis*-1,2-DCE was found to be lowest among DCE isomer, this resulted in little formation of ethene, the primary non-toxic end product desired. With the reaction scheme considered (not shown), the ideal species therefore would be 1,1-DCE, which directly yields ethene. However, 1,1-DCE has the lowest solubility in water among the three DCE isomers. Moreover, it also has the highest Henry's law constant and the most stringent drinking water standards compared to two other

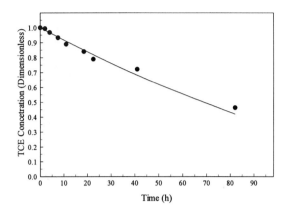

FIGURE 3. Comparison between predicted and experimental TCE profiles with 40 mesh zero-valent iron (points- experimental, line- model)

DCE isomers and hence it is the least desirable. Between the other two isomers, the rate constant of *trans*-1,2-DCE is almost an order of magnitude higher than *cis*

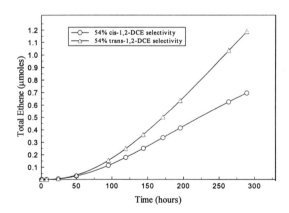

FIGURE 4. Effect of selectivity of DCE isomers on Ethene Production

-1,2-DCE. Hence it would be prudent to develop an electron mediator, which would change the selectivity of TCE daughter products to *trans*-1,2-DCE. This is also evident from the Figure 4, where the amount of ethene production (simulated) is plotted as a function of selectivity of TCE to the two DCE isomers.

SUMMARY AND CONCLUSIONS

The reductive dehalogenation of TCE with zero-valent iron was found to be independent of initial TCE concentration, feed pH in the range of parameters studied. A two-fold increase in the rate constant was obtained when iron particle size was decreased by a factor of 2.5. A two-phase kinetic model was developed to quantify the behavior of system. For two different studies (vitamin B_{12} and zero-valent iron), the modeling showed good agreement with the experimental data.. A new concept of fractional active site concentration is introduced to incorporate the differences in the sorption on the iron surface. Using vitamin B_{12} case , it is shown how model can also be used to arrive at future research directions in order to improve the overall effectiveness of the process. Furthermore, such models can be combined with traditional transport theories to determine monitoring protocols and evaluate remediation alternatives for specific contaminated sites.

ACKNOWLEDGEMENTS

This work was partially supported by a Memorandum of Agreement between Kentucky Natural Resources and Environmental Protection Cabinet and the University of Kentucky through the Kentucky Water Resources Research Institute.

REFERENCES

Burris, D.R., T.J. Campbell, and V.S. Manoranjan. 1995. 'Sorption of Trichloroethylene and Tetrachloroethylene in a Batch Reactive Metallic Iron-Water System', *Environ. Sci. Technol.*, 29(11): 2850-2855.

Burris, D.R., C.A. Delcomyn, M.H. Smith, and A.L. Roberts. 1996. 'Reductive Dechlorination of Tetrachloroethylene and Trichloroethylene Catalyzed by Vitamin B_{12} in Homogeneous and Heterogeneous Systems', *Environ. Sci. Technol.*, 30(10): 3047-3052.

Gotpagar, J., E. Grulke, T. Tsang, and D. Bhattacharyya. 1997. 'Reductive Dehalogenation of Trichloroethylene using Zero-valent Iron', *Environmental Progress*, 16(2): 137-143.

Gillham, R.W., and S.F. O'Hannesin. 1994. 'Enhanced degradation of Halogenated Aliphatics by Zero-Valent Iron', *Ground Water*, 32(4): 958-967.

ELECTRON TRANSFER TO ORGANOCHLORINE COMPOUNDS BY AN IRON(II) PORPHYRIN-DITHIONITE SYSTEM.

Richard A. Larson and Javiera Cervini-Silva (University of Illinois, Urbana-Champaign, Illinois).

ABSTRACT: Heme, an iron(II) porphyrin, reacted with 1, 1, 1-trihalomethanes such as carbon tetrachloride (CCl_4), chloroform ($CHCl_3$), chloropicrin (CCl_3NO_2), hexachloroethane (CCl_3CCl_3) and pentachloroethane (CCl_3CHCl_2) in the presence of sodium dithionite. Heme was found to act as an electron mediator in this system. Kinetics and mechanisms are discussed.

INTRODUCTION

The role of iron (II) is of importance in the reduction of chlorinated compounds using iron containing clays and zero-valent iron (Sivavec and Horney, 1997; Johnson et al., 1998). Iron(II) was shown to act as an electron transfer mediator while bound to the mineral surface and facilitate the electron transport from the solid surface to the bulk solution. In this paper we focus our attention on the role of iron (II) in solution. For this purpose, we study the reaction of an iron (II) containing porphyrin and selected chlorinated compounds.

Iron porphyrins in some abiotic systems transfer electrons to reducible compounds. Schwarzenbach et al. (1990) studied the reduction of quinone by cysteine in the presence of an iron (II) porphyrin and found that its presence enhanced the reduction rate of the quinone. They proposed this abiotic system to behave similarly to those occurring in nature where porphyrins act as catalysts to promote electron transfer in biochemical cycles. Furthermore, the participation of the reduced form of porphyrins as electron mediators has been corroborated by studies of reduction of nitrobenzene, azobenzenes, aromatic brominated compounds, and halogenated aliphatic compounds (Castro and Wade., 1973; Schanke and Wackett, 1992).

In this paper we report studies of the kinetics and mechanisms of the reaction between 1,1,1-trichlorocompounds, $RCCl_3$, and heme in aqueous media. We discuss the experimental conditions that promote heme for serving as electron mediator in the system dithionite-hematin-$RCCl_3$. Also, we discuss the importance of R electron properties in describing $RCCl_3$ reactivities toward heme.

MATERIALS AND METHODS

Hematin stock solution. A modified procedure to prepare the stock solution of hematin (protoporphyrin IX (2, 4-divinyldeuteroporphyrin IX) iron(III)) was followed as reported elsewhere (Castro and Wade). A solution 1:1 $NaHCO_3$ (3.4x 10^{-3}M)-H_2O/EtOH was prepared, contained in a closed vessel and stirred overnight. Later, it was purged with argon (99.9%) for 15 minutes and used to prepare heme (60 mg hematin/100ml of solution; final pH=8).

Heme preparation. A 5ml aliquot of 8e-5M hematin was placed in contact with 0.02g of sodium dithionite ($Na_2S_2O_4$) in a 6ml oxygen free vessel and shaken manually. A rapid change of color from dark green (hematin) to bright red (heme, Figure 1) indicated the change in the oxidation state from Fe^{3+} to Fe^{2+} (Solution A). The concentration of hematin and heme were determined spectroscopically (λ=566nm, and λ=575nm respectively; Castro and Wade).

Solutions of $RCCl_3$ were prepared in the alkaline-ethanolic solution. The containers were shaken manually up to a point where no emulsion was visibly detected (final concentration of 0.01M; Solution B).

Aliquots of solution A and B were added to a sealed cuvette to adjust the initial concentration of sodium bicarbonate and $RCCl_3$ to 5e-3M and 8.3e-4M respectively. Heme initial loss rate was determined with a Beckman 7400 DU spectrometer.

Additional experiments in the absence of sodium dithionite were performed by extracting heme with octanol from the alkaline-ethanolic solution. An aliquot of the resultant extract was withdrawn with a valve syringe and placed in the cuvette. The concentration of heme in the cuvette was adjusted to 8e-5M.

Alkaline aqueous solutions of hematin (>4e-4M, $NaHCO_3$ +Na_2CO_3 =0.005 M, pH_0=8.6) and of $RCCl_3$ (2e-3M, $NaHCO_3$ =0.005 M pH_0=8) were prepared and placed in oxygen free flasks, and stirred overnight. Aliquots of these solutions were added to a 5ml sealed vessel containing 50 mg of dithionite such that the total liquid volume added was 3ml and the resulting initial concentrations were 8.3 e-4 M and 8 e-5M, respectively. The experiment was repeated in absence of hematin. The samples were diluted 1:1000 after 25 minutes of contact time and extracted 1:1 with pentane for their identification. Reactants and products were injected into a 3740 Varian GC, equipped with an electron capture detector and a J & W Scientific column (DB5 phase, 30m X 0.322 mm ID, and 0.25 μm film thickness). A 27 minute program with an initial temperature of 30°C (5 min), and a final temperature of 150°C (5 min), with a heating rate of 7°C/min was used.

RESULTS AND DISCUSSION

The reaction of hematin (protoporphyrin IX (2, 4-divinyldeuteroporphyrin IX) iron (III)) with sodium dithionite was observed to be fast and of high yield in the absence of oxygen. Hematin loss and heme formation were corroborated with the spectra found elsewhere (Castro and Wade).

In alkaline solutions of pH ≥ 8, a mole of dithionite disproportionates to form two moles of sulfoxyl radicals, $SO_2 \bullet^-$ (Equation 1) (Cassatt et al., 1977). Once formed, $SO_2 \bullet^-$ (E_h= -0.27V, Bard and Lund, 1979) is able to transfer an electron to hematin (E_h= 0.1V; Schwarzenbach and Gschwend, 1990) and promote change in the oxidation state of the iron atom (Fe^{3+} to Fe^{2+}) to form heme (Equation 2). Hence, the rate of hematin loss and heme formation ($\Re_f$) will depend on the initial dithionite and hematin concentration in solution (Equation 3).

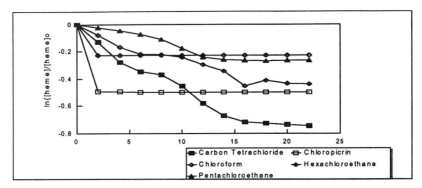

FIGURE 1. Heme loss in presence of 1,1,1-chlorocompounds (RCCl₃) vs time (min). Conditions:[Heme]₀ = 10⁻⁴M, [RCCl₃]=[HCO₃⁻]=5e-3M, pH₀=8, dithionite added.

$$S_2O_4^{2-} \leftrightarrow 2\ SO_2^{\bullet-} \tag{1}$$

$$\text{hematin} + SO_2^{\bullet-} \rightarrow \text{heme} + SO_2\ (g) \tag{2}$$

Hematin, $[Hm]_T$, in solution can be protonated (pKa=5.89, Cassatt et al. (1977) and behave as an acid ($[H_2O\text{-}Hm\text{-}OH_2^+] = \alpha_o^*[Hm]_T$), or be deprotonated and to a base, ($[H_2O\text{-}Hm\text{-}OH] = \alpha_1^*[Hm]_T$) (Equation 3). Because both acid base forms react with sulfoxyl radicals, k' may be expressed in terms of dissociation constant, α_o, α_1, k_1, k_2, $[SO_2^{\bullet-}]$ and $[Hm]_T$ as $k' = \alpha_o k_1 + \alpha_1 k_2$ (Equation 6).

$$H_2O\text{-}Hm\text{-}OH_2^+ \leftrightarrow H_2O\text{-}Hm\text{-}OH + H^+ \qquad K_a = 10^{-5.89} \tag{3}$$

$$H_2O\text{-}Hm\text{-}OH_2^+ + SO_2^{\bullet-} \leftrightarrow \text{heme} \qquad k_1 = 6 \pm 0.7 \times 10^6 M^{-1}s^{-1} \tag{4}$$

$$H_2O\text{-}Hm\text{-}OH + SO_2^{\bullet-} \leftrightarrow \text{heme} \qquad k_2 = (5 \pm 2) \times 10^3 M^{-1}s^{-1} \tag{5}$$

$$\Re_f = (\alpha_o k_1 + \alpha_1 k_2)\ [SO_2^{\bullet-}][Hm]_T \tag{6}$$

From equation (6) we estimated k' to be ~ $10^{4.65} < k' < 10^{4.76}$ L/mol min. We expect that by increasing the pH of the solution, the consumption of $SO_2^{\bullet-}$ by hematin to produce heme will be slower and therefore, we predict heme formation to be the rate-limiting step for its reaction towards RCCl₃. On the other hand, hematin solubility also increases as the pH of the solution does. Therefore, pH is critical in order for hematin to serve as electron mediator in the reduction of RCCl₃.

The mechanism by which heme reacts with halocompounds of low molecular weight is controversial (Castro and Wade; Gonsior and Klecka, 1984; Schanke and Wackett, 1992). Recently, Schanke et al. studied the reaction of 1,1 1-trichloroethanes with heme in water (5e-9M heme/1e-4M dithiothreitol / anaerobic conditions at pH 8.2) and proposed a dechlorination involving a transference of two electrons in a single step (Equation 7).

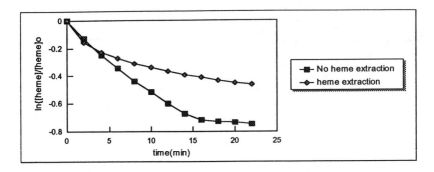

FIGURE 2. Heme loss when reacted with carbon tetrachloride in presence (no heme extraction) and absence (heme extraction) of dithionite.

$$CCl_3CH_3 + 2 e^- \rightarrow CCl_2CH_3^{\cdot -} + Cl^- \tag{7}$$

In this respect, we noticed that the reactivity of $RCCl_3$ towards heme was greatly dependent on R electronic properties (Table 1). Compounds with substituents (R) of stronger electron withdrawing properties (EW) showed faster reduction rates than those with good electron donating (ED) groups relative to hydrogen (R=H): CCl_3NO_2 and CCl_3CCl_3 reacted faster followed by CCl_3CHCl_2, CCl_4 and $CHCl_3$. This order of reactivity agreed with previous studies of polarity and relative rates of chlorine abstraction of $RCCl_3$ type compounds (Coates and Tedder, 1978) for equation (8). They suggested that EW groups are able to stabilize radical intermediates $RCCl_2\cdot$ to a greater extent than ED groups do, hence contribute to faster chlorine abstraction rates ($\Re_{chlorine}$), ("reverse polarity" effect). Folli et al. studied the reaction of ferrous chloride ($FeCl_2\cdot4H_2O$) towards selected $RCCl_3$ (R= substituted aromatic ring). They proposed the formation of $RCCl_2\cdot$ as first step and they concluded that the relative reactivity on these compounds towards Fe (II) rely on R electronic contribution for stabilizing $RCCl_2\cdot$. Our results suggested that this trend of reactivity towards iron(II) containing species also exist for aliphatic halocompounds (Table 1). Our observations of the reaction between heme and the selected $RCCl_3$ type compounds pointed to the formation of $RCCl_2\cdot$ species *via* one electron transfer as to be the mechanism for dechlorination.

$$(CH_3)_3Sn \cdot + RCCl_3 \rightarrow (CH_3)_3SnCl + RCCl_2\cdot \tag{8}$$
$$heme + CCl_3R \rightarrow chlorohemin + \cdot CCl_2R \tag{9}$$

The heme solution was noticed to 'age' and passed from a bright to a dull red ~7-8 minutes after preparation. Heme served efficiently as an electron mediator before aging with dithionite present in excess. These observations were corroborated by extracting heme from the alkaline solution with octanol, and reacting it with CCl_4 (Figure 3).We found a faster heme loss rate ($\Re_1$) when dithionite was present.

Further studies in aqueous solution of the reactions CCl_3CHCl_2-hematin-dithionite and CCl_3CHCl_2-dithionite suggested that while dithionite does not contribute to the reduction of CCl_3CHCl_2 in the first 25 minutes, this process does take place when hematin is present. Further analyses of the system CCl_3CHCl_2-dithionite after an hour of contact time and in excess of dithionite (200mg) showed $CHCl=CCl_2$ and $CCl_2=CCl_2$ as the main components in solution.

Dithionite has a dual role: serving as a continuous source of electrons for hematin for $RCCl_3$ reduction (Equation 9) and participating as direct electron transfer agent towards $RCCl_3$ (Equations 10-13). Hence, heme participates in the initial step of chlorine abstraction to form $•CCl_2CHCl_2$ followed by further electron transfer to this intermediate by sulfoxyl radicals present in solution.

$$CCl_3\text{-}CHCl_2 + SO_2•^- \rightarrow SO_2(g) + • CCl_2\text{-}CHCl_2 \quad (10)$$

$$•CCl_2\text{-}CHCl_2 + SO_2•^- \rightarrow SO_2(g) + {}^-CCl_2\text{-}CHCl_2 \quad (11)$$

$${}^-CCl_2\text{-}CHCl_2 \rightarrow CCl_2=CHCl + Cl^- \quad (12)$$

$${}^-CCl_2\text{-}CHCl_2 \rightarrow CCl_2=CCl_2 + H^- \quad (13)$$

Preliminary studies of the reaction between CCl_3CHCl_2 with zero-valent iron at pH_o=6 in aqueous solution led also to the formation of these products, suggesting similarities in the mechanisms of dechlorination.

TABLE 1. (i) Initial loss rate of heme in presence of chlorinated compounds $(k^{heme}_{obs}$, min$^-$); *Data estimated from $[CCl_4]$ = $[CCl_3NO_2]$ = $[CCl_3CCl_3]$= 3.3e-4M; (ii) $RCCl_3$ loss rate in presence of zero-valent iron, (k^{Fe}, min$^-$); (iii) Chlorine abstraction rate in CCl_3R relative to $CHCl_3$. ($\Re_{chlorine}$= $rCCl_x$/$rCHCl_3$ where r $CHCl_3$ is defined as 1;Coates and Tedder, 1978).

Compound	R	k^{heme}_{obs}	k^{Fe}	$\Re_{chlorine}$
CCl_3NO_2	NO_2	1.58	3.23e-3	-
CCl_3CCl_3	CCl_3	0.43	7.24e-3	3.7(*CCl_3CF_3)
CCl_3CHCl_2	$CHCl_2$	0.34	1.81e-3	4.7
CCl_4	Cl	0.0371	6.6e-3	2.8
$CHCl_3$	H	0.0275	2.8e-3	1

CONCLUSIONS

Our results suggest that the couple $Na_2S_2O_4$-hematin promotes the dechlorination of selected $RCCl_3$ type compounds. We observed that the reactivity of 1,1,1-trihalomethanes towards heme is strongly dependent on R electronic properties and propose the formation of $RCCl_2•$ as to be the first step towards their dechlorination. Further analysis of products for the reaction of dithionite and $RCCl_3$ type compounds also suggest the mechanism of dechlorination to be a two step one-electron transfer.

REFERENCES

Bard, A.J. and M. Lund (Eds).1979. *Encyclopedia of Electrochemistry of the Elements*; Vol 12 and 13, 5th ed., Marcel Dekker, NY.

Cassatt, J.C., M. Kukuruzinska, J.W. and Bender. 1977. "Kinetics of Reduction of Hemin and the Hemin Bis(pyridine) Complex by Dithionite." *Inorganic Chemistry.* 16: 3371-3372.

Castro, C.E. and R.S.Wade. 1973. "Oxidation of Iron (II) Porphyrins by Alkyl Halides." *J. Amer.Chem. Soc.* 95: 226-234.

Coates, D.A. and J.M.Tedder. 1978. "Free Radical Substitution in Aliphatic Compounds. Part 33. Halogen Abstraction from Alkyl Halides by Trimethylgermanium Radicals in the Gas Phase." *J.Chem.Soc.Perkin Trans 2.* 8:725-728.

Folli, U., F. Goldoni, D. Iarossi, S. Sbardellati, and F. Taddei 1995. "Selectivity towards Hydrodehalogenation and Dehalocoupling in the Reduction of Trichloromethyl Derivatives with Iron(II) Chloride." *J. Chem. Soc. Perkin. Trans. 2.* 1017-1019.

Gonsior, S.J. and G.M.Klecka. 1984. "Reductive Dechlorination of Chlorinated Methanes and Ethanes by Reduced Iron (II) Porphyrins." *Chemosphere.* 3: 391-402.

Johnson, T.L., W. Fish, Y.A.Gorby, and P.G.Tratnyek. 1998. "Degradation of Carbon Tetrachloride by Iron Metal: Complexation Effects of the Oxide Surface." *J. Contam. Hydrol.* 29:377-396.

Shanke, C.A., and L.P. Wackett.1992. "Environmental Reductive Elimination Reactions of Polychlorinated Ethanes Mimicked by Transition-Metal Coenzymes." *Environ. Sci. Technol.* 26:830-833.

Schwarzenbach, R.P., R. Stierli, K. Lanz, and J. Zeyer.1990. "Quinone and Iron Porphyrin Mediated Reduction of Nitroaromatic Compounds in Homogeneous Aqueous Solution." *Environ. Sci. Technol.* 24:1566-1574.

Sivavec, T.M. and D.P. Horney. 1997. "Reduction of Chlorinated Solvents by Fe (II) Minerals." *Prep. Pap. ACS Natl. Meet., Am. Chem. Soc., Div. Environ. Chem.,* 37(1):115-117.

REDUCTIVE DECHLORINATION OF TCE BY DITHIONITE-TREATED SEDIMENT

Edward C. Thornton, James E. Szecsody, Kirk J. Cantrell, Christopher J. Thompson, John C. Evans, Jonathan S. Fruchter, and Alex V. Mitroshkov (Pacific Northwest National Laboratory, Richland, Washington)

ABSTRACT: Laboratory flow-through column tests have been performed to evaluate the degradation of TCE by dithionite-treated sediment. These tests involved packing sand-sized samples obtained from the Hanford formation into columns and reducing the iron fraction of the sediment with a pH-buffered dithionite solution. TCE solutions were subsequently pumped through columns at various flow rates and the effluent analyzed for TCE and byproducts. Column experimental data collected to date indicate reduction of TCE can be approximated as a pseudo-first-order reaction with half-life values ranging from 19 to 66 hours. Analysis of column effluent samples indicates that most of the TCE is converted to acetylene, though minor amounts of ethylene and chloroacetylene were also observed.

INTRODUCTION

Laboratory and field investigations have demonstrated the viability of constructing in situ permeable barriers for passive treatment of groundwater contaminant plumes through the injection of pH-buffered dithionite solutions. A portion of the iron oxide components of the aquifer sediment is reduced to ferrous iron following reaction with dithionite, thus providing a significant reducing capacity for treatment of contaminants subsequently passing through the treated zone.

Field-scale dithionite treatment of aquifers has been successfully undertaken at two sites, demonstrating that the reduction of hexavalent chromium in groundwater is readily accomplished by this approach (Fruchter et al., 1996). Laboratory studies have shown that 50 to 250 pore volumes of oxygen-saturated water are required to oxidize the reduced sediment, suggesting that the reduced zone can be maintained for a considerable length of time at the field scale (Amonette et al., 1994).

The potential of treating chlorinated solvent groundwater plumes also needs to undergo investigation in view of the important environmental problems associated with this class of contaminants. While other studies have shown that zero-valent iron can dechlorinate solvents (e.g., Orth and Gillham, 1996), the effectiveness of dithionite treatment has not been adequately evaluated. The purpose of this study is to quantify the rates and capacity of dithionite-reduced sediment for abiotically degrading trichloroethylene (TCE). This information will provide a basis for determining if this approach can be used at the field scale for chlorinated solvents.

Column experiments were used to evaluate TCE degradation following dithionite treatment of sediment and to evaluate subsequent reoxidation of the sediment. Because the dechlorination of TCE and degradation intermediates involves multiple solutes and multiple reactions, several column experiments at different velocities (i.e., residence times) were undertaken to quantify the different reaction rates. Analysis of the effluent also allows identification of degradation byproducts and permits verification of the degradation process through mass balance considerations. A long-term column experiment was used to evaluate how long the TCE degradation can be maintained while the sediment is reoxidized. The information obtained in these tests provides a basis for designing a barrier for treatment of TCE-contaminated groundwater and for estimating barrier longevity.

METHODS AND MATERIALS

Columns were packed with silty sand (<10 mm) from the Hanford formation of the unconfined aquifer at the Hanford site. The sediment was reduced by injecting 0.1 molar sodium dithionite (buffered to pH 11 with K_2CO_3 and $KHCO_3$) into the columns over a period of several days. Dithionite concentration was monitored in the column effluent during the second test by UV spectrophotometry as a means of verifying reduction of the sediment.

A solution of 0.1 to 1 mg/L TCE was subsequently injected into the columns at flow rates required to achieve 0.2- to 6-day residence times. Effluent samples were collected and analyzed for TCE and degradation byproducts by gas chromatography and gas chromatography/mass spectroscopy. Dissolved oxygen, pH, and electrical conductivity were also measured during the study to monitor geochemical conditions during the dechlorination and reoxidation reactions.

RESULTS AND DISCUSSION

Test Results. Two tests were conducted in the course of this study. The first was designed to maximize column residence time to ensure measurable degradation of TCE, which was present in the influent at a concentration of about 1 mg/L. The flow rate was held at 0.02 mL/min during most of this test to achieve a column residence time of 147.5 hours, but was run at 0.04 mL/min (residence time = 73.75 hours) for part of the test. The longer residence time was sufficient to essentially destroy all measurable TCE (<10 µg/L), as indicated by analysis of effluent samples collected early in the experiment (Figure 1). This indicates a maximum half-life of about 19 hours, assuming a pseudo-first-order dechlorination rate. Analysis of the effluent samples also indicated that the primary degradation product was acetylene, though minor amounts of ethylene were also observed. The observed mass of acetylene in effluent samples was found to be roughly equal to the influent mass of TCE, thus indicating that byproducts other than acetylene were not generated in significant concentrations.

Monitoring of oxygen concentrations in the effluent using a microelectrode indicated that the system remained anoxic for the duration of the first test, which ran for a total of 43 column pore volumes. Initial breakthrough of TCE, however, appears to have occurred after only about 14 pore volumes (see

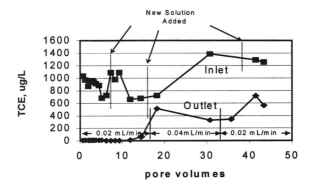

FIGURE 1. TCE inlet and outlet concentrations measured during the first test as a function of column pore volumes.

Figure 1). It is possible that adequate reduction of the sediment may not have been achieved in this test, however, because the effluent was not monitored for dithionite breakthrough during the dithionite-treatment step.

Based on positive evidence of TCE degradation observed in the first test, a second test was designed to obtain quantitative information regarding the degradation rate of TCE and additional information regarding the identity and proportion of byproducts generated. In addition, greater effort was undertaken to ensure maximum reduction of the sediment during the dithionite-treatment step, thus obtaining a greater potential for TCE degradation. To accomplish these objectives, the test was conducted in four steps: 1) dithionite-treatment of the sediment, 2) injection of TCE influent at a flow rate required to achieve a column residence time of about 3.5 days, 3) injection of TCE influent at a slower flow rate (corresponding to a column residence time of about 5.5 days), and 4) injection at a very high flow rate to achieve a short residence time (5 hours). Step 1 was monitored for dithionite in the outlet by UV spectrophotometry to ensure complete treatment during the sediment reduction phase. The primary objective of steps 2 and 3 was to obtain information related to TCE degradation rates, while step 4 (in progress) will be run for a sufficient number of pore volumes to achieve breakthrough of oxygen and TCE. The latter information can be used as a basis for predicting barrier longevity.

The concentration of TCE injected into the inlet during step 2 of the second test averaged about 1.17 μmolal (0.15 mg/L), though an apparent loss with time was observed (Figure 2). This is attributed to a slow diffusion of TCE from the influent reservoir. The column residence time in step 2 was about 87 hours, though some variation in flow rate was noted. The organic mass effluent data were relatively constant, however, allowing the TCE disappearance rate to be approximated by a first-order function with a half-life of 38 hours. Acetylene and ethylene were the only products present in measurable quantities (see Figure 2).

The inlet concentration of TCE also decreased slowly with time during step 3 (Figure 3), but averaged about 2.0 μmolal (0.26 mg/L). The flow rate was steady, however, and the column residence time was 132 hours. A pseudo-first-order function for degradation yields a half-life for TCE of about 66 hours during this step, though effluent TCE concentrations were still increasing at the end of step 3 and the degradation rate may actually be somewhat faster. Acetylene and ethylene were again observed as the principal byproducts of TCE degradation.

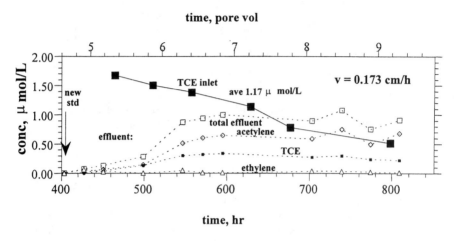

FIGURE 2. Inlet TCE and outlet TCE and byproduct concentrations during step 2 of the second test as a function of time and column pore volumes.

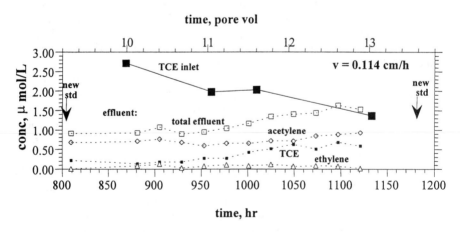

FIGURE 3. Inlet TCE and outlet TCE and byproduct concentrations during step 3 of the second test as a function of time and column pore volumes.

Step 4 of the second test was still in progress at the time this paper was prepared, but information is available for the initial portion. This step was undertaken at a higher flow rate, with a residence time of only about 0.2 days. As indicated above, the goal of this step was to run the test until breakthrough of oxygen and TCE is observed. Thus, this step will provide an indication of the total reductive capacity of the reduced sediment, as indicated by oxygen breakthrough, and the longevity of TCE degradation. To date, analytical data indicate that TCE degradation persists to at least 230 pore volumes with a half-life of 40.7 hours and is accompanied primarily by acetylene. Other byproducts observed include trace DCE, chloroethane, and chloroacetylene, however, which may be intermediates in the degradation of TCE to acetylene and ethylene. Oxygen outlet concentrations are also currently low, verifying anoxic conditions are being maintained.

Summary and Implications for Barrier Design. Testing results to date indicate that dithionite-reduced Hanford formation sand has the capability to degrade TCE in dilute aqueous solutions and can be approximated as a pseudo-first reaction with a half-life of about 40 hours. These rates are similar in magnitude to those observed in other studies with Fe(II), but slower than in studies associated with Fe or Fe-Ni metal barriers (Roberts et al., 1996; Sivavec et al., 1997). The primary end byproduct is acetylene, which is readily destroyed in the natural environment by microbial activity. Achieving maximum reductive capacity of the dithionite-treated sediment requires diligent monitoring of the treatment process, as illustrated by a comparison of the results of the two tests conducted during this study. It is concluded that TCE degradation in excess of 200 pore volumes can be achieved if thorough treatment of sediment similar to that used in this study is performed. This approach may be advantageous over iron metal walls in deep applications.

Assuming a degradation half-life for TCE of 40 hours, it is possible to present predictions related to barrier performance for a simple example. If 10 half-lives are sufficient to lower TCE concentrations below a specified target concentration, the barrier must be thick enough so that groundwater residence time in the barrier is at least 400 hours. For a groundwater velocity of 1 ft/d (0.3 m/d), for example, a barrier thickness of 20 ft (6 m) would thus be sufficient. In small scale field tests, barriers of 50 ft (15 m) thickness have been easily achieved.

ACKNOWLEDGMENT

This work was conducted at Pacific Northwest National Laboratory, operated by Battelle for the U.S. Department of Energy under Contract DE-AC06-76RLO 1830.

REFERENCES

Amonette, J. E., J. E. Szecsody, H. T. Schaef, J. C. Templeton, Y. A. Gorby, and J. S. Fruchter. 1994. "Abiotic Reduction of Aquifer Materials by Dithionite: A

Promising In-Situ Remediation Technology." In G. W. Gee and N. R. Wing
(Eds.), *In-Situ Remediation: Scientific Basis for Current and Future
Technologies*, pp. 851-881. Battelle Press, Columbus, OH.

Fruchter, J. S., J. E. Amonette, C. R. Cole, Y. A. Gorby, M. D. Humphrey, J. D.
Isok, F. A. Spane, J. E. Szecsody, S. S. Teel, V. R. Vermeul, M. D. Williams, and
S. B. Yabusaki. 1996. *In Situ Redox Manipulation Field Injection Test Report –
Hanford 100-H Area.* PNNL-11372, Pacific Northwest National Laboratory,
Richland, WA.

Orth, W. S., and R. W. Gillham. 1996. "Dechlorination of Trichloroethene in
Aqueous Solution Using Fe°." Environ. Sci. Technol. 30(1): 66-71.

Roberts, A. L., L. A. Totten, W. A. Arnold, D. R. Burris, and T. J. Campbell.
1996. "Reductive Elimination of Chlorinated Ethylenes by Zero-Valent Metals."
Environ. Sci. Technol. 30(8): 2654-2659.

Sivavec, T. M. , P. D. Mackenzie, D. P. Horney, and S. S. Baghel. 1997.
"Redox-Active Media for Permeable Reactive Barriers." In *Conference
Proceedings: 1997 International Containment Technology Conference and
Exhibition*, pp. 753-759, February 9-12, 1997, St. Petersburg, FL.

EFFECT OF IRON CORROSION INHIBITORS ON TRICHLOROETHYLENE REDUCTION

Baolin Deng, Shaodong Hu (New Mexico Tech, Socorro, NM)
David R. Burris (Air Force Research Laboratory, Tyndall, FL)

ABSTRACT: A laboratory study has been performed to examine the inhibition effects of 14 organic and inorganic chemicals on anaerobic iron corrosion and trichloroethylene (TCE) degradation. Anaerobic iron corrosion is characterized by measuring hydrogen (H_2) production, and TCE degradation was evaluated by the formation of its reduction products, ethylene and ethane. Cysteine and adenine significantly decrease the amounts of H_2 produced from iron corrosion. Chemicals that inhibit H_2 production also decrease the rates of TCE reduction.

INTRODUCTION

An *in situ* permeable barrier, with metallic iron serving as a reactive material for the degradation of chlorinated compounds, has emerged as an important approach for groundwater remediation (Gillham and O'Hannesin, 1994; Matheson and Tratnyek, 1994). In order to optimize design parameters and predict the long-term performance of a reactive barrier, we must understand the kinetics and mechanisms of dechlorination and other relevant reactions.

The reactive iron barrier technology is based on the reduction of chlorinated compounds (RCl) to form non-toxic hydrocarbons and chloride:

$$Fe^0 + RCl + H_3O^+ = Fe^{2+} + RH + 2Cl^- + H_2O \qquad (1)$$

In terms of reaction mechanism, the reduction could potentially occur through reactions with: 1) zero-valent iron on the surfaces, 2) Fe^{2+} generated by iron corrosion, and 3) H_2 generated by iron corrosion. Experiments with alkyl halides (Matheson and Tratnyek, 1994; Scherer et al., 1997) and vinyl chloride (Deng et al., 1997) suggest that the first process (i.e, direct reaction on metallic iron surfaces) be most likely. The dechlorination reaction, however, is just one of the corrosion reactions occurring in metallic iron-water systems. Metallic iron may corrode through additional reactions with water itself and dissolved oxygen in water according to the following stoichiometries:

$$Fe^0 + 2H_2O = Fe^{2+} + H_2 + 2OH^- \qquad (2)$$
$$2Fe^0 + O_2 + 2H_2O = 2Fe^{2+} + 4OH^- \qquad (3)$$

Excess oxygen may further oxidize Fe^{2+} to form iron oxides such as goethite and hematite. Although all the reactions lead to iron corrosion, we use dechlorination here to denote Reaction 1, while iron corrosion for the other corrosion reactions.

The interaction between dechlorination and iron corrosion is unclear. If they occur in parallel, it would be beneficial to block the undesirable side corrosion reactions, while promote the dechlorination reaction. Less iron corrosion not only preserves its reducing capacity for dechlorination, but also avoids formation of iron oxides which, as is generally believed, may clog the permeable iron barriers. Surface reactions however occur normally on active surface sites and competition among various processes is common. Blocking iron corrosion may inhibit the dechlorination process if both are utilizing the same set of surface sites. The objective of this research is to seek a better understanding of the relationship between dechlorination and corrosion reactions by altering corrosion rates with various chemicals.

MATERIALS AND METHODS

Imidazole, adenine, and various amino acids were from Sigma Chemical and had at least 98% purity. Inorganic regents and triethanolamine were from Fisher Scientific. Metallic iron used was a powdered material produced by pentacarbonyl iron reduction by H_2, which had a purity > 97%, an average size of 4.5 - 5.2 μm and a bulk density of 4.4 g/cm^3 (Sigma). The carbon content of the iron is 0.83% (w/w) as measured on a Leco WR-112 carbon analyzer, and the N_2-BET specific surface area is 0.47 m^2/g as measured using Micrometrics FlowSorb 2300. The iron sample was used without pretreatment.

The effect of various chemicals on the rates of iron corrosion in the absence of chlorinated solvents was tested in 60 ml bottles, with each containing 1.000 g iron and 30.00 ml solution of triethanolamine (5.0 mM, as a buffer to control pH at 7.5) and a potential inhibitor (1.00 mM). The iron corrosion was evaluated by headspace H_2 analysis. Trichloroethylene degradation by metallic iron in the presence of several corrosion inhibitors was examined in 160 ml bottles containing 3.00g iron, 100.00 ml solution (with triethanolamine, an inhibitor, and TCE), and a 60-ml headspace. The degradation reaction was monitored as a function of time by headspace analysis of H_2, TCE, and TCE reduction products. All experiments were prepared under anaerobic conditions with initial headspace being pure N_2. Reaction bottles were mixed on a rotor drum at 8 RPM in a 20.0°C incubator.

Hydrogen was analyzed on a Shimadzu GC-8AIT gas chromatograph, equipped with a thermal conductivity detector and molecular sieve columns. TCE and its degradation products were analyzed by a dual-column sequence-reversal technique employing flame ionization detection (Campbell et al., 1996).

RESULTS AND DISCUSSION

Iron Corrosion Inhibition: Various methods can be used to evaluate iron corrosion, such as directly measuring the loss of metallic iron from corrosion, recording the electrochemical current-voltage relationship, and characterizing H_2 evolution (Schaschl, 1973). We used the hydrogen production as an indicator for the rate of iron corrosion.

Table 1 shows the structures of organic compounds examined. The amounts of hydrogen produced after 70 hours of reaction are listed in Table 2. In the controls which contain only iron and triethanolamine, headspace H_2 concentration is 2750 ppmv (average of triplicates). H_2 productions in the presence of imidazole, histidine, and methionine are very close to the control, indicating

TABLE 1. Structures of Organic Compounds Examined in This Study

Imidazole	Histidine	Methionine
Tryptophan	Phenylalanine	Glutamic Acid
Cysteine	Adenine	Triethanolamine

that they do not significantly alter iron corrosion rate. In contrast, phenylalanine, tryptophan, and glutamic acid decrease the amount of H_2 production by more than 50%. More dramatically, H_2 productions in the presence of cysteine and adenine are only 2 to 3% of the hydrogen produced in the control.

Some amino acids such as histidine and methionine do not inhibit iron corrosion while others (e.g. phenylalanine, tryptophan, and glutamic acid) do, indicating that the amino acid group is not the only functional group responsible for the corrosion inhibition. Other reactive moieties in the amino acids must be involved. Cysteine strongly inhibits iron corrosion in comparison with other amino acids, suggesting that the -SH group in cysteine is effective in blocking iron corrosion. The -S- group (as in methionine), however, could not significantly prevent iron corrosion. The significant inhibition effect of adenine likely results from the -NH_2 group attached to the ring structure, since the compounds with nitrogen fused in aromatic rings (i.e. imidazole and histidine) do not show a strong effect.

Table 3 shows H_2 production in the presence of six inorganic compounds. Sodium silicate and sodium molybdenate do not significantly affect iron corrosion. Borate and phosphate however show an inhibiting effect, as indicated by the approximately 50% decrease in H_2 production in the presence of these two compounds. In contrast, potassium sulfate and aluminum sulfate do not inhibit, but rather enhance iron corrosion. The H_2 concentration in the presence of 1.0 mM potassium sulfate is three times more than that in the control, and in the presence of 1.0 mM aluminum sulfate ($Al_2(SO_4)_3$), it is 38 times more than that of the control. The unique enhancement for iron corrosion appears to be from sulfate.

TABLE 2. Hydrogen Production in the Presence of Various Organic Compounds Under Anaerobic Conditions (pH 7.5)

Inhibitor (Replicates)	H$_2$ Production (ppm)	Average (ppm)
Control (w/Triethanolamine)	1721, 4737, 1801	2753
Imidazole	2697, 1266	1981
Histidine	1689, 1796	1743
Methionine	1523, 1166	1345
Phenylalanine	911, 779	845
Tryptophan	434, 937	686
Glutamic acid	81, 646	364
Adenine	88, 86	87
Cysteine	63, 52	58

TABLE 3. Hydrogen Production in the Presence of Various Inorganic Compounds Under Anaerobic Conditions (pH 7.5)

Inhibitor (Replicates)	H$_2$ Production (ppm)	Average (ppm)
Control (w/Triethanolamine)	1066, 1350	1208
Na$_2$SiO$_3$	2234, 860	1547
Na$_2$MoO$_4$	1006, 754	880
Na$_2$B$_4$O$_7$	182, 626	404
K$_2$HPO$_4$	575, 358	467
K$_2$SO$_4$	4139, 4295	4217
Al$_2$(SO$_4$)$_3$	51923, 42753	47338

Effect of Iron Corrosion Inhibitors on TCE Degradation: Trichloroethylene reduction by metallic iron was examined in the presence of each of the following compounds: cysteine, sodium silicate, sodium borate, histidine, and potassium phosphate. H$_2$ production as a function of time is measured by GC with headspace injection. Based on H$_2$ solubility (Readon, 1995) and the current experimental design, dissolved H$_2$ in water accounts for less than 3% of the total hydrogen in the system. Thus H$_2$ in the headspace is used to calculate the total amount of H$_2$ produced from anaerobic iron corrosion. TCE degradation is demonstrated by the production of various degradation intermediates and products including acetylene, ethylene, ethane, and C$_3$-C$_6$ hydrocarbons. Since high initial amount of TCE (23.5 µmole/160ml bottle) is used, the TCE degraded is less than 10% of the total in the experimental time period. Therefore the production of ethylene and ethane, rather than the decrease in TCE concentration, is used to evaluate dechlorination rates.

As shown by Figure 1, hydrogen and TCE degradation products are not observed in the control containing only triethanolamine and TCE. In the system with TCE, triethanolamine and iron, but without corrosion inhibitor, significant amount of H$_2$ is produced along with significant ethylene and ethane production.

Figure 1. Hydrogen Production in Systems with TCE

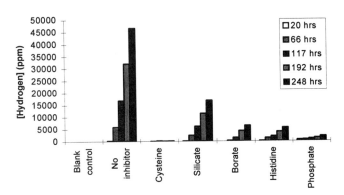

Figure 2. Ethylene Production From TCE Degradation

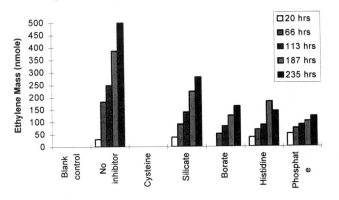

Figure 3. Ethane Production From TCE Degradation

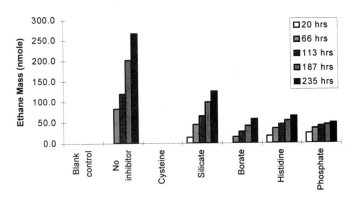

When one of the corrosion inhibitors is added into the system, hydrogen production is decreased with the effect of cysteine >> phosphate > histidine ~ borate > silicate (Figure 1A). The inhibitors similarly influence TCE degradation as indicated by the decreased ethylene and ethane production (Figure 1B and 1C). Cysteine totally blocks TCE degradation. Phosphate, borate, histidine, and silicate decrease ethylene and ethane production by about 50 - 70%. We can not, however, unambiguously establish that the effect on H_2 production is exactly proportional to the effect on TCE degradation .

In summary, this study has clearly shown that iron corrosion can be totally blocked by some chemicals such as cysteine and adenine. Other compounds, including phenylalanine, tryptophan, glutamic acid, sodium borate, and potassium phosphate, can partially inhibit iron corrosion. Chemicals that inhibit iron corrosion also decrease the rate of TCE degradation. Sulfate uniquely enhance the rate of iron corrosion, and how it affects the reduction of chlorinated compounds in groundwater needs to be examined. Further research is also needed to relate the current study to the potential effect of natural organic materials on dechlorination and iron corrosion reactions, and to explore the application of corrosion inhibition to the design of better reactive iron barriers.

REFERENCES

Campbell, T. J., and D. R. Burris. 1996. "Analysis of Chlorinated Ethene Reduction Products in Vapor/Water Phase Systems by Dual-Column, Single-Detector Gas Chromatography." *Intern. J. Environ. Anal. Chem. 63*: 119-126.

Deng, B., T. J. Campbell, and D. R. Burris, 1997. "Reduction Kinetics of Vinyl Chloride by Metallic Iron in Zero-headspace System." Preprints of Papers Presented at 213th ACS National Meeting, San Francisco, *35*(1): 453-456.

Gillham, R. W., and S. F. O'Hannesin. 1994. "Metal Enhanced Abiotic Degradation of Halogenated Aliphatics." *Ground Water, 32*(6): 958-967.

Matheson, L. J., and P. G. Tratnyek. 1994. "Reductive Dehalogenation of Chlorinated Methanes by Iron Metal." *Environ. Sci. Technol. 28*: 2045-2053.

Reardon, E. J. 1995. "Anaerobic Corrosion of Granular Iron: Measurement and Interpretation of Hydrogen Evolution Rates." *Environ. Sci. Technol. 29*: 2936-2945.

Schaschl, E. 1973. "Methods for Evaluation and Testing of Corrosion Inhibitors." In C. C. Nathan (Ed.), *Corrosion Inhibitors*, pp. 28-41. National Assoc. Corrosion Engineers, Houston.

Scherer, M. M., J. C. Westall, M. Ziomek-Moroz, and P. G. Tratnyek. 1997. "Kinetics of Carbon Tetrachloride Reduction at an Oxide-Free Iron Electrode." *Environ. Sci. Technol. 31*: 2385-2391.

ABIOTIC TRICHLOROETHYLENE DEHALOGENATION USING SODIUM HYDROSULFITE: LABORATORY AND FIELD INVESTIGATION

Ralph Ludwig, Alina Rzeczkowska, and Sandra Dworatzek
Water Technology International Corporation
Burlington, Ontario

ABSTRACT: A microcosm study and field study were undertaken to investigate the feasibility of using the reducing agent sodium hydrosulfite to induce abiotic degradation of selected chlorinated hydrocarbons in groundwater. Aqueous microcosm tests were conducted with trichloroethylene (TCE), 1,1,1-trichloroethane (TCA), and carbon tetrachloride. Microcosm tests indicated abiotic dehalogenation of each of the compounds tested as evidenced by the production of the by-products 1,1-dichloroethylene, 1,1-dichloroethane, and chloroform, respectively. Parent compound losses of at least 20% relative to controls not containing sodium hydrosulfite were observed. An in situ field study involving injection of sodium hydrosulfite into an aquifer contaminated with TCE and 1,1,1-TCA was conducted at a Landfill site near Ottawa. The field test did not indicate significant degradation of TCE and 1,1,1-TCA in the aquifer. The results suggest that although an abiotic degradation process may be induced through the addition of sodium hydrosulfite, the process is likely too insignificant to have a measurable impact in an aquifer restoration application.

INTRODUCTION

Technologies continue to be sought for remediating aquifer systems contaminated with chlorinated solvents. Recent efforts have focused on the use of zero-valent metals as a means of abiotically inducing dehalogenation of these compounds. These efforts have proved very promising. Little work to date has been conducted using reducing agents. Previous work conducted by Klecka and Gonsior (1984) indicated that selected chlorinated solvents could be abiotically degraded in the presence of iron porphyrins and a reducing agent. Carbon tetrachloride, chloroform, and 1,1,1-trichloroethane were all reduced to lower chlorinated homologs in aqueous solutions containing sulfide redox buffer and the iron porphyrin, hematin. Abiotic degradation was, however, not observed for TCE or tetrachloroethylene (PCE).

A microcosm study and field study were conducted to evaluate the potential of sodium hydrosulfite to induce abiotic dehalogenation of TCE, 1,1,1-TCA, and carbon tetrachloride. If viable, the use of reducing agents such as sodium hydrosulfite to manipulate redox conditions and induce dehalogenation processes in aquifer systems would represent a potentially flexible and cost-effective means of remediating chlorinated compounds in groundwater.

MATERIALS AND METHODS

Microcosm Study. The microcosm study was conducted in aqueous solutions contained within 125 milliliter (ml) glass hypo-vials. An aqueous stock solution of 0.05M phosphate buffer (K_2HPO_4) in de-ionized water autoclaved twice at $120^\circ C$ for a total of 30 minutes (to sterilize and to remove oxygen) was first introduced into all microcosms. Once introduced into microcosms, the phosphate buffer solution was purged with nitrogen for a period of 20 minutes to further expel any residual oxygen and to remove oxygen from the headspace.

Immediately following purging, microcosms vials were capped with a combination butyl rubber septum and aluminum crimp seal. Microcosms were subsequently amended with sodium hydrosulfite and chlorinated hydrocarbon test compounds aseptically prepared and introduced into microcosms by syringe injection through the butyl rubber septa. TCE, 1,1,1-TCA, and carbon tetrachloride were introduced into all microcosms (in a methanol carrier) at concentrations ranging from of 1-4 mg/L. All tests were conducted over a 96-hour period. The specific treatments considered in the microcosm study were:

1. 0.05M K_2HPO_4 + TCE (4 mg/L) + 0.05M sodium hydrosulfite
2. 0.05M K_2HPO_4 + 1,1,1-TCA (3 mg/L) + 0.05M sodium hydrosulfite
3. 0.05M K_2HPO_4 + carbon tetrachloride (1 mg/L) + 0.05M sodium hydrosulfite

All tests were conducted at room temperature. The final liquid volume in all microcosm vials was 100 mls (leaving 25 mls of headspace) and all tests were conducted in triplicate. Microcosm vials were covered with aluminum foil, stored in the dark, and continually shaken during the course of the study in an inverted position using a mechanical shaker. An additional replicate was prepared for each treatment to allow for monitoring of redox potential and pH over the course of the study.

Microbiological anaerobic and aerobic presence/absence tests were conducted to ensure that sterile conditions were being maintained in the microcosm vials during the course of the study. The anaerobic presence/absence test was conducted using a fluid thioglycollate medium formulated with Bacto Yeast Extract and Bacto Casitone or pancreatic digest of casein (Difco Manual, 10th Edition). The liquid medium was inoculated with test solution from all triplicates both at initiation and termination of tests. Test samples for inoculation were extracted from microcosm vials using a sterile syringe. The absence of any visually observed turbidity in the liquid medium after 48 hours at $35^\circ C$ was interpreted to indicate the absence of anaerobic microbial activity. Microbiological aerobic presence/absence tests were conducted using heterotrophic plate counts with R2A agar (Standard Methods 18th Edition, Method 9215C). The R2A agar was inoculated with test solution from all treatment triplicates and subsequently monitored over a 96-hour period at $35^\circ C$ for colony formation. As in the case of the anaerobic study, test sample used for inoculation was extracted from microcosm vials using a sterile syringe.

Redox and pH conditions in replicate treatment microcosm vials were monitored during the course of the study using a Corning Model 350 pH/ion meter with a gel-filled combination electrode for pH measurements and a combination platinum Ag:AgCl reference electrode for redox measurements.

Field Test. A field study was undertaken to evaluate the feasibility of using sodium hydrosulfite for the in situ abiotic degradation of chlorinated compounds in groundwater. The study was undertaken at the Gloucester Landfill site near Ottawa, Ontario where two aquifers have been extensively impacted by chlorinated hydrocarbons including TCE and 1,1,1-TCA. The objective of the field study was to determine whether the levels of chlorinated compounds in the groundwater could be significantly reduced by addition of sodium hydrosulfite. An existing multi-level monitoring well completed to a depth of approximately 15 m at the site was used to conduct the study. The multi-level well consisted of nine lengths of 1.27 cm internal diameter (ID) high density polyethylene (HDPE) tubing with nytex screen affixed to a 1.9 cm ID PVC Schedule 40 center stalk at nine discrete depths. Initial concentrations of TCE and other chlorinated hydrocarbons were first determined at the nine depths (3.0, 4.5, 6.0, 7.5, 9.0, 10.5, 12.0, 13.5, and 15.0 m) within the multi-level well. Groundwater samples from the discrete depth increments were collected on three separate occasions prior to conducting the test to establish a baseline time-zero concentration for target chlorinated compounds including TCE and 1,1,1-TCA. 500 mls of a 0.125M solution of sodium hydrosulfite were subsequently injected under nitrogen gas at a pressure of 165 kilopascals (24 pounds per square inch) at each of the nine depth intervals through the nine lengths of HDPE tubing. Twenty-four hours of equilibration time were allowed following injection after which time groundwater samples were again collected from each of the nine depth increments using a peristaltic pump.

RESULTS

Microcosm Study. The results of the anaerobic and aerobic microbial presence/absence tests indicated no evidence of microbial activity over the 96-hour period the tests were conducted. No evidence of turbidity after 48 hours was noted in the thioglycollate media used for the anaerobic absence/presence tests and no evidence of colony formation after 96 hours was noted on the R2A agar used for the aerobic absence/presence test. These results suggested that any observed differences in chlorinated hydrocarbon losses between controls and treatments could not be attributed to microbial degradation.

Redox potentials measured in microcosm triplicates during the course of the study indicated equilibrium values of -380 to -400 mV. Equilibrium pH values in all microcosms ranged from 6.5 to 7.0.

The results of the study involving TCE are graphically presented in Figure 1. The results are presented as percentage TCE loss over time. Figure 1a indicates significantly increased loss of TCE at room temperature within

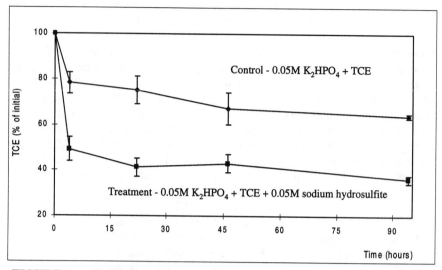

FIGURE 1a   TCE loss with time as mean of triplicates for initial concentration of 4 mg/L in presence and absence of sodium hydrosulfite.

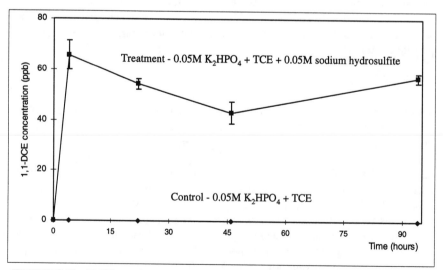

FIGURE 1b   DCE production with time as mean of triplicates for initial TCE concentration of 4 mg/L in presence and absence of sodium hydrosulfite.

microcosm triplicates in the presence of sodium hydrosulfite relative to triplicate controls not containing sodium hydrosulfite. Most if not all of the increased TCE loss relative to the control occurs within the first few minutes of treatment. The difference in TCE loss between the treatment and control is observed to be approximately 25%. After the initial few minutes, there is no evidence of increased TCE loss with time in the treatment relative to the control. Vertical bars shown in Figure 1 indicate the standard deviations observed for triplicates at each measurement point. (Note: The initial drop in the TCE concentration observed for the control is attributed to loss of TCE into the microcosm headspace during and following injection.)

Figure 1b indicates the production of 1,1-DCE as a by-product of TCE following the injection of sodium hydrosulfite into the microcosms. The production of 1,1-DCE is not observed for the control to which sodium hydrosulfite was not added. 1,1-DCE concentrations up to 65 µg/L (for an initial TCE concentration of 4 mg/L) were observed in the microcosms following injection of the sodium hydrosulfite. The production of 1,1-DCE was further evidenced by the detection of 1,1-DCE in the headspace of treatment microcosms but not in controls. (The apparent increase in the concentration of 1,1-DCE noted between the 45-hour sampling period and the 96-hour sampling period in Figure 1b is likely due to analytical variability rather than an actual increase in 1,1-DCE production.)

Abiotic degradation in the presence of sodium hydrosulfite was also observed for 1,1,1-TCA and carbon tetrachloride in the microcosm tests conducted. For 1,1,1-TCA, a loss of over 20% relative to controls was observed and for carbon tetrachloride, a loss of over 50% relative to controls was observed. By-products detected for these compounds included   1,1-DCA and chloroform, respectively. Abiotic degradation of all compounds was observed in the absence of dissolved iron suggesting that iron is not a necessary component in the degradation process.

Field Test. Following injection of sodium hydrosulfite into the multi-level test well at the Gloucester Landfill site, significant drops in redox levels were observed.   The greatest redox drop was observed at the deepest multi-level location (15 m below ground surface) where the redox level dropped from –45 mV to –560 mV following injection of the sodium hydrosulfite. The initial TCE and 1,1,1-TCA concentrations at this depth were 364.3 µg/L and 49.9 µg/L, respectively prior to injection of the sodium hydrosulfite. Twenty four hours following addition of the sodium hydrosulfite, the concentrations of TCE and 1,1,1-TCA measured at this depth were 360.1 µg/L and 45.7 µg/L, respectively. 1,1-Dichloroethane, also present in the aquifer, was measured at a concentration of 60.5 µg/L prior to injection and at a concentration of 55.1 µg/L following sodium hydrosulfite injection at this depth.

Similar observations were made at the other eight depths evaluated. At no depth was there a marked decrease in TCE or 1,1,1-TCA concentrations observed following   addition   of   the   sodium   hydrosulfite.   TCE   and   1,1,1-TCA measurements conducted one week after sodium hydrosulfite injection also did not indicate any change despite the reduced redox levels having been maintained throughout the one week period.

CONCLUSIONS

The results of the microcosm study indicate that chlorinated aliphatics such as TCE and 1,1,1-TCA can undergo abiotic degradation in the presence of sodium hydrosulfite. However, based on the field test results, the process is apparently too insignificant to have a measurable impact in an aquifer restoration application. It is speculated that chlorinated aliphatics such as TCE that may be degraded in the aqueous phase following sodium hydrosulfite addition in an aquifer setting are likely rapidly replaced by TCE desorbed from the aquifer solids as a result of equilibrium partitioning.

ACKNOWLEDGEMENTS

The authors would like to thank Dr. Carol Moralejo for her assistance in the chemical analyses of compounds and by-products during the course of the study and Benoit Lalonde, David Janes and Andrew Patton for their assistance in carrying out the field study.

REFERENCES

American Public Health Association. 1985. *Standard Methods for the Examination of Water and Wastewater.* 16th ed. American Public Health Association, Inc., Washington, D.C.

DIFCO Manual. 1994. *Dehydrated Culture Media and Reagents for Microbiology.* 10th ed. DIFCO Laboratories, Detroit, Michigan 48232 USA.

Fruchter, J.S., C.R. Cole, M.D. Williams, V.R. Vermeul, S.S. Teel, J.E. Amonette, J.E. Szecsody, and S.B. Yabusaki. 1997. "Creation of a Subsurface Permeable Treatment Barrier Using In Situ Redox Manipulation". *Proceedings International Containment Technology Conference.* St. Petersburg, Florida February 9-12, 1997

Gillham, R.W., and S.F. O'Hannesin. 1994. "Enhanced Degradation of Halogenated Aliphatics by Zero-Valent Iron". *Ground Water* 32(6):958-967.

Klecka, G.M., and S.J. Gonsior. 1984. "Reductive Dechlorination of Chlorinated Methanes and Ethanes by Reduced Iron (II) Porphyrins". *Chemosphere.* 13(3):391-402.

Ludwig, R.D., A. Rzeczkowska and R. Booth. 1995. "Investigation of Abiotic Degradation of Trichloroethylene in the Presence of Dissolved Iron: A Microcosm Study." *Proceedings Fifth Annual Symposium on Groundwater and Soil Remediation.* Toronto, Ontario October 2-6, 1995

FIELD DEMONSTRATION OF IN SITU FENTON'S DESTRUCTION OF DNAPLS

Karen M. Jerome (Westinghouse Savannah River Company (WSRC), Aiken, SC USA)
Brian B. Looney (WSRC, Aiken, SC USA)
Jim Wilson (Geo-Cleanse International, Inc. Kenilworth, NJ USA)

ABSTRACT: A collaborative effort between Geo-Cleanse International, Inc. and WSRC in 1997 resulted in demonstration of in situ oxidation based on Fenton's chemistry. The technology treatment yielded an approximate 90% destruction of a 272 kilogram dense non-aqueous phase liquid (DNAPL) pool located 47 m below ground surface. Factors to be considered in selecting this technology are volume of DNAPL, other compounds which may be oxidized under similar conditions, geochemical makeup of the treatment zone, and tightness of treatment zone. Economic evaluation of this technology indicated the costs are sensitive to both depth and DNAPL quantity. Based only on economic considerations, large DNAPL volumes (e.g. 1000's of kilograms) are needed to make the technology viable. Positive attributes of this technology are the end products are innocuous materials, no waste is generated from the treatment process and no material is brought to the surface.

INTRODUCTION

At large industrial sites where contamination resulted from release of chlorinated solvents to the ground, a significant barrier to clean up is treatment of the undissolved DNAPL in the soil and groundwater. Chlorinated solvents, such as trichloroethylene (TCE) and tetrachloroethylene (PCE), are more dense than water and migrate downward through the vadose zone and into the groundwater. The migration path is influenced by geology, hydrology, and solvent physical/chemical characteristics. Data collected at the Savannah River Site (SRS) over the past 15 years indicate that DNAPL below the water table occurs as relatively diffuse ganglia and/or a thin layer on the top of aquitards. DNAPL acts as a reservoir that will continue to generate contaminant levels far above remediation concentration goals well into the future. In an effort to achieve remediation goals and reduce future costs, technologies that will recycle or destroy DNAPL are being developed and demonstrated.

A demonstration of one such destruction technology was conducted at the Savannah River Site in April of 1997. This demonstration evaluated an in situ oxidation technology based on Fenton's chemistry to destroy DNAPL below the water table. Westinghouse Savannah River Company (WSRC) worked in cooperation with Geo-Cleanse International, Inc. in conducting this test, sponsored by the Department of Energy (DOE).

Objective. The objective of this demonstration is to evaluate a technology in the general class of DNAPL destruction technologies. This was accomplished by determining a destruction efficiency and preparing a cost analysis of the demonstration. The goal is to identify a DNAPL recycling or destruction technology to incorporate into the existing suite of technologies being used in the A/M Area of SRS to remediate a 600 hectare groundwater plume resulting from four major DNAPL sources.

Site Description. A/M Area, a former metals fabrication facility within the DOE's Savannah River Site, located in Aiken, South Carolina, was chosen for the location of this demonstration. One and a half million kilograms of solvents, primarily TCE and PCE, were discharged to the soils and groundwater in A/M-Area from 1958 until 1985. Typical of the Atlantic Coastal Plain, sediments beneath A/M-Area are interbedded sands, silts and clays deposited during periods of fluctuation in sea level and modified by erosion during intervening times. Clay rich confining intervals are interspersed with more transmissive, sandier intervals. The uppermost significant clay beneath the water table, termed "Green Clay", is at an approximate elevation of 61 m msl (10 m below the water table and 46 m bgs).

MATERIALS AND METHODS

The demonstration was conducted in three phases: pre-test characterization, technology test, and post-test characterization. Pre- and post-test characterization consisted of collecting continuous soil cores from which soil samples were taken and analyzed by GC headspace method to measure TCE and PCE concentrations. Rotosonic drilling was used to collect soil cores and to install all wells for the test. By comparing post-test characterization data to pre-test characterization data the destruction efficiency of the treatment technology was determined. The test of the Geo-Cleanse® process, based on Fenton's chemistry, was conducted over a 6 day period.

Fenton's Chemistry. H. J. H. Fenton developed a chemistry which oxidized malic acid through use of hydrogen peroxide and iron salts in the 1890s. This chemistry has been, and is still widely used by the waste water industry for treatment of organic wastes. Hydrogen peroxide is the active ingredient in oxidation of organic compounds by this methodology. The hydroxyl radical is the reactive species in this process.

The chemistry of Fenton's reagent (1) is well documented as a method for producing hydroxyl radicals by reaction of hydrogen peroxide and ferrous iron (Fe^{+2}). Hydroxyl radicals are very powerful, effective and nonspecific oxidizing agents, approximately 10^6 to 10^9 times more powerful than oxygen or ozone alone.

$$H_2O_2 + Fe^{2+} \Rightarrow Fe^{3+} + OH^- + OH^\bullet \qquad (1)$$

During the optimum reaction sequence and when the catalyst is iron, ferrous iron (Fe^{+2}) is converted to ferric iron (Fe^{+3}). Ferrous iron is soluble in water at pH 5 to 6 and is necessary for generation of the hydroxyl radical, but ferric iron will not generate the hydroxyl radical and is less soluble at pH 5 to 6. Under properly controlled and buffered conditions, ferric iron can be regenerated back to ferrous iron by a subsequent reaction with another molecule of hydrogen peroxide (2).

$$H_2O_2 + Fe^{3+} \Leftrightarrow Fe^{2+} + H^+ + HO_2^{\bullet} \qquad (2)$$

The iron will remain available in ferrous form as long as pH is properly buffered and there is sufficient hydrogen peroxide. As hydrogen peroxide is consumed, some iron will precipitate out as ferric iron (if pH is moderate).

Many reactions occur during the oxidation of a contaminant, but as shown by equation (3) a contaminant (RHX), hydrogen peroxide, and ferric iron, as a catalyst, are consumed to produce water and carbon dioxide. RHX represents an organic compound and X represent a halide (such as chloride). If the compound is non-halogenated (no X), then the hydrogen ion and halide anion are not formed in the overall reaction.

$$RHX \;^{Fe+2} + H_2O_2 \Leftrightarrow H_2O + CO_2 + H^+ + X^- \qquad (3)$$

Geo-Cleanse® Technology Description. Geo-Cleanse® technology, an in situ destruction technique, utilizes Fenton's reagent to convert organic contaminants to water and carbon dioxide. Hydrogen peroxide and catalyst are injected into the groundwater zone where DNAPL contamination is located. The injection process is proprietary and Patents #5,525,008 and #5,611,642 have been issued.

The injector contains a mixing head which is utilized for mixing reagents and has components to stimulate circulation of groundwater to promote rapid reagent diffusion and dispersion. Thus, all reagents are injected into the subsurface through the injectors. Upon start of the injection process, air with catalyst solution is injected to ensure the injector is open to the formation prior to injection of peroxide and catalyst solution. When an acceptable flow has been established, peroxide and catalyst will be injected simultaneously. This ensures that catalyst and peroxide will not mix together in a sealed system. The injector is designed with a check valve and constant pressure delivery system ensures safety in field operations.

The number of injectors installed and volume of injectate is based on the source area size. Injection of catalyst solution adjusts the groundwater pH to between 4 and 6, where metals, specifically iron, will be at the optimal electron state, +2. This is followed by the simultaneous injection of hydrogen peroxide and catalyst. Mixing of catalyst and hydrogen peroxide in the subsurface will generate heat as the reaction with organic contaminants progresses. Monitoring is conducted during the treatment phase for water vapor, carbon dioxide gas, hydrogen peroxide, the contaminants to be destroyed, pH, conductivity, and dissolved oxygen.   Catalyst solution may be added throughout the injection process to maintain groundwater pH within the range of 4 to 6.

Test Site.   The selected test site is in a bowl shaped surface depression approximately 15 meters square.   It is located within a suspected subsurface trough in the Green Clay along which DNAPL is migrating. Soil concentrations of PCE in the test area ranged from 10 to 150 µg/g.   Highest concentrations were found in a zone at approximately 43 m below ground surface (bgs).   An injector was installed in the center of the test zone with 3 additional injectors and 3 monitoring wells installed radially outward at distances of 5 m and 8 m, respectively (Figure 1).   Also installed were 4 vadose zone piezometers.

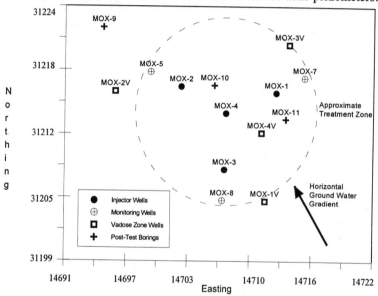

FIGURE 1.   Schematic of In Situ Oxidation Field Demonstration Site Layout (coordinates are a local grid in meters).

RESULTS AND DISCUSSION

Success of the demonstration is based on destruction of DNAPL in the treatment zone.   The best measure of destruction is based on measurement of DNAPL globules in the sediment before and after the treatment process. Results of pre- and post- test characterization indicated a significant decrease in DNAPL concentration after the technology test, as represented in Figure 2.   The estimated pre-test mass of DNAPL in the treatment zone was 272 kgs and the estimated post-test mass of DNAPL was 18 kgs.   Resulting in a destruction rate of approximately 90 percent for the treatment zone.   The treatment zone is defined as the vertical distance between the water table

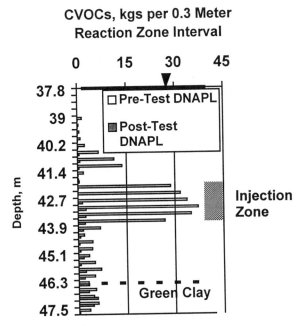

FIGURE 2. Pre- and Post- Test DNAPL Mass for the In Situ Oxidation Demonstration

(38 m bgs) and the Green Clay (46 m bgs) and a 8 m radius from the center injector. Contaminant mass was estimated by averaging sediment concentrations at 0.3 m depth intervals and assuming a treatment zone of 1800 m³. Total destruction of all DNAPL was not achieved and can be attributed to the process not contacting all DNAPL globules in the fine grained sediments. Injected hydrogen peroxide takes the path of least resistance through areas of higher permeability, which in this case is through sandy regions of the treatment zone.

Cost Evaluation of Demonstration. This evaluation examined costs from two perspectives. First, overall cost of the demonstration is discussed in relationship to the influence of each component of the demonstration. Second, cost on a per kilogram of DNAPL removed basis is compared to the cost per kilogram of DNAPL removed for the baseline system of pump and treat using air stripping.

In evaluating the influence of each component of the demonstration on the overall demonstration cost, six categories were identified: site preparation, pre-test drilling and characterization, technology test, post-test drilling and characterization, demobilization, and documentation/project management. The majority of costs are related to the technology test and the pre-test drilling, 36% and 30% of total costs, respectively. Breaking these two categories down into their respective activities showed that drilling activities and peroxide will vary greatly dependent on depth to contamination and amount of DNAPL present.

In an effort to determine the cost effectiveness of this technology, a unit cost based on a kilogram of DNAPL treated or destroyed was determined and compared to the unit cost of the baseline technology. For A/M Area, the baseline technology is pump and treat using airstripping. The baseline cost is $192/kg DNAPL treated (note this is related to groundwater concentration, and the unit cost will increase over time as concentrations decrease). The break even point for in situ oxidation is dependent on depth to contamination. For the A/M Area this occurs at volumes ranging from 2900 kgs to 4300 kgs of DNAPL as depth to contamination increase from 18 m to 47 m. Unit cost of in situ oxidation at sites with small volumes of DNAPL, less than 1800 kgs, is greater than $220/kg of DNAPL. Unit costs escalate to greater than $1500/kg of DNAPL for sites with approximately 450 kgs of DNAPL.

CONCLUSIONS

During this demonstration approximately 90% of 272 kgs of DNAPL was destroyed in a six day operating period, leaving a residual of 18 kgs of DNAPL in the target zone. On a unit cost basis, this technology becomes cost competitive with pump and treat using airstripping ($192/kg, SRS costs) for a DNAPL pool of approximately 4300 kgs at a depth of 47 m bgs. Depth is a major contributor to the overall costs when this technology is employed. For a DNAPL pool of equal volume, remediation costs will increase as depth to the DNAPL pool increases. Thus, both size of the DNAPL pool and depth to the DNAPL pool must be considered in determining when this technology becomes cost competitive with the chosen baseline technology.

Other factors contributing to the decision to use this technology include duration of treatment, and end products of treatment. Treatment is conducted in a period of days to weeks. Factors effecting the duration of the treatment would include: other compounds which may be oxidized under similar conditions, geochemical makeup of treatment zone, and tightness of treatment zone (i.e., access to DNAPL). End products of in situ oxidation are very appealing. No waste is generated from the treatment process, and no material is brought to the surface. End products of this process are carbon dioxide, water, and chloride ions. All of these compounds are considered innocuous materials.

Additional questions raised during the demonstration concerned effects of the treatment on the geochemistry and microbiology in the target zone. Work underway in 1998 will attempt to determine these effects.

REFERENCES

Jerome, K. M., B. Riha, B. B. Looney. 1997. Final Report for Demonstration of In Situ Oxidation of DNAPL Using the Geo-Cleanse® Technology. WSRC-TR-97-00283. Prepared for the U. S. Department of Energy

AQUEOUS OXIDATION OF TRICHLOROETHENE (TCE): A KINETIC AND THERMODYNAMIC ANALYSIS

Kevin G. Knauss, Michael J. Dibley, Roald N. Leif, Daniel A. Mew and Roger D. Aines (Lawrence Livermore National Laboratory, Livermore, CA)

ABSTRACT: An empirical kinetic rate law was determined for the aqueous oxidation of trichloroethene (TCE). By measuring both the rate of disappearance of TCE and the rate of appearance of carbon dioxide and chloride ion, mass balance was monitored to confirm that "mineralization" was the ultimate reaction. Dilute buffer solutions were used to fix pH and stoichiometrically sufficient amounts of dissolved oxygen to make the reactions zero-order in oxygen. Using standard chemical kinetic methods, two orders of magnitude were spanned in initial TCE concentration and used in the resulting double-log plot vs. initial rate (regressed using both linear and polynomial fits) to determine the rate constant and "true" reaction order (i.e., with respect to concentration, not time). By determining rate constants over the temperature interval 343-373K, an Arrhenius activation energy was determined for the reaction. The potential effects of buffer ligand concentration and type (phosphate, borate, acetate, carbonate, sulfate), ionic strength, specific electrolytes, and pH on the rate of TCE aqueous oxidation were studied. The aqueous oxidation reaction rate was found to be pH dependent over the pH range pH 2 to pH 10 and strongly inhibited by high bromide concentration. The equilibrium aqueous solubility of TCE was determined by making reversed measurements from 294-390K.

INTRODUCTION:

Under oxidizing conditions, DNAPL components are thermodynamically unstable. Both ex situ and in situ methods, collectively known as "wet oxidation" using powerful oxidants like hydrogen peroxide, potassium permanganate, peroxydisulfate and ozone, have been developed capitalizing on this fact. However, when applied in situ these methods involve injection of fluids that, due to the very nature of plug flow, displace the contaminant. Efficient mixing can be problematic.

Another approach to in situ aqueous oxidation (Knauss et al., 1995) involves acceleration of rates using heat and is referred to as HPO (Hydrous Pyrolysis/Oxidation). The oxidant may include the dissolved oxygen already present in groundwater, the oxide minerals (e.g., MnO_2) already present in soils, or additional O_2 gas injected with steam into the subsurface. Benefits of the heated groundwater environment include enhancement of contaminant aqueous solubility and mobility by the in situ heating. Desorption and mobilization of DNAPLs from less permeable sediments will also be enhanced to achieve DNAPL destruction in areas which were commonly the source of problems for traditional pump and treat technologies.

Unfortunately, there is a lack of knowledge regarding the basic chemical mechanisms and rates of the aqueous/organic oxidation reactions for heated groundwater conditions. Basic thermodynamic data as a function of temperature required for transport modeling (e.g., solubilities, Henry's Law constants, etc.) are also scarce. This work will begin to address the chemical kinetic and thermodynamic data information needs.

MATERIALS AND METHODS

The laboratory experiments were performed in flexible pure Au bag reactors employing a reusable pure Ti closure contained in a water-filled steel pressure vessel heated inside a large tube furnace. This hydrothermal system is ideal for studying the aqueous phase oxidation of pollutants because it allows repeated sampling of the reacting fluids while precisely controlling temperature and pressure for the duration of the experiment. In all experiments the gold bag was completely liquid filled; there was no headspace. A confining hydrostatic pressure of 1 MPa outside the gold bag was used for most runs, because this pressure was sufficiently high to keep the system single phase (liquid) within the gold bag at all temperatures studied.

Glass gas-tight syringes were used to collect separate samples for analysis of TCE (and any other hydrocarbons), pH, Cl^-, CO_2, and dissolved oxygen. Prior to collecting the TCE sample, the syringe was pre-loaded with an equal volume of HPLC grade methanol to act as a cosolvent. Analysis of samples for TCE (and other hydrocarbons) was made using GC and GC/MS. The aqueous samples were analyzed for inorganic anions (Cl^-, Br^-, SO_4^{-2}, etc.) using HPLC/IC. Dissolved total CO_2 (i.e., $H_2CO_3 + HCO_3^- + CO_3^{-2}$) was determined using an IR CO_2 analyzer. The sample pH was determined using glass electrodes calibrated following the NIST convention.

RESULTS & DISCUSSION

Preliminary Experiments. In order to identify alternate mechanisms to complete mineralization, a few runs were made with very high TCE concentrations (0.095 mmolal) and a stoichiometric excess of dissolved oxygen to make the reaction zero-order in O_2. After approximately 16 days at 373K the concentration of TCE had dropped a thousand-fold and by 29 days was below the EPA maximum contaminant level of 0.000038 mmolal, representing a 2500-fold reduction. The analyses of total inorganic carbon and chloride ion indicated mass balance suggesting that complete mineralization had occurred.

With the initial TCE concentration set at 0.053 mmolal, initial dissolved oxygen concentrations were set at 0.094, 2.94 and 5.97 mmolal in experiments at 373K designed to investigate the reaction dependency on O_2. Even at the lowest oxygen concentration, a stoichiometric excess of dissolved oxygen sufficient to completely oxidize the TCE was present. The results showed that no dependence on the dissolved oxygen concentration was apparent for the aqueous phase oxidation rates of TCE under these conditions. The results suggest that the only constraint at 373K for complete mineralization of TCE with respect to dissolved oxygen concentration is to have at least a stoichiometric amount of dissolved oxygen present to completely oxidize the TCE.

The molar ratio of H^+ production to TCE destruction is 3 to 1, so in unbuffered systems where the pH of the solution is not fixed, a downward drift in the solution pH is observed as the reaction proceeds. Preliminary results suggest that the aqueous oxidation reaction rate is pH dependent (both acid and base catalyzed) over the pH range pH 2 to pH 10. For this reason the kinetic experiments were fixed to pH 7, the pH of normal ground water, using a dilute (1 mmolal) phosphate buffer. Results of kinetic runs made at the same pH using phosphate and carbonate buffers suggested that at these dilute concentrations there were no buffer effects.

In a series of runs to investigate ionic strength and other electrolyte effects, runs were made using Na salts of Cl^-, Br^-, SO_4^{-2}, etc. No salt effects were noted, with the exception that high Br^- ($\geq$ 10 mmolal) significantly inhibited aqueous oxidation of TCE.

Kinetic Analysis. The results of the preliminary screening experiments just described suggest that in the presence of at least a stoichiometrically equivalent amount of dissolved oxygen (O_2(aq)), the rate of TCE aqueous oxidation is zero order in oxygen. The experimental matrix of kinetic runs was designed to take advantage of this so-called method of isolation, in which an unmeasured reactant is present in excess. In these runs, the O_2(aq) was actually measured periodically, but only to confirm that this reactant remained in excess throughout the entire run. As an experimental convenience, it was decided to fix initial O_2(aq) in all runs to the same value (≈ 0.25 mmolal) by equilibrating the fluids with the atmosphere before starting. This effectively meant that the experiments were limited to an initial TCE concentration below ≈ 0.18 mmolal in order to maintain a stoichiometric abundance of oxygen. Because the reactant H_2O is also the solvent, it is present in excess and the aqueous oxidation rate equation reduces to the form:

$$r_{TCE} = dC/dt = -kC_o^n \qquad (1)$$

where r_{TCE} is the rate, k is the rate constant, C is the concentration of TCE (C_o being the initial concentration) and n is the reaction order. This rate law may be linearized by taking the log of both sides to give:

$$\log r_{TCE} = n \log C_o + \log k \qquad (2)$$

In the preferred differential method of kinetic analysis, separate runs are carried out at different initial concentrations, C_o. Initial rates are determined by measuring initial slopes. By using only concentration data collected at early time, potential catalysis or inhibition by reaction products is minimized. A double-log plot of log initial rate vs. log initial concentration has a slope of n_c and an intercept equal to the rate constant, k. In order to span sufficient concentration space, it is desirable to vary initial TCE concentration by at least 2 orders of magnitude. These runs spanned the range 0.15 to 0.0015 mmolal TCE.

The influence of temperature on the rate of aqueous oxidation can be interpreted using the Arrhenius equation. This is commonly expressed in the form:

$$k = Ae^{-E_a/RT} \qquad (3)$$

where k has been defined, A is the pre-exponential factor (assumed to be independent of temperature), E_a is the activation energy (assumed to be independent of temperature), while R and T have their usual meanings. By determining rate constants at a number of temperatures, one can use the linearized version of this equation:

$$\log k = \log A - (E_a/R)(1/T) \qquad (4)$$

where a plot of rate constant, k, vs. 1/T will have slope (E_a/R) and intercept log A. These runs spanned the temperature range 343-373K.

Rate Constant, Reaction Order & Activation Energy. A regression analysis of the results at pH 7 and 90°C (Figure 1) yields the rate constant, k, and true reaction order with respect to TCE, n_c, for the aqueous oxidation of TCE. The results are $k = 5.77 \pm 1.06 \times 10^{-7}$ mols/kg-s and $n_c = 0.85 \pm 0.03$. The simplified rate expression for TCE aqueous oxidation under these conditions then becomes:

$$r_{TCE} = dC/dt = -(5.77 \pm 1.06 \times 10^{-7} \text{ mols/kg-s})C_o^{(0.85 \pm 0.03)} \qquad (5)$$

at 90°C and pH 7 in the presence of stoichiometrically sufficient dissolved oxygen. A regression analysis of the results over the temperature interval 343-373K

(Figure 2) yields the activation energy, $E_a = 108.0 \pm 4.5$ kJ/mol. Assuming that the pre-exponential factor is independent of temperature, the rate constant at any temperature may be calculated using:

$$\ln k_2 = \ln k_1 + (E_a/R)[(1/T_1) - (1/T_2)] \qquad (6)$$

where: the subscript 1 refers to the first temperature in K and subscript 2 refers to the second temperature in K and the other terms either have been defined or have their usual meaning. For example, the calculated rate at 25°C is about 2460 times slower than the 90°C rate of aqueous oxidation. At the average annual temperature of most ground waters contaminated with TCE, the rate would be significantly slower than at 25°C. This is why TCE degradation via aqueous oxidation is not a significant sink at normal temperature compared to other processes, except when microbially mediated.

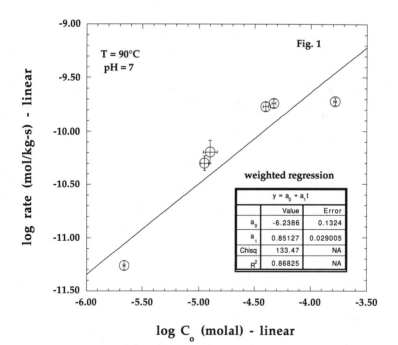

FIGURE 1. Double-log plot used to determine of log initial rate vs. log initial concentration having a slope of n_c (true reaction order, i.e., with respect to concentration, not time) and an intercept equal to the rate constant, k.

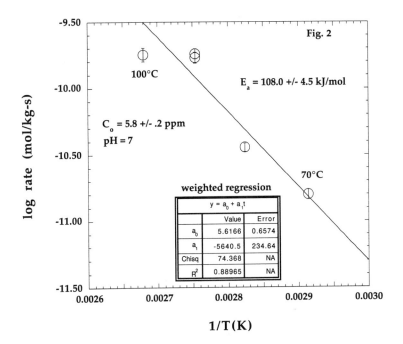

FIGURE 2. Arrhenius plot having a slope equal to E_a/R, where E_a is the activation energy for the aqueous oxidation reaction

Solubility of TCE & PCE. Equilibrium data (e.g., solubility data) measured over a wide temperature range allow calculation of the temperature dependency of thermodynamic quantities at constant pressure following the method of Clarke and Glew (1966). This method involves a linear multivariate regression analysis of the expression:

$$R \ln K = a_0 + a_1 \mu_1 + a_2 \mu_2 + a_3 \mu_3 + a_4 \mu_4 + a_5 \mu_5 \qquad (7)$$

$$= \sum_{i=0}^{n} a_i \mu_i \qquad n \le 5 \qquad (8)$$

where: $\mu_0 = 1$ and the a_i are the partial regression coefficients of $R \ln K$ on the temperature variables μ_i. The temperature variables μ_i are defined by:

$$\mu_i = x^i \sum_{n=1}^{\infty} \frac{n}{n+i+1} (-x)^{n-1} \qquad 1 \le i \le 5 \qquad (9)$$

where: $x = (T-\theta)/\theta$, $T = T(K)$ and θ is the reference temperature (= 298.15 K in this case). The relation of the first three regression coefficients to the thermodynamic parameters at $T = \theta$ are:

$$a_0 = -\frac{\Delta \overline{G}(\theta)}{\theta} \qquad (6)$$

$$a_1 = \frac{\Delta \overline{H}(\theta)}{\theta} \qquad (10)$$

$$a_2 = \Delta \overline{C}_p(\theta) \qquad (11)$$

and by substituting Eqns. 6 and 7 into the expression $\Delta \overline{G} = \Delta \overline{H} - T\Delta \overline{S}$:

$$a_0 + a_1 = \Delta \overline{S}(\theta) \qquad (12)$$

Applying the method of Clarke and Glew (1966) to the weighted (weighting factors proportional to the inverse of the measured variance - σ^2) R ln K data measured here, the following thermodynamic values are calculated at 298 K for TCE: $\Delta \overline{G}_{soln} = 11.282\ (\pm.003)$ kJ/mol, $\Delta \overline{H}_{soln} = -3.35\ (\pm.07)$ kJ/mol, $\Delta \overline{S}_{soln} = -49.07\ (\pm.24)$ J/mol-K, and $\Delta \overline{C}_{p\ soln} = 385.2\ (\pm3.4)$ J/mol-K.

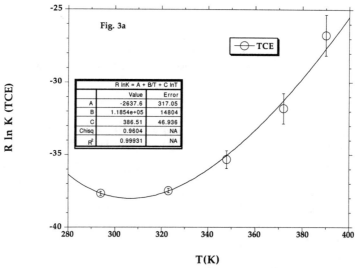

T(K)

FIGURE 3. Plot of R ln K for TCE vs. T (K), where: K = equilibrium constant for the dissolution reaction: $TCE_{(l)} = TCE_{(aq)}$.

REFERENCES

Clarke, E.C. and D.N. Glew. 1966. "Evaluation of thermodynamic functions from equilibrium constants." *Trans. Farady Soc.* 62: 539-547.

Knauss K. G., M. Alai, D.A. Mew, S. A. Copenhaver and R. D. Aines. 1995 "TCE : Thermodynamic measurements and destruction via Hydrous/Pyrolysis Oxidation." *Geol. Soc. Am. Abstr.* 27: 249.

VOX DESTRUCTION BY PEROXIDE/ULTRAVIOLET

Ronald E. Hutchens, P.E.(Environ International Corporation, Buffalo Grove, IL)
Larry A. Ganzel, P.E. (ERM-North Central, Inc., Vernon Hills, IL)
Carl Raycroft, P.E. (Environ International Corporation, Buffalo Grove, IL)

ABSTRACT: Two Superfund sites in Indiana, are being remediated in accordance with a Consent Decree pursuant to the Comprehensive Environmental Responses, Compensation and Liability Act of 1980. The sites, operated during the 1970s, are being remediated with a ground water collection system, a hydrogen peroxide/ultraviolet (HP/UV) treatment system and deep well injection (required due to the saline nature of the ground water). The system has been in operation since August 1995, and has successfully treated over 20 million gallons (75,700 cubic meters) of ground water contaminated with halogenated volatile organic compound (VOX) and other compounds to meet Superfund delisting standards. The design flow rate is 50 gallons per minute (gpm) (3.15 liters per second [lps]).

The operational data for the two treatment plants is presented and analyzed (the ground water at each site is separately treated, and transferred by pipeline to the deep well). The data include untreated ground water, treatment plant operational and performance data, and final effluent data.

The variation in ground water quality and the presence of small amount of hydrocarbons and other less photo-oxidizable compounds has reduced the economic advantages of the HP/UV treatment system. The combined influent data are poorly correlated with individual well data, and the ability to maintain efficient photo-oxidation is limited due to ionic instability of the ground water and subsequent coating of the oxidation lamps.

Our data demonstrate that the use of HP/UV treatment is appropriate for low concentrations of VOX, but that more effort should be allocated to issues of ground water variability, the presence of less photo-oxidizable compounds and ionic instability before selection of this process. In other respects, the operation has been reliable and the effluent quality has met all requirements, but at a greater than predicted cost.

TREATMENT SYSTEM DESCRIPTION

In general, ground water is extracted from six extraction wells at the Site II, treated and pumped through a pipeline to the post-treatment tank at Site I. Ground water at Site I is extracted from six extraction wells, treated, and combined with the Site II water in the post-treatment tank. The combined ground water flow is then pumped from the post-treatment tank to the deep injection well. The description of the treatment system, is noted below.

Site II. Ground water is collected from six extraction wells using pneumatic pumps. The individual extraction well flows are combined at the treatment system. The design flow rate is 30 gpm (1.89 lps). The combined ground water is mixed with a small amount of HP to convert the soluble ferrous iron to insoluble ferric hydroxide (to reduce scaling in the HP/UV unit). Flow is then discharged into an oil/water separator, which removes free product and settable organic matter (to reduce the organic loading to the HP/UV unit). The ground water overflows into an equalization tank. The contents of the equalization tank are pumped through filters to remove oils and large particles and then through a fine filter to remove the remaining suspended solids (to reduce fouling of the HP/UV unit). Chemical metering pumps are then used to pump sulfuric acid into the ground water flow for pH adjustment down to a pH of 6.5 for optimum oxidation. The ground water then flows into the HP/UV unit. The HP/UV unit consists of 12 lamps, each rated at 15 kilowatts. Hydrogen Peroxide is metered and blended with water prior to entering the first oxidation chamber. Air release valves are used on the HP/UV unit to release carbon dioxide generated during oxidation.

The treated ground water discharges to a tank with a portion of the flow being analyzed by an on-line gas chromatograph (GC). Based on characteristics of site waste data, the GC analyzer has been calibrated for vinyl chloride (VC) as an indicator for sufficient treatment of wastewater through the HP/UV system. If the GC detects a VC concentration exceeding the MAC, the Site II extraction wells and treatment system are shut down. A chemical metering pump is used to pump sodium hydroxide into the tank to raise the pH to 7. The contents of the tank are pumped through a filter for further removal of remaining suspended solids (to reduce fouling of the underground pipeline) and discharged to the underground pipeline.

Underground Pipeline. The 4-inch-diameter high density polyethylene underground pipeline conveys the treated, delisted ground water from Site II to Site I where it is combined with the Site I treated ground water in the post-treatment room tank. The 21,000-foot-long (6,400-meter) pipeline between the two sites is buried with directional drilled casings below the Calumet River crossing and all road and railroad crossings.

Site I. The treatment process is identical to the Site II system with the following modifications: (1) the design flow rate is 20 gpm (1.26 lps); (2) there is no oil removal equipment; (3) the pH is adjusted to 5.5; (4) the HP/UV treatment unit consists of 3 lamps, each rated at 30 kilowatts; (5) an air stripper unit with off gas treatment through granulated activated carbon is used for its polishing treatment; and (6) the GC analyzer is calibrated for MeCl. The air stripper with off gas treatment is included to further remove methylene chloride, a difficult to oxidize compound, from the waste stream.

Site I Post-Treatment. Treated ground water from Site II, combined with the treated ground water from Site I combines in the Site I post-treatment tank The contents of the tank are pumped through final filters (with one micron filter media) to the Deep Well Site injection tank.

Deep Well Site. The contents of the injection tank are pumped by a positive displacement pump to the deep well. Injection flow, pressure, and temperature are monitored. The deep well annulus pressure is also monitored. Deviations from specified critical operating values result in the shutdown of both Sites I and II ground water extraction and treatment systems.

SITE I EVALUATION

Operating Data

Operating Condition	Value
Influent pH, unitless	7.8+/-
HP iron oxidation dosage, mg/L	5
Treatment flow rate, gpm (lps)	16.5 (1.04)
Primary solids filter media size, microns	3
Sulfuric acid setpoint, unitless	5.5
HP oxidation dosage, mg/L	700
HP/UV unit lamps, kw	90
Tube cleaning cycle, minutes per hour (recycle minutes)	7
Net treatment flow rate, gpm (lps)	14 (0.88)
Air stripper air rate, scfm (scms)	140 (3.95)
GC analyzer shutdown (MeCl), $\mu g/l$	12

Parameter Concentrations ($\mu g/l$)

Parameter	Influent Average	Maximum	Effluent	MAC
1,1-Dichloroethane	61	63	2.7	880
1,1,1-Trichloroethane	25	30	2.8	1,260
cis-1,2-Dichloroethene	175	210	2.5	441
Methylene chloride	**73.5**	**100**	3.6	31.5
Toluene	189	280	2.3	6,300
Vinyl chloride	**38.5**	**59**	2.8	12.6
Xylenes (total)	220	280	12.4	63,000

All other volatile organic compounds (VOCs) were less then detection limits and the MAC. MeCl after the HP/UV treatment unit and prior to the air stripper unit averaged 27.3 $\mu g/l$ with a maximum of 46 $\mu g/l$, exceeding the MAC.

SITE II EVALUATION

Operating Data. Site II operating data that deviates from Site I is listed below:

Operating Condition	Value
Influent pH, unitless	8.5 +/-
Treatment flow rate, gpm (lps)	24.5 (1.55)
Sulfuric acid setpoint, unitless	6.5
Tube cleaning cycle, minutes per hour (recycle minutes)	10
Net treatment flow rate, gpm (lps)	20.8 (1.31)
GC analyzer shutdown (VC), µg/l	32

Parameter Concentrations (µg/l)

Parameter	Influent		Effluent	MAC
	Average	Maximum		
1,1-Dichloroethane	102.2	170	74.9	880
1,1,1-Trichloroethane	66.5	150	51.6	1,260
cis-1,2-Dichloroethene	74.7	130	9.4	441
Methylene chloride	27.3	**52**	20.5	31.5
Toluene	439	850	10.1	6,300
Vinyl chloride	**55.5**	**100**	10	12.6
Xylenes (total)	258	560	35.6	63,000

All other VOC parameters were less then detection limits and the MAC.

HP/UV TREATMENT UNIT OPERATING ISSUES

To maintain the HP/UV treatment unit efficiency, the influent wastewater is filtered with disposable particulate cartridges. Solids in the water reduce light transmission and treatment efficiency. The initial filter media was a plastic polypropylene material with a 10-micron rating and an anticipated 5-day filtration capacity. To maintain treatment efficiency, the media has been changed to a cotton fabric filter cartridge with a 3-micron rating and a 2-day filtration capacity. This change of filter media has increased cartridge material costs, but reduced the influent color and suspended organic matter with an improvement in HP/UV treatment efficiency.

During initial startup testing at each site, the operating parameters were established for optimal treatment efficiency of the HP-UV systems by adjusting the following conditions:

1)   pH range between 5.0 and 7.0
2)   HP oxidation dosage rate between 300 mg./l to 1,000 mg/l
3)   Partial dosage feeding of HP between UV lamp chambers versus feeding full dosage up front.

To maintain the HP/UV treatment unit efficiency, cleaning of the oxidation lamps with an internal scraper on a defined schedule is required. The scraper is mounted in a shuttle assembly in each oxidation chamber that is forced by water pressure across each tube to scrape residual material off the tube during a tube cleaning cycle. Upon initiation of a tube cleaning cycle, the water flow through the HP/UV system is reversed and water flow pressure builds behind the shuttle and forces the device to the opposite end of the chamber. Upon complete movement of all shuttles, the flow is reversed again to return the shuttle to the original position. A build up of calcium and solid deposits on the lamp surface reduces light transmission and treatment efficiency, inhibits scraper movement, and can result in accumulation of heat in the tube resulting in lamp failure. During a tube cleaning, the HP/UV unit internal flow receives partial treatment and requires recycling to the system feed tank for re-treatment through the HP/UV system following the tube cleaning. During initial operations, the tube cleaning was set at 2 cycles every 48 hours. One cycle is designated as a back and forth movement of the shuttle assembly across the tube. To maintain treatment efficiency, the tube cleaning was revised to one cycle every hour. Due to the quantity of 12-chambers at Site II and 3-chambers at Site I, the tube cleaning cycle time is 10-minutes and 7-minutes respectively, to complete a cycle. Consequently, the net treatment flow rate is reduced by 14.3% and 10.5%, respectively.

Even with the above system operating changes, to maintain HP/UV treatment unit efficiency, a scheduled monthly maintenance program was implemented to manually clean the individual tubes. This program requires shut down of the treatment system and draining of the HP/UV unit. The tubes are removed and hand cleaned, the quartz tubes and scrapers are removed, inspected, rebuilt or replaced, and the end chamber ports inspected and repaired. A 1- or 2-day shutdown for each site is required for this scheduled maintenance.

Subsequent to the start up of the Site I treatment system, greater then anticipated influent parameter concentrations, specifically MeCl, were encountered. To meet the MAC, an air stripper system with off-gas treatment was added. Greater then anticipate influent parameter concentrations were also experienced at Site II, specifically total dissolved solids (10,000 + mg/l) and chloride concentrations (5,000 + mg/l), which accelerated corrosion rates of stainless steel parts. No additional treatment equipment was added at Site II.

The higher then anticipated and variable influent concentrations are attributed to: (1) utilizing samples from investigation wells, not actual extraction wells, to determine influent concentrations; (2) varying Lake Michigan water level, which affect water table elevation, gradients, and extraction rates; and (3) periodic individual extraction pump failures, which changes influent characteristics. To meet the MAC limits, the UV lamp requires replacement approximately every 3,000 hours of operation to maintain UV light efficiency.

CONSTRUCTION COST

Construction costs are site specific and are influenced by (1) site access limitation, (2) utility requirements, (3) complexity of the ground water extraction system, (4) pre-treatment requirements, (5) enclosure requirements, (6) climate conditions, and (7) monitoring requirements. The construction costs (1994 cost unless noted), including engineering and construction oversight are presented below:

Treatment Component	Costs
Site II treatment system	$ 2,220,000
Site I treatment system	$ 1,530,000
Site I air stripper system (1996)	$ 90,000
Pipeline from Site II to Site I	$ 1,420,000
Deep well injection system	$ 240,000
Deep well (1993)	$ 1,300,000
Total Construction Costs	$ 6,800,000

The above costs do not include ground water monitoring wells, site fencing, and construction reports.

ANNUAL OPERATION AND MAINTENANCE COSTS

The annual operation and maintenance costs are site specific and are influenced by (1) degree of automation; (2) personnel site attendance requirements; (3) monitoring and sampling requirements; (4) water characteristic variations; (5) equipment trouble shooting, maintenance, or replacement; and (6) system testing requirements. The 1994 projected annual operation and maintenance cost was estimated at $618,000. This cost does not include site influent and effluent sampling, ground water sampling, air monitoring, and reports. At an anticipated combined treatment system flow rate of 41 gpm (2.59 lps), the treatment cost per 1,000 gallons (3,785 liters) is $28.66.

The 1997 annual operation and maintenance cost was $728,000. The higher cost is attributed to (1) increased HP dosage, (2) increased filter media usage, and (3) increased equipment maintenance due to the corrosive nature of treated water. The 1997 average combined treatment system flow rate was 28 gpm (1.77 lps). The flow rate is lower due to HP/UV unit tube cleaning cycles, scheduled monthly shutdown to conduct HP/UV unit maintenance, and unscheduled shutdowns to repair or replace equipment components. The treatment cost per 1,000 gallons (3,785 liters) is $49.47. The cost breakout is summarized below:

Item	Cost ($) per 1,000 Gallons (3.785 Liters)
Acid and caustic	0.88
Hydrogen peroxide	7.61
Filter media	6.80
HP/UV repair parts	2.89
Equipment repair/replacement	1.97
Consumable supplies	4.11
Electricity	10.94
Labor	14.27
Total	49.47

IN SITU TREATMENT OF ORGANICS BY SEQUENTIAL REDUCTION AND OXIDATION

Paul G. Tratnyek and Timothy L. Johnson (Oregon Graduate Institute)
Scott D. Warner (Geomatrix Consultants, San Francisco, CA)
H. Steve Clarke (Chemical Waste Management of the Northwest, Arlington, OR)
John A. Baker (Waste Management Technology Center, Geneva, IL)

ABSTRACT: We have investigated the possibility of treating mixtures of VOCs by adding aqueous solutions of chemical oxidants, reductants, and sequential combinations of oxidants and reductants. Dithionite reduces carbon tetrachloride rapidly, but has little effect on most other VOCs. Permanganate oxidizes chlorinated ethylenes, but has little effect on halogenated methanes or ethanes. Fenton reagent (hydrogen peroxide and ferrous iron) rapidly oxidizes all VOCs except the perhalogenated alkanes. In our experiments, the chemical oxidants and reductants were more effective in laboratory water systems, less effective in site water, and least effective in the presence of aquifer solids. This is presumably due to competition between the VOCs and other aquifer constituents for the oxidant or reductant. We also found that it is difficult of achieve the benefits of sequential treatment by reductants (to reduce perhalogenated VOCs) followed by oxidants (to mineralize the rest), due to reaction between the oxidant and residual reductant.

INTRODUCTION

Mixtures of volatile organic compounds (VOCs) are difficult to remediate when the components of the mixture are diverse in their susceptibility to available treatments. This study was motivated by the need to remediate groundwater that is contaminated with such a mixture. The contamination is associated with the construction of a monitoring well, which penetrates a confining layer and ends in a low-permeability zone ~200 feet below ground. Since other wells in the area are clean, the contamination is believed to be localized in the vicinity of the well casing.

The chemicals that have been found in the well water include chlorinated ethenes, ethanes, and methanes; chlorofluorocarbons; and petroleum hydrocarbons. Maximum reported concentrations (C_{max}) are shown in Table 1 along with a risk based ranking based on an analysis of human health effects data, federal drinking water standards (where applicable), and ecological risk information for each of the contaminants detected at the site. The purpose of this study was to develop an in situ remediation strategy to treat this VOC mixture, giving priority to those VOCs that represent the highest risk. We have investigated the possibility of treating mixtures of VOCs by adding aqueous solutions of chemical oxidants, reductants, and sequential combinations of oxidants and reductants.

Chemical Oxidants and Reductants. The chemical oxidants and reductants used in this study are summarized in Table 2 along with the corresponding redox half-reactions (for low and high pH conditions) and standard reduction potentials.

Hydrogen peroxide (H_2O_2) and H_2O_2 with Fe^{2+} (Fenton reagent) have been studied extensively for in situ oxidation of groundwater contaminants (e.g., Walton et al. 1991; Vella and Veronda 1993). Two alternative reagents that are gaining recognition are the oxidant potassium permanganate ($KMnO_4$) and the reductant sodium dithionite ($Na_2S_2O_4$).

TABLE 1. Ranking VOCs detected at the site.

Contaminant	C_{max} (mg/L)	Conc Rank	Risk Rank	Comb Rank
1,1-Dichloroethene	0.56	4	1	1
Trichloroethene (TCE)	3.2	2	4	2
1,2-Dichloroethene	0.26	7	3	3
Tetrachloroethene (PCE)	0.28	6	6	4
Chloroform	0.32	5	7	5
1,1,1-Trichloroethane	3.4	1	11	6
Toluene	1.4	3	9	7
Carbon tetrachloride	0.015	15	2	8
Methylene chloride	0.18	9	8	9
Benzene	0.032	14	5	10
Freon 11	0.21	8	13	11
1,1-Dichloroethane	0.098	11	10	12
Ethylbenzene	0.035	13	12	13
Freon 113	0.12	10	14	14
Xylenes	0.039	12	15	15

TABLE 2. Chemical oxidants and reductants.

Treatment	Redox Half-Reaction	E° (V vs SHE)
Hydrogen peroxide	$H_2O_2 + 2\,H^+ + 2\,e^- \rightleftharpoons 2\,H_2O$	+ 1.77
	$HO_2^- + H_2O + 2\,e^- \rightleftharpoons 3\,OH^-$	+ 0.88
Permanganate	$MnO_4^- + 4\,H^+ + 3e^- \rightleftharpoons MnO_2 + 2\,H_2O$	+ 1.70
	$MnO_4^- + 2\,H_2O + 3e^- \rightleftharpoons MnO_2 + 4\,OH^-$	+ 0.59
Dithionite	$2\,H_2SO_3 + H^+ + 2e^- \rightleftharpoons HS_2O_4^- + 2\,H_2O$	− 0.08
	$2\,SO_3^{2-} + 2\,H_2O + 2e^- \rightleftharpoons S_2O_4^{2-} + 4\,OH^-$	− 1.12

METHODS

Batch experiments were performed in 12-mL serum vials that were crimp-sealed using Hycar septa and allowed to react at 22°C with gentle mixing. Vials were extracted by removing 2 mL of water, adding 1 mL of hexane, and recrimping the septa. After the vials were shaken vigrously, vortexed, and allowed to equilibrate for 1 hr, 0.5 mL of hexane was transfered to autosampler vials for analysis by gas chromatography.

RESULTS AND DISCUSSION

Dithionite. $Na_2S_2O_4$ reduced CCl_4 rapidly in laboratory water (Figure 1). The disappearance of CCl_4 was accompanied by appearance of nonstoichiometric amounts of $CHCl_3$ and a trace of CH_2Cl_2, suggesting that simple reductive dechlorination was not the only pathway of CCl_4 degradation. A previous study of products in this reaction (Rodríguez and Rivera 1997), showed that the reaction products include the trichloromethanesulfinate anion ($CCl_3SO_2^-$), presumably due to coupling of the trichloromethyl radical (intermediate in reductive dehalogenation) and the sulfonyl anion radical ($SO_2^{\bullet-}$ from the dissociation of $S_2O_4^{2-}$). The subsequent fate of $CCl_3SO_2^-$ is not known.

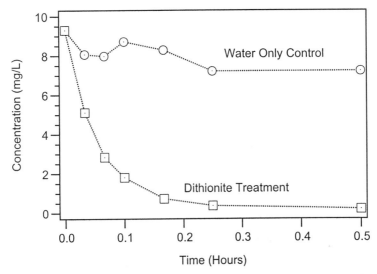

FIGURE 1. Reduction of CCl_4 by 0.09 M dithionite (pH 6.6 unbuffered).

For the data in Figure 1, the initial kinetics of CCl_4 reduction by dithionite are pseudo first-order with $k_{obs} = 0.27$ min^{-1}. Assuming the reaction is also first-order in dithionite, a second-order rate constant can be estimated. The value, $k = 0.05$ M^{-1} s^{-1}, is in remarkably good agreement with the value ($k = 0.03$ M^{-1} s^{-1}) we estimate from previously published data for a similar system that was buffered at pH = 7.5 with 0.1 M phosphate (Rodríguez and Rivera 1997). In binary mixtures with CCl_4, other chlorinated VOCs (PCE, 1,1,1-TCA, and $CHCl_3$) were not significantly degraded by dithionite under these conditions.

Site water gave slower degradation rates but similar selectivity among the contaminants. In the presence of crushed aquifer material, $Na_2S_2O_4$ had no detectable effect on contaminant concentrations over several hours. However, this treatment caused the aquifer solids to change color from yellow to gray, suggesting reduction of iron oxides and formation of "structural" Fe(II). Structural Fe(II) can mediate the reduction of CCl_4 by dithionite (Amonette et al. 1994), but this apparently requires much longer contact times than were tested in this study.

Permanganate. KMnO$_4$ oxidized the chlorinated ethenes (PCE, TCE) in site water (Figure 2). No products were observed, but the expected products are hydroxyketones, aldehydes, and carboxylic acids (Walton et al. 1991; Schnarr and Farquhar 1992) that would not be detected by our method.

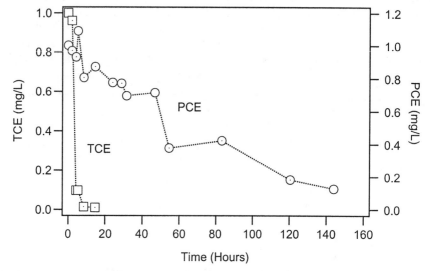

FIGURE 2. Reduction of PCE and TCE by 0.7 mM KMnO$_4$ (site water).

Although the data in Figure 2 does not demonstrate that disappearance of PCE and TCE follows first-order kinetics, the data are consistant with a psuedo first-order model (linear regression of ln concentration vs. time gives R^2 > 0.94 for both analytes). Applying the model gives k$_{obs}$ = 0.01 hr^{-1} for PCE and 0.54 hr^{-1} for TCE over the conditions of this experiment. As expected, oxidation of the less chlorinated alkene (TCE) is more rapid (in this case by about 50-fold).

A series of experiments performed in deionized water with PCE and permanganate concentrations varying from 11 to 111 mg/L (data not shown), confirms that the reaction is very close to first order in permanganate concentration (the log k$_{obs}$ vs. log [MnO$_4$] plot has slope = 1.3±0.1). From the latter data, we can estimate the second-order rate constant k = 0.0084 M^{-1} s^{-1} for oxidation of PCE by permanganate in site water (at pH 8).

Oxidation of chlorinated ethenes by permanganate involves attack at the double bond, and it is rearrangement of the adduct that leads to dechlorination. Since there is no mechanism for direct dechlorination by attack of permanganate at C-Cl bonds, there is no reason to expect oxidation of halogenated alkanes. In fact, we found permanganate has little effect on the ethanes (1,1,1-TCA, 1,1-DCA), and no effect on the methanes (CH$_2$Cl$_2$, CHCl$_3$), or CFC-11.

Hydrogen Peroxide. Treatment with H$_2$O$_2$ removed TCE and PCE from site water (containing ~0.4 mg L^{-1} Fe), but had little effect on the other VOCs (Figure

3). No products were detected by gas chromatography, consistant with oxidation to give polar intermediates and some mineralization.

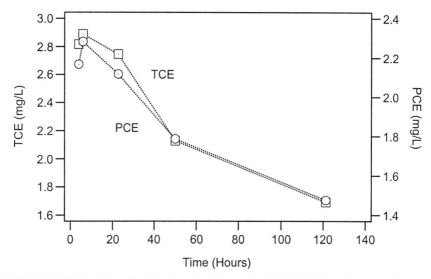

FIGURE 3. Reduction of chlorinated ethenes by 0.1 M H_2O_2 (site water).

Assuming pseudo first-order kinetics apply, the data in Figure 3 give k_{obs} = 0.0056 hr^{-1} for PCE and 0.0072 hr^{-1} for TCE. These correspond to half-lives of about 4 to 5 days, which seem slow considering that the chlorinated ethenes react with hydroxyl radical at nearly diffusion-controlled rates. Therefore, we believe that H_2O_2 appears to be less effective than permangante (c.f. Figures 2 and 3), due to inefficient generation of hydroxyl radical by the Fenton reaction, and not because permanganate is intrinsically more reactive than Fenton reagent with PCE and TCE. An important advantage of H_2O_2 treatement is that it is less selective than permanganate, giving comparable degradation rates for many of the site water constituents (data not shown).

Reductant/Oxidant Sequences. The data presented above confirm that chemical reductants are most effective with the perhalogenated components of the site water (recall Table 1) and chemical oxidants are more effective at degrading the less halogenated (more hydrogenated; i.e., reduced) components. Thus, it would seem attractive to pretreat with a reductant to convert all the components of the mixture into compounds that are readily treatable with chemical oxidants. Sequences consisting of dithionite followed by H_2O_2, with various amendments to adjust pH, were tested on the site water. Figure 4 shows that dithionite partially degraded CCl_4, but that making the solution alkaline with NaOH gave little benefit (aside from dilution) and adding peroxide to the site water containing alkaline dithionite also gave no significant benefit.

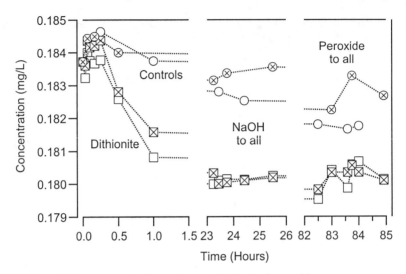

FIGURE 4. CCl₄ concentrations after a dithionite/peroxide sequence.

Data (not shown) for other constituents in the site water also showed little benefit from sequential treatments of dithionite and peroxide. Preliminary efforts to adjust pH (up for treatment with dithionite and down for optimal Fenton chemistry) have not been successful. These results suggest that the anticipated benefits of chemical reduction and oxidation in sequence could be difficult to achieve in practice. Column experiments are underway that may yield more promising results by allowing longer contact times with the chemical reagents.

REFERENCES

Amonette, J. E., J. E. Szecsody, H. T. Schaef, J. C. Templeton, Y. A. Gorby, and J. S. Fruchter. 1994. "Abiotic reduction of aquifer materials by dithionite: A promising in-situ remediation technology". *Proceedings of the 33rd Hanford Symposium on Health & the Environment.* Pasco, WA, Battelle. pp. 851-881.

Rodríguez, J. C. and M. Rivera. 1997. "Reductive dehalogenation of carbon tetrachloride by sodium dithionite". *Chem. Lett.* pp. 1133-1134.

Schnarr, M. J. and G. J. Farquhar. 1992. "An in situ oxidation technique to destroy residual DNAPL from soil". *Subsurface Restoration Conf.*, Dallas, TX.

Vella, P. A. and B. Veronda. 1993. "Oxidation of trichloroethylene: A comparison of potassium permanganate and Fenton's reagent". *3rd Int. Symp. Chemical Oxidation: Technology for the Nineties*, Nashville, TN, Technomic. pp. 62-78.

Walton, J., P. Labine, and A. Reidies. 1991. "The chemistry of permanganate in degradative oxidations". *1st Int. Symp, Chemical Oxidation: Technology for the Nineties*, Nashville, TN, Technomic. pp. 205-221.

IN SITU OXIDATION OF TRICHLOROETHYLENE USING POTASSIUM PERMANGANATE: PART 2. PILOT STUDY

Daniel McKay and Alan Hewitt (U.S. Army CRREL, Hanover, New Hampshire)
Stanley Reitsma (University of Windsor, Windsor, Ontario)
John LaChance and Ralph Baker (ENSR, Acton, Massachusetts)

ABSTRACT: Pilot-scale testing is being performed at the U.S. Army Cold Regions Research & Engineering Laboratory to evaluate the feasibility of using a concentrated solution of potassium permanganate ($KMnO_4$) to promote in-situ oxidation of trichloroethylene (TCE) in low-permeability layers of the vadose zone. The permanganate is injected into higher-permeability strata embedded within silt and clay zones zones where residual-phase TCE has been found. Monitoring of pore water indicates oxidation of TCE is occurring, while data from soils analyses are presently inconclusive.

INTRODUCTION

For 27 years, the Cold Regions Research & Engineering Laboratory utilized trichloroethylene (TCE) extensively for such purposes as refrigeration, ice coring, and degreasing. Since the initial discovery of TCE at this site in 1990, residual levels of subsurface TCE contamination have been identified in unsaturated soils between 15 and 35 ft below the surface. The geological composition within this zone consists largely of fine sands with some silts, interbedded with veneer-thin stringers of sand and thicker (one to several inches) layers of nearly saturated silts and clays. Because of the depth of contamination and presence of an extensive network of underground utilities, only in-situ cleanup alternatives may be considered. Preliminary selection of a chemical oxidant in solution was based on the deliverability of an aqueous carrier to otherwise inaccessible, dense, moist strata via induced pressure, gravity, capillary suction, and diffusion. This paper presents initial results of a chemical oxidation pilot test utilizing potassium permanganate ($KMnO_4$) that is delivered in solution via vertical injection wells.

Potassium permanganate is a strong oxidizer that that reacts readily in the presence of many organic compounds. Reaction with TCE in water (Equation 1) cleaves the double carbon bond to yield carbon dioxide (CO_2), manganese dioxide (MnO_2), potassium chloride (KCl), and hydrochloric acid (HCl).

$$C_2Cl_3H(TCE) + 2KMnO_4 \Rightarrow 2CO_2(aq) + 2MnO_2(s) + 2KCl + HCl. \qquad (1)$$

Dissolution of the resulting CO_2 in pore water results in formation of a weak carbonic acid while CO_2 in excess of carbonate saturation levels is released as vapor [Schnarr, 1992]. One of the concerns with generation of CO_2 vapor during treatment, if it occurs, is that it may impair deliverability by reducing the relative permeability to pore water. Solubilized KCl and HCl dissociate into ionic constituents yielding potassium (K^+), hydrogen (H^+) and chloride ions (Cl^-) whereas precipitates of MnO_2 exist as a thin coating on surfaces. While this encrustation onto soil surfaces may play an important role in oxidizing or adsorbing products of an incomplete reaction, subsequent reduction in permeability was observed in column experiments [Schnarr, 1992]. Delivery and distribution may thus be effected by generation of MnO_2 as well as CO_2 vapors during treatment. Field-scale pilot tests may help to resolve these and other issues.

METHODS

Two sites (Figure 1) were selected for the pilot tests, representing moderate and high levels of TCE contamination. Factors to be assessed include the extent of TCE destruction, changes to soil chemistry and permeability, and permanganate distribution. Prior to injection, samples were collected at two locations within each site to establish baseline conditions. Cores were collected at discrete depths between 16 and 24 ft beneath the surface using a Geoprobe® Large Bore™ soil sampler and 2-ft clear polymer liners.

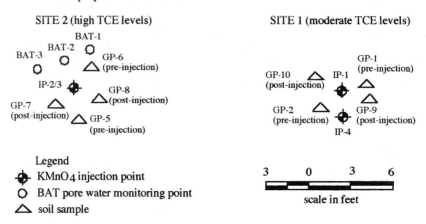

Figure 1. Injection and sample locations.

Prior to sub-sampling, each 2-ft core was visually inspected to note stratigraphy and identify layers of high moisture and zones of silts and clays that likely contained the highest levels of TCE. Sub-samples for TCE analysis were obtained by dividing the core at the desired depth and collecting approximately 5 g of undisturbed soil with a modified 10-ml disposable syringe. Each sample was immediately placed into a pre-weighed 40 ml glass vial containing 10 ml of methanol. The percentage of solids for each TCE sample was determined by placing an adjacent undisturbed sample into a pre-weighed glass vial. These were then weighed before and after oven-drying at 105°C to determine the moisture fraction. Other analyses consisted of potassium, manganese, chloride, pH, cation exchange capacity (CEC), and porosity. Soil porosity was determined by mercury porosimetry using approximately 1 g of intact sample. The analytical methods used for soil chemistry are listed in Table 1:

Table 1. Analytical methods for soil samples.

volatile organics (TCE)	EPA Method 8010
potassium	EPA Method 3050/6010
manganese	EPA Method 3050/6010
chloride	EPA Method 9253
pH	EPA Method 9045
cation exchange capacity	EPA Method 9081

Addition of KMnO$_4$ solution to the subsurface has thus far been accomplished using two types of direct-push wells to enable injection at multiple discrete depths. Initially, a 6-in Solinst® piezometer threaded to 3/4-in (nominal) iron pipe was installed at each site. These were later replaced with 8-in length, 2-in diameter stainless steel wire-wrap screens installed using a cone penetrometer truck. Initial installation depths were selected to allow injection into a predominantly sandy zone above a less permeable silt layer. Thus, the bottom of the well screens were driven to 20.5 ft at site 1 and to 21 ft at site 2. A 1.5% $KMnO_4$ solution (15 g/l) prepared from tap water was delivered from a 55 gal. barrel using either a rotary lobe or a centrifugal pump at typical injection pressures of 6–12 psi.

At site 2, three BAT® groundwater samplers (Hogentogler & Co., Columbia, MD) were placed near the injection well to collect pore water samples during the treatment process. These samplers consisted of either ceramic (BAT-1) or stainless-steel (BAT-2 and -3) lysimeter blocks installed to depths of 21, 22, and 22.7 ft (BAT-1,-2, and -3 respectively). Pore water was collected prior to and during $KMnO_4$ injection to monitor changes in TCE and chloride concentrations using gas and ion chromatography.

Similar to the baseline sampling, post-injection soil samples were collected at two locations within each site using a Geoprobe® Large Bore™ sampler. One-inch diameter cores, 2 ft in length, were collected from 18 to 24 ft below the ground surface. Samples were again analyzed as described for the baseline study with one exception—moisture levels were determined by obtaining directly the wet and dry weight of samples placed in methanol rather than collecting adjacent samples.

RESULTS

Selection of the initial injection depth was aided by visual inspection of stratigraphy and the pre-injection distribution of pore water/TCE as shown in Figure 2. As expected, the level of moisture and TCE was observed to correspond generally well with soil type such that silts and clays contained higher amounts of water and TCE than the fine sand intervals.

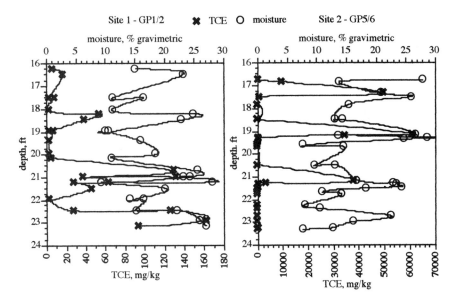

Figure 2. Pre-injection moisture and TCE concentrations.

Analysis of pre-and post injection soil porosity is continuing, but data may eventually provide insight with respect to permeability changes. Pre-injection soil chemistry data for both sites are given in Table 2.

Table 2. Pre-injection soil chemistry.

site	location	depth, ft	K, mg/kg	Mn, mg/kg	Cl⁻, mg/kg	pH	CEC, meq/100g
1	GP1	18.25	930	300	470	7.34	0
1	GP1	18.75	970	320	510	7.26	0
1	GP1	19.25	970	280	600	7.26	0
1	GP1	19.75	780	280	660	7.06	0
1	GP1	20.75	1040	260	550	7.26	0
1	GP1	21.25	780	250	730	7.32	0
1	GP1	21.75	890	260	360	7.23	322.8
1	GP1	22.75	970	270	480	7.17	0
1	GP1	23.75	800	230	250	7.56	16.6
1	GP2	21	1180	290	500	7.64	181.8
2	GP5	19.55	2300	330	1000	7.26	41.7
2	GP5	20.25	1100	370	730	7.3	22.8
2	GP5	20.75	880	310	500	7.48	29
2	GP5	21.09	970	350	830	7.56	24.5
2	GP5	21.63	1700	350	660	7.53	35.9
2	GP5	22.67	980	290	550	7.09	19
2	GP5	23.84	1200	330	920	7.51	20.8
2	GP6	18.65	960	350	2300	7.72	15
2	GP6	18.96	1060	375	1700	7.87	16.4
2	GP6	19.46	1120	300	570	7.81	14.4
2	GP6	20.84	1170	360	640	7.85	16.9
2	GP6	21.34	1500	360	470	7.8	24.5
2	GP6	21.84	700	280	550	7.07	21.9
2	GP6	22.21	1200	370	750	7.59	17.3
2	GP6	23.59	1300	340	450	7.14	33

Approximately 200 gal. of $KMnO_4$ solution was injected in several batches at site 1 over a 53-day period while 358 gal. was delivered to site 2 during a 21-day interval. Pore water samples collected prior to and during the injection period from BAT® lysimeters were analyzed for TCE and chloride. Table 3 lists data averages.

Table 3. Pre- and post-injection monitoring of pore water.

sample location	cumulative $KMnO_4$, gal.	TCE, mg/l	chloride, mg/l
BAT1	0	770	19.5
BAT1	159	940	75
BAT1	212	800	4200
BAT1	265	730	6200
BAT1	318	786	6420
BAT1	358	733	5367
BAT2	0	245	30
BAT2	53	240	29
BAT2	159	230	35
BAT2	212	120	83
BAT2	318	72	175
BAT3	0	48.5	20
BAT3	212	74	21
BAT3	318	30.5	78.5

Concentrations of TCE remain at nearly saturated levels at sample location BAT-1 despite an increase in chloride following $KMnO_4$ injection. This is likely a consequence of the residual levels of TCE in the surrounding soil at equilibrium with TCE-saturated pore water. As only aqueous TCE is dechlorinated by $KMnO_4$, reduction of non-aqueous TCE is limited by the rate of mass transfer of residual TCE into solution. Slight decreases in TCE corresponding to increases in chloride were observed in BAT-2 and BAT-3. As noted, the samplers at these locations were installed beneath the fine sand interval in and below a zone of lower permeability silts.

Comparison of post-injection soil chemistry data (Table 4) with the pre-injection values in Table 2 provides a means to trace paths of $KMnO_4$ distribution. In addition, increases in cation exchange capacity may indicate the buildup of Mn oxides which have a CEC value of 240 meq/100 g [Golden et al., 1986]. Based on the injection depths and the marginal changes observed in BAT-3 samples at 22.7 ft, the primary zone of interest may be identified as being from 19.7–22.7 ft. Inspection of the data provides little evidence for chemical transport beyond this range as expected due to the silt lens beneath the injection zone. This evidence is inconclusive however, as average post-injection values within the zone of interest are statistically comparable to average background levels. Some individual exceptions may be noted however. Near 21 ft at site 2 (bottom of injection well screen), potassium and manganese levels increased approximately twofold (GP8-21.17) while, at site 1, increases in chloride are apparent at 20.83 ft (GP9 and GP10). No notable changes in pH or CEC were observed.

Table 4. Post-injection soil chemistry.

site	location	depth, ft	K, mg/kg	Mn, mg/kg	Cl⁻, mg/kg	pH	CEC, meq/100g
2	GP7	20.5	1810	385	407	7.17	44.7
2	GP7	20.83	1990	685	573	7.44	50.8
2	GP7	21.17	800	346	450	7.63	52.1
2	GP7	21.5	780	344	1370	7.60	58
2	GP7	22.5	1390	339	569	7.84	41.6
2	GP7	22.83	369	146	573	7.83	42.8
2	GP8	19.17	710	255	1160	7.62	48.7
2	GP8	19.5	853	294	1630	7.69	44.7
2	GP8	19.83	760	322	1200	7.68	52.1
2	GP8	20.83	122	355	1181	7.76	35.2
2	GP8	21.17	2130	609	518	7.82	55.7
2	GP8	21.83	652	290	591	7.56	53.2
1	GP9	19.5	979	361	670	7.39	55.4
1	GP9	19.83	1200	334	310	7.90	49.4
1	GP9	20.5	996	354	518	7.80	49.1
1	GP9	20.83	983	321	916	7.78	50.0
1	GP9	21.17	1020	401	306	7.88	47.7
1	GP9	21.5	803	328	500	7.92	49.0
1	GP9	21.83	2013	290	–	8.06	49.4
1	GP9	22.5	1100	414	353	7.94	44.8
1	GP9	22.83	715	296	408	7.95	45.7
1	GP10	20.83	1280	398	824	7.81	49.8
1	GP10	21.17	1110	359	666	7.78	77.5

Table 5 lists the concentrations of TCE in soil before and after $KMnO_4$ injection for depths ranging from 19–21.5 ft below the ground surface. As noted, the injection depths were from 19.7–20.5 ft at site 1 and 20.3–21 ft at site 2. Although elevated TCE concentrations remained in the treatment area, evidence of attenuation may be argued by noting that all post-treatment concentrations at site 2 were about 1 order of magnitude less than the highest measured pre-treatment values. Evidence of cleanup at site 1 is inconclusive.

Table 5. Pre- and post-injection TCE concentrations in soil.

site	location	depth, ft	pre-injection TCE, mg/kg	site	location	depth, ft	post-injection TCE, mg/kg
1	GP1	19	3.3	1	GP10	19.83	0
1	GP2	19	6.46	1	GP9	20.5	5.8
1	GP2	19.42	3	1	GP10	20.5	20.5
1	GP1	20	3.21	1	GP9	20.75	31.5
1	GP2	20.17	4.89	1	GP10	20.75	8.2
1	GP2	20.75	130	1	GP9	21	51.2
1	GP1	21	37.9	1	GP10	21	10.8
1	GP2	21	133	1	GP9	21.25	47.5
1	GP1	21.25	64.5	1	GP10	21.25	88.3
1	GP2	21.25	28	1	GP9	21.5	110
1	GP1	21.5	46.3	1	GP10	21.5	70.3
2	GP6	19.13	61000				
2	GP5	19.21	34800	2	GP7	19.5	45
2	GP6	19.29	517	2	GP8	19.5	146.3
2	GP5	19.38	459	2	GP8	19.75	8231.3
2	GP6	19.63	16	2	GP7	19.92	3741.8
2	GP5	19.71	23	2	GP7	20.42	9202.8
2	GP5	20.5	42	2	GP7	20.67	85.4
2	GP6	20.5	34	2	GP8	20.83	87.4
2	GP5	21.17	38000	2	GP7	20.92	32.9
2	GP5	21.29	3210	2	GP8	21	33
2	GP6	21.33	311	2	GP7	21.17	24.2
2	GP5	21.46	392	2	GP8	21.25	17.2
2	GP6	21.5	92	2	GP8	21.5	23.8

CONCLUSIONS

Two TCE-contaminated sites are undergoing pilot-scale treatment by injection of a 1.5% $KMnO_4$ solution into unsaturated soils. Increases of chloride in pore water indicate oxidation of TCE is occurring. Analyses of post-injection soil samples also indicate cleanup may be occurring, but confirmation requires additional treatment and sample collection. In any case, significantly larger volumes of $KMnO_4$ solution or higher concentrations of $KMnO_4$ are required if complete cleanup is to be achieved.

REFERENCES

Golden, D.C., J.B. Dixon, and C.C. Chen. 1986. "Ion Exchange, Thermal Transformation and Oxidizing Properties of Birnessite." *Clays and Clay Minerals* 34:(5): 511–520.

Schnarr, M.J. 1992. An In-Situ Oxidative Technique to Remove Residual DNAPL from Soils. *M.S. Thesis. University of Waterloo*, Waterloo, Ontario, Canada.

IN-SITU CHEMICAL OXIDATION OF PENTACHLOROPHENOL AND POLYCYCLIC AROMATIC HYDROCARBONS: FROM LABORATORY TESTS TO FIELD DEMONSTRATION

Bruce K. Marvin, M.S., Fluor Daniel GTI, Martinez, California, USA
Christopher H. Nelson, M.S., P.E., Fluor Daniel GTI, Golden, Colorado, USA
Wilson Clayton, Ph.D., P.G., Fluor Daniel GTI, Golden, Colorado, USA
Kevin M. Sullivan, P.E., Fluor Daniel GTI, Martinez, California, USA
George Skladany, M.S., Fluor Daniel GTI, Trenton, New Jersey, USA

Abstract
A detailed demonstration of in-situ chemical oxidation was performed using ozone to treat pentachlorophenol (PCP) and polycyclic aromatic hydrocarbons (PAHs). The technology demonstration included a slurry-phase laboratory treatability study and an intensively monitored in-situ ozone injection study.

The laboratory study was designed to demonstrate the ability of ozone to oxidize the PCP and PAHs and to evaluate competing reactions with organic and inorganic ozone sinks. The laboratory results demonstrated 97% reduction of PCP and PAHs in site soils. In addition, it was shown that competing ozone reactions with site soils would not limit treatment effectiveness.

The field study was designed to evaluate subsurface ozone transport and the effectiveness of remediating contaminants in place. The ongoing field test includes multi-level saturated and unsaturated-zone injection points and an intensive monitoring array.

Field results after one month of operation showed an average of 82% reduction in vadose and saturated-zone dissolved concentrations of PCP and PAHs. Soil gas and vadose and saturated-zone dissolved ozone data show that a relatively uniform distribution of ozone is achieved in the subsurface. Continuing field operations will further investigate ozone distribution, soil concentration reductions and biological activity.

Introduction
Historic operations, between 1950 and 1970, at a site in Sonoma county, California included wood treating and cooling tower and water tank manufacturing. These operations involved the use of pentachlorophenol and creosote. As a result, both PAHs and PCP are contaminants of concern (COC) at this site. Impacts are distributed in shallow soils and extend down to the water table. The majority of the COC is localized in two areas: the former dip tank (5 spot) and product drying (3 spot) areas, as shown in Figure 1.

Several remedial options were considered including capping with institutional controls, excavation and disposal off-site, excavation and disposal to an on-site Class I facility, and in-situ chemical oxidation using ozone. In-situ

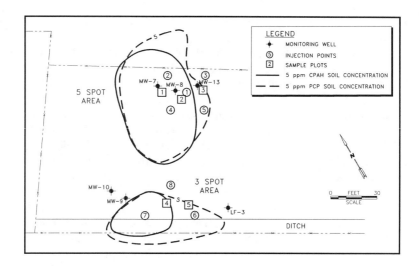

FIGURE 1. Site plan

ozonation was selected as the preferred remedial alternative because it was most protective of human health and the environment, it minimized the need for long-term monitoring and maintenance, and it offered significant cost savings over the other remedial options. This paper presents final laboratory treatability study results and preliminary results from an ongoing field demonstration on in-situ ozonation.

Laboratory Study

Fluor Daniel GTI, Inc. was contracted to evaluate ozonation of site soils in its Remediation Technology Testing Facility in December 1996. The primary objective of this test was to assess the ability of ozone addition to destroy PCP and PAHs in impacted soils from the site.

Methods. Representative samples of site soils and groundwater were collected for the laboratory study. Initial characterization of theses soils was performed using EPA Method 8270 for PCP and PAHs. In addition, the samples were analyzed for total organic carbon (TOC) and inorganic parameters. These additional parameters were evaluated to allow quantification of competing ozone reactions such as reaction with naturally organic matter and carbonates.

The laboratory test was conducted in slurry vessels to minimize potential mass transfer limitations. A slurry of 20% site soils and 80% groundwater was prepared and placed in a 5 liter reactor vessel. A concurrent control vessel was prepared for sparging with nitrogen. The ozonation reactor was subjected to approximately 2 grams of ozone per hour delivered in an oxygen carrier gas. The control reactor was subjected to the same flow rate of nitrogen. Sampling of both reactors was preformed periodically throughout the laboratory study for period of 50 hours.

Results. Ozone treatment of a slurry of site soils and groundwater resulted in greater than 97 percent reduction in contaminant concentrations as shown in Table 1. The results also indicated that PAHs and PCP were preferentially oxidized over TOC. Contaminant concentrations were reduced an order of magnitude after six hours of treatment as compared to unchanged levels of TOC.

TABLE 1. Laboratory soil/groundwater slurry results

	Control			Ozonation		
	Initial	Final	% Reduction	Initial	Final	% Reduction
Soil Results (mg/kg)						
PCP	13	7.5	42	27	<1.7	97
CPAH	140	160	--	210	2.6	99
NCPAH	1500	1800	--	2600	2.4	100
TOC	12	11	8	10	<0.7	97
Aqueous Results (mg/L)						
PCP	6.7	7.1	--	5.7	<0.05	100
CPAH	0.20	0.067	67	0.21	0.08	62
NCPAH	4.5	3.9	13	4.3	0.063	99
TOC	0.015	0.054	--	0.021	0.1	--

Laboratory study results indicated that after 50 hours of ozonation the majority of COC were destroyed. In addition, the results from the control reactor clearly indicated that ozonation caused the reductions of COC in the ozonation vessel. Increases in aqueous TOC in both the ozonation and control reactors was likely the result of vigorous mixing. The preferential oxidation of COC over TOC showed that competing reactions with TOC will not impede reaction with COC. The results of the laboratory study indicated that oxidation of COC by ozone may be feasible. As a result, design of a field demonstration system was initiated.

Field Study

Upon completion of the laboratory study, goals for the field study were identified. These goals were to monitor the effectiveness of in-situ ozonation, assess subsurface ozone transport, determine appropriate well spacing, and evaluate microbial response.

Design. The geology of the site is semi-continuous layers ranging from fine sands to clays. This geology resulted in both highly stratified contamination and variable gas entry pressures. As a result, multi-level ozone injection was deemed necessary. Figure 2 is a cross section through the former dip tank (5 spot) area showing site geology (as determined by downhole conductivity surveys), injection well intervals, and in-situ instrumentation.

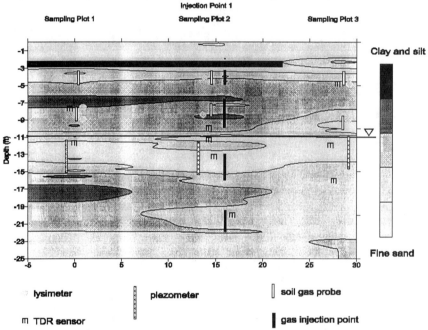

FIGURE 2. Geologic cross section through 5 spot treatment plot

Instrumentation includes: soil gas monitoring points, peizometers, lysimeters, monitoring wells, thermocouples, and time-domain reflectometry instrumentation to measure soil moisture content. Soil vapor extraction wells were placed outside the treatment areas to ensure that fugitive ozone emissions were minimized.

Results. After one month of continuous ozone injection a full round of sampling was completed. Table 2 shows that dissolved COC concentrations in both the saturated and unsaturated-zones have been reduced more than 82%, on average, to distances greater than 15 feet from the injection point. PAH reductions ranged from 67% and 99.5% while PCP concentrations declined 39 to 98%.

Figure 3 shows that subsurface gaseous ozone concentrations are relatively uniform and decrease with increasing distance from the injection point. Soil gas data also suggests that ozone utilization of greater than 90% is achieved.

The lateral distribution of dissolved ozone in soil waters, shown in Figure 4, was relatively uniform across the 5-spot treatment area. Dissolved ozone concentrations were less than concentrations predicted assuming equilibrium partitioning and measured soil gas ozone concentrations.

TABLE 2. Groundwater results after one month of in-situ ozonation

Location	Total PAHs			PCP		
	Initial	One month	% Reduction	Initial	One month	% Reduction
MW-13	1200	<10	99.5	2600	200	88
LY-1A	<100	<30	--	2200	230	90
LY-1B	<600	<200	--	8800	880	90
LY-2A	14000	3500	75	8400	670	92
LY-3A	560	72	87	200	<10	98
LY-3B	1000	70	93	1700	250	85
LY-4A	960	240	75	330	200	39
PZ-1C	<200	<300	--	3500	1300	63
PZ-2C	<200	<50	--	3300	960	71
PZ-3C	<200	<100	--	2300	1000	57
Average	**1800**	**530**	**86**	**3300**	**570**	**77**
MW - Monitoring well		LY - Lysimeter		PZ - Piezometer		

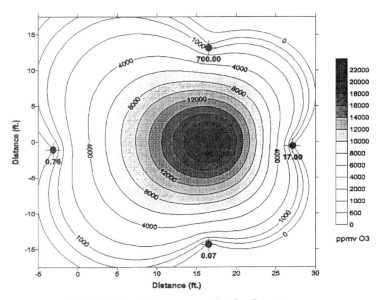

FIGURE 3. Soil gas ozone in the 5 spot area

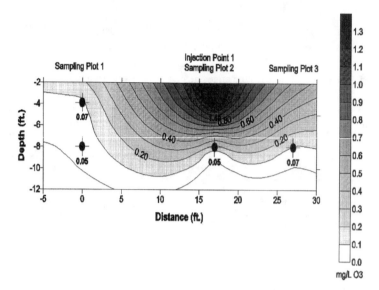

FIGURE 4. Dissolved ozone cross section through 5 spot area

Summary and Conclusions

The laboratory study showed that ozone preferentially oxidizes both PCP and PAHs over TOC in site soils. In addition, the laboratory study indicated that destruction of COC occurs rapidly under optimal ozone loading conditions. A field study was then designed to further evaluate the operational effectiveness of in-situ ozonation.

Field data after one month of in-situ ozonation show significant reductions in dissolved PCP and PAHs in both soil moisture and shallow groundwater. An average of 82% reduction in saturated and unsaturated zone concentrations was observed.

The uniform distribution of dissolved ozone across the treatment area indicates that ozone mass transfer into the aqueous-phase is effective. The observed dissolved ozone concentrations are below equilibrium partitioning predicted values given the soil gas ozone concentrations. This implies rapid ozone reaction with contaminants and non-equilibrium mass transfer.

The field demonstration will continue for six months. During this time additional data will be acquired to assess the effectiveness of in-situ ozonation. Soil borings will be advanced after three and six months of treatment, respectively. In addition, respirometry will be performed to determine if in-situ ozonation represses or enhances microbial activity. Finally, engineering issues such as distribution and well spacing will be further investigated.

OZONE AND CONTAMINANT TRANSPORT
DURING IN-SITU OZONATION

Wilson S. Clayton, Ph.D., P.G., Fluor Daniel GTI, Inc., Golden, Colorado, USA

Abstract: Simple analytical and numerical models were used to assess several important properties of subsurface transport, mass transfer, and reactions of ozone and contaminants during in-situ ozonation. The analysis and discussion is relevant to both vadose zone injection and ozone sparging. An upper limit on ozone transport is imposed by first order ozone decay in the presence of native subsurface material. During treatment, second order ozone reactions with contaminants limits ozone gas transport. As contaminants close to the injection point are oxidized, ozone gas is transported farther. Therefore, as treatment proceeds, the radius of effective ozone treatment expands, and is eventually limited to the first order estimate. Comparison of ozonation of trichloroethylene and naphthalene indicates that less volatile compounds have greater incremental benefit in ozonation, since contaminant volatilization also occurs during in-situ ozonation.

INTRODUCTION

Ozone (O_3) is a gas with strong oxidizing properties, which has recently been applied for the in-situ oxidation and enhanced biodegradation of recalcitrant organic contaminants (Marvin et al., 1998, Brown et al. 1997, Nelson et al., 1997, Leahy et al., 1997). Among the primary advantages of in-situ ozonation is that large gas flow rates can be delivered, which can potentialy drive rapid contaminant oxidation. Successful in-situ ozonation depends on ozone mass transfer into groundwater. Ozone is 12 times more soluble than oxygen and is therefore more readily transferred into the aqueous phase. Since oxygen mass transfer is an established mechanism for biosparging, we may expect at least as favorable mass transfer of ozone. Another potential advantage of in-situ ozonation is that contaminant volatilization may also occur.

While in-situ ozonation has numerous advantages, there can be several limitations, primarily related to subsurface ozone transport and mass transfer. For example, in the presence of oxidizable compounds, the very fast reaction rates of the ozone can potentially limit its transport over large distances. Oxidizable compounds can include organic contaminants, soil humic matter, or carbonates. Also, the fundamental diffusion limitations of air sparging can be more problematic for mass transfer of a highly reactive gas such as ozone. Subsurface heterogeneity can lead to channelized flow of injected gases in high permeability zones, such that ozone transport into low permeability zones is limited by penetration diffusion.

Characterization of the mechanisms of subsurface ozone transport and mass transfer is critical to assess the applicability of in-situ ozonation. This paper uses results from an analytical radial transport model and a one-dimensional multiphase multicomponent numerical model to describe ozone and contaminant transport and mass transfer. The analysis is applicable to vadose zone injection or ozone sparging. In-situ ozonation is compared for two different organic contaminants: naphtahalene a polynuclear aromatic hydrocarbon (PAH), and trichloroethylene (TCE), a chlorinated solvent.

ANALYTICAL RADIAL TRANSPORT MODEL

Subsurface ozone transport is limited by ozone reaction as it moves through the soil. The importance of ozone reaction rates on ozone transport is illustrated by considering simplified radial transport of ozone from an injection well.

Model Development. The radial transport of a gaseous chemical subject to first order degradation can be solved analytically as follows. The flow velocity (v) during radial flow is expressed as,

$$v = \frac{dR}{dt} = \frac{Q}{A} = \frac{Q}{2\pi R Z \phi S_g} \tag{1}$$

Where, R = radial distance from well, t = time, Q = injection rate, A = flow cross sectional area, Z = height of flow zone, ϕ = soil porosity, S_g = gas saturation. Solving for t in eq (1),

$$t = \int dt = \int \frac{1}{V} dR = \int \frac{2\pi R Z \phi S_g}{Q} dR = \frac{\pi Z \phi S_g R^2}{Q} \tag{2}$$

Combining eq (2) with the standard first order decay equation yields,

$$C = C_0 e^{-kt} = C_0 e^{-[k\pi Z \phi S_g R^2 / Q]} \tag{3}$$

Where, k = degradation constant = 0.693/half life, C_0 = initial concentration. Equation (3) is an analytical solution for steady state radial gas transport subject to first order decay.

Analytical Model Results and Discussion. Application of equation (3) shows that steady state subsurface ozone transport is highly sensitive to reaction rate (Figure 1). For 5 percent ozone injection into a 5 foot thick zone at 5 cubic feet per minute, ozone concentrations are depleted to 1 ppm at distances of 11, 18, and 32 feet for ozone half lives of 5, 15, and 45 minutes, respectively. These results represent the steady state ozone concentration distribution at a constant degradation rate. However, this analysis yields an upper limit estimate of maximum transport distances, because it does not account for ozone mass transfer into groundwater, and the assumption of first order kinetics is simplified. Actual ozone reaction rates are second order, i.e. a function of ozone and contaminant concentrations, which means they vary spatially and temporally. The numerical model described below simulates unsteady state ozone transport which accounts for mass transfer and second order reactions.

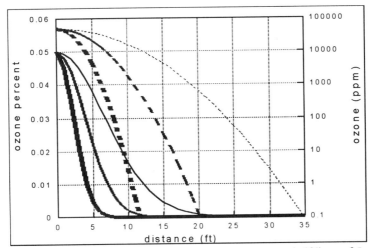

Figure 1. Steady state radial ozone transport at ozone half lives of 5, 15, and 45 minutes (coarse to fine lines). Ozone concentrations shown in percent and ppm (dashed)

MULTICOMPONENT, MULTIPHASE NUMERICAL MODEL

Ozone and contaminant transport, mass transfer, and reactions during ozone sparging are depicted conceptually in Figure 2. Ozone is transported by advection within gas-filled pores in the subsurface. Gas fingering occurs at the pore-scale during sparging (Clayton, 1998), and ozone mass transfer into groundwater is diffusion limited. Contaminant diffusion and mass transfer are in the opposite directions from ozone. Dissolved ozone comes into contact and reacts with dissolved and sorbed contaminants. Ozone gas is lost to mass transfer and reaction, so that ozone concentrations in the gas stream decrease along the flow path.

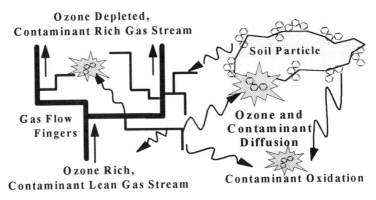

Figure 2. Conceptual Model for diffusion limited mass transfer of ozone and contaminants between pore scale gas fingers, groundwater, and soil.

Model Development. Based on the above conceptual model, the simultaneous mass transfer and transport of contaminants and ozone, is simulated an unsteady state three-phase, two component numerical model, which accounts for (a) one dimensional transport of ozone and contaminants by advection in the gas phase, (b) diffusion-limited interphase mass transfer of ozone and contaminants between gaseous and aqueous phases, (c) diffusion-limited contaminant desorption, and (d) second order reaction of ozone and contaminants in gas and aqueous phases.

The transport and reaction of ozone and contaminant are described in terms of solute fugacity (Mackay, 1991). Diffusive mass transfer rates are a function of the interfacial area, diffusion path lengths, concentration gradients, and chemical diffusion coefficients. Clayton and Nelson (1994) provided details of the mathematical derivation of the mass balance equations for diffusion-limited interphase mass transfer, based on pore scale air finger geometry during air sparging. The mathematical modeling results presented here are based on those mass balances, with the addition of a second chemical component (ozone) and second order reactions between contaminants and ozone in gas and aqueous phases.

Unsteady state numerical solution of the mass balances in time and one-dimensional space was achieved using the Runge-Kutta technique. The results should be interpreted as preliminary, since the numerical model was not subjected to mathematical verification or calibration to measured data. In fact, extensive research will be required to precisely quantify many of the important parameters such as mass transfer coefficients and second order reaction rates. The results are primarily intended to illustrate several important transport characteristics of in-situ ozonation.

Numerical Modeling Results and Discussion. Numerical modeling was carried out to evaluate the sensitivity of ozone sparging transport and treatment time frame to ozone injection concentration, injection rate, and gas saturation. Results for TCE and naphthalene are compared to illustrate differences between highly and moderately volatile and soluble compounds. Model conditions were configured to represent one-dimensional flow in a 10 cm diameter, 20 cm long soil column. While the model solves for concentrations of both components in all phases, the results presented here depict dissolved and sorbed contaminant concentrations, and dissolved ozone concentrations over time and distance.

Ozonation Model Results and Discussion for Naphthalene. Naphtalene has low vapor pressure (10.4 Pa) and solubility (37 mg/l). These low values are reflected by slow treatment in the absence of ozone (Figure 3). The addition of ozone to the injected gases increases treatment rates several orders of magnitude, as a result of chemical oxidation of naphthalene. Increasing the ozone concentration results in faster treatment times and more uniform treatment over distance, since ozone gas concentrations are less depleted by loss to mass transfer and reaction.

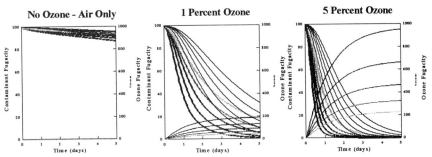

Figure 3. Influence of ozone concentration on naphthalene treatment. Each graph shows three sets of curves for dissolved ozone (increasing over time), dissolved naphthalene (bold lines), and sorbed naphthalene (finer lines). Each set of curves has 5 lines, representing increasing distance. Contaminants are reduced faster and dissolved ozone levels are higher at closer distances.

At low gas injection rates (Figure 4), the subsurface gas velocity is lower, and much ozone gas is lost to mass transfer and reactions near the injection point. Treatment at distance is delayed until contaminant levels near the injection point are reduced. At later times, ozone reaction rates are less near the injection point, thus minimizing interphase mass transfer rates, and allowing ozone gas to be transported farther. At higher flow rates, treatment near the injection point is unaffected. However, gas velocity is increased, ozone transport occurs over larger distances sooner, and treatment at distance occurs faster.

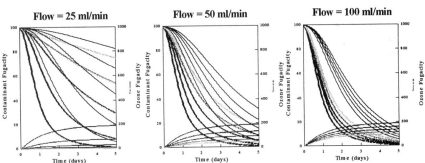

Figure 4 Influence of gas flow rate on naphthalene treatment. Each graph shows three sets of curves for dissolved ozone (increasing over time), dissolved naphthalene (bold lines), and sorbed naphthalene (finer lines). Each set of curves has 5 lines, representing increasing distance. Contaminants are reduced faster and dissolved ozone levels are higher at closer distances.

At high gas saturation values (Figure 5) diffusion limitations are lessened, and interphase mass transfer rates are higher. Therefore, treatment closer to the injection point is faster. However, the larger mass transfer rate depletes ozone gas concentrations, and treatment at distance is delayed. This is indicated by the high dissolved ozone levels near the injection point and reduced dissolved ozone farther from the injection point. At lower gas saturation values, treatment near the injection point is slower, but treatment is more uniform over distance. Optimal treatment at intermediate and large distances is at an intermediate gas saturation.

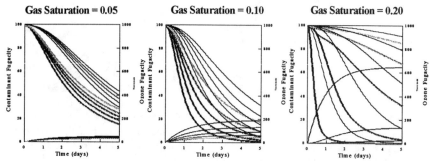

Figure 5 Influence of gas saturation on naphthalene treatment. Each graph shows three sets of curves for dissolved ozone (increasing over time), dissolved naphthalene (bold lines), and sorbed naphthalene (finer lines). Each set of curves has 5 lines, representing increasing distance. Contaminants are reduced faster and dissolved ozone levels are higher at closer distances.

<u>Ozonation Model Results and Discussion for TCE.</u> TCE has much higher vapor pressure (7,700 Pa) and solubility (1,100 mg/l) than naphthalene. These higher values are reflected in the higher treatment rates in the absence of ozone (Figure 6), which are attributable to TCE volatilization. The addition of ozone to the injected gas increases the TCE treatment rate, to a larger degree than naphthalene (Figure 3), as a result of the increased solubility and volatility and reduced absorption coefficient (log K_{ow} = 2.3 vs. 3.5) of TCE. The sensitivity of in-situ ozonation of TCE to gas flow rate and air saturation is not shown, but is similar to the naphathalene results in Figure 4 and 5.

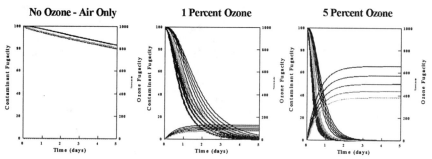

Figure 6 Influence of ozone injection concentration on TCE treatment. Each graph shows three sets of curves for dissolved ozone (increasing over time), dissolved TCE (bold lines), and sorbed TCE (finer lines). Each set of curves has 5 lines, representing increasing distance. Contaminants are reduced faster and dissolved ozone levels are higher at closer distances.

CONCLUSIONS
Subsurface ozone transport is limited by ozone reactions. An upper limit on ozone transport distances is imposed by first order ozone degradation. During the early phase of treatment, ozone reactions with contaminants or other oxidizable material near the injection well depletes ozone gas concentrations, and limits ozone transport. As contaminants and other materials near the injection well are oxidized, ozone gas is transported farther. Changes in ozone gas distribution over time are controlled by the balance between gas velocity and mass transfer and reaction rates. Increased injection flow rates and ozone concentrations help transport ozone farther.

Interphase mass transfer rates are diffusion limited. Higher gas saturation results in a higher mass transfer rate, causing (1) faster remediation close to the injection well, and (2) greater depletion of ozone gas concentrations, and therefore more limited ozone transport and slower remediation at distance. Optimal treatment effectiveness is achieved at an intermediate gas saturation.

Geologic Heterogeneity, which effects gas flow paths, may define separate regions of advective gas transport from regions of aqueous penetration diffusion transport of ozone. The modeling here applies to the region of advective gas flow. Rates of contaminant destruction in the region of penetration diffusion depend on the thickness of the low permeability layer, and on the dissolved ozone concentrations which can be achieved at the boundary of the layer.

REFERENCES
Clayton, W.S., 1998. "A Field and Laboratory Investigation of Air Fingering During Air Sparging" *Ground Water Monitoring and Remediation*, in press.

Clayton, W. S., and C. H. Nelson, 1995, "In-Situ Sparging: Managing Chemical Transport and Mass Transfer", in *proceedings of the 1995 Superfund Conference and Exhibition*, E. J. Krause and Assoc., Bethesda, Maryland.

Brown, R.A., Nelson, C., and Leahy, M.C. 1997, "Combining Oxidation and Bioremediation for the Treatment of Recalcitrant Organics", *in In-Situ and On-Site Bioremediation: Volume 4(4). Battelle Press*, pp. 457-462. Columbus, Ohio.

Leahy, M.C., Nelson, C.H., Fiorentine, A.M., and Schmitz, R.J., 1997. "Ozonation as a Polish Technology for In-Situ Bioremediation", *in In-Situ and On-Site Bioremediation: Volume 4(3). Battelle Press*, pp. 479-4483. Columbus, Ohio.

Marvin, B.K., Clayton, W.S., Nelson, C.H., Sullivan, K.M., and Skladany, G. 1998, "In-Situ Chemical Oxidation of Pentachlorophenol: From Laboratory Tests to Field Demonstration", in (this volume), presented at First International Conference on Remediation of Chlorinated and Recalcitrant Compounds. Battelle, Columbus, Ohio.

Mackay, D. 1991, Multimedia Environmental Models: The Fugacity Approach, Lewis Publishers, Chelsea, Michigan,

Nelson, C.H., Seaman, M., Peterson, D., Nelson, S., and Buschbom, R., 1997, "Ozone Sparging for the Remediation of MGP Contaminants", *in In-Situ and On-Site Bioremediation: Volume 4(3). Battelle Press*, pp. 457-462. Columbus, Ohio.

In Situ Oxidation of Trichloroethene using Potassium Permanganate
Part 1: Theory and Design

John C. LaChance, ENSR, Acton, Massachusetts
Stanley Reitsma, ENSR, Acton, Massachusetts
Daniel McKay, US Army Cold Regions Research and Engineering Laboratory,
Hanover, New Hampshire
Ralph Baker, ENSR, Acton, Massachusetts

ABSTRACT: Trichloroethene (TCE) exists as immiscible phase in low permeability layers in the unsaturated zone at the US Army Cold Regions Research and Engineering Laboratory (CRREL) in Hanover, New Hampshire. The low permeability layers remain saturated or nearly saturated while higher permeability layers are unsaturated, making remedial methods that rely on flushing the subsurface with fluids (air and/or water) to remove contaminants, ineffective for removal of the TCE due to capillary effects and mass transfer limitations. Recent research indicates that potassium permanganate ($KMnO_4$) will completely oxidize chlorinated ethenes and may potentially improve mass transfer rates due to the in situ chemical reaction with the chlorinated solvent.

A pilot-scale in situ oxidation demonstration using $KMnO_4$ has been designed and implemented at the CRREL facility. The results of this work are described in a companion paper presented at this conference titled: In Situ oxidation of Trichloroethylene Using Potassium Permanganate: Part 2. Pilot Study (McKay et. al., 1998). This paper provides an overview of the $KMnO_4$ in situ oxidation process, background information on site conditions and contaminant distributions, and the conceptual design of the $KMnO_4$ in situ oxidation demonstration.

INTRODUCTION

The Cold Regions Research and Engineering Laboratory (CRREL) located in Hanover, NH, used trichloroethylene (TCE) from 1960 to 1987 for such purposes as refrigeration, ice coring, and degreasing. The extensive use of TCE has led to releases to the environment and contamination of soil and groundwater at the facility. One well-documented release occurred in 1970, when repairs to an above ground storage tank (AST) resulted in the discharge of approximately 3,000 gallons of TCE to surrounding paved areas and soil.

Currently available methods of addressing chlorinated hydrocarbon contamination are generally limited to those that contain offsite migration and/or treat dissolved constituents in groundwater. Approaches capable of addressing source areas, especially those in heterogeneous geologic settings, are often prohibitively expensive or not feasible (NRC, 1994). The chief limitations plaguing existing technologies, can be summarized as follows: 1) preferential flow and partial treatment of the subsurface due to variations in the capillary properties of the medium; 2) tailing and rebound of groundwater concentrations due to rate limited mass transfer of contaminants out of inaccessible deadened pores and untreated low permeability zones; and 3) potential remobilization and spreading of DNAPL due to alteration of hydraulic gradients and interfacial tension.

In situ oxidation using potassium permanganate (KMnO$_4$) represents an innovative and promising approach to cleaning up sites contaminated with chlorinated hydrocarbons, existing both as dissolved and immiscible phase (DNAPL), in groundwater, soils, and bedrock. KMnO$_4$ oxidation technology is particularly applicable at sites having heterogeneous soils (this includes most sites) or fractured systems where the rock or clay matrix is porous.

The following describes the design of a pilot demonstration to evaluate the applicability and feasibility of using KMnO$_4$ for the removal of TCE from the subsurface beneath the CRREL facility. Background information on site conditions, KMnO$_4$ oxidation chemistry, and implementation issues are also provided.

OVERVIEW OF KMnO$_4$ OXIDATION PROCESS

KMnO$_4$ has been found to readily oxidize chlorinated alkenes such as PCE and TCE. The oxidative reaction between KMnO$_4$ and PCE or TCE cleaves the double bond between the two carbon atoms and subsequently dechlorinates the molecule. The products of complete destruction include carbon dioxide, potassium chloride, chlorine, hydrochloric acid and manganese dioxide. The complete reactions for TCE and PCE, respectively, are given as (Schnarr et al., 1997):

$$C_2Cl_3H(TCE) + 2KMnO_4 \rightarrow 2CO_2(aq) + 2MnO_2(s) + 2KCl + HCL \quad (1)$$

$$C_2Cl_4(PCE) + 2KMnO_4 \rightarrow 2CO_2(aq) + 2MnO_2(s) + 2KCl + Cl_2(aq) \quad (2)$$

In the reactions between PCE/TCE and KMnO$_4$, various chemicals will be released into the groundwater as unavoidable by-products (Schnarr, 1992). The primary by-products of concern include: HCL, Cl$_2$, CO$_2$, and MnO$_2$, Carbonate soils will typically have sufficient buffer capacity to neutralize the HCl. The reactivity of the chlorine gas (Cl$_2$) is quite high and it will immediately combine with water to from hypochlorous acid (HOCl) and hypochlorite (OCl$^-$), releasing chloride ions. HOCl and OCl$^-$ are themselves very strong oxidants and will in turn be readily reduced by available organic material in the aquifer. CO$_2$ will combine with water to form the carbonate series and lower the pH of the water. If the reaction rate exceeds the carrying capacity of the water, a separate CO$_2$ vapor phase will form within the soil matrix. Increases in the concentration of the gas phase within the soil matrix will result in a corresponding decrease in the relative permeability with respect to water of the matrix. The build up of CO$_2$ vapor phase within the soil pores could adversely affect the ability to inject KMnO$_4$ solution into the subsurface. The manganese dioxide (MnO$_2$) and other forms of manganese oxides that are formed during the KMnO$_4$ oxidation process, will remain in the aquifer as coatings on the soil grains. These compounds have oxidative and absorptive properties and may help to control any incomplete oxidative organic by-products. The build up of manganese oxides on the soil grains may result in reductions in soil permeability over time. The release of

KMnO$_4$ into the groundwater beyond the treatment area should not be a problem as it will react with aquifer material that is readily oxidized (Schnarr, 1992).

TECHNICAL ISSUES OF KMnO$_4$ IN SITU OXIDATION

Several factors related to the treatment of soils containing chlorinated solvents using KMnO$_4$ influence the potential of achieving typical remediation objectives. These include: 1) Ability of KMnO$_4$ to oxidize chlorinated solvents present without production of daughter products; 2) Insolubility of KMnO$_4$ in the immiscible phase (DNAPL) present in the soil and dependence of reaction rate on solubility and mass transfer rates; and 3) Ability to deliver KMnO$_4$ solution to the zones containing dissolved and/or immiscible phase (DNAPL) contamination.

Oxidation Ability of KMnO$_4$. KMnO$_4$ will readily and completely oxidize chlorinated alkenes (Schnarr, 1992; Schnarr et al., 1997). However, chlorinated solvents having single carbon bonds, such as trichloroethane (TCA), are not readily oxidized and therefore may not be treated during KMnO$_4$ treatment. There is some indication that high concentrations of KMnO$_4$ or combinations of KMnO$_4$ and catalysts may oxidize these compounds (Schnarr, 1992). Thus characterization of contamination present is very important in determining remedial strategy.

Effects of KMnO$_4$ on Mass Transfer Rates. KMnO$_4$ is essentially insoluble in organic liquids such as PCE or TCE. Therefore the oxidation reaction takes place in the aqueous phase. This requires that immiscible phase (DNAPL) first dissolve into the aqueous phase before oxidation of the organic will take place. It is well known that the dissolution of DNAPL may be a slow process, limited by inter-phase mass transfer (e.g. Powers et al., 1994; Lamarche, 1991; Imhoff et al., 1994). As DNAPL saturation (fraction of pore space) decreases during treatment, the rate of inter-phase mass transfer slows, resulting in lengthy treatment times as has been demonstrated during DNAPL dissolution experiments (e.g. Powers et al., 1994). This fact is one reason for the very poor performance of pump-and-treat operations.

KMnO$_4$ technology is also faced with the same rate-limited inter-phase mass transfer. However, the inter-phase mass transfer rate during oxidation treatment is greatly improved because chemical gradients are increased significantly. Figure 1 depicts an idealized single film boundary-layer model illustrating the chemical concentrations occurring in close proximity to the DNAPL - water interface. The solubility of the DNAPL in water is given by Cs and the dashed lines illustrate the change in concentration profile with decreased oxidant concentration in the bulk phase. Modified after Schnarr et al. (1997).

Increased KMnO$_4$ concentration in the aqueous phase also leads to increased mass transfer since concentration gradients are increased for both KMnO$_4$ and the organic as is illustrated in Figure 1. Conceptually, further increases in reaction rate may be achieved by introducing an alcohol or surfactant that will increase the solubility of the chlorinated compound in water. KMnO$_4$ may oxidize the alcohols prior to reaching the DNAPL of interest and therefore this approach requires testing in the laboratory.

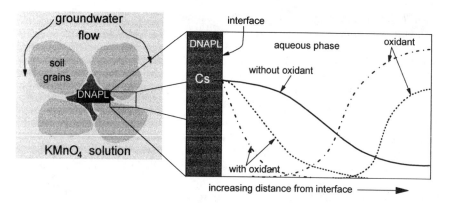

Figure 2. Effect of Oxidant on Aqueous Concentration in Close Proximity to DNAPL-Water Interface. (Modified after Schnarr et al., 1997)

KMnO$_4$ Delivery in Heterogeneous Settings. The delivery of KMnO$_4$ solution to zones containing dissolved and/or immiscible phase contamination is a crucial issue for successful soil remediation. Approximate location of the contamination must be established before delivery is possible. As with inter-phase mass transfer described above, the oxidation reaction causes increased chemical gradients as shown in Figure 1 except on a larger macro-scale. KMnO$_4$ solution will diffuse from higher permeability lenses into the lower permeability lenses and react with dissolved contaminants in the low permeability layers. The reaction leads to higher concentration gradients for both the oxidant and the contaminant. Also, since KMnO$_4$ is delivered in aqueous solution, delivery is not seriously inhibited by variation in soil capillary properties, which can not be said for remedial technologies that rely on delivery of a gas phase. Thus, KMnO$_4$ oxidation technology provides several advantages over many remedial technologies whose objective it is to address DNAPL source areas or dissolved phase present in low permeability lenses.

SITE CONDITIONS

The following summarizes the subsurface conditions beneath the KMnO4 pilot area: 1) the area is underlain by 65 m of unconsolidated glaciolacustrine and glaciofluvial deposits; 2) the depth to the water table is 40 m below the ground surface (BGS); 3) the upper 16 m of the unconsolidated deposits consist of layers of interbedded silty sands, 10 cm to 3 m thick, separated by layers of dense and cohesive silt, 0.5 to 50 cm thick; 4) below 16 m, the unconsolidated deposits consist of interbedded sands; 3) the moisture contents of the sand and silt layers were between 10 and 20 %, and 20 and 30%, respectively; and 5) immiscible phase TCE (DNAPL) is present in the dense silt layers between 6 and 8 m BGS.

Figure 2 presents a summary of data collected in the pilot area in 1996 and 1997 work. The 1997 investigations confirmed the findings of the 1996 work and provided additional stratigraphic and contaminant detail useful for design of the pilot demonstration. Interestingly, the 1997 contaminant data, which were horizontally located within 1.5 meters of the 1996 investigations, were generally lower than the 1996 data. This is likely an indication of the heterogeneous distribution of immiscible phase TCE in the silt zones.

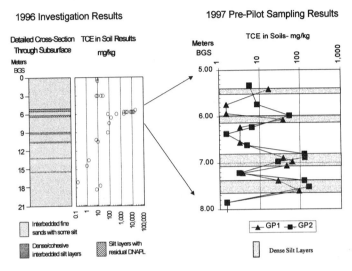

Figure 1. Summary of 1996 and 1997 Soil Data

CONCEPTUAL DESIGN OF KMnO₄ PILOT DEMONSTRATION

Figure 3 presents the conceptual design of the KMnO₄ in situ oxidation pilot demonstration. The sandy zones provide pathways for the delivery of KMnO₄ to the targeted treatment zones. Following injection into the sandy zones, the KMnO₄ will move into the silt layers via advective flow and diffusion and oxidize the TCE.

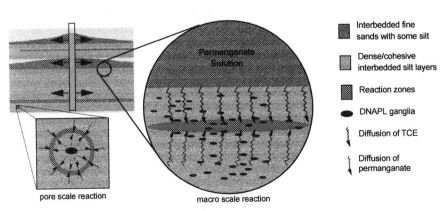

Figure 1. Conceptual Design of KMnO4 Pilot Demonstration

SUMMARY

The following summarizes the salient features of the CRREL $KMnO_4$ pilot demonstration: 1) $KMnO_4$ has been demonstrated to destroy chlorinated solvents in situ, making surface treatment of soil, water, or off-gas unnecessary; 2) $KMnO_4$ in situ oxidation is not hindered by the presence of utilities or buildings; 3) The materials and technology are cheap to design, build, and operate; 2) $KMnO_4$ increases the rate of remediation in low permeability media and dead-end pores by increasing the chemical potential gradients for both $KMnO_4$ and the chlorinated solvent; 3) High permeability zones provide delivery pathways for $KMnO_4$ from which it can diffuse into the contaminated low permeability zones; 4) Unlike most other treatment technologies (e.g., steam or surfactant flooding), $KMnO_4$ causes no significant decrease in interfacial tension properties that could lead to remobilization of DNAPL; 5) $KMnO_4$ is highly soluble in water and may be delivered in a water solution, providing a means for easy delivery to zones of high permeability; 6) Unlike technologies that rely on delivery of a gas phase (e.g. air sparging or vapor extraction), $KMnO_4$ treatment is not hindered by restrictive capillary properties (e.g., air-entry pressures) of the layered system; 7) $KMnO_4$ will not persist in the subsurface environment.

REFERENCES

Hood, E., and Farquhar, G.J., 1994. In Situ permanganate oxidation of PCE and TCE in soil, The ATV Conference on Groundwater Contamination, The Danish Academy of Technical Sciences. Vejle, Denmark, March.

Imhoff, P. T., Jaffé, P. R., and Pinder, G. F., 1994. An experimental study of complete dissolution of a nonaqueous phase liquid in saturated porous media, Water Resources Research, 30(2):307-320.

Lamarche, P., 1991. Dissolution of immiscible organics in porous media, Ph.D. Thesis, University of Waterloo, Waterloo, Ontario, Canada.
National Research Council, 1994. Alternatives for Ground Water Cleanup, June.

Powers, S. E., Abriola, L. M., and Weber, W. J., Jr., 1994. An experimental investigation of nonaqueous phase liquid dissolution in saturated subsurface systems: Transient mass transfer rates, Water Resources Research, 30(2):321-332.

Schnarr, M., 1992. An in situ oxidative technique to remove residual DNAPL from soils, M.A.Sc. Thesis, University of Waterloo, Waterloo, Ontario, Canada.

Schnarr, M., Truax, C., Farquhar, G.J., Hood, E., Gonullu, T., and Stickney, B., 1997. Laboratory and controlled field experiments using potassium permanganate to remediate trichloroethylene and perchloroethylene DNAPLs in porous media, Journal of Contaminant Hydrology, in press.

OXIDATION OF CHLORINATED SOLVENTS BY PERMANGANATE

Y. Eugene. Yan and Franklin.W. Schwartz
(The Ohio State University, Columbus, Ohio)

Abstract: This paper presents the results of various batch experiments to elucidate reaction pathways and the kinetics of chlorinated ethylene degradation. Degradation is rapid, but rates decrease with increasing numbers of chlorine substituents. Chlorinated ethylenes degrade to CO_2 via intermediates, including various carboxylic acids.

INTRODUCTION

Both laboratory and field experiments have shown that potassium permanganate is able to oxidize common chlorinated ethylenes like trichloroethylene (TCE) and tetrachloroethylene (PCE) (e.g., Vella and Veronda, 1992; Gates et al., 1995; Gonullu et al., 1998). Most of these experiments can be considered as a proof-of-concept that demonstrates the efficacy of the remedial concept rather than pathways and kinetics.

The goal of our investigations is to provide a detailed process-level understanding of the oxidative destruction of chlorinated ethylenes by permanganate. The specific objectives here is to present results concerning reaction order, degradation rate and kinetic behavior of chlorinated ethylenes in reactions with permanganate, and to elucidate the detailed reaction pathways based on product analysis.

There has been little direct process-oriented work on the oxidation of chlorinated ethylenes by permanganate. Work on alkene (C=C) oxidation in chemical synthesis provides a general understanding of the oxidation of chlorinated ethylenes. The reaction can be described as:

$$\alpha \, C_2Cl_nH_{4-n} + \beta \, MnO_4^- \xrightarrow{k_1} \eta \, I \xrightarrow{k_2} \gamma \, CA + \delta \, MnO_2 + \zeta \, Cl^- \quad (1)$$

where α, β, η, γ, δ, and ζ are stoichiometric coefficients, $C_2Cl_nH_{4-n}$ represents various chlorinated ethylenes, I is a cyclic complex, CA is some carboxylic acid, and k_1 and k_2 are rate constants. This study explores this reaction, particularly the identity of carboxylic acids, their pH dependencies, and the extent to which carboxylic reactions are converted to CO_2.

MATERIALS AND METHODS

Kinetic experiments were undertaken with a closed reaction vessel to establish chemical transformation rates of various chlorinated ethylenes present as a dissolved phase in aqueous solutions. The compounds of interest included PCE (C_2Cl_4), TCE (C_2HCl_3), cis-DCE ($C_2H_2Cl_2$), trans-DCE ($C_2H_2Cl_2$), and 1, 1-DCE ($C_2H_2Cl_2$) (Aldrich Chemical Co., Milwaukee, WI). The degradation rates of these chlorinated ethylenes were measured in Milli-Q water, and phosphate-buffered Milli-Q water. Table 1 describes four of the different experiments conducted as part of this study.

TABLE 1. Summary of experiments conducted

Type	Design	Monitoring Species
(1) oxidation kinetics TCE	MnO_4^- - 1 mM TCE - 0.031 to 0.083 mM	pH, Cl^-, TCE
(2) oxidation kinetics TCE	TCE – 0.078 MnO_4^- - 0.37 – 1.2 mM	pH, Cl^-, TCE
(3) oxidation kinetics PCE and DCE isomers	MnO_4^- - 1 mM PCE, DCEs < 0.1 mM	PCE or DCE isomers
(4) reaction pathway characterization	TCE - 0.09 mM MnO_4^- - 0.63 mM	TCE, various carboxylic acids, CO_2

Page constraints preclude a detailed discussion of the experimental design and methods for analyzing reactants and products. Readers can refer to Yan (1998) for details. The chlorinated ethylenes were analyzed using a Fisons Instruments 8060 gas chromatograph equipped with a Ni^{63} electron capture detector and a DB-5 capillary column (J&W Scientific, Rancho Cordova, CA). Cl^- activity in the reaction vessel was measured using either an Orion ion selective electrode (ISE) with glass body (Model 9617) or a Buchler Digital Chloridometer.

The last group of kinetic experiments (4) (Table 1) were designed to identify and quantify the various products over time, as a function of pH (4, 6 and 8). Chemical synthesis studies determined that various carboxylic acids should form as intermediates, so the experiments were geared to their identification. Our kinetic tests were conducted in 50-ml glass vials with PTFE/silicone septum-lined screw-top caps. Acidic products were analyzed on a Waters high performance liquid chromatography (HPLC) fitted with a Bio-Rad Aminex HPX-87H strong cation-exchange resin column (300 × 7.8 mm I.D.).

RESULTS AND DISCUSSIONS

Reaction Order. From Eq. 1, chlorinated ethylenes degrade according to this rate equation

$$r = -\frac{1}{\alpha}\frac{d[C_2Cl_nH_{4-n}]}{dt} = k[C_2Cl_nH_{4-n}]^\alpha[MnO_4^-]^\beta \qquad (2)$$

In order to find unknown kinetic orders, an isolation technique was used in kinetic experiments by using an excess of MnO_4^- to isolate $C_2Cl_nH_{4-n}$. Determination of reaction order usually involve a logarithmic form of Eq. 2

$$\log r = \log k_{obs} + \alpha \log [C_2Cl_nH_{4-n}] \qquad (3)$$

$$\log k_{obs} = \log k + \beta \log [MnO_4^-]_0 \qquad (4)$$

where k_{obs} is a pseudo-order rate constant estimated from experiments and $[MnO_4^-]_0$ is held essentially constant throughout the period of experimental observation. To avoid complications from subsequent reactions or catalysis, an initial rate method (Casado, 1986) was used here and Eq. 3 can be expressed as:

$$\log \, r_0 \; = \; \log \, k_{obs} \; + \; \alpha \log \, [\,C_2 Cl_n H_{4-n}\,]_0 \qquad (5)$$

By varying the concentration of $C_2Cl_nH_{4-n}$ and measuring the initial rate, the order α with respect to $C_2Cl_nH_{4-n}$ can be determined by Eq. 3. Having determined α, we repeated experiments with a fixed initial concentration of $C_2Cl_nH_{4-n}$ to determine k_{obs} at several high concentrations of MnO_4^-, respectively. The order β with respect to MnO_4^- can then be determined by Eq. 5.

Experiments (1) and (2) (Table 1) provide estimates of α and β values for TCE. The initial reaction rates were estimated as the tangent to the TCE concentration-time curve. As shown in Figure 1a, the slope $\alpha = 1.01 \pm 0.02$ was calculated through a linear regression of the logarithm of initial rates versus the logarithm of initial TCE concentration (r^2=0.998). The reaction order with respect to TCE is unity and k_{obs} is a pseudo-first-order rate constant.

Based on data from experiment (2), a slope of $\beta = 1.05 \pm 0.03$ was determined (Figure 1b). Thus, the reaction order with respect to MnO_4^- is also unity. Overall then, the reaction between TCE and MnO_4^- is a second-order reaction with α=1 and β=1. The second-order rate constant, $0.66 \pm 0.02 \, M^{-1}s^{-1}$, was estimated from data measured in experiment (2) using eq. 4 with $\beta = 1$.

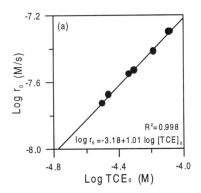 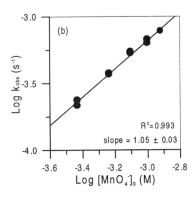

FIGURE 1. (a) Initial rates versus initial TCE of 0.031-0.083 mM oxidized by 1 mM MnO_4^-. (b) k_{obs} versus initial MnO_4^- of 0.37-1.2 mM reacting with 0.078 mM TCE.

Degradation of Chlorinated Ethylenes. Five chlorinated ethylenes were investigated in Experiment (3) (Table 1) with concentrations of chlorinated ethylenes at least 10 times less than MnO_4. The disappearance of chlorinated ethylenes (Figure 2) can be described by a pseudo-first-order model with values ranging from 0.45 to 300 x 10^{-4} s^{-1}. The degradation is most rapid for the less

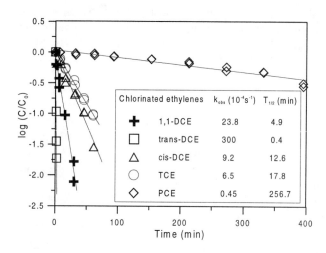

Chlorinated ethylenes	k_{obs} $(10^{-4}s^{-1})$	$T_{1/2}$ (min)
+ 1,1-DCE	23.8	4.9
□ trans-DCE	300	0.4
△ cis-DCE	9.2	12.6
○ TCE	6.5	17.8
◇ PCE	0.45	256.7

FIGURE 2. Degradation of chlorinated ethylenes by 1 mM MnO$_4$⁻.

chlorinated compounds with half lives generally less than 20 minutes. PCE oxidation is much slower with a half life of 257 minutes. This kinetic behavior is consistent with the idea of electrophilic addition proposed by Freeman (1975). In effect, the high deficiency of electrons in the carbon double bond, induced by four chlorine substituents in PCE, reduces the rate of electrophilic attack. Thus, PCE degradation is slow.

In general for alkenes, the trans isomers are more stable than the corresponding cis isomers. However, we observed the opposite with trans-DCE oxidizing more rapidly than cis-DCE (Figure 2). This result points to a significant steric effect in the reaction. The steric interaction of cis substituents is caused by the change in bond angles in the addition reaction involving the large cyclic activated complexes, such as five- and six-membered cyclic complexes (Freeman, 1975).

Products. The last experiment (4) was concerned with identifying some of the intermediate products early in the reaction. The system was phosphate-buffered to inhibit precipitation of solids. In all, three separate experiments were conducted (pH 4, 6, and 8) to discover any pH dependencies. Samples were collected one hour after reaction started in 8-hour kinetic experiments. Four carboxylic acids, formic, oxalic, glyoxylic, and glycolic acids, were identified in the system as intermediate products (Yan, 1998). Either formic or oxalic acid predominated, depending on pH. In general, a maximum of 43% initial TCE was converted to either formic or oxalic acid and up to 25% of TCE was transformed to glyoxylic acid. Glycolic acid was measured in very small quantities (<2%) in TCE transformation.

Other data, not presented here, indicates that the majority of TCE was transformed to CO$_2$ as a final product. Assuming TCE is stoichiometrically converted to CO$_2$, 57-88% of initial TCE, varying depending on pH, was converted to CO$_2$ at the time when the experiments were terminated at 8 hours.

Proposed Reaction Scheme for TCE Oxidation. Reaction scheme for alkene

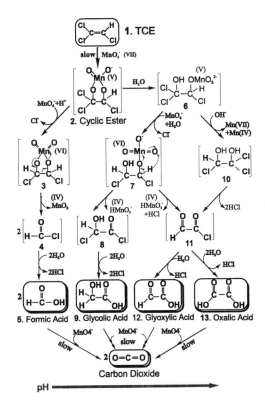

FIGURE 3. Proposed TCE oxidation scheme

oxidation by permanganate has come out of the field of organic synthesis chemistry (Wiberg and Saegebarth, 1957; and Stewart, 1965). Based on this work and our experiments we propose the chemical transformation pathways specifically for TCE oxidation in Figure 3. In particular, the products and intermediates in the shadowed boxes were identified in our experiments. TCE oxidation begins with an attack of the permanganate ion, as an electrophile, on carbon-carbon double bond (Yan, 1998). An organometallic compound, cyclic hypomanganate ester (**2**) in Figure 3 forms. The decomposition of cyclic ester (**2**) occurs rapidly via oxidative hydrolysis and hydrolysis along several potential pathways.

Oxidative hydrolysis transforms cyclic hypomanganate ester (**2**) rapidly to cyclic manganate (VI) ester

(**3**) and to formic acid (**5**) via **4**. Two other possible pathways involve the hydrolysis of cyclic ester (**2**) to acyclic hypomanganate (V) ester (**6**). The acyclic manganate (VI) ester (**7**) oxidized from (**6**) transforms to glycolic acid (**9**) via **8**,or glyoxylic acid (**12**) and/or oxalic acid (**13**) via **11**. In alkalinic solution, acyclic ester (**6**) may simply hydrolyze to trichloroglycol (**10**), which again hydrolyzes to either glyoxylic acid (**12**) or oxalic acid (**13**). We measured a gradually increasing quantity of CO_2 with time in kinetic experiments indicating that carboxylic acids are oxidized to carbon dioxide at a relatively slow rate.

CONCLUSIONS

This study has shown that chlorinated ethylenes can be oxidized rapidly by MnO_4^- in aqueous solution. Half-lives for TCE, cis-1,2-DCE, trans-1,2-DCE and 1,1-DCE ranged from 0.4 to 18 minutes under our experimental conditions. PCE reactions were slowest with a half-life of 257 minutes. The extensive kinetic study on TCE oxidation suggests that the reaction is second order and that the second order rate constant for TCE is $0.66 \pm 0.02 M^{-1}s^{-1}$.

Indications from the chemical synthesis literature are that the reaction pathways are complex and involve several intermediate compounds and pathways. Our work has verified the formation of various carboxylic acids and the ultimate transformation to CO_2. Overall, the transformations are sufficiently rapid to facilitate further study of oxidation schemes for site remediation.

ACKNOWLEDGEMENTS

This material is based on work supported by the Department of Energy under Grant No. DE-FG07-96ER14735.

REFERENCES

Casado, J., López-Quintela, M. A. and Lorenzo-Barral, F. M., 1986. The initial rate method in chemical kinetics. *Journal of Chemical Education*, 63:450-452.

Freeman, F., 1975. "Possible criteria for distinguishing between cyclic and acyclic activated complexes and among cyclic activated complexes in addition reactions." *Chemical Reviews*. 75:439-491.

Gates, D. D, R.L. Siegrist, and S.R. Cline. 1995. "Chemical oxidation of volatile and semivolatile organic compounds in soil." *Proceedings of 88th annual meeting and exhibition*, San Antonio, Texas.

Gonullu, T., G. Farquhar, C. Truax, M.J. Schnarr, and B. Stickney. 1998. "Studies on the use of permanganate to oxidize chlorinated solvents in soil." *Journal of Contaminant Hydrology*, in press.

Stewart, R., 1965. Oxidation by permanganate. In *Oxidation in organic chemistry*, Wiberg, K. B., ed., Academic Press, New York, Part A, Charpter 1, pp 1-68.

Vella, P. A. and B. Veronda, 1992. "Oxidation of trichloroethylene: comparison of potassium permanganate and Fenton's reagent." In *Chemical oxidation technologies for the nineties*, Eckenfelder, W. W., ed., Technomic publishing, Lancaster, Basel.

Wiberg, K. B. and K.A. Saegebarth. 1957. "The mechanisms of permanganate oxidation. IV. Hydroxylation of olefins and related reactions." *Journal of American Chemical Society*. 79: 2822-2824.

Yan, Y. E., 1998. "Abiotic remediation of ground water contaminated by chlorinated solvents." *Unpublished Ph.D. Dissertation*, The Ohio State University.

EVALUATION OF FLAMELESS THERMAL OXIDATION FOR VAPOR-PHASE VOC TREATMENT

Peter R. Guest (Parsons Engineering Science, Inc., Denver, Colorado)
Steven R. Archabal (Parsons Engineering Science, Inc., Phoenix, Arizona)
Mark J. Vessely (Parsons Engineering Science, Inc., Denver, Colorado)
E. Kinzie Gordon, (Parsons Engineering Science, Inc., Denver, Colorado)
Dan Kraft (Booz•Allen & Hamilton, Inc., San Antonio, Texas)
Jim Gonzales (AFCEE/ERT, Brooks AFB, Texas)
Brady Baker (AFBCA/OL3A, Plattsburgh, New York)
Luke Gilpin (Lockheed Martin, Fort Worth, Texas)

ABSTRACT: A performance/cost evaluation of flameless thermal oxidation (FTO) vapor-phase treatment of extracted soil vapors was conducted at two Air Force sites. One site is a former fire training area where soils are contaminated with petroleum (aromatic) and chlorinated hydrocarbons. Influent volatile organic compound (VOC) vapor concentrations ranged from 12 to 6,000 parts per million, by volume (ppmv). The destruction/removal efficiency (DRE) evaluation indicated that the FTO unit was 99.96% efficient at removing total VOCs, and >99.98% efficient at removing benzene, trichloroethene (TCE), and tetrachloroethene (PCE). Treatment costs ranged from $6.00/lb to $9.05/lb. The other demonstration site was a TCE-contaminated aircraft production plant. Influent TCE concentrations ranged from 67 to 170 ppmv at this site, and the DREs for TCE were >99.97%. Treatment costs ranged from $84.61/lb to $139.72/lb TCE removed. The FTO system shut down once during each demonstration due to internal system problems. The system was 75% and 60% available at Site FT-002 and Building 181, respectively. Vendor information and the results of the two site demonstrations indicate FTO technology is a cost-competitive vapor treatment method for a full-scale application at each site.

INTRODUCTION

The Air Force Center for Environmental Excellence (AFCEE) sponsors an ongoing program to promote the use of cost-effective soil vapor treatment technologies to be used in conjunction with soil vapor extraction (SVE) for remediation of fuel- and solvent-impacted sites. Parsons Engineering Science, Inc. (Parsons ES) was contracted by the Air Force to provide services to evaluate flameless thermal oxidation (FTO) vapor-phase treatment of SVE system off-gas at Air Force sites, including a former fire training area (Site FT-002) at Plattsburgh AFB, NY, and a trichloroethene (TCE) release site (Building 181) at Air Force Plant 4 (AFP 4), Fort Worth, TX. Thermatrix, Inc. of Knoxville, TN provided their GS Series FTO treatment system to be evaluated during the demonstrations. The FTO technology demonstration at Plattsburgh

AFB was performed over a 30-week period. The FTO demonstration at AFP 4 was conducted over an approximate 26-week period.

This paper presents an evaluation of the effectiveness of the FTO system during the two demonstrations; a summary of FTO system performance, operational costs, and reliability; and an evaluation of full-scale treatment system application for Site FT-002 and Building 181. More detailed site-specific technical reports were prepared for each demonstration site (Parsons ES, 1997 and 1998).

Site Background. Site FT-002 is a fire training area that was used from the mid-1950s through 1989. Training activities involved the release of waste fuels and solvents into unlined pits, where the fuels were ignited and extinguished. Uncombusted fuels and solvents percolated into the soils, resulting in contamination of soils and groundwater. The depth to groundwater at the site averages approximately 35 feet below ground surface (bgs).

At AFP4, parts degreasing operations were performed in the northwestern corner of Building 181. Soil and groundwater contamination at the site resulted from TCE leaking from an aboveground storage tank. The leak was discovered in May 1991, and on July 15, 1991, the degreasing tanks were removed from service. The depth to perched groundwater beneath the site ranges from 1 to 5 feet bgs, and the unconfined aquifer occurs approximately 23 feet bgs.

MATERIALS AND METHODS

Description of Technology. Thermatrix has developed a proprietary technology for FTO of volatile organic compounds (VOCs) in vapor streams. The Thermatrix GS Series FTO system employs a "packed-bed" ceramic matrix that is resistant to moisture and acid, noncatalytic, and has a temperature rating of up to 2,500°F. The oxidation of VOCs in the influent vapor stream occurs in a reaction zone within the ceramic matrix. Typical operating temperatures are between 1,600 and 1,850°F. If the British thermal unit (BTU) value of the influent soil gas is not sufficient to maintain reactor bed operating temperatures, supplemental fuel (e.g., propane) is required. System exhaust gases are discharged directly to the atmosphere, or can be routed through a caustic scrubber to remove hydrochloric acid (HCl) if the influent vapors contain chlorinated VOCs. The caustic scrubber was not used during operation at Site FT-002 because estimated mass emission rates for HCl were below the New York State Department of Environmental Conservation (NYSDEC) annual guideline concentration for air emissions (7 µg/m^3) The scrubber was used at the Building 181 site to ensure compliance with Texas Natural Resource Conservation Commission (TNRCC) HCl allowable emission rate (0.0247 lb/hr).

The FTO demonstration unit is designed to extract and treat contaminated vapors at flow rates of 20 to 120 cfm, and to reduce the influent

VOC concentrations by not less than 99.99%. Thermatrix testing has demonstrated the 99.99% destruction/removal efficiency (DRE) of the FTO system for a wide variety of VOCs, including chlorinated hydrocarbons (Meltzer, 1992; Thermatrix, 1992).

Capital Equipment Costs. The Thermatrix GS Series FTO treatment system was purchased by the Air Force on a "shared cost" basis for these demonstrations. After the Thermatrix share, the cost to the Air Force was $235,265, versus a commercial cost of $275,265.

To determine the prorated capital cost for the 210-day Plattsburgh AFB demonstration, the total capital cost was averaged over an estimated 3-year life of the FTO system. Because use of the quench/scrubber was not necessary at this site, the capital cost for the Plattsburgh AFB demonstration excluded $62,000 for the quench/scrubber [($213,265/1,095 days) x 210 days = $40,900]. The prorated capital cost (including the quench/scubber) for the 180-day AFP 4 demonstration was $45,249 [($275,265/1,095 days) x 180 days].

FTO Operations. At Site FT-002, testing of the FTO system was conducted over a 30-week period from August 27, 1996 through March 25, 1997, and soil vapors were extracted from 14 site wells. Wells were tested for periods of up to 2 weeks to determine the optimum vacuum/extraction flow rate and the VOC concentration. Demonstration of the FTO system at AFP 4 was conducted over an approximate 26-week period from April 19 through October 15, 1997. SVE occurred from one deep and seven shallow (perched-zone) 2-inch-diameter monitoring wells were tested.

RESULTS AND DISCUSSION

At each site, monthly sampling of influent and effluent FTO system vapor streams was conducted to evaluate the performance of the FTO system. The vapor streams were sampled using 1-liter SUMMA® canisters, and samples were analyzed by Air Toxics, Ltd. of Folsom California for total volatile hydrocarbons (TVH) and VOCs using USEPA Method TO-14. In addition, at Building 181, analysis for HCl using NIOSH Method 7903 was conducted to evaluate the effectiveness of the quench scrubber.

SVE Rates and Soil Vapor Concentrations. The primary chemicals of concern at Site FT-002 are benzene, TCE, and tetrachloroethene (PCE). The SVE flow rates for individual extraction wells ranged from 40 to 90 cfm. The influent vapor flow rate to the FTO unit was maintained at 100 cfm. The concentrations of total VOCs in the influent vapor samples ranged from 12 to 6,000 ppmv. During the field demonstration a total of 8,162 lb of VOC vapors was recovered from the soil during 139 days of SVE.

The primary chemical of concern at Building 181 is TCE. The influent vapor flow rate to the FTO unit was held constant at 105 cfm by using an automatic air bleed-in valve. The concentrations of TCE detected in the post-dilution influent vapor stream ranged from 67 to 170 ppmv. During the field demonstration, an estimated 572 lb of TCE was recovered from the soil over 109 days of extraction.

Contaminant Destruction/Removal Efficiency. At Site FT-002, the FTO unit was 99.96% efficient at removing total VOCs, and >99.98% efficient at removing benzene, TCE, and PCE from extracted soil vapors. At Building 181, the FTO unit was >99.97% efficient at removing TCE. The effluent caustic scrubber was effective in removing HCl to a discharge rate of <0.040 lb/hr; the maximum allowable emission rate for HCl is 0.0247 lb/hr.

Operating Costs. FTO technology demonstration operating costs included Thermatrix mobilization and startup, treatment unit transportation, propane, electricity, and demobilization. Excluded from these costs are Parsons ES labor costs and the cost of vapor and air emission sampling, which would be relatively consistent for other vapor treatment technologies.

At Site FT-002, the total cost for the FTO demonstration for 210 days was $73,934, or $352/day. During the field demonstration, a total of 8,162 lb of VOC vapors was recovered from site soils during 139 days of SVE. The treatment costs ranged from $6.00/lb (based on 139 days of SVE) to $9.05/lb (based on 210 days on site).

At Building 181, the total cost for the FTO demonstration for 180 days was $79,912 ($444/day). A total of 572 lb of TCE was recovered from site soils during 109 days of SVE. The TCE treatment costs ranged from $84.61/lb ($444 x 109 days/572 lb) to $139.72/lb ($444 x 180 days/572 lb).

Low influent TCE concentrations (67 to 170 ppmv) were observed during the Building 181 demonstration due to SVE from groundwater wells with limited screen exposure to the vadose zone. Therefore, the FTO system operated at only 5 to 10% of the designed loading rate, dramatically increasing the treatment cost per pound. The maximum designed loading rate of the FTO system is 3.67 lb/hr TCE. This influent maximum loading rate is equal to 1,500 ppmv TCE at 120 cfm. This theoretical maximum loading rate would have resulted in approximately 9,590 lb of TCE recovered from site soils during 109 days of SVE. Assuming this maximum loading rate, the TCE treatment costs would range from $5.05/lb ($444 x 109 days/9,590 lb) to $8.33/lb ($444 x 180 days/9,590 lb).

Onsite system monitoring includes checking oxidizer and scrubber temperatures, supplemental fuel consumption, caustic solution flow rates, and liquid levels in the scrubber caustic and quench tanks and inlet moisture separator. Generally, each visit takes 30 minutes. Monthly sampling of the system influent and effluent vapor samples takes approximately 2 hours per event.

Reliability and Maintainability. The FTO unit shut down once during each of the two demonstrations due to system problems. At Site FT-002, heavy rains caused damage to the variable frequency drive (VFD) controller for the SVE blower on two separate occasions. A redesigned VFD was installed, and the system was operational 96% of the remainder of the demonstration period, including 100% during February 1997. In January 1997, the FTO system shut down twice due to low supplemental fuel pressure readings caused by very cold (-27°F) ambient temperatures. Increasing the propane pressure-regulator setting alleviated this problem. The FTO system was operationally available for 156.5 days, or 75% of the 210 days onsite. The majority of the down time was due to heavy rains causing damage to the VFD controller.

At Building 181, the FTO system operated for 109 of 180 onsite days (i.e., was 60% available). A quench tank thermocouple was replaced after its malfunction caused a system shut down. External causes of FTO shut downs included 1) loss of water supply to the quench scrubber; 2) failure of a float switch in the groundwater treatment system air stripper sump (the discharge point for the scrubber effluent), which caused a high-water level in the quench tank; and 3) propane exhaustion due to the fuel tank not being filled on schedule. Increasing the scrubber water inlet flow rate on September 10, 1997, addressed the water-pressure fluctuation problem. The system then was 100% operational for the remainder of the demonstration period.

Cost Comparison of Vapor Treatment Technologies. Vendor information was used to compare the FTO technology to thermal oxidation, catalytic oxidation, and resin-bed vapor treatment technologies for full-scale application at each site. For full-scale application at Site FT-002 (assuming mass recovery of total VOCs at 204 lb/day, and vapor flow rate of 500 cfm), the estimated costs of treating VOC vapors using the four technologies range from approximately \$1.98 to \$3.46/lb over a 1-year operating period, and from \$1.40 to \$1.57/lb over 3 years. The costs of the Thermatrix FTO technology at this site were estimated at \$3.41/lb over a 1-year period of operation, and \$1.45/lb over 3 years. Based on vendor information, the most cost-effective vapor treatment technologies for full-scale application at Site FT-002 were thermal oxidation for a 1-year period and catalytic oxidation for a 3-year period.

For a full-scale application at Building 181 (assuming mass recovery of TCE at 203 lb/day, and vapor flow rate of 1,140 cfm), the costs of treating TCE vapors using thermal, catalytic, or resin bed treatment technology may range from approximately \$0.93 to \$1.95/lb over a 5-year operating period, and approximately \$0.56 to \$1.46/lb over 12 years. The costs of the Thermatrix FTO technology at this site were estimated at \$1.60/lb over a 5-year period, and \$0.91/lb over 12 years. The most cost-effective vapor treatment technology for full-scale application at Building 181 was catalytic oxidation.

CONCLUSIONS

The FTO system has proven to be a reliable VOC vapor treatment technology that is capable of achieving >99.96% DRE. Based on vendor information and the results of the two site demonstrations, the Thermatrix FTO technology is a cost-effective method for treating a mixed aromatic/chlorinated hydrocarbon vapor stream, or a TCE vapor stream. Based on comparative vendor quotes, the capital costs for the Thermatrix FTO full-scale system were the highest; however, the total annual operating costs were among the lowest of the three technologies considered. Therefore, the longer the period of vapor-phase treatment, the more cost-competitive the Thermatrix FTO technology becomes. Thermatrix was the only vendor claiming a VOC DRE of 99.99%; other vendor-estimated DREs ranged from 90 to 99%. The Thermatrix FTO technology is best suited for high-concentration (>3,000 ppmv) VOC vapor streams that contain chlorinated compounds and where DREs greater than 99.9% are required. The most appropriate vapor treatment technology will be a function of the site-specific system operating period and the expected changes in soil vapor VOC concentrations over that time period.

ACKNOWLEDGEMENTS

Parsons ES performed this work under contract to the AFCEE Technology Transfer Division (ERT), Contract F41624-94-D-8136, Delivery Order 28. The authors would like to thank the Parsons ES field engineers who collected the site data and performed FTO system operation and monitoring: Mr. Dave Brown, Ms. Kim Makuch, Mr. John Mackey, and Mr. John Mastracchio (Plattsburgh AFB), and Mr. Tom Dragoo (AFP 4).

REFERENCES

Meltzer, J.S. 1992. *Flashback Testing of Thermatrix ES-60H Oxidizer.* Report No. SSR-1628, Fenwal Safety Systems, Inc., Marlborough, MA.

Parsons Engineering Science, Inc. (Parsons ES). 1997. *Site-Specific Technical Report for the Evaluation of Thermatrix GS Series Flameless Thermal Oxidizer for Off-Gas Treatment of Soil Vapors with Volatile Organic Compounds at Site FT-002, Plattsburgh Air Force Base, New York.* Prepared by Parsons ES. June.

Parsons ES. 1998. *Site-Specific Technical Report for the Evaluation of Thermatrix GS Series Flameless Thermal Oxidizer for Off-Gas Treatment of Trichloroethene Vapors at Building 181, Air Force Plant 4, Texas.* Prepared by Parsons ES. February.

Thermatrix, Inc. 1992. *Destruction of Organic Compounds in the Thermatrix Flameless Thermal Oxidizer.* San Jose, California.

INNOVATIVE AIR STRIPPING AND CATALYTIC OXIDATION TECHNIQUES FOR TREATMENT OF CHLORINATED SOLVENT PLUMES

Stephen H. Rosansky, Arun R. Gavaskar, Byung C. Kim, Eric Drescher, Charles A. Cummings (Battelle, Columbus, Ohio), and Say-Kee Ong (Iowa State University, Ames, Iowa)

ABSTRACT: An innovative field pilot-scale pump-and-treat system was evaluated at Dover Air Force Base using groundwater containing mainly 1,2-dichloroethane. A crossflow air stripper design, a concept initially developed by Louisiana State University, was used in an effort to enhance the air-water contact. Baffles were installed in the air-stripping tower to increase the flow path of the air as it traveled up the column. Two 17-ft-long packed towers, one with the crossflow design and the other with a conventional countercurrent design, were tested side-by-side. A 10-week-long continuous test was performed to study long-term performance and the effects of iron and biological fouling of the stripping towers. The towers were operated at a water flowrate of 20 gal/min each. The off-gas from the towers was treated using an innovative photocatalytic oxidizer as well as three conventional fixed-bed catalytic oxidizers. The photocatalytic unit used a combination of metal oxide catalyst and ultraviolet lights to destroy the chlorinated organics at a much lower temperature (around 175 °F) than that required by conventional catalysts (around 800 °F). The operational data from this study was used to develop a full-scale design and to determine if crossflow air stripping and photocatalytic oxidation are economically viable treatment alternatives.

INTRODUCTION

Groundwater contamination by chlorinated compounds has occurred at various Air Force Bases (AFB) as a result of spills, leaks, and seepage from routine operations. Air stripping is an accepted technology for removing these dissolved phase chlorinated compounds from groundwater. Typically, air-stripping towers are designed in a countercurrent arrangement; water gravity flows downward through the tower and air is pumped from the bottom to the top of the tower. A packing material is used to provide surface area to increase the mass transfer of contaminants from the liquid to the vapor phase. The off-gas generated by the air stripping process can contain high concentrations of volatile contaminants. In many cases the off-gas stream requires some form of treatment. Oxidation and carbon adsorption technologies are commonly employed to treat the off-gas stream.

The objective of this study was to field demonstrate two innovative technologies, crossflow air stripping and photocatalytic oxidation, for cleanup of groundwater contaminated with chlorinated volatile organic compounds (VOCs). Results of short-term tests (Gavaskar et. al., 1995) were used to optimize tower design and operating parameters prior to implementing a 10-week extended test.

The crossflow air stripping concept was first introduced by Baker and Shyrock (1961) and was developed by researchers at Louisiana State University (Thibodeaux, 1969; Wood et al., 1990; Mertooetomo et al., 1993; Verma et al., 1994). Operation of the crossflow tower is still countercurrent; however, a number of partial baffles are spaced at equal distances along the length of the tower. The airstream is deflected by the baffles causing it to pass through the packing at approximately 90 degrees from the vertical. Bench-scale experiments performed by LSU (Mertoooetomo et al., 1993) indicated that this arrangement reduced the gas velocity and pressure drop across the tower, creating a potential for reduced energy cost for the air blower. Also, higher air-to-water ratios than were possible with conventional strippers were achieved with the crossflow design without causing flooding.

The photocatalytic oxidizer employs a technology called the Adsorption-Integrated-Reaction™ (AIR™) process. It uses a proprietary catalyst in conjunction with banks of ultraviolet (UV) lights. The lights activate the oxygen present in the air stream, which in turn oxidizes the contaminants in the vapor phase.

The field demonstration was performed at Dover Air Force Base (AFB) Dover, Delaware. A 1/2-acre plot next to the runway at Dover AFB was chosen as the test site because of its past history of groundwater contaminated with chlorinated compounds including 1,2-dichloroethane (DCA), 1,2-dichloroethylene (DCE), trichloroethylene (TCE), and perchloroethylene (PCE). A 10-week continuous test was conducted to evaluate the performance of the crossflow air stripping (CFAS) and the countercurrent air stripping (CCAS) towers and the photocatalytic oxidizer under steady-state conditions. Results were used to develop a full-scale design and to determine if crossflow air stripping and photocatalytic oxidation are economically viable treatment alternatives.

MATERIALS AND METHODS

Design. The CCAS and CFAS towers were constructed of aluminum and were 22.1 ft high. Both towers were packed with 1-inch-diameter polyethylene pall rings to a 17-ft vertical packing depth. Each tower contained the same volume of packing. The inner diameter (I. D.) of the CCAS tower was 1.5 ft; the I. D. of the CFAS tower was 2.0 ft. Two cylindrical sections (8.3 ft and 8.5 ft high) were used to contain the packing in the CCAS tower. The packing in the CFAS tower was contained using six 2-ft-long sections and two 2.4-ft-long sections, all with flanged ends. Baffles were made of thin aluminum sheets of size similar to the flanged ends. Nine baffles were installed in the CFAS tower. The details of the CFAS tower are shown in Figure 1. The packed portion of each tower was installed on a 2.9-ft-long base piece mounted to the trailer. A 1.5-ft distributor section was placed on the top of each tower to complete the construction.

The air stripping process is illustrated in Figure 2. The air stripping/catalytic oxidation system was constructed on a trailer for easy transport from site to site. Groundwater was extracted from 3 wells at 40 gpm. The groundwater was spiked with DCA to obtain an approximate inlet concentration of 1,000 ppb. Spiking was necessary because of the unexpectedly low concentra-

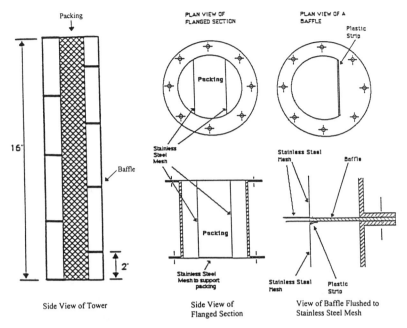

Figure 1. Crossflow Air-Stripping (CFAS) Tower Details

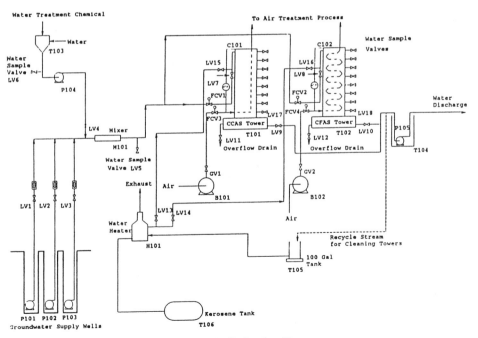

Figure 2. Air Stripping Process

tions of contaminants in the groundwater. A 1,000 ppb inlet concentration was needed so that the concentration of DCA in the effluent stream from the towers would be in a detectable range. The stripping towers were operated in parallel; each tower received 20 gpm of water during the 10-week test. The treated water leaving the towers was collected in a surge tank located at the base of the towers and was discharged to a drainage ditch. Percent removal results calculated from short-term air stripping tests previously performed (Gavaskar et al., 1995) were used to select air flowrates for the 10-week test. It was desired to achieve 95 percent removal of DCA in each tower. The CFAS tower, therefore, was operated at 250 scfm (air-to-water ratio of 94), and the CCAS tower was operated at 160 scfm (air-to-water ratio of 60). These air flowrates yielded identical removal efficiencies in both towers.

Off-gas generated by the two stripping towers was combined and directed to a 50 scfm fixed-bed catalytic oxidizer and to three bench-scale catalytic oxidizers equipped with conventional fixed-bed catalysts. A schematic illustration of the oxidation system is shown in Figure 3.

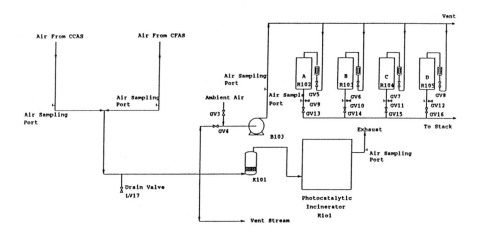

Figure 3. Off-Gas Treatment Process

The bench-scale catalysts were operated at 0.1 scfm which corresponds to a space velocity of 5,000 hr^{-1} (space velocity is defined as the volumetric flowrate of air divided by the volume of the catalyst bed). Catalyst A, B, and C, were operated at 880, 880, and 800 °F, respectively as recommended by their manufacturers.

The pilot-scale photocatalytic oxidizer was designed to achieve 95 percent destruction of DCA at an air flowrate of 50 scfm. The unit employs a preheater to heat the inlet air stream, thereby reducing the relative humidity of the air stream. The preheater was operated in the temperature range of 165 to 180 °F. The heated air passes through the reactor, consisting of a catalyst bed illuminated with UV lamps. Thirty-four UV lamps were needed to obtain approximately 95 percent conversion.

Sampling and Analysis. All water and air analyses were performed on site. A minimum of two sets of samples were collected and analyzed during each week of operation. A typical set consisted of inlet and outlet water and air samples from each tower, a common inlet air sample before the catalysts, and outlet air samples after each catalyst. Water samples were collected into pre-cleaned 40-mL VOA vials, while air samples were collected using 100-mL glass syringes.

Water analyses were performed using EPA Method 601 as a general guide. The analytical system employed an OI 4460 purge-and-trap device equipped with a two-phase sorbent trap (Carbopack™ B/Carbosieve™ S-III) for collecting the organics purged from the water samples. Analytical separation was performed with a Varian 3700 gas chromatograph using an HP-1 (Hewlett-Packard) fused silica capillary column (60 meters x 0.53 millimeter I.D.) with a flame-ionization detector (FID). A Nafion™ dryer reduced the moisture loading on the sorbent trap to enhance the stability of the FID.

The analytical system for off-gas analyses employed a two-phase sorbent trap (Tenax TA/Carboxen 1000) to collect the organic components in the off-gas samples. A sampling pump with mass flow controller system was installed down-stream of the trap so that the sample would be pulled into the trap at a uniform rate. The analytical separations were accomplished with an HP 5890 GC using an HP-1 fused silica capillary column (60 meters x 0.53 millimeter I.D.) with an FID.

A single-point calibration standard and blank samples were analyzed on both aqueous and off-gas analytical systems each day samples were collected. Duplicate sample analyses and matrix spike analyses were performed periodically.

RESULTS AND DISCUSSION

Air Strippers. The air strippers were operated continuously for 10 weeks at a 20 gpm water flowrate and an air flowrate of 250 scfm (CFAS) and 160 scfm (CCAS). The percent removal of DCA is presented in Figure 4. It is seen that the stripping efficiency of each tower remained relatively constant throughout the duration of the test. The average stripping efficiency was 96.4 percent for the CCAS tower and 94.4 percent for the CFAS tower. The pressure drop across each tower was measured periodically during the 10-week period. The results are presented in Figure 5. The pressure drop across the CFAS tower was about one magnitude of order less than that observed across the CCAS tower.

The consistent nature of the operation, with regard to stripping efficiency and pressure drop, indicates that microbial fouling was not a major concern at this site. The tower packing was removed from the stripping towers at the completion of the extended test. This packing was covered with a very thin film of reddish-brown oxide. Over a longer time (6 months or more), it is likely that the iron fouling would affect the performance of the stripping towers.

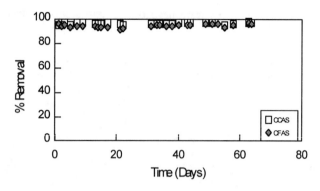

Figure 4. Removal of DCA Using the CFAS and CCAS Towers

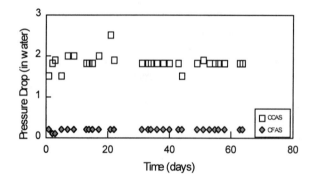

Figure 5. Pressure Drop Across Packing

Catalytic Oxidation. The photocatalytic oxidizer had an average conversion of 95.6 percent for dichloroethane. The results are shown in Figure 6. This destruction efficiency was achieved using 34 UV lamps in conjunction with an air pre-heater temperature of 180 °F. During the first 5 weeks of operation, the oxidizer occasionally overheated and shut off. It was believed that the high ambient summer air temperature was interfering with the heating control mechanism of the oxidizer, causing the thermal overload switch to activate. To compensate, the preheater temperature was decreased to 175 °F on very hot days. This resulted in slightly lower conversions (94.1 percent average); however, the oxidizer no longer shut down. To compensate for the reduced operating efficiency, an additional row of seven lamps was activated. The average conversion using these operating conditions was 96.1 percent. During the last week of operation, an additional 8 lamps were activated (total of 49), increasing the average conversion to 98.3 percent.

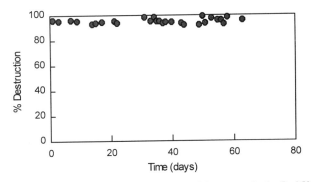

Figure 6. Percent Destruction of DCA in Photocatalytic Oxidizer

The destruction efficiency of the photocatalytic oxidizer was comparable to the three conventional catalysts that were used. The destruction efficiencies of the 3 conventional catalysts observed during the 10-week test are plotted in Figure 7. The average destruction efficiencies of catalysts A and B were greater than 99 percent throughout the duration of testing. Catalyst C had an average destruction efficiency of about 92 percent during the test.

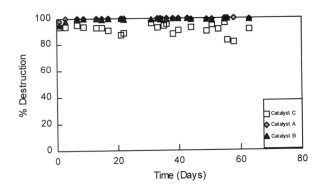

Figure 7. Percent Destruction of DCA in Fixed-Bed Catalytic Oxidizers

ECONOMIC ANALYSIS

Air Stripping. Assuming that the Dover site were to be remediated at a rate of 250,000 gpd, a treatment plant that treats approximately 180 gpm has to be designed. The contaminant is DCA at a concentration of 1,000 ppb. Assuming a 50 ppb cleanup level gives a stripping efficiency requirement of 95 percent removal. Based on these design requirements and the results of the extended test, design calculations were performed to scale up both the CCAS and CFAS towers. Based on these calculations, each stripping tower will require approximately 17 ft of packing. The CCAS tower will require a 4.2-ft diameter, whereas the CFAS tower will require a 6.0-ft diameter. The approximate capital costs for the towers are presented in Table 1. The water pump was sized based on pumping 180 gpm of water to a height of 25 ft. The air blower (1.5 hp for the CCAS tower and 1.0 hp for the CFAS tower) was sized to deliver 1,440 scfm of air in the CCAS tower and 2,250 scfm of air in the CFAS tower, against pressure drops of 2.0 and 0.2 inches of water, respectively.

If the air-stripping process is examined in isolation from catalytic oxidation of the airstreams, the operating costs for both stripping towers will be identical, with the exception of the cost to operate the blower. As a result of the lower pressure drop across the CFAS tower, there is a saving of approximately 0.5 hp by using the CFAS tower. Assuming $0.08 per kW-hour, this translates into a savings of about $400 per year.

Table 1. Capital Costs for Full-Scale Air Strippers

Item	CCAS Cost	CFAS Cost
Tower/Ductwork	$25,000	$37,500
Installation	$10,000	$10,000
Water Pump	$2,000	$2,000
Blower	$5,000	$4,000
Total:	**$42,000**	**$53,000**

Catalytic Oxidation. The data collected during the extended test were used to develop a design for the full-scale oxidizers. Scale-up is based on an inlet DCA concentration of 2,000 to 3,000 ppbv with a desired destruction efficiency of 99 percent. A space velocity of 5000 hr^{-1} and an operating temperature of 880 °F was selected for the fixed-bed catalytic oxidizers. The photocatalytic oxidizer was designed using a preheater temperature of 160 °F. The design assumes an air flowrate of 1,440 scfm from the CCAS tower and an air flowrate of 2,250 scfm from the CFAS tower.

Capital costs for the catalytic oxidation units are based on several factors such as air flow rate, catalyst bed volume, type of catalyst, heat recovery from exhaust air, and pressure drop across the catalyst bed. The catalyst manufacturers who participated in this study where asked to provide estimates for scaled-up catalytic oxidizers that could treat between 1,440 and 2,250 scfm of air. Initial estimates provided by the manufacturers of both conventional and photocatalytic units ranged from $110,000 to $250,000. Exact cost comparisons are not included because of the varying features offered by different manufacturers. Also, some manufacturers suggested that most units had enough turndown capacity, so that the same size of unit probably would be provided for both the 2,250- and 1,440-scfm requirement.

The main operating cost item associated with catalytic oxidation is the energy input required to heat the incoming air to the catalyst operating temperature. For the photocatalytic oxidizer, there are two forms of energy input. One is the energy required in the preheater to heat the incoming air to 160 °F. The other is the electrical energy required by the UV lamps.

The operating costs for the conventional catalytic oxidizers are presented in Table 2. The energy costs do not include the energy requirements of any blowers required to overcome the pressure drop across the catalyst bed systems. The blower size and energy requirement vary with catalyst type and system design. From these tables, it is apparent that the photocatalytic oxidizer has a much lower operating cost than the conventional oxidizers. It also is apparent that catalytic oxidation in conjunction with the CFAS tower has a higher operating cost than catalytic oxidation with the CCAS tower. This is because a larger airstream is generated in the CFAS tower, and this air stream requires a larger energy input to bring it to the catalyst operating temperature.

CONCLUSIONS

The crossflow tower was capable of the same stripping efficiency as the conventional countercurrent tower, but at a higher air-to-water ratio. However, the higher air-to-water ratio required in the CFAS tower is achieved at a pressure drop that is an order of magnitude lower than in the CCAS tower. Therefore, higher air-to-water ratios are achievable in the CFAS tower allowing compounds having low Henry's Law constants (e.g., DCA) to be removed. During the 10-week test, both the CFAS and CCAS tower consistently demonstrated greater than 95 percent removal efficiency of DCA at air-to-water ratios of 60 (CCAS) and 94 (CFAS).

There is some energy savings that results from the lower pressure drop across the CFAS tower. However, the resulting reduction in operating cost does not appear to be high enough to offset (over a reasonable period of time) the additional capital cost of building a CFAS versus a CCAS tower. Also, because the CFAS tower is operated at a much higher air-to-water ratio, a larger volume of contaminated air is generated. If contaminant levels in the air stream are such that the air can be discharged directly to the atmosphere, then the larger air volume is not a concern. However, if the air stream requires treatment by catalytic oxidation, then total operating costs are higher for a treatment system incorporating a CFAS tower because a larger airstream has to be heated to the catalyst operating temperature.

The photocatalytic oxidizer performed consistently at the design condition of 95 percent destruction efficiency during the test period. Approximately 99 percent destruction efficiency was achieved under selected test conditions. These destruction efficiencies are comparable to the destruction efficiencies of the conventional catalysts that were tested. The photocatalytic oxidizer operates at lower temperatures compared to conventional catalysts, resulting in lower operating costs.

In general, a combination of air stripping and catalytic oxidation was found to be a good method for remediation of organics in groundwater. The crossflow air stripping concept did not appear to offer any significant cost advantages over conventional air stripping. There may be other advantages, however, such as effective stripping of other compounds with low Henry's Law constants that could be evaluated in the future.

Table 2. Operating Costs for Catalytic Oxidizers

Item	Using CCAS	Using CFAS
Conventional Fixed-Bed Catalytic Oxidizers		
Heat input (Btu/hour)	1.62×10^6	2.53×10^6
Fuel Cost ($/$10^6$ Btu)	$5.50	$5.50
Annual Energy Cost ($/year)	$77,900	$121,800
Photocatalytic Oxidizer		
Heat Input (Btu/hour)	0.215×10^6	0.336×10^6
Fuel Cost ($/$10^6$ Btu)	$5.50	$5.50
Annual Heating Cost	$10,400	$16,200
Lamp Electricity Requirements (kW)	52	82
Annual Electrical Energy (kW-hours)	455,520	718,320
Electrical Cost	$0.08	$0.08
Annual Electricity Cost ($/year)	$36,400	$57,500
Total Annual Energy Costs ($/year)	$46,800	$73,700

REFERENCES

Baker, D.R., and H.A. Shyrock, 1961. "A Comprehensive Approach to the Analysis of Cooling Tower Performance," *J. Heat Transfer, Trans. AIME,* 83:339.

Gavaskar, A.R., Kim, B.C., Rosansky, S.H., Ong, S.K., and Marchand, E.G. "Crossflow Air Stripping and Catalytic Oxidation of Chlorinated Hydrocarbons from Groundwater," *Environmental Progress*, 14(1):33-40.

Mertooetomo, E., K.T. Valsaraj, D.M. Wetzel, and D.P. Harrison, 1993. "Cascade Crossflow Air Stripping of Moderately Volatile Compounds Using High Air-to-Water Ratios." *Water Research*, 27(7):1139-1144

Thibodeaux, L.J., 1969. "Continuous Crosscurrent Mass Transfer in Towers." *Chemical Engineering*, 17(6):165-170.

Verma, S., K.T. Valsaraj, D.M. Wetzel, and D.P. Harrison, 1994. "Direct Comparison of Countercurrent and Cascade Crossflow Air Stripping under Field Conditions," *Water Research*.

BATCH AND COLUMN TESTING TO EVALUATE CHEMICAL OXIDATION OF DNAPL SOURCE ZONES

Eric Drescher, Arun R. Gavaskar, Bruce M. Sass, Lydia J. Cumming, Melody J. Drescher, and Travis K.J. Williamson
(Battelle, Columbus, Ohio, USA)

ABSTRACT: Bench-scale batch and column tests were conducted to simulate the in-situ application of chemical oxidation to remediate a DNAPL source zone. A trichloroethylene source zone emplaced in a column of aquifer soil was successfully remediated with a potassium permanganate solution.

INTRODUCTION

Soil and groundwater contaminated with chlorinated solvents represent a major environmental restoration problem. Manufacturing and use of these industrial chemicals often has resulted in the migration of these contaminants into subsurface aquifers. Since most chlorinated solvents are heavier than water, they sink through the water table to the bottom of the aquifer, leaving a trace of free-phase solvent. Some mobile solvent is carried away by the groundwater flow. The residual solvent forms a source zone that may persist for hundreds of years as the solvent dissolves slowly into the groundwater flow. The free-phase chlorinated solvent in the source zone is commonly referred to as dense non-aqueous phase liquid (DNAPL). The dissolved phase solvent migrates along with the groundwater flow and forms a plume that persists as long as the source remains.

Permeable barriers are an innovative alternative that is being applied for passive remediation of chlorinated solvent plumes. However, alternatives for source zone remediation are still being researched. Among source zone technologies, resistive heating and surfactant flushing have been demonstrated on a field pilot scale. Both these technologies mobilize and extract the DNAPL to the surface along with large volumes of water that require further treatment. On the other hand, in-situ chemical oxidation is an option that has the potential to destroy the DNAPL in the aquifer itself.

Chemical oxidation has been used mostly for aboveground treatment of organic and inorganic contaminants in drinking water and wastewater. In one study (Vella et al., 1990), the ability of potassium permanganate to oxidize phenols in wastewater to below regulatory requirements (< 19 ug/L) was reported. These researchers observed the formation of manganese dioxide (MnO_2) and Mn (II) as by-products of the oxidation reaction. Manganese has a secondary drinking water action level of 0.05 mg/L because of taste and water discoloration. Another study by Gates et al. (1995) evaluated the destruction of chlorinated solvents and other organics incorporated in soil matrices.

In situ aquifer treatment requires the study of the physical and chemical interactions between the oxidant and the aquifer matrix, in addition to those between the oxidant and the contaminants. In the bench-scale study described in this paper, the chemical oxidizer tested was potassium permanganate ($KMnO_4$). Permanganate has some advantages over other common oxidizers such as hydrogen peroxide and ozone. Permanganate is more stable in an aquifer environment and is effective over a broad range of pH. Trichloroethylene (TCE) was used as the target contaminant. A series of batch and continuous column tests were conducted during 1996-97 to examine the potential for field applicability of potassium permanganate to chemically oxidize a DNAPL source zone. The results from this laboratory study are presented in this paper.

MATERIALS AND METHODS

First, batch tests were conducted to verify the dosage of potassium permanganate required to oxidize TCE. Second, a continuous column test was conducted in two columns (see Figure 1). The "control column" was filled with a sandy aquifer soil. The "DNAPL column" was filled with the same water-saturated soil, into which a TCE source was emplaced. Permanganate solution was passed through both columns. The objective was to study the oxidant demand of the natural aquifer matrix (inorganic soil and organic matter constituents) and contaminants (TCE).

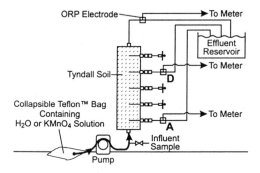

Figure 1. One of two columns used to simulate in-situ chemical oxidation

Batch Tests. In the batch tests (Table 1), potassium permanganate was added to glass vials containing combinations of TCE-water, TCE-sand-water, and TCE-sand-humic acid-water. The sand used for these experiments was Ottawa sand. Humic acid was added to simulate aquifer organic matter. The batch tests were conducted using 125-mL glass serum vials with Teflon^{TM} stoppers. The vials were filled to eliminate headspace. Each sample vial was injected with $KMnO_4$ solution and placed on an orbital shaker for 48 hours. The vials were sampled and analyzed on a gas chromatograph equipped with a flame ionization detector (GC/FID).

Table 1. Batch test series for verifying oxidant dose

Test Series	KMnO₄ Range (mg/L)	TCE (mg/L)	Sand (g)	Water (mL)	Humic Acid (mg)
1	1 to 100	7.2	0	110	0
2	1 to 100	7.2	15	110	0
3	1 to 400	7.2	15	110	10

Column Tests. The glass columns were each 24 inches (56 cm) in length and 2 inches (5 cm) in diameter. The column ends were sealed with Teflon^{TM} screw-caps. Each column had five sample ports approximately 3.5 inches (9 cm) apart.

The uncontaminated test soil for the column experiments was obtained from Tyndall Air Force Base, Florida. Native aquifer soil was removed from the base at a depth of 4 to 6 feet (1.2 to 1.8 m). This sandy soil was homogenized, passed through a 10-mesh sieve, and packed into the two columns. Each column was packed with 1,684.7 g of dry soil to obtain a total porosity of 0.214. Therefore, one pore volume in the column is equivalent to 225 mL. About 50 pore volumes of water were initially passed through each column to saturate it and stabilize flow.

TCE was emplaced into the DNAPL column by using a 6-inch syringe. A 2-mL aliquot of TCE was injected at each of three depths (2, 4, and 6-inches from the top of the column) and five points along each depth cross-section. A total of 30-mL of TCE was emplaced in this manner. After emplacement, the TCE was allowed to settle into the soil matrix for 24 hours. Deionized water was then flushed through the column to remove mobile TCE. Water flooding was conducted for about 90 pore volumes until it appeared that the TCE concentration in the column effluent had stabilized.

A 1,000-mg/L KMnO₄ solution was passed upward through both columns at a flow rate of 3.5 mL/min each, generally until the oxidation/reduction potential (ORP), a measure of residual permanganate concentration, reached steady-state at Ports A and D and in the column effluent (see Figure 2). For the DNAPL column, this point in time corresponded approximately with the time beyond which no dissolved TCE could be detected in the column effluent. The ORP and pH were measured with a Corning pH/ion analyzer Model 350. The ORP probes were installed in-line with the test columns, and the pH probe was used to measure pH in discrete effluent samples of ports A and D. Residual KMnO₄ concentrations were estimated from the ORP readings by calibration against a set of standard KMnO₄ solutions.

After TCE disappeared from the DNAPL column effluent, permanganate addition in both columns was stopped and the columns were flushed with deionized water. Effluent water samples were collected from the DNAPL column and analyzed for manganese. Samples of the column soil were taken after permanganate treatment to determine the concentration of manganese byproduct that was formed during the reaction. Duplicate 500-g samples were taken from the two soil columns that were run. A 500-g sample of untreated soil was also taken for background evaluation. Samples were taken by removing 1.5-inch layers of soil from the top, middle and bottom portions of the

column. The samples were thoroughly homogenized, dried, and shipped off-site
for analysis.

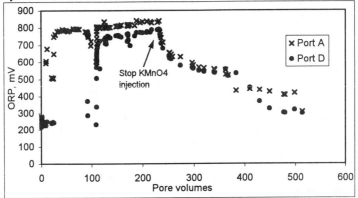

Figure 2. ORP readings indicate residual KMnO₄ levels in DNAPL column

Tracer Tests. Two tracer tests were performed on the soil column spiked with
TCE to determine any change in porosity that the KMnO₄ flushing may have
generated during TCE destruction. The tracer tests were performed by adding
potassium bromide to the inlet of the column as a conservative tracer. The
tracer concentration was measured using a bromide-selective probe
manufactured by Microelectrodes, Inc. The first tracer test was done after TCE
emplacement and prior to the injection of permanganate solution. The second
tracer test was done at the end of the column test after the KMnO₄ injection was
stopped and the column had been rinsed with 50 pore volumes of deionized
water to flush out residual permanganate.

RESULTS AND CONCLUSIONS
        The results of the batch, column, and tracer tests for evaluating
application of an oxidant to a simulated aquifer setting are described below.

Batch Tests. The results of the first series of batch tests in which TCE
dissolved in water was treated with varying concentrations of KMnO₄ are
shown in Figure 3. The results appear to follow the following stoichiometry
fairly well, with TCE destruction observed in the test matching the destruction
predicted by the equation.

$$2 \text{ KMnO}_4 + \text{C}_2\text{HCl}_3 \rightarrow 2 \text{ MnO}_2 + 2 \text{ CO}_2 + 2 \text{ KCl} + \text{HCl}$$

Based on this stoichiometry, 1 mg of TCE requires 2.4 mg of KMnO₄ for
oxidation. However, in a natural aquifer, the aquifer matrix itself is likely to
contain oxidizable matter that could consume some of the permanganate.

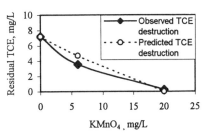 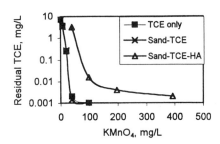

Figure 3. KMnO₄ stoichiometry   Figure 4. Evaluating KMnO₄ consumption

Figure 4 shows the results of all three series of tests. There was no noticeable difference in the permanganate consumption when sand was added to the TCE-water matrix. However, when humic acid was added to the matrix, consumption of permanganate increased significantly. This indicated that, in a field application, the organic matter constituent of the aquifer could exert a substantial permanganate demand. Suitable column tests were designed to study the destruction of TCE in a simulated aquifer setting.

Column Tests. Figure 5 shows the effect of permanganate treatment on the dissolved TCE levels in the DNAPL column effluent, and in Ports A (near the inlet of the column) and Port D (near the outlet of the column). After emplacement of the TCE source zone in the column soil, water flooding removed all the mobile TCE. Residual TCE generated elevated dissolved TCE levels in the water flow. For this particular emplaced source, mass transfer limitations appear to have stabilized the TCE concentration at around 600 mg/L in the effluent and at just below 1,000 mg/L in Port D. TCE concentration was very low in Port A. This indicated that the residual TCE source was mostly above Port A in the column. TCE has a solubility of 1,100 mg/L.

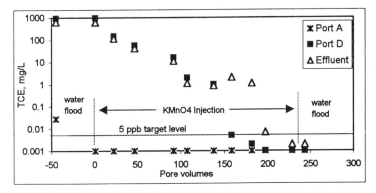

Figure 5. TCE concentrations in the effluent from the DNAPL column

When injection of the 1,000-mg/L permanganate solution was started, TCE levels in Port D and effluent started dropping. The drinking water clean up target of 5 ug/L was reached at Port D after 158 pore volumes (36 L) of permanganate solution were run through the column. The 5-ug/L target was met in the effluent after 229 pore volumes (52 L). The total $KMnO_4$ loadings corresponding to these flow volumes are 36,000 mg (to oxidize all reduced species up to Port D) and 52,000 mg (to oxidize all reduced species in the column), respectively. For a total of 1,684.7 g of soil in the column, the $KMnO_4$ required works out to be 30,866 mg/kg of aquifer soil containing a residual TCE source. Because it was not clear how much TCE remained in the column after the initial water flood, it was difficult to say how much of the permanganate was consumed by the TCE versus by the organic matter in the soil.

In order to obtain some idea of the relative oxidant demands of the TCE and the native organic matter, the control column (containing no TCE) was run alongside. Figure 6 shows the estimated residual $KMnO_4$ measured at Port D in the two columns. The residual $KMnO_4$ concentrations were estimated by measuring the ORP at Port D in both columns. The residual $KMnO_4$ levels were only slightly higher in the control column. This indicates that more $KMnO_4$ was consumed by the native organic matter in both columns than by the residual TCE source in the DNAPL column.

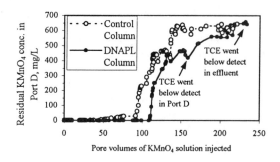

Figure 6. Residual $KMnO_4$ levels at Port D in the two columns

The residual $KMnO_4$ levels at Port D stabilized after around 144 pore volumes (32 L) in the control column. This corresponds to 32,000 mg of $KMnO_4$ consumed by the native organic matter. Assuming 50% more permanganate would be required to oxidize the remaining 8 inches (beyond the 16 inches to Port D) of the column, about 48,000 mg of $KMnO_4$ were consumed by native organic matter. Therefore, out of the 52,000 mg of $KMnO_4$ injected into the DNAPL column, 48,000 mg (roughly 90% of the $KMnO_4$ injected) may be assumed to have been consumed by the native organic matter and, possibly, by other soil constituents. Thus a large portion of the permanganate injected into a residual DNAPL source zone may be consumed by the aquifer matrix itself.

Table 2 shows that the native Tyndall soil contained 16,000 mg/kg (1.6%) total organic carbon (TOC). After permanganate treatment, 6,000 mg/kg of TOC were left that do not seem to have been mineralized by permanganate. The TOC content of the soil, an easily measured parameter, would be a good indicator of the aquifer's permanganate demand.

TABLE 2. Contaminant By-Product Analysis of Column Soil and Effluent Water

Sample	Manganese	TOC
Soil	*mg/kg*	*%*
Untreated soil	0.80	1.60
Control Column Soil after KMnO$_4$ treatment	7,535	0.60
DNAPL Column Soil	7,635	0.65
Water Effluent (After permanganate treatment)	*µg/L*	*µg/L*
Control column, 0 pore volumes rinse	11,300	57
Control column, 1-7 pore volumes rinse	590	34
Control column, 7-15 pore volumes rinse	300	20
DNAPL column, 0 pore volumes rinse	121,000	120
DNAPL column, 1-7 pore volumes rinse	4,540	56
DNAPL column, 7-15 pore volumes rinse	170	6

The build up of manganese in both columns (see Table 2) also indicates that the native organic matter in the soil exerts a greater demand on the oxidant. Both columns had similar manganese levels in the soil. This probably represents manganese as MnO$_2$ solid in the oxidation state Mn(IV). In the aquifer, insoluble Mn(IV) is likely to convert to soluble Mn(II) over time as it comes into contact with more reducing groundwater flow following the end of permanganate injection. It would be important to continue flushing the aquifer with water (and recovering it) until manganese levels drop. In this study, flushing with another 15 pore volumes of water brought manganese levels down to 300 and 170 ug/l in the effluent from the control and DNAPL columns, respectively. Additional flushing would be required to reduce the manganese levels to 50 ug/L, which is the secondary drinking water target.

Tracer Tests. To evaluate the hydrologic impact of manganese dioxide buildup in the aquifer, a conservative tracer (potassium bromide) was injected into the DNAPL column. The first tracer injection was done following the emplacement of the TCE zone, just before permanganate solution was introduced into the column (towards the end of the initial water flood). The second tracer injection was conducted after permanganate treatment and after the final 15-pore volume water rinse. The tracer profiles in Port A, Port D, and the effluent are shown in Figure 7. At all three locations in the column, the tracer peaks after permanganate treatment arrive earlier as compared to the tracer peaks before treatment. Because the water flow rate was kept constant by adjusting the pump throughout the experiment, the shorter residence time indicates a lower porosity in the column after permanganate treatment, probably due to a buildup of manganese dioxide solid.

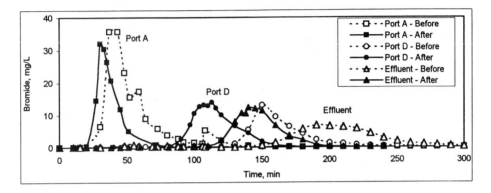

Figure 7. Tracer injections in DNAPL column

DISCUSSION

This study shows that in-situ chemical oxidation is a promising alternative for remediation of chlorinated solvent source zones. A relatively low $KMnO_4$ concentration of 1,000 mg/L was used in the column tests mainly to control formation of manganese dioxide solid, which could potentially affect flow through the source zone. Lower $KMnO_4$ concentrations also could afford better control over eventual Mn(II) migration in the groundwater flow. However, this approach requires several pore volumes of flow to remove the source zone. Higher permanganate concentrations could allow faster clean up. Verifying the byproducts of permanganate versus native organic matter reactions and controlling MnO_2 and Mn(II) products are the main research needs for this technology.

REFERENCES

Gates, D. D., R.L. Siegrist, and S.R. Cline. 1995. "Chemical Oxidation of Volatile and Semi-Volatile Organic Compounds in Soil." *Air & Waste Management Association, 88th Meeting & Exhibition.* 95-TP66.03

Vella, P.A., and B. Veronda. 1996. "Oxidation of Trichloroethylene: A Comparison of Potassium Permanganate and Fenton's Reagent." *Chemical Oxidation: Technologies for the Nineties, Proceedings First International Symposium.* Aqualine Abstract No. 94-5074.

OXIDATIVE DECHLORINATION OF CHLORINATED ORGANICS
BY SOLUBLE RUTHENIUM CATALYSTS

Mario Bressan, Nicola D'Alessandro and Lolita Liberatore (Università "G.
D'Annunzio", Pescara, Italy)
Antonino Morvillo (Università di Padova, Italy)

ABSTRACT: Water-soluble ruthenium(II) complexes are effective catalysts for
the oxidation of chlorinated organics in the presence of hydrogen peroxide and other
peroxidic oxidants at ambient temperature. Reactions are conducted in water (or
water-acetonitrile mixtures) with ca. 0.1 mM catalyst: chlorinated phenols and α-
chlorinated olefins are converted within hours to various oxygenated and
chlorinated organic products, and also, depending on the reaction conditions, to
hydrochloric acid and carbon dioxide.

INTRODUCTION

Liquid-phase catalysts are close models to the mono-oxygenase enzymes
used by the biological systems for the oxidation of a wide variety of xenobiotics,
included chlorinated organics, and may become a gentle alternative method of
destruction of these recalcitrant substrates (Sheldon, 1997), especially when they
are dissolved in small, but significant amounts, in waters. The feasibility of this
approach has been demonstrated in the recent years by a number of reports dealing
with succesful metal-catalyzed oxidations of chlorinated substrates in the liquid
phase (Sorokin and Meunier, 1996; Labat et al. 1990; Sorokin et al., 1996, 1995;
Bailey et al., 1993; Bressan et al., 1994, 1995). As a continuation of our previous
studies, we report here on the effective oxidation of a number of chlorinated olefins
and phenols in the presence of water-soluble ruthenium catalysts, i.e.:
$[Ru(H_2O)_2(dmso)_4](BF_4)_4$ (dmso = dimethylsulfoxide) and [RuPcS] (PcS = sodium
tetra-sulpho-phthalocyaninate), and of potassium monopersulfate (Oxone) or
hydrogen peroxide.

MATERIALS AND METHODS

$[Ru(H_2O)_2(dmso)_4](BF_4)_2$ was prepared by published procedures (Bressan
et al., 1995) and [RuPcS] were prepared by template synthesis starting from
ruthenium(II) salts, sodium 4-sulfophthalate and urea, following the early
procedure of Weber and Busch (1965).

The reactions were carried out at 20°C in a 10 mL vial, by stirring
magnetically an aqueous solution of the substrate (5-20 mM) and the catalyst (0.1-
0.5 mM), to which suitable amounts of 30% hydrogen peroxide or oxone were
added, corresponding to 100-600 mN H_2O_2 or HSO_5^-, as determined by iodometric
titrations. In the experiments with the surfactant agent, 2.5%
cetyltrimethylammoniun hydrosulfate was added; the experiments in alkaline media
were carried out in the presence of the desired buffer (0.1 M solutions) or of added
0.1 M NaOH.

Carbon dioxide evolved during the reaction was captured by an aqueous
solution of $Ba(OH)_2$ 0.1 M, and aliquots from the surnatant solution were
backtitrated with HCl 0.1 N. Chloride ions produced upon dechlorination of the
substrates were analyzed by the mercury thiocyanate method. Organic analyses
were performed a HP 5890-II GLC instrument;. mass spectra were obtained on a
VG 16F spectrometer operating in the electron ionization mode at 70 eV; NMR
spectra were measured on a Bruker Avance 300 MHz spectrometer.

RESULTS AND DISCUSSION

Trichloroethylene is affected by the conventional RuO_4-NaClO system in water-CCl_4 double phase, with ca. 50% of the substrate being oxidized within hours, whereas replacement of hypochlorite by monopersulfate does not improve reaction rates. We also tried hydrogen peroxide (1÷30% aqueous solutions), but in the presence of the ruthenium catalysts a very rapid dismutation took place with little oxidation of the substrates. Other oxidizing agents, such as iodosobenzene, magnesium peroxophtalate and potassium peroxodisulfate, were inactive. When the reaction is conducted in the water phase only and in the presence of a surfactant agent, which effectively 'dissolves' the substrate, a dramatic increasing in the oxidation rate is observed in the presence of the soluble ruthenium(II) complex $[Ru(H_2O)_2(dmso)_4](BF_4)_2$, with complete disappearance of the substrate within minutes and formation of chloride ions (>85%), together with major amounts of CO_2 and HCOOH and minor amounts of $CHCl_3$. Although oxidation also proceeds at lower rate in the absence of catalyst, it should be noted that: the presence of catalyst critically enhances the oxidation of the more recalcitrant substrates, like tetrachloroethylene, which is quantitatively converted into to CO_2 and HCl. 1,2-cis-dichloroethylene and other chlorinated olefins, are also effectively degraded to HCl and CO_2 (and other organic products, like HCOOH). No evidence of CO is found. It is likely that oxidation of chloro-olefins proceed via an initial oxidative cleavage of the double bond, often accompanied by epoxidation and/or ketonizeation as minor pathways, and followed by further oxidation and/or hydrolysis of the formed oxygenated derivatives.

The above catalytic system is also able to effectively degrade various polychlorophenols. The amount of inorganic chlorine produced indicates extensive dechlorination, whereas only 50% of carbon was transformed into CO_2: analyses indicate the presence of large quantities of a wide variety of water-soluble organic products, among which the para-dihydroxylate and the corresponding chloroquinone derivatives and various isomers of the product of radical coupling.

TABLE 1. Oxidation of chloroalkenes and chlorophenols by monopersulfate[a]

Compound	time[b]	Cl[c]	CO_2[c]
2,6-dichlorophenol[d]	90 min	90%	60%
2,4,6-trichlorophenol[d]	30 min	80%	50%
pentachlorophenol[d]	5 h	80%	45%
1,2-dichloroethylene[e]	1 min	85%	45%
trichloroethylene[e]	1 min	85%	40%
tetrachloroethylene[e]	6 h	100%	100%

[a]$[Ru(H_2O)_2(dmso)_4](BF_4)_2$ catalyst in water and 2.5% cetyltrimethylammoinum hydrosulfate; 20°C. [b]Time for complete disappearance of substrate. [c]Determined after 6h reaction and calculated on the basis of the amount of substrate disappeared. [d]Catalyst 0.5 mM; molar ratio catalyst : substrate : $KHSO_5$ = 1 : 10 : 500. [e]Catalyst 0.1 mM; molar ratio catalyst : substrate : $KHSO_5$ = 1 : 100 : 500.

In the presence of the simple $[Ru(H_2O)_2(dmso)_4](BF_4)_4$ or $RuCl_3$ catalysts, hydrogen peroxide undergoes extensive dismutation, by being the substrate recovered practically unchanged. Replacement of the above 'naked' ruthenium ion by the water-soluble RuPcS, where a ruthenium(II) metal ion is strongly complexed by a tetrasulfophthalocyaninate macrocycle (likely with two water molecules in apical positions) results in a definite improvement of the reaction course:

dismutation is significantly lowered, thus making possible an effective oxidation of the substrates (Table 2).

TABLE 2. Oxidation of 2,4-dichlorophenol by hydrogen peroxide[a]

Catalyst	Cl^{-} [c]	CO_2 [c]	H_2O_2 [d]
$[Ru(H_2O)_2(dmso)_4](BF_4)_2$	15%	8%	90%
	(=)	(nd)	(100%)
RuPcS	20%	9%	20%
	(20%)	(nd)	(65%)

[a]Catalyst 0.1 mM; molar ratio catalyst : substrate : H_2O_2 = 1 : 250 : 3500; in water at pH:5-6 (within parentheses, pH: 10-11); 20°C. [b]Time for complete disappearance of substrate. [c]Determined after 24h reaction and calculated on the basis of the amount of substrate disappeared. [d]Consumption of the oxidant.

Oxidation of various chlorophenols was conducted in water solution and resulted to be highly sensitive to the pH values. Experiments at the natural, slightly acidic pH values (pH 5-6), show a fast and irreversible degradation of the catalyst (the color of the reaction mixtures changes sharply from green to light yellow), apparently due to the oxidative action of hydrogen peroxide: in these conditions, significant but slow dismutation of the oxidant takes place with reduced, even if still significant, degradation of the substrate (see Table 3). Addition of acetonitrile, to enhance the solubility both of the substrates and of the reaction intermediates, does not improve the reaction course.

The catalyst is fairly stable in strongly alkaline conditions (up to pH 13), but in these conditions oxidation is strongly inhibited, likely because of extensive and fast dismutation of the oxidant, whose concentration approaches zero within a few hours. Indeed, intermediate pH conditions, attained by conducting the reactions in buffered solutions, definitely favor the oxidative dechlorination of the substrates, while in the same time limiting to a reasonable level the dismutation of hydrogen peroxide. For example, in phosphate buffer at pH 7.6, almost 50% of 2,4,6-trichlorophenol is dechlorinated within 24 h at ambient conditions and in the presence of a 2-3-fold excess of hydrogen peroxide, as calculated on the equivalents required for the complete mineralization of the substrate to CO_2 and HCl.

TABLE 3. Oxidation of trichlorophenol by hydrogen peroxide[a]

Solvent	Cl^{-} [b]	CO_2 [b]
Water:acetonitrile (2:1 vol) (	4%	5%
Water[c]	15%	(nd)
Water (pH 7.6)[d]	45%	26%
Water (pH 9.5)[e]	17%	13%
Water (pH 9.5)[f]	23%	(nd)
Aqueous NaOH 0.1 M[g]	13%	4%

[a][RuPcS] catalyst in water, 0.1-0.2 mM; molar ratio catalyst : substrate : H_2O_2 = 1 : 100 : 1500; 20°C. [b]Determined after 24h reaction and calculated on the basis of the amount of substrate disappeared. [c]Saturated solutions (ca 8 mM): slightly acidic (pH 5-6). [d]Phosphate buffer. [e]Na_2HPO_4 solution. [f]$Na_4B_4O_7$ solution. [g]pH progressively decreases approaching neutrality.

CONCLUSION

The results indicate that a catalytic system comprised of simple ruthenium derivatives and aqueous monopersulfate (Oxone) is effective for the defunctionalization of polychlorophenols and polychloroolefins. Polar media, among which water itself, markedly increase the oxidation rates. A definite improvement is achieved by using complexed ruthenium(II) derivative, like RuPcS, which promote the same oxidations with the more valuable hydrogen peroxide and at very low concentrations (ca 1%). Practical applications of the systems described in this paper may be envisaged for the degradation of chlorinated organics dissolved in small amounts in groundwaters.

ACKNOWLEDGMENTS: The authors acknowledge the Ministry of University of Italy (MURST, 40% fundings) and the Regione Abruzzo, Italy (ARSSA-POM fundings) for financial support.

REFERENCES

Bailey, A.J., W.P. Griffith, S.I. Mostafa, and P.A. Sherwood. 1993. "Studies on transition-metal oxo and nitrido complexes. 13. Perruthenate and ruthenate anions as catalytic organic oxidants." *Inorg. Chem. 32* (3): 268-271 and references therein.

Bressan, M., L. Forti, and A. Morvillo. 1994. "Effective ruthenium-catalyzed oxidation of chlorinated olefins by monopersulfate in aqueous medium." *J. Chem. Soc. Chem. Commun.* : 253-254.

Bressan, M., L. Forti, and A. Morvillo. 1995. "Oxidation of nitrobenzene, chlorobenzene and chlorophenols using liquid-phase ruthenium catalysts." *New J. Chem. 19* (8-9): 951-957.

Labat, G., J.-L. Seris, and B. Meunier. 1990. "Oxidative degradation of aromatic pollutants by chemical models of ligninase based on porphyrin complexes." *Angew. Chem. Int. Ed. Engl. 29* (12): 1471-1473.

Sheldon, R.A. 1997. "Catalysis: the key to waste minimization." *J. Chem. Tech. Biotechnol. 68* (2): 381-388.

Sorokin, A., J.-L. Seris, and B. Meunier. 1995. "Efficient oxidative dechlorination and aromatic ring cleavage of chlorinated phenols catalyzed by iron sulphophthalocyanine." *Science 268* :1163-1166.

Sorokin, A., S. De Suzzoni-Dézard, D. Poullain, J.-P. Noël and B. Meunier. 1996. "CO_2 as the ultimate degradation product in the H_2O_2 oxidation of 2,4,6-trichlorophenol catalyzed by iron tetrasulphophthalocyanine." *J. Am. Chem. Soc. 118* (31): 7410-7411.

Sorokin, A., and B. Meunier. 1996. "Oxidative degradation of polychlorinated phenols catalyzed by metallosulfophthalocyanines." *Chem. Eur. J. 2* (10): 1308-1317 and references therein.

Weber, J.H., and D.H. Busch. 1965. "Complexes derived from strong field ligands. XIX. Magnetic properties of transion metal derivatives of 4, 4', 4'', 4'''-tetrasulfophthalocyanine." *Inorg. Chem. 4* (4): 469-475.

FULL-SCALE SOIL REMEDIATION OF CHLORINATED SOLVENTS IN CLAY SOILS BY *IN SITU* CHEMICAL OXIDATION

Richard Levin, Edward Kellar (QST Environmental Inc., Gainesville, Florida)
James Wilson (Geo-Cleanse International, Inc., Kenilworth, New Jersey)
Leslie Ware [Anniston Army Depot (SIOAN-RK), Anniston, Alabama]
Joseph Findley, John Baehr (U.S. Army Corps of Engineers, Mobile, Alabama)

ABSTRACT: In July 1997, the first large scale use of the patented Geo-Cleanse process for removal of DNAPL chlorinated solvents and hydrocarbons was begun over an area of approximately 2 acres at a former clay backfilled lagoon site on Anniston Army Depot (ANAD), Anniston, Alabama.

This rapid *in situ* technology is being used to reduce volatile organic compounds to approved risk based cleanup levels on over 43,125 cubic yards (yd^3) of soils containing up to 31-percent trichloroethene, methylene chloride, daughter products, and BTEX. Soil injection extends from 6 feet (ft) to at least 25 ft below grade. Groundwater pockets of high strength DNAPL in residual sludge layers and associated clay smear zone are being eliminated selectively from 29 to 71 ft (top of bedrock) to prevent overlying soil recontamination.

This *in situ* chemical oxidation relies on the well documented Fenton's Chemistry for creation of hydroxyl radicals from hydrogen peroxide to non-specifically "mineralize" solvents, oil and grease, and naturally occurring organics to water, carbon dioxide, oxygen, and inorganic salts through an aggressive exothermic reaction. Treatment proceeds from shallow to deep zones through an array of patented steel injectors screened at specifically designed depths to treat organics present in both saturated and unsaturated zones. Injection quantities are based on pollutant mass present.

The full-scale 120-day injection of up to 109,000 gallons (gal) of peroxide into 255 injectors is complete, and over 99 percent of the area has been initially treated. Soil concentrations of up to 1,760,000 micrograms per kilogram (mg/kg) have been reduced to below detection. Operating data to date indicate no adverse migration of organics to surrounding soils or groundwater. Project pilot test data, remedial design and operations summaries, *in situ* monitoring, and pre- and post-treatment results are presented.

BACKGROUND

The U.S. Environmental Protection Agency (EPA) has designated 44 sites at Anniston Army Depot (ANAD), Anniston, Alabama, as Resource Conservation and Recovery Act (RCRA) Solid Waste Management Units (SWMUs). The former Facility 414 Old Lagoons have been designated as SWMU12.

SWMU12, consisted of three unlined industrial waste lagoons that cover an area of approximately 470 by 300 ft. Industrial wastes placed into the lagoons included: abrasive dust waste containing cadmium and lead, petroleum hydrocarbons, solvents, including trichloroethene (TCE) degreasers and methylene chloride paint stripper sludges, alkaline cleaners, grease trap wastes, and Industrial Wastewater Treatment Plant (IWTP) sludge.

During closure of the lagoons in 1978, the majority of sludge were removed from the lagoons and local clays were used as fill; however, the closure was incomplete and a significant amount of solvent and waste oil remained in the site soils and aquifer matrix, particularly in the middle lagoon. Subsequent RI/FS activities led to a decision to perform the emergency removal of 2,000 yd³ of contaminated soils from three isolated areas.

Additional delineation was conducted increasing the area of treatment needed. The majority of VOCs were found at depths of 8 ft and greater. High groundwater levels have been encountered at levels of 14 feet below land surface (ft-bls). TCE accounts for approximately 85 percent of the 71,000-pounds VOC total. The total surface area of impacted soil is approximately 45,000 square feet (ft²) (220 ft x 120 ft and 120 ft x 120 ft). The total affected volume (exceeding SSLs for VOCs) was increased from the original 2,000 yd³ to 7,200 yd³ above the mean high water table (14 ft-bls) and 6,500 yd³ below groundwater. The highest concentrations of TCE, 20,100 mg/kg, occurred in samples taken from the 10-ft depth interval. However, concentrations greater than 9,000 mg/kg TCE at depths down to 24 ft.

PURPOSE AND OBJECTIVES
The emergency removal (ER) objective at SWMU12 was to use *in situ* chemical oxidation on soils to treat (remove) waste chemical constituents, volatile organic compounds (VOC), that may be contributing significantly to exceedences of health-based concentration limits in onpost soils and area groundwater. Specific objectives of the removal included:
- Destruction of organics (VOCs) in soils contributing to groundwater contamination,
- Containment of chemical constituents within the zone of treatment until destruction was complete,
- Prevention of mobilization of inorganics, and
- Assurance that treatment was complete by monitoring groundwater and soils chemical concentrations.

Based on a detailed site assessment performed according to current EPA Soil Screening Levels (SSL) Screening Guidance using all of the newly developed relevant site data, the site specific approved soil SSLs, as soil cleanup criteria were developed. These were calculated to be protective of groundwater at the Depot boundary.
- Trichloroethene—41 mg/kg,   • Methylene Chloride—63 mg/kg, and
- 1,2-dichloroethene—51 mg/kg,   • Tetrachloroethene — 5.2 mg/kg.
- Vinyl Chloride—23 mg/kg,

Following extensive site characterization, risk assessment, and a pilot demonstration program, full-scale soil remediation injection is proceeding under an Emergency Removal Plan.

METHODOLOGY
Process Description. *In situ* chemical oxidation of organic contaminants is achieved by injection of hydrogen peroxide and a catalyst formulation into the affected media under carefully controlled conditions. The Geo-Cleanse® Process used, is a patented technology in which hydrogen peroxide and trace quantities of metallic salts are injected into the impacted media. This *in situ* oxidation system is capable of rapid complete, non-selective oxidation of organic compounds in soil and groundwater. The basic reaction in the Geo-Cleanse® Process is simplified below:

| Hydrogen Peroxide | + | Organic Contaminant | ----> | Carbon Dioxide | + | Water |

The Geo-Cleanse® Process delivers a calculated charge of hydrogen peroxide and catalyst to the contaminated region via a patented injection methodology and equipment. This process maximizes the dispersion and diffusion of the reagent through the soil and/or the affected aquifer. The patented injectors are specially designed to withstand the elevated temperatures and pressures resulting from Fenton's reaction, while achieving maximum dispersion of the reagents through the subsurface.

The injection of hydrogen peroxide and the catalytic system results in an exothermic subsurface reaction that generates heat, pressure, oxygen, and carbon dioxide. During the reaction sequence, the organic compounds are successively converted to shorter chain mono- and di carboxylic (fatty) acids. These compounds are non-hazardous, naturally occurring substances, and are further degraded into carbon dioxide, chlorides, and water by subsequent reactions.

The actual oxidation is driven by formation of a free hydroxyl radical via Fenton's reaction chemistry. This methodology for the treatment of organic compounds in wastewater has been widely studied, utilized, and proven effective by the wastewater industry.

The preferred Fenton's Reaction is:

$$H_2O_2 \quad + \quad Fe^{2+} \quad ----> \quad OH. \quad + \quad OH^- \quad + \quad Fe^{3+}$$

hydrogen	ferrous	hydroxyl	hydroxyl	ferric
peroxide	iron	radical	ion	iron

The hydroxyl free radical (OH.) is an extremely powerful oxidizer organic compounds. Residual hydrogen peroxide, due to its unstable characteristics, rapidly decomposes to water and oxygen in the subsurface environment. Soluble iron amendments added to the subsurface during the Geo-Cleanse® Process in trace quantities are precipitated out during conversion to ferric iron.

Within SWMU12, the organic-contaminated waste was treated *in situ* to achieve health-based remedial action goals protective of onsite groundwater as a drinking water source. Treatment consisted of *in situ* organic destruction accomplished by a controlled sequence of catalyzed peroxide injections into an engineered array of approximately 250 injectors. Chemical oxidation of the contaminated soils proceeded over a 5-month period (120 injection days) and entailed the closely monitoring of the establishment of Fenton's chemistry conditions during injection of an estimated 109,000 gal of 50-percent hydrogen peroxide.

Injection System Design. Three types of injectors were used targeting three distinct depth intervals. Single shallow injectors (screened from 8 to 14 ft) were installed in areas where contamination is above 15 ft. The top of the injection interval for the shallow injectors was established to provide a minimum hydraulic confinement to induce acceptable horizontal migration of the reagents and minimizing the potential of reagents migrating to the surface. Cluster injectors consisting of a shallow injector paired in the same borehole with a deep soil injector (screened from 20 to 26 ft) were constructed in areas where contamination was found at both deep and shallow depths. Single intermediate injectors were placed in areas where contamination existed between 14 and 20 ft. A total of 255 injectors were installed.

Shallow and cluster injectors were installed approximately 20 ft apart. Intermediate injectors were placed between the shallow injectors approximately 20 ft from each other and 10 ft from the nearest shallow or cluster injector.

In addition, 25 deep groundwater injectors screened in the saturated zone were installed as monitor wells during chemical oxidation. The groundwater monitor wells (deep injectors) were located throughout the contaminated area approximately 60 ft apart. These injector wells were used to detect effects on the groundwater from the *in situ* treatment. A line of groundwater monitor wells were located on a line paralleling the eastern margin of the SWMU to detect any downgradient effect of the treatment immediately outside the SWMU boundary. Finally, one monitor well was located within the SWMU, but upgradient of the contaminated area. All monitor wells were used regularly and frequently to define the potentiometric gradient at the site to monitor injection conditions.

Each of the deep groundwater injector wells were constructed in the uppermost water-bearing strata. In most locations, this strata occurs in the weathered zone at the top of the bedrock or within a few feet of the top of the rock. The wells penetrated a minimum of 5 ft into the bedrock or approximately 50 ft.

PROGRESS AND RESULTS

Injection Control and *In Situ* Monitoring. Catalyst solution (ferrous sulfate) and sulfuric acid (for pH control) were injected to prepare the soils for peroxide injection. Surrounding wells were monitored for the following parameters to determine readiness for peroxide:
- pH (maintain near 5.5),   • Fe^{3+} (catalyst ion availability).
- Conductivity, and

Reaction of peroxide with ferrous sulfate and organics yields oxygen, carbon dioxide, chlorides, and water as products. After commencement of peroxide injection, the following parameters were monitored in the vent well off-gas to determine the progress of the reaction:
- O_2 (reaction product),   • H_2O_2 (excess reactant).
- CO_2 (reaction product), and

Injection proceeded at a rate (0.25 gal per minute) and pressure (1 - 5 psi) that maximizes the radius of influence of the reaction without causing preferential flow or undue venting. The injection rate and pressure was recorded continuously to ensure accurate records of peroxide volume at each injector. Hourly reading of carbon dioxide and breathing zone volatiles using an OVM were monitored from the surrounding injectors to monitor the progress of the reaction. If during monitoring it is determined that there is a preferential flow direction of peroxide reaction, surrounding injectors were capped to prevent the release of pressure, causing the peroxide to disperse in another direction.

As the rate of reaction with available organics peaks, carbon dioxide, and oxygen levels in the off-gas also peaked. During monitoring, carbon dioxide concentrations peaked at 10 to 14 percent and oxygen levels peaked at approximately 40 percent. As the available organics was consumed, the concentration of carbon dioxide and oxygen in the off-gas began to decline and the concentration of excess peroxide increased.

Each sub-array of injectors receiving peroxide at one time were treated with the minimum target chemical quantities and/or the *in-situ* monitoring indicates adequate endpoints for the operational parameters (oxygen, carbon dioxide, and hydrogen peroxide) in the involved and neighboring vents and injectors. Following treatment of a sub-array and pre-treatment catalyst adjustment of the subsurface environment, peroxide connections was moved to the succeeding sub-array position.

Monitoring Groundwater. In addition to carbon dioxide monitoring, the monitor wells located in SWMU12 were monitored daily for peroxide, pH, iron, conductivity, temperature, and chlorides. The monitoring data was collected and tabulated to determine effects on the groundwater from soil treatment.

Vent Flow Balance. A vent flow balance (VFB) system was installed to aid in maintaining an effective radial dispersion of catalyst and peroxide. During catalyst and peroxide injection, air lines were connected to up to twelve injector/monitors adjacent to the injectors being used. By monitoring the pressures and carbon dioxide measurements of the adjacent injectors, it could be determined if the injected fluids are dispersing into the soils radially. Ideally, the injection should disperse from the injector radially up to approximately 15 to 20 ft from the injector. If monitoring indicates that there was a preferential direction of flow, the VFB system was used to effectively disperse the peroxide in a radial direction by creating a slightly negative pressure differences near the injector.

Post-Treatment Sampling and Confirmation. To determine the effectiveness of the oxidation process, post-treatment sampling commenced while the full-scale treatment program was active. The injection program started in the far west portion of the middle lagoon and proceed to the east. Once the injection program reached the "O" line on the injection grid, post-treatment sampling began.

The locations of the post-treatment samples were chosen to ensure that the edges of the contaminated area remain clean and that the injection treated soils were below the SSLs. By beginning the post-treatment sampling during treatment, site condition changes could be effected based on sampling results. Soil samples were collected using a Geoprobe®. Samples will be collected at intervals corresponding to the intervals with the highest concentrations in the nearest delineation boring and/or injector locations. Samples were sent to QST's laboratory in Gainesville and analyzed for VOCs using Method 8010/8020. When the sample analyses show that the soil concentrations are below the SSLs, the area was deemed clean (Table 1). The injection has reduced the total VOC concentration by up to <90 percent from original concentrations. Confirmatory samples were taken at selected location and sent to the Gainesville laboratory for complete VOC and metals analysis. If the screening samples showed that contamination concentrations remain above SSLs, the location was re-treated with peroxide for a polishing treatment. After polishing treatment, another set of screening samples were taken until the analyses showed concentrations of VOCs below SSLs.

CONCLUSIONS

In situ chemical oxidation using hydrogen peroxide was proven to be effective in reducing the contaminant concentrations in clays and groundwater at the Anniston Army Depot to below the SSLs. The use of in-situ chemical oxidation at ANAD was the first large scale use of hydrogen peroxide to treat chlorinated solvents. The objectives of the project were met by destruction of the organics (VOCs) in soils which were contributing to groundwater contamination, being able to contain the chemical constituents within the zone of treatment until destruction was complete, prevented the mobilization of inorganics, and documented that treatment was complete by monitoring groundwater and soils chemical concentra-tions. The cost data is being evaluated for the project, however, it appears that the

Table 1. Soil Analysis (Pre and Initial Post Sampling) - Trichloroethene (TCE)

Boring	Depth (ft)	Pre-Sampling Sampling (μg/kg)	Post Sampling sampling (μg/kg)	Percent Reduction	Actual Injected Volume (gal)	Target Volume Gal
E60	14	62100	<1000	>99	303	200
E48	16	22700	<1000	>99	220	250
E44	10	38900	<1000	>99	80	200
G40	14	51050	<1000	>99	220	200
I60	8	258000	90000*	65		
	10	60600	5000	92	518	200
I52	8	1760000	<1000	>99	511	250
	10	36500	<1000	>99		200
K56	8	152000	<1000	>99	276	200
M60	12	164000	<1000	>99	402	200
M52	8	85100	<1000	>99	160	200
O40	12	198000	370	>99	494	200
Q60	8	480000	42800*	91	244	200
	10	370000	25500	93		
Q52	4	190000	2270	99		
	8	12300000	563000*	95	638	500
	14	226000	25000	89		
S48	10	3270000	174000*	95		
	14	5780000	64600*	99	515	500
	16	2170000	250000*	88		
	18	1198500	329000*	73		
	22	486000	6900	99	371	350
	26	827000	212000*	74		
U20	8	10100000	108000*	99	376	1500
U52	4	484000	44000*	91	502	500
U48	18	263000	21500	92	410	350
W20	8	5410000	17900	>99	139	200
	10	2040000	94300*	95		
W24	16	2960000	169000*	94	460	1100
W28	10	15900000	596000*	96	357	3400

Source: QST   SSL TCE 41,000 μg/kg

* Polishing Treatment Required.

cost of the *in situ* chemical oxidation treatment is one-fourth the cost of excavation and disposal of the contaminated soils.

To date there has been no evidence of adverse pollutant mobility produced by the process. Conversely, adequate dispersion in the tight clays, while difficult in several locations, has proceeded a reasonably productive rates and low pressures. The target volumes of peroxide have been adequate to treat the organics except where localized dispersion is affected. Approximately 10 to 20 percent of the injectors locations needed polishing through either existing injectors or the construction of additional injectors. The process has proven to capability to reduce soil concentrations to below detection limits. The development of the treatment sequence and site specific injection technique is continuing.

PHOTOCATALYTIC DEGRADATION OF ORGANICS IN AQUEOUS SUSPENSIONS

Ajay K. Ray and Dingwang Chen

National University of Singapore, 10 Kent Ridge Crescent, Singapore 119260

ABSTRACT: Photocatalytic degradation of phenol, 4-chlorophenol and 4-nitrophenol catalyzed by Degussa P25 TiO_2 in aqueous suspension has been studied in laboratory scale. Experiments were conducted to investigate the effects of parameters such as UV light intensity, catalyst dosage, pollutant concentration, temperature, and partial pressure of oxygen in suspensions. Simple model for predicting the optimal catalyst dosage for different photosystems was proposed. Pseudo first-order kinetics with respect to all the parent compounds was observed. Experimental data obtained under different conditions were fitted with mathematical equation to describe the dependency of degradation rate as a function of all the above mentioned parameters. Adsorptive properties of all the organics were also experimentally measured and fitted with Langmuir equation. The extremely low surface coverage of the organics on the catalyst may be one of the main factors that result in the low efficiency of the photocatalytic process. Mass transfer of organics and oxygen in the TiO_2 suspensions has also been discussed.

INTRODUCTION

Heterogeneous photocatalysis is one of the AOPs and has proven to be a promising method for the elimination of toxic and bioresistant organic and inorganic compounds from wastewater by transforming them into innocuous species (Ollis and Al-Ekabi, 1993). Many examples of complete photooxidation of organic compounds have been reported. Matthews studied the photocatalytic degradation of 22 organics both in TiO_2 aqueous suspensions (1990) and over immobilized TiO_2 thin films (1988). Pruden and Ollis (1983) carried out the photomineralization of halogenated organic pollutants in TiO_2 suspensions. Langmuir-Hinshelwood model was extensively used to describe the degradation rate expressions. It has been demonstrated catalyst dosage, initial concentration of pollutants, UV light intensity, dissolved oxygen, temperature and pH of the solution are the main parameters that affect the degradation rate. However, no kinetic expression that correlates above parameters has been reported. Optimal TiO_2 catalyst dosage in aqueous suspensions reported had a wide range from 0.5 to 10 g/L for different photosystems and reactor configurations, and no quantitative explanation was reported. Most commonly proposed mechanisms were based on the reaction between the adsorbed species, however, few adsorptive results of organic compounds on the TiO_2 catalyst have been reported, which may be helpful to elucidate the process. In this paper, the photodegradation of phenol, 4-chlorophenol (4-CP) and 4-nitrophenol (4-NP) in P25 TiO_2 aqueous suspensions has been investigated. A degradation kinetic expression describing the influences of the main operational parameters was obtained and can satisfactorily predict the photodegradation behaviour under different conditions. A simple model that can predict the optimal catalyst dosage for different systems was proposed and has been verified by the photodegradation of phenol, 4-CP and 4-NP. In order to elucidate the process, the adsorptive experiments of above compounds on TiO_2 have been studied. Besides, the mass transfer of organics and O_2 in TiO_2 aqueous suspensions has also been discussed quantitatively although it is not the governing step.

EXPERIMENTAL DETAILS

Materials. Degussa P25 catalyst provided by Degussa Company (Germany) was used throughout this work without further modification. 4-NP (98+%) was obtained from BDH chemicals, phenol (99.5+%) from Merck chemicals, 4-CP (99+%) from Fluka chemicals, acetonitrile (for HPLC) from Fisher. All chemicals were used as received. Water used to make up solutions was Milli-Q water.

Apparatus and Analyses. The batch photoreactor used in the present study is the same as mentioned elsewhere (Chen and Ray, 1998). A Shimadzu TOC-5000A analyzer with an ASI-5000 autosampler was used to analyze the TOC in samples. Analyses of 4-NP, 4-CP and phenol were carried out by using a Shimadzu UV 3101 PC spectrophotometer and HPLC (Perkin Elmer). Mobile phase for HPLC was acetontrile (60%) and ultrapure water (40%) at the flowrate of 1.5 ml/min, the column was C-18 (Chrompack). The pH value and chloride ion concentration of the reaction solution were measured by the Cyberscan 2000 pH meter and Orion 720A ion meter, respectively.

RESULTS AND DISCUSSIONS

Photodegradation of Organics. In this paper degradation of 4-CP, 4-NP and phenol over illuminated P25 TiO_2 was investigated and the results demonstrated that all compounds could be completely degraded into CO_2, water and corresponding mineral acids. Figure 1 illustrates the degradation course of 4-CP, formation and subsequent degradation of intermediates, chloride ion concentration and pH variation during the photocatalyzed reaction. The decrease of pH value in solution resulted from the formation of HCl during the degradation, which was reflected by the increase of the chloride ion concentration. In literature up to six intermediates have been detected during the degradation of 4-CP in TiO_2 aqueous slurry, and all of which could be degraded further to final products CO_2 and HCl. In the present work, the total concentration of intermediates measured during the reaction was quite low, therefore no effort has been made to identify them. But we can conclude that there is almost no Cl⁻ group in intermediates because the measured chloride ion concentration can be well balanced by the decrease of 4-CP during the reaction. According to the manufacturer's report, the P25 TiO_2 used in this study contains a little bit HCl, this is why the initial chloride ion concentration is not zero (ca. 2 mg/L).

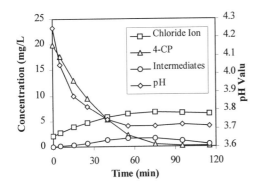

FIGURE 1. Typical diagram of degradation of 4-CP in P25 TiO₂ suspensions

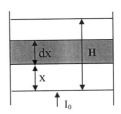

FIGURE 2. Schematic diagram of light absorption in reactor

Catalyst Dosage. In slurry system catalyst dosage is an important parameter that has been extensively studied. The optimal catalyst dosage reported in literature was in a wide range from 0.5 to 10 g/L for different photocatalyzed systems and photoreactors, and so far no explanation has been made quantitatively.

Figure 2 is the schematic diagram of the photoreactor used in this study. Because the photodegradation takes place on the surface of catalyst, the degradation rate in a differential volume is proportional to the illuminated catalyst surface and can be written as:

$$dr_i = k_1 f(C_i) I_x^\beta A C_{cat} dx \tag{1}$$

and

$$I_x = I_0 \exp(-\varepsilon C_{cat} x) \tag{2}$$

where C_i is substrate concentrate in bulk solution, I_0 is incident light intensity, A is illuminated area of the reactor window, C_{cat} is the catalyst concentration, ε is light absorption coefficient of catalyst in suspensions and k_1 is a proportional constant. Substituting Eq. (2) into Eq. (1) then integrating Eq. (1) at given C_i and I_0 yields:

$$r_i = K[1 - \exp(-\varepsilon \beta C_{cat} H)] \tag{3}$$

where

$$K = k_1 A f(C_i) I_0^\beta \tag{4}$$

The β values for different photocatalyzed systems in above equations can be determined experimentally by carrying out the experiments at different light intensities and herein are listed in Table 1. Eq. (3) can well correlate the degradation rates obtained under different catalyst dosages from 0.2 to 4 g/L for all the three photocatalyzed systems. The ε values for the three systems obtained from Eq.(3) by best fitting experimental data were very close (see in Table 1).

This result supports the reliability of the above simple model because the ε value should be same for the same catalyst in dilute solutions of different organics.

Eq. (3) can also successfully explain the dependence of degradation rate on the TiO_2 dosage at low and high light intensities. Degradation rate is usually proportional to $I_0^{0.5}$ and $I_0^{1.0}$ at high and low intensities, respectively. Thus, according to Eq. (3) we can make the conclusion that the optimal catalyst dosage at high light intensity is higher than that at low light intensity and this is in accord with the results obtained by Okamoto et al. (1985).

Degradation Kinetics. Besides the catalyst dosage, effects of other kinetic factors on the photocatalytic degradation including initial concentration of pollutants, UV light intensity, dissolved oxygen and temperature have also been carefully investigated in this work. Based on the derivation made in the previous paper (Chen and Ray, 1998), the photocatalytic degradation kinetics for 4-NP, 4-CP and phenol have the following expression:

$$r_s = -\frac{dn_s}{Sdt} = k_o \exp(-\frac{E}{RT})I_a^\beta \frac{K_{O2}p_{O2}}{(1+K_{O2}p_{O2})} \frac{K_s C_s}{1+K_s C_{s0}} \tag{5}$$

In above equation, the photodegradation rate (mol/m²s) is defined as the mole (n_s) reduction of pollutant per irradiated reactor window area (S). All experiments were performed at natural pH value and constant TiO_2 dosage of 2 g/L, while other parameters were varied as follows: T = 288-323 K, I = 1.5-24 mW/cm², C_{S0} = 0.07-0.9 mM, p_{O2} = 0.05-1.0 atm. Kinetic parameters in Eq. (5) were determined by best fitting above equation with the experimental data obtained under different conditions, and are listed in the following Table.

Table 1. Kinetic parameters in Eq. 5 and light absorption coefficient of TiO_2

Organics	k_0 (mol/m²/s)	E (kJ/mol)	β	K_{O2} (1/atm)	K_S (1/Mm)	ε (L/g/cm)
4-NP	1.87×10^{-3}	7.42	0.84	9.98	10.43	1.58
4-CP	3.79×10^{-2}	13.72	0.72	17.64	6.19	1.71
Phenol	1.23×10^{-2}	11.80	0.82	12.71	4.26	1.76

Adsorptive Characteristics on P25 TiO_2. The key steps of the photocatalytic degradation are the reactions between adsorbed species, therefore, the photocatalytic process should strongly depend on the adsorptive properties of the organic compounds. In this study the adsorption experiments of 4-CP, 4-NP, and phenol have been investigated in TiO_2 suspensions at natural pH value and room temperature. Results demonstrated that Langmuir model could well describe the adsorption processes of above four compounds and the adsorptive parameters are listed in Table 2. Although the K_S in Eq. (5) is generally taken as the equilibrium adsorptive constants of organics on TiO_2, however, the results clearly showed that the adsorption constants obtained from kinetic experiments are discrepant with those obtained from dark adsorption for all 4-NP, 4-CP and phenol by comparing column 6 in Table 1 with column 3 in Table 2.

Table 2. Adsorptive parameters and the ratio of C_S to C_b

Compound	Q_{max} (μmol/g)	K_S (1/mM)	θ_{max} (%)	θ (%)	$(C_S/C_b)_{min}$
4-NP	5.94	1.06	1.7	0.30	0.990
4-CP	4.20	2.83	1.2	0.45	0.989
Phenol	3.28	0.82	0.95	0.14	0.989
Oxygen					0.992

Assuming that a complete monolayer of organic compound is formed on the surface of TiO_2, the maximum surface coverages based on Q_{max} of above compounds can be calculated by the following equation and are listed in column 4 of Table 2.

$$\theta = \frac{Q_{max} N_A \sigma_0}{A_{SP}} \times 100\% \tag{6}$$

where N_A is Avogadro's number, σ_0 is the average area of organic molecule. Obviously, θ values are extremely low and it indicates that most of the potential sites of the catalyst are not accessible to these species. This may result from the presence of high concentration of water. Alternatively, it also may be that many H_2O molecules attach to the organics when they absorb onto the catalyst. Under the experimental condition where the concentration of organics is around 0.2 mM, the surface coverages are even much low (listed in column 5). The reaction between hydroxyl radical and absorbed organics is the main step of the photocatalytic process, and the degradation rate is proportional to the surface coverage. Therefore, we can conclude that the extremely low coverage of organics on TiO_2 catalyst is one of the main reasons that result in the low efficiency of the photocatalytic process.

Mass Transfer of Organics in TiO_2 Suspensions. P25 TiO_2 is a non-porous catalyst and its external mass transfer rate can be determined by the following equation:

$$r_m = 10^3 k_m (C_b - C_s) A_{SP}^m W \text{ (mmol/L/s)} \tag{7}$$

where the factor 10^3 is required to render the units consistent in the equation. C_b and C_s are the bulk and local concentration (mM) of organics, respectively. W is the catalyst dosage (g/L). A_{SP}^m is the modified specific surface area of the catalyst, because catalyst particle in aqueous solution was reported to be up to 5μm in diameter due to the aggregation of the elementary particles. Mass transfer coefficient for a motionless spherical particle in stagnant solution can be estimated by the following equation:

$$k_m^0 = \frac{D}{R} \tag{8}$$

where D and R are the diffusivity of organics in water and actual radius of catalyst particle. But in this study the solution was magnetically stirred, k_m in Eq. (7) should be larger than k_m^0. Assuming that at steady state the mass transfer rate is equal to the photodegradation rate that was determined by the kinetic experiments, the ratio of C_s to C_b can be calculated for all organics and oxygen and is listed in Table 2. The results indicate that the mass transfer resistance of organics and oxygen in TiO_2 suspensions can be neglected definitely.

CONCLUSIONS

Heterogeneous photocatalysis as a promising candidate for water purification has been extensively investigated. Catalyst dosage is one of the important kinetic factors. Eq. (3) can well predict the optimal catalyst dosage for different photosystems and guide the design of photoreactor configuration. Other operational parameters have also been carefully studied and their influences on the photocatalytic degradation can be correlated by Eq. (5) for 4-CP, 4-NP, and phenol in P25 TiO_2 suspensions. The kinetic parameters in Eq. (5) have been experimentally determined. The low efficiency of the photocatalytic process may result from the extremely low surface coverage of organics on the catalyst. The mass transfer resistance of organics and oxygen in TiO_2 suspensions is negligible.

REFERENCES

Ollis D. F., and H. Al-Ekabi (Eds.). 1993. *Photocatalytic purification and treatment of water and air.* Elsevier Science Publichers B. V. Amsterdam, Netherlands.

Matthews, R. W. 1990. "Purification of water with near-UV illuminated suspensions of titanium dioxide" *Water Research*, 24: 653-660.

Matthews, R. W. 1988. "Kinetics of Photocatalytic Oxidation of Organic Solutes over Titanium Dioxide." *J. Catal.* 111: 264-272.

Pruden, A. L., and D. F. Ollis. 1983. "Photoassisted Heterogeneous Catalysis: The Degradation of Trichloroethylene in Water." *J. Catal.* 82: 404-417.

Chen, D. W., and A. K. Ray. 1998. "Photodegradation kinetics of 4-nitrophenol in TiO_2 aqueous suspensions." In press, *Water Research.*

Okamoto K., and Y. Yamamoto, H. Tanaka, and A. Itaya. 1985. "Kinetics of heterogeneous photocatalytic decomposition of phenol over anatase TiO_2 powder." *Bull. Chem. Soc. Jpn.* 58: 2023-2028.

DEVELOPMENT OF TECHNOLOGY FOR ELECTROCHEMICAL REDUCTION OF CHLORINATED AROMATIC COMPOUNDS

Nigel J. Bunce, Simona G. Merica, Wojceich Jedral, Claudia Banceu, and Jacek Lipkowski (University of Guelph, Guelph, Ontario, Canada N1G 2W1)

ABSTRACT: The electrochemical reduction of hexachlorobenzene (HCB) has been studied in methanol and in micellar aqueous solutions using Triton-SP 175® which, unlike conventional surfactants, is acid-labile. Electrolysis at a wide range of cathodes leads to dechlorination to less chlorinated benzenes, in each case with a quantitative material balance. In the system 1 mM HCB/methanol/0.05 M tetraethylammonium chloride/Hg pool cathode, a maximum current efficiency of 60% was obtained when the potential was maintained at the first voltammetric reduction peak (-1.2 V *vs.* SHE). In a mainly aqueous medium of 0.1% Triton-SP 175, 1% heptane, and 98.9% water, amperostatic electrolysis of 30 μM HCB proceeded with current efficiency of 5% at a Pb cathode, using sodium sulfate as an environmentally friendly electrolyte.

INTRODUCTION

Highly chlorinated aromatic compounds owe their environmental persistence to their low chemical and biological reactivity. Besides being recalcitrant to conventional biological treatment, aqueous waste streams or leachates containing these compounds are frequently toxic to the microorganisms in a biological reactor. Electrochemical technologies show promise for treating these materials due to the relative simplicity of the equipment, environmental friendliness, and the possibility of high energy efficiency compared with thermal and photochemical processes, but the low chemical reactivity of chlorinated aromatic compounds is reflected in very negative potentials for reduction and very positive potentials for oxidation.

Electroreduction of polychlorobenzenes involves two one-electron transfers for each chlorine atom removed as chloride ion. A radical anion ($ArCl^{\cdot-}$) formed initially expels Cl^- to give an aryl radical ($Ar^{\cdot}$) that rapidly accepts a second electron to yield the reduced species Ar^-. Protonation completes the reduction to ArH.

The target molecule in this research is the pesticide intermediate hexachlorobenzene (HCB); our goal was to evaluate electrolytic methods for its removal from aqueous waste streams.

MATERIALS AND METHODS

Triton-SP175 was donated by Union Carbide Canada; other chemicals were purchased commercially (Aldrich or Fisher). For most of the experiments, we used a two compartment cell constructed of high density polypropylene, with interior channels 4 mm $\times$ 15 mm $\times$ 55 mm, and the anode and cathode separated at 0.2 cm distance by a Nafion 417 membrane in the K^+ form. Cathodes included a mercury pool, metal wire ($A = 30$ cm^2) or polished glassy carbon; the auxiliary electrode was made of Ti/IrO$_2$; for most experiments, the reference electrode was internal Ag/AgCl.

The cell was controlled by an EG&G PAR model 273 potentiostat/galvanostat interfaced to a PC and controlled using Head Start electrochemical software. Experiments were carried out in two modes: continuous flow electrolyses involved recirculating the catholyte, whereas in single flow experiments the catholyte was passed only once through the cell before being directed to waste. In all cases the anodic solution, consisting of solvent plus electrolyte, was recirculated. In methanol solution, the reaction products could be analyzed directly by HPLC using a C_{18} reverse phase column and methanol-water as the mobile phase. In the Triton SP-175 microemulsion, the products were obtained by breaking the emulsion through hydrolysis at pH<2 and extraction into heptane.

RESULTS AND DISCUSSION

Differential pulse voltammetry was carried out in methanol solution containing 0.05 M tetraethylammonium chloride (TEACl) as supporting electrolyte, using a sessile drop Hg cathode, at which HCB showed four reduction peaks (Figure 1).

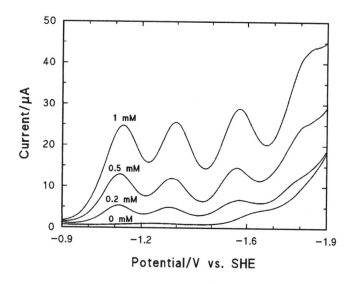

FIGURE 1. Differential pulse voltammograms of HCB

These were attributed to successive reduction of HCB to penta-, tetra-, tri-, and di-chlorobenzenes, and confirmed by showing that the latter three maxima corresponded to the first reduction potentials of these reduction products. In a parallel experiment using 0.05 M NaCl as supporting electrolyte, only a single maximum was seen, at E ~ -1.35 V *vs.* SHE, and in addition, hydrogen evolution was shifted to less negative potentials, diminishing the useful potential range for electroreduction. Tetraethylammonium cations are known to modify the electrode surface by adsorption from non-aqueous solutions, thereby changing the composition of the double layer and the potential distribution across the interface.

Bulk electrolyses at a mercury pool cathode confirmed the identities of the reduction products, and allowed us to obtain the mass balance, which was in all cases > 98% based on phenyl residues. Products were identified by HPLC retention time and by GC/MS. Depending on the potential, the reduction could be interrupted fairly cleanly after the removal of one (at -1.27 V), two (at -1.50 V) or three (at -1.75 V) chlorine atoms (Figure 2).

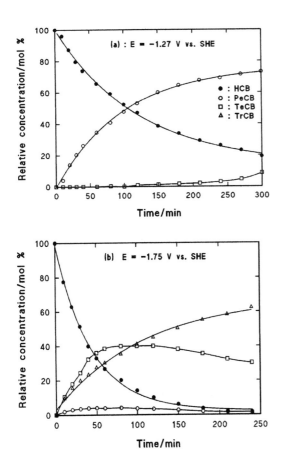

The loss of HCB followed first order kinetics; the apparent rate constant increased only weakly with applied potential, but linearly with flow rate, indicating that the rate of reduction is limited by the rate of mass transfer to the cathode. Single flow experiments were employed for the purpose of determining the current efficiency in methanol, near the maximum potential for the first reduction peak ($\sim$ -1.3 V), the current efficiency for removal of the first chlorine atom from a 1 mM solution of HCB was ~60%, after correction for IR drop. Competing H_2 evolution decreased the current efficiency at more negative potentials, and lowering the concentration of HCB to 0.1 mM reduced the efficiency to ~25% at -1.3 V. In continuous flow experiments carried out with 1 mM HCB

FIGURE 2. Kinetics of electrolysis of HCB in the recirculating flow cell

under amperostatic conditions (i = 20 mA), the rate of formation of dechlorinated products during the first hour of electrolysis occurred with 13% current efficiency (the cathodic potential in the amperostatic experiments was ~ -2.1 V and competing hydrogen evolution was copious).

The foregoing experiments provided "proof of concept" for electrolytic dechlorination of HCB, but were not carried out under technologically useful conditions. Three specific requirements were next addressed: (a) the need to work in aqueous solution, even though HCB is poorly ($\sim 10^{-8}$ M) soluble in water; (b) the need to replace mercury as the cathode, given the negative experience with this material in the chlor-alkali industry; (c) the need for a supporting electrolyte that is less costly and more environmentally friendly than tetraethylammonium chloride.

Water Solubility. The addition of small (< 20% v/v) amounts of methanol to aqueous solutions of HCB was ineffective at raising the solubility even to the μM range. We therefore turned our attention to surfactant media, which have been used successfully in other systems (Iwunze, 1989; Couture et al., 1992; Zhang, 1993). A difficult problem has been to find a method to break the emulsion and recover the electrolysis products before further biological treatment (Gao et al., 1996), since surfactants are themselves toxic towards microorganisms.

The recent introduction of the Triton-SP® series of surfactants manufactured by Union Carbide offered a prospective solution to this problem. These non-ionic surfactants consist of a proprietary hydrophobe attached to a polyethoxylate hydrophilic chain. The significant advance is that the hydrophobe can be cleaved from the polyethoxylate chain by brief treatment with dilute acid (pH < 3), upon which the cleaved Triton-SP loses its surfactant properties irreversibly and becomes compatible with subsequent biological treatment.

Concentrations of HCB up to at least 50 μM were attained in 0.1% Triton-SP175 with cosolvent 1% heptane: the solutions were milky white and stable towards breaking the emulsion for months. HCB was successfully reduced in 0.1% Triton-SP 175/1% heptane/0.05 M TEACl at the mercury pool cathode in the recirculating flow-through cell under amperostatic conditions that closely duplicated our previous experiments in methanol. The experiments were analyzed by HPLC, after removing aliquots of solution from the electrolyte, breaking the emulsion with dilute hydrochloric or sulfuric acid, and partitioning the products into heptane. Penta-, tetra-, and tri-chlorobenzenes were formed, as in methanol, and a quantitative material balance (> 98%) was achieved.

Supporting Electrolyte. TEACl is unsuited to practical technology on grounds of cost, and because it would afford Cl_2 in the anodic reaction. We considered sodium sulfate to be an ideal supporting electrolyte for this application; sodium ions and sulfate ions are neither electroactive nor pose an environmental threat (no limits are placed on their concentrations in aqueous discharges). The HCB/Triton-SP175 microemulsion was stable in the presence of 0.05 M Na_2SO_4 and reduction to the same dechlorination products occurred with similar current efficiency as with TEACl. By contrast, in methanol, the current efficiency was lower when NaCl replaced TEACl as the supporting electrolyte; this effect was ascribed to preferential adsorption of Et_4N^+ ions at the surface of the cathode, thereby altering the electrical double layer so as to increase the useful potential range of the electrode (Fawcett 1982). We speculate that in the microemulsion the surfactant itself is the chief adsorbent to the cathode surface.

All subsequent experiments were carried out using 0.05 M Na_2SO_4 as the supporting electrolyte in the aqueous medium of 0.1% Triton-SP 175 and 1% heptane.

Choice of Cathode. Because mercury is undesirable as an electrode for industrial processes, we investigated a series of other metals as possible cathodes for comparison against Hg. The most important property sought was a high overvoltage for hydrogen evolution, otherwise the current efficiency is severely reduced. Voltammetric measurements showed that Hg, Pb, and Sn exhibited only residual current beyond the potential required for the first reduction of HCB; a small hydrogen evolution current was already apparent for Zn, and a large residual current was seen for stainless steel, Fe, Ni, Cd, and glassy carbon. Bulk electrolyses were carried out in the flow-through cell under amperostatic polarization. In all cases the product distributions were similar to those recorded in Figure 2, and a quantitative material balance was observed throughout the reactions.

The results of these experiments confirmed the voltammetric measurements. The rate of disappearance of HCB decreased in the approximate order of cathodes Hg > Pb > Cd, Zn, Sn > stainless steel > Fe > Ni > glassy carbon. Hydrogen evolution was copious and the current efficiency was therefore very low (~ 1%) due to competing evolution of hydrogen. We then carried out single flow experiments at the more promising cathodes, and obtained the results shown in Table 1.

TABLE 1. Relative current efficiencies for various cathode materials (reference point : Pb cathode with current efficiency ~ 5 %)

Electrode	Current Density (mA/cm²)	-E (V vs. SHE)	HCB: PeCB: TeCB: TrCB (mol %)	Relative Current Efficiency
Hg	0.06	1.47	87.0: 5.7: 5.2: 2.1	53
Pb	0.01	1.40	85.3: 0.6: 4.8: 9.3	100
Zn	0.15	1.25	81.0: 2.8: 7.8: 8.4	10
Sn	0.05	1.58	91.8: 2.1: 3.7: 2.4	10

Lead was the best cathode material, based on both current efficiency (~ 5%) and extent of dechlorination. Its superior performance was explained because the potential of -1.4 V vs SHE was attained at very low current density, allowing successive reduction of HCB to penta-, tetra- and tri-chlorobenzenes. This can be stated conversely, that low current density minimizes the extent of cathodic polarization and therefore limits the amount of competing reduction of hydrogen.

Knowing that current efficiency is optimized at low current density allowed us to investigate the effect of cathode area, since we reasoned that a practical technology will require the cathode to have the largest possible surface area. The most successful approach was to deposit Pb cathodically as long bright needles from 0.15 M Pb(NO₃)₂, 0.15 M sodium citrate solution at pH 6.5 on to Pb wire. Comparing electrolysis at 75 mA at this cathode with electrolysis at a plain Pb wire of area 30 cm^2, the reaction was over an order of magnitude faster, but we could not determine its surface area. One difference noted was in product distribution; whereas the major product at the plain wire was trichlorobenzene, the reaction progressed only to tetrachlorobenzene at the high surface area electrode, because the lower current density led to lower cathodic polarization.

CONCLUSIONS

We have shown that practically useful concentrations of HCB can be electrolytically dechlorinated in aqueous solution using Triton-SP175 and 1% heptane as surfactant and cosolvent. The reaction occurs at a variety of cathodes, of which lead was the best under our conditions. The electrolysis products are recovered following mild hydrolysis of the surfactant. Unlike previous studies with ionic surfactants, our methodology requires the use of a supporting electrolyte, of which 0.05 M Na$_2$SO$_4$ was found to be both effective and environmentally friendly. Our results indicate that cathodes of the highest possible surface area are needed in order to combine rapid reaction rate with high current efficiency.

ACKNOWLEDGEMENTS

We thank the Natural Sciences and Engineering Research Council of Canada for financial assistance and Union Carbide Canada for a gift of Triton-SP175.

REFERENCES

Couture, E. C., J. F. Rusling, and S. Zhang. 1992. "Mediated Electrolytic Dechlorination of Pollutants in Water using Surfactants." *Trans I Chem E. 70*(B): 153–157.

Fawcett, W. R., and J. S. Jaworski. 1982. "Electroreduction of Alkaline-earth Metal Cations at Mercury in Aprotic Media." *J. Chem. Soc. Faraday Trans. 1 78*: 1971.

Gao, J., J. F. Rusling, and D. Zhou. 1996. "Carbon-Carbon Bond Formation by Electrochemical Catalysis in Conductive Microemulsion." *J. Org. Chem. 61*: 5972–5977.

Iwunze, M. O., and J. F. Rusling. 1989. "Aqueous Lamellar Surfactant System for Mediated Electrolytic Dechlorination of Polychlorinated Biphenyls." *J. Electroanal. Chem. 266*: 197–201.

Zhang, S., and J. F. Rusling. 1993. "Dechlorination of Polychlorinated Biphenyls by Electrochemical Catalysis in a Bicontinuous Microemulsion." *Environ. Sci. Technol. 27*: 1375–1380.

OZONE/ELECTRON BEAM IRRADIATION REMOVES PERCHLOROETHYLENE AND GENOTOXIC COMPOUNDS FROM GROUNDWATER

Peter Gehringer (Austrian Research Centre Seibersdorf, Austria)

ABSTRACT: With regard to hydroxy free radical generation the ozone/electron beam irradiation treatment process is most efficient among the ozone based Advanced Oxidation Processes (AOPs). This process is able to mineralize trace amounts of perchloroethylene in groundwater in a single stage process without formation of any by-product to be disposed of. Moreover, experiments performed just recently have demonstrated that the ozone/electron beam process is also apt for total removal of genotoxic compounds detected in groundwater contaminated with perchlorothylene. The design of the world-wide first ozone/electron beam full scale plant to purify groundwater contaminated with traces of perchlorothylene and additionally genotoxic compounds of unknown origin and structure is presented. The capacity of the plant which will be located in Bad Fischau-Brunn - a small community south of Vienna - is 108 m^3/h; the treated water will be supplied into the tap water distribution system of the community. In the present case the genotoxic components cannot be detected by chemical methods. The Tradescantia micronucleus assay used for the proof of genotoxicity needs about one month to yield the result. Under these conditions activated carbon filtration alone is not apt for remediation of such a contaminated water.

INTRODUCTION

Hydroxy free radicals are the strongest oxidants known to occur in water. Processes which generate hydroxy free radicals for the subsequent use of pollutant decomposition are generally referred to as "Advanced Oxidation Processes" (AOPs).

It has been found that the addition of ozone in combination with UV radiation or hydrogen peroxide can promote the decomposition of ozone and generate hydroxy free radicals. Combining hydrogen peroxide and UV radiation alone will achieve similar results to the former mentioned ozone based processes by generating hydroxy free radicals with a considerable less yield though ("Glaze et al., 1987"). All these AOPs have in common that the hydroxy free radicals originate from one single source only (ozone or hydrogen peroxide, respectively), i.e. the hydroxy free radical concentration obtained is comparatively low. On the other hand, groundwater very often is contaminated by rather small amounts of highly toxic compounds. To be effective in the treatment of groundwater containing trace amounts of pollutants high hydroxy free radical concentration is necessary. Combination of ozone with electron beam irradiation - which represents ionizing radiation - is able to produce much higher hydroxy free radical concentrations than the AOPs mentioned before because in this combination the hydroxy free radicals originate from two different but simultaneous working sources: (1) directly from the water to be purified by the action of the electron beam irradiation (so-called water radiolysis) and (2) from ozone decomposition accelerated by the reducing species formed during water radiolysis. ("Gehringer et Eschweiler, 1996").

Table 1 shows the most important steps of all ozone based AOPs including also the combination of ozone and hydrogen peroxide. It clearly illustrates that O_3/UV and O_3/H_2O_2 processes are one and the same regarding hydroxy free radical generation: in the former, one is merely forming hydrogen peroxide in situ, rather than adding it from the outside. From this table it comes out clearly that two completely different mechanisms for OH free radical generation occur. UV-radiation is absorbed by ozone (and if present by other organics absorbing UV at 254 nm) while the energy of the fast electrons is totally absorbed by water and not by any solute. In other words: Under the conditions given UV-irradiation represents the direct effect of radiation while electron beam irradiation stands for the indirect effect of radiation.

TABLE 1: Ozone based AOPs with regard to their most essential reactions for hydroxy free radical formation in water

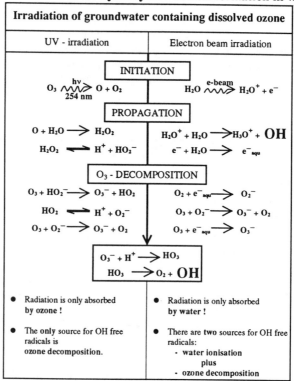

The latter produces hydroxy free radicals directly from water and simultaneously the promoters for ozone decomposition (by formation of the solvated electrons

e_{aqu}^-) while the former only produces the promoter for ozone decomposition (by formation of hydrogen peroxide and its deprotonated anion, respectively). Accordingly electron beam irradiation must be more efficient regarding hydroxy free radical generation than any other ozone based process. Since UV irradiation of hydrogen peroxide is less effective than ozone/UV ("Glaze et al., 1987") it is obvious that ozone/electron beam irradiation is the most attractive AOP for groundwater remediation.

The present paper deals with the use of the combined ozone/electron beam irradiation process for groundwater remediation on a technical scale. Such a plant is at present constructed in Bad Fischau - Brunn, Austria to clean up a groundwater contaminated with traces of perchloroethylene (PCE) and additionally, with some genotoxic compounds of unknown origin and concentration. The technical design of the plant will be introduced.

MATERIALS AND METHODS

Groundwater from the polluted well was transported from Bad Fischau-Brunn to Seibersdorf and treated in the existing 3 m^3/h pilot plant for continuous ozone/electron beam irradiation treatment of water. This facility has been already described elsewhere ("Gehringer et al., 1993"). The groundwater contained 251 mg/L bicarbonate, 8 mg/L chloride, 15 mg/L nitrate, < 0.4 mg/L DOC; it was polluted with about 61 µg/L PCE and some genotoxic substances detected with the Tradescantia micronucleus (Trad-MCM) assay (Tradescantia clone # 4430) described by "Ma (1983)". In general, five slides were prepared per sample and 300 tetrads per slide were scored. PCE was measured by gas chromatography using cold on – column injection and electron capture detection.

RESULTS AND DISCUSSION

It was found that a radiation dose of 200 Gy is sufficient to reduce the PCE in the water from about 61 µg/L to about 1 µg/L provided the ozone concentration in water before irradiation is ≥ 6 mg/L. Moreover, all the genotoxic components present in water before the treatment were completely removed by the ozone/electron beam irradiation process (see Table 2).

TABLE 2: **Results of Trad-MCN test with water from the well in Bad Fischau – Brunn before and after treatment with ozone/electron beam irradiation.**

Sample	Clastogenic concentrations in micronuclei (MCN)/100 tetrads	
	individual values	mean value ± s
groundwater before treatment	53.6; 19.2; 9.6; 53.9; 29.7	33.2 ± 20.0
groundwater after treatment	5.8; 5.4; 11.0; 14.9; 6.3	8.7 ± 4.1
tap water (negative control)	2.9; 6.9; 2.9; 12.1; 10.1	7.0 ± 4.1
tap water with 0.25 mM As$_2$O$_3$/L (positive control)	31.7; 19.7; 17.0; 16.3; 16.2	20.2 ± 6.5

The unit of the radiation dose 1 Gray (Gy) is equivalent to an absorbed energy of 1 Joule/kg. Accordingly a radiation dose of 200 Gy means a very small energy transfer into the water. According to these dose and ozone requirements a 9 kW electron beam accelerator and 650 g O_3/h are required in theory to purify 108 m^3/h of the polluted groundwater.

Figure 1 shows now the block diagram of the planned 108 m^3/h water treatment plant in Bad Fischau-Brunn. Ozone is made from pure oxygen, an ozone concentration of 13 % per weight is planned. The ozone/oxygen stream is then compressed and mixed with the polluted groundwater at elevated pressure in a static mixing unit. Under the conditions given almost all ozone will be dissolved in the water but not all the oxygen. Therefore, most of the gaseous oxygen still present is released into the environment by means of a gas separator. Trace amounts of ozone which will be by that also removed from the water are destructed in an ozone destructor after the gas separator.

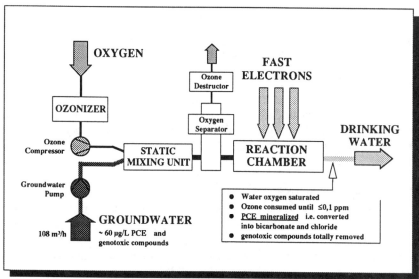

FIGURE 1. Process schematic for Bad Fischau-Brunn's treatment plant

Under the conditions given almost no stripping of PCE takes place. Despite of that inevitable ozone losses it is assumed that about 90 % of the ozone produced remains in the water after oxygen release. The ozone containing polluted groundwater is then irradiated with 500 keV electrons in a closed irradiation chamber.

The electron beam accelerator is the most essential part of the process because it supplies the energy necessary for PCE decomposition. Please note! Fast electrons form part of ionizing radiation. Accordingly an electron beam accelerator is a machine for the production of ionizing radiation i.e. ionizing radiation can be switched on and off, respectively and most importantly it is produced without using any radioactive material! In a process using electron beam accelerators up to about 4.5 MeV (what is supposed to be the upper limit in radiation processing of water) no radioactivity can be induced. Therefore, in such a process there is **no radioactivity involved at all!** (For details see "Gehringer et al., 1994").

As already mentioned the PCE contained in groundwater is almost totally mineralized by the action of the electron beam irradiation in the presence of ozone ("Gehringer et al., 1992"). As can be seen from Table 3 the additional amount of bicarbonate and chloride as well to the natural solute content of these components as consequence of pollutant mineralisation is so small that it would be impossible to detect it by any analytical procedure. It is, therefore, of no influence to the water treated; just the contrary, as a consequence of the ozone introduction the oxygen content of the water is increased what can be classified as improvement of the water quality.

TABLE 3: Change in concentration of some natural solutes in Bad Fischau-Brunn groundwater as a consequence of the treatment process

Solute	Concentration in mg/L		
	before the treatment	change by the treatment	after the treatment
bicarbonate	251.0	+ 0.06	251.06
chloride	8.0	+ 0.068	8.068
oxygen	8.7	$\approx$ + 20.00	$\approx$ 28.7 *
genotoxic compounds	clearly present	totally removed	

* can be roughly estimated only because of some degassing after treatment

CONCLUSIONS

It is obvious that the combined ozone/electron beam irradiation process is the most efficient AOP in water regarding concentration of hydroxy free radicals formed. Moreover, it has been demonstrated that the ozone/electron beam process is also apt for total removal of genotoxic compounds detected in groundwater contaminated with perchloroethylene. Since the genotoxic compounds could not be detected by chemical methods but only by the rather time consuming Trad-MCN test – which need about one month to yield the results – activated carbon filtration alone is not apt for remediation of a groundwater containing such genotoxic compounds.

Moreover, there are some data available from the literature ("Helma at al., 1994") concerning the groundwater stream of the region the community of Bad Fischau-Brunn is located. 6 km down-stream close to a big waste disposal site genotoxic and chlorinated compounds were detected in the groundwater, too, using the same Trad-MCN test for the genotoxicity proof. It was found that activated carbon filtration could not control the genotoxic compounds. Furthermore, when UV-irradiation was applied after activated carbon filtration the genotoxicity of the water treated increased exponentially with the UV irradiation dose applied. Therefore, application of UV-irradiation for disinfection of such a contaminated water is not possible. Beyond that, it is even questionable whether AOPs based on UV-radiation in combination with oxidants can be applied when the water to be purified contains such genotoxic substances or their precursors.

To date only the ozone/electron beam irradiation treatment process has demonstrated to be apt for remediation of such a contaminated groundwater. The first full scale plant for 108 m^3/h is under construction in Bad Fischau-Brunn in Austria with SIEMENS Austria as industrial partner. The project is supported by the European Union LIFE Programme (LIFE 96 ENV/A/314).

REFERENCES

Gehringer, P., and H. Eschweiler. 1996. "The use of radiation-induced Advanced Oxidation for water reclamation." *Wat. Sci. Tech.* 34: 343-349.

Gehringer, P., H. Eschweiler, and H. Fiedler. 1994. "Advanced Oxidation Process Based on Ozone/Electron Beam Irradiation for Treating Groundwater." Proc. 1994 Annual Conf. Am. Water Works Assoc., *Water Quality.* New York, June 1994: 633-639.

Gehringer, P., H. Eschweiler, W. Szinovatz, H. Fiedler, R. Steiner, and G. Sonneck. 1993. "Radiation-induced OH radical generation and its use for groundwater remediation." *Radiat. Phys. Chem.* 42: 711-714.

Gehringer, P., E. Proksch, H. Eschweiler, and W. Szinovatz. 1992. "Remediation of groundwater polluted with chlorinated ethylenes by ozone-electron beam irradiation treatment." *Appl. Radiat. Isot.* 43: 1107-1115.

Glaze, W.H., J.-W. Kang, and D.H. Chapin. 1987. "The chemistry of water treatment processes involving ozone, hydrogen peroxide and ultraviolet radiation." *Ozone Sci&Eng.,* 9: 335-352.

Helma C., R. Sommer, R. Schulte-Hermann, and S. Knasmüller. 1994."Enhanced clastogenicity of contaminated groundwater following UV irradiation detected by the Tradescantia micronucleus assay." *Mutation Res. Letters* 323: 93-98.

Ma, T. H. 1983. "Tradescantia micronuclei test (Trade-MCN) for environmental clastogens." In Kolber, A. R., T. K. Wong, L. D. Grant, R. S. Dewoskin and T. J. Hughes (Eds.). *In Vitro Toxicity Testing of Environmental Agents,* Part A, Plenum, New York pp. 191-214.

ELECTROKINETICS AS A SUPPORTING TECHNOLOGY FOR *IN SITU* TCE BIODEGRADATION

John H. Pardue, W. Andrew Jackson, Robert Gale and M. Fazzle Rabbi,
(Louisiana State University, Baton Rouge, LA, USA)
Boyce Clark and Elif Ozsu-Acar (Electrokinetics, Inc., Baton Rouge, LA, USA)

ABSTRACT: Bench-scale experiments have been conducted to determine the use of electrokinetics to deliver amendments (electron donors) to support *in situ* anaerobic reductive dechlorination of trichloroethylene (TCE) in the subsurface environment. Electrokinetics has the potential to effectively deliver amendments such as electron donors and nutrients regardless of porosity and heterogeneities in the porous media. Experiments were conducted in two systems: a sub-soil sampled from Vicksburg, MS, USA and aquifer material removed from an oxic zone of a TCE-contaminated aquifer at the Petro-Processors, Inc. site in Baton Rouge, LA, USA. In bench scale studies using the Vicksburg subsoil, sections of the cell demonstrated strong evidence of TCE dechlorination, stimulated by benzoate addition by electrokinetics. This evidence included apparent loss rates of TCE virtually identical to batch microcosm studies. In addition, lower chlorinated metabolites of TCE dechlorination, primarily cis-dichloroethylene (cis-DCE), were observed equivalent to ~`25% of the available TCE mass. By contrast, bench scale experiments using the Petro Processors material were inconclusive. Electrokinetics appears to have potential in the delivery of amendments to support TCE dechlorination. However, optimization of the rate of delivery is required.

INTRODUCTION

In situ bioremediation is an important technology for the remediation of chlorinated organics in the subsurface. In particular, natural attenuation of chlorinated solvents such as trichloroethylene (TCE) has been proposed as a remedial alternative for many sites. In some cases, however, sufficient quantities of electron donors do not exist and amendments of donors has been utilized to create anaerobic conditions in the aquifer. Additions of these amendments is difficult. Recent surveys have concluded that the ineffective transport of amendments have been the primary cause of bioremediation system inefficiencies and failures (Cookson, 1995). These problems include the existence of preferential flow paths, heterogeneities, adsorption or chemical reactions in the soil and biofouling. Electrokinetic injection has the potential to effectively deliver amendments such as electron donors and nutrients regardless of porosity and heterogeneities in the porous media (Acar et al., 1997). Studies have demonstrated that charged species can be introduced into the porous soil media by electrokinetic processing regardless of the heterogeneities present (Rabbi, 1997). Mechanisms of electrokinetics applicable to its use at *in situ* biodegradation sites include electroosmosis and electromigration.

Bench-scale experiments have been conducted to determine the use of electrokinetics to deliver amendments (electron donors) to support *in situ* anaerobic reductive dechlorination of trichloroethylene (TCE) in the subsurface environment. The results of these studies are presented below.

Objective. Experiments were conducted to examine the use of electrokinetics to deliver benzoate as a donor to support TCE dechlorination. The goal was to demonstrate that electrokinetics could be a useful tool for delivering amendments to support the bioremediation of chlorinated solvents such as TCE. Specifically, the approach targets oxic aquifers where insufficient electron donors exist to drive the aquifer anaerobic. Studies were conducted using two materials: a subsoil from Vicksburg, MS and aquifer material from the Petro Processors site in Baton Rouge, LA.

Site Description. Petro Processor's Inc. operated an unengineered disposal facility for industrial wastes from 1962-1981 near Baton Rouge, La (Constant et al., 1995). The facility consisted of several unlined pits in two separate disposal sites, "Scenic" and "Brooklawn", named for the adjacent surface roads. An estimate of the original volume of waste disposed in the two sites is 330,000 m^3. The material deposited included a wide range of industrial solid waste and free-phase organic chemicals. Primary organic waste components were hexachlorobutadiene (HCBD) and hexachlorobenzene (HCB), halogenated solvents, and polynuclear aromatic hydrocarbons. The total estimated volume of material to be remediated, including contaminated soils, is 990,000 m^3. A consent decree for site closure was developed in 1983 and entered into court record in 1984. A remedial plan, approved by the U.S. Environmental Protection Agency in 1987, includes the hydraulic containment of the waste and recovery and incineration of organics. Remedial action is currently being implemented. Portions of the aquifer beneath the site are candidates for natural attenuation as indicated by an ongoing field monitoring program, however, oxic regions of the aquifer may require amendment of electron donors.

MATERIALS AND METHODS

Studies were conducted using two types of experimental protocols: batch microcosm studies and bench-scale cells constructed of glass and Teflon. Experiments were conducted using both the Vicksburg subsoil and the Petro Processors aquifer material.

Microcosm studies. Microcosm studies were conducted by dispensing subsoil and aquifer material into glass serum vials (30-125 mL) in an anaerobic glove box. Slurries were constructed with groundwater from the Petro Processors site for the aquifer material and deionized water for the Vicksburg subsoil. Soils were uniformly contaminated with neat TCE (Vicksburg subsoil) and PCE in acetonitrile as a solvent (Petro Processors aquifer material). Killed controls (autoclaved or Hg^{2+} added as a biocide) were included for both experiments.

Treatments included controls (no benzoate added) and samples amended with benzoate. Initial concentrations of TCE were 50 mg/kg soil in the subsoil while PCE concentrations were 3 mg/kg in the Petro Processors soil, equivalent to the highest observed concentrations in the aquifer. First-order apparent loss coefficients for TCE and PCE were calculated using non-linear regression using a Marquardt algorithm.

Bench-scale studies. Bench-scale studies were conducted using cells (0.3 or 1 m long) constructed of glass with Teflon end caps. Subsoils and aquifer materials were packed into the cells, saturated with a TCE-spiking solution and sealed. Several treatments were used in the studies. In the Vicksburg soil, 2 cells were used: one in which electrokinetics was used to amend the soil with benzoate, another in which the electrical field was applied but no benzoate added. For the Petro Processors experiment, two additional treatments were used. One in which the cell was uniformly mixed with benzoate in addition to the TCE. The other cell did not receive benzoate or was it subject to an electric field. TCE and breakdown products were monitored throughout the experiment by monthly sampling of cores through ports with subsequent extraction and GC-MS analysis..

RESULTS AND DISCUSSION

Microcosm studies. Batch studies using the Vicksburg subsoil demonstrated rapid loss of TCE in treatments amended with benzoate as compared with killed controls and controls without benzoate amendments. First order degradation rate constants for the control treatment (no benzoate added) was 0.006 day^{-1} ($\pm$ 0.001) and an r^2 of 0.711. By contrast the rate constant for the benzoate amended treatments was 0.038 day^{-1} ($\pm$ 0.001) and a r^2 of 0.99. Very minor losses were observed in the killed controls. By contrast, no statistically significant effect could be determined for benzoate on PCE degradation using Petro Processors aquifer material. These results can be explained by several factors: the higher concentrations of TCE in the Vicksburg experiment likely overwhelmed the existing concentrations of carbon donors. Benzoate was necessary for stimulation of dechlorination of high levels of TCE used in this experiment. Degradation of PCE in the Petro Processors appears to progress naturally once the initial oxygen present in the groundwater is consumed.

Bench-scale studies. In bench scale studies using the Vicksburg subsoil, sections of the cell demonstrated strong evidence of TCE dechlorination, stimulated by benzoate addition by electrokinetics. This evidence included apparent loss rates of 0.039 day^{-1} ($\pm$0.007 and r^2 of 0.998) for cells with benzoate delivered by electrokinetic injection. While cells that received only the electrokinetic current had apparent loss rates of 0.002 day^{-1} ($\pm$0.001 and r^2 of 0.51). This rate of TCE degradation is virtually identical to the batch microcosm studies (0.038 day^{-1}) (Figure 1). In addition, lower chlorinated metabolites of TCE dechlorination,

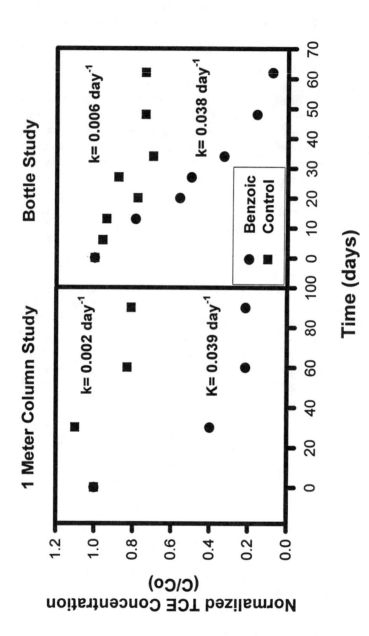

FIGURE 1. Apparent first-order loss rates of TCE in microcosm bottle studies and electrokinetic bench scale studies with and without the addition of benzoate.

primarily cis-dichloroethylene (cis-DCE) and trans-dichloroethylene (trans-DCE), were observed equivalent to ~`25% of the available TCE mass. TCE loss was minimal in the unamended cells and only trace metabolites were detected. By contrast, bench scale experiments using the Petro Processors material were inconclusive. These experiments were conducted in the shorter 0.3 m columns. While metabolites were observed in some samples, it could not be determined if benzoate added using electrokinetics had a positive effect. Because of electroosmosis, TCE is constantly moving out of the cell into the wells at each end of the column. Because of this loss, closing the mass balance in the bench scale experiments is difficult.

Implications. Evidence is presented that electrokinetics can deliver benzoate to porous media, effectively stimulating TCE dechlorination to dichloroethylenes. Much remains to be understood about the process including methods of optimizing delivery to insure a uniform level of benzoate reaching the entire cell. Because of electroosmosis, closing the mass balance is difficult in studies of this design. With further development, electrokinetics may be the method of choice for delivering ionic amendments to support *in situ* bioremediation in heterogeneous media. However, the need for donor amendments needs to be established on a case-by-case basis.

REFERENCES

Acar, Y.B., M.F. Rabbi and E.E. Ozsu-Acar. 1997. "Electrokinetic injection of ammonium and sulfate ions into sand and kaolinite beds". J. Geotech. & Geoenviron. Engr. 123:239-249.

Cookson, J.T. 1995. *Bioremediation Engineering: design and application.* McGraw-Hill Book Company, Inc., New York, N.Y.

Rabbi, M.F. 1997. Injection of Species Under Electrical Field to Enhance Bioremediation. Ph.D. Thesis. Louisiana State University.

LOW-PROFILE STRIPPING OF ORGANIC POLLUTANTS AND DESTRUCTION BY GAS-PHASE PHOTOLYSIS

Yusuf G. Adewuyi, Vicky L. Harrison and Riyad A. Jadalla
North Carolina Agricultural and Technical State University
Greensboro, North Carolina 27411, U.S.A.

ABSTRACT: We have demonstrated the feasibility of effectively removing organic pollutants in aqueous solutions using a Shallow-tray, low-profile air stripper. Experiments were conducted to determine the influence of air flowrate, water flowrate (2-20 GPM), volumetric air-to-water ratios (100-700), temperature, and inlet contaminants' concentrations (500-1500 ppm) on the stripping rate constants and steady-state removal efficiencies of benzene, toluene, xylene (BTX), trichloroethylene (TCE), methyl tert-butyl ether (MTBE), and acetone. Steady-state stripping efficiencies were determined spectrophotometrically by measuring the concentrations of the pollutants in the inlet and outlet streams of the stripper. At room temperatures and varying air-to-water ratios, high stripping efficiencies of 97-100% were obtained for highly volatile compounds such as TCE and BTX. However, the mass transfer rates of acetone and MTBE were limited by their high water solubilities, producing steady-state removal efficiencies of 30-65% and 70-80% respectively.

INTRODUCTION

Contamination of soil and groundwater is a serious health and environmental problem. Contamination sources include manufacturing plants, petroleum refineries, fuel and chemical storage facilities, gas service stations, underground storage tank leaks, poor disposal practices, chemical wastes from agricultural practices, and accidental spills of contaminants. The majority of the contaminants are volatile organic compounds (VOCs). VOCs commonly found in water include BTX, naphthalene, acetone, and a wide range of chlorinated hydrocarbons such as TCE, tetrachloroethane and trichloroethane.

Traditional technologies for treatment of contaminated soil and groundwater include landfilling, air stripping/carbon adsorption, incineration, biological, and chemical treatments. Incineration, adsorption and landfilling merely transfer the contaminant to another phase or location and produce a potentially dangerous and toxic secondary disposal requirement. Biodegradation provides a promising method for environmental remediation. However, currently available processes are slow and produce unpredictable results. Much research is needed to develop efficient processes through coupling of physical and biological approaches with chemical oxidative techniques such as advanced oxidation processes (AOPs). Examples of AOPs include the use of Fenton's reagent (H_2O_2/Fe^{2+}), UV-catalyzed oxidation (e.g., H_2O_2/UV, O_3/UV, TiO_2/UV), ultrasonic irradiation, wet air oxidation and supercritical water oxidation. UV-catalyzed oxidative techniques are effective for mineralizing organic pollutants in aqueous waste streams. However, these methods are sometimes hampered by competition for light or radicals by such solutes as humic acid or nitrate and bicarbonate ions in water. Metals such as iron and magnesium are oxidized and precipitated in the process units. These problems could be avoided by developing sequential techniques involving the physical disengagement of contaminants from the groundwater or soil matrix and subsequent destruction in the gas phase. The destruction of organics in air by UV-catalyzed photolysis is faster than in water due to the following advantages: 1) lower UV absorption of competitive organics in air than in water, 2) scaling due to precipitation is minimized, 3) higher mobility of dissociation species in gas phase which prevents the reverse process of toxics recombination, and 4) small portion of oxygen (O_2) in air is converted by UV to ozone (O_3), which contributes to UV-aided photolysis.

Unlike the conventional packed-tower air strippers, which create a number of operating and maintenance problems due to clogging, the tray-type, low-profile strippers have no packing materials and offer easy access for removal of scale and other fouling agents. They are also capable of operating at very high air-to-water ratios (greater than 1,000:1 on a volume basis), which is desirable for high treatment efficiencies, particularly for more-difficult-to-strip compounds. This feature of the tray-type air stripper is also an advantage when treating waters that contain high concentrations of soluble inorganics.

Objective. The objectives of this project is to determine the feasibility of effectively decontaminating polluted water by coupling air stripping (using Shallow-tray, low-profile air stripper) with UV-catalyzed photolysis directly in the gas phase (using xenon flashlamps and UV pulsed lamps). This research sought to determine the capabilities and limitations of tray-type (Shallow Tray™) air-stripping technology.

MATERIALS AND METHODS

The stripping experiments were accomplished in a Shallow Tray Low Profile Air Stripper (Model 2311, Northeast Environmental Products, Inc.). The stripper is designed to have a maximum flowrate of 30 GPM. The Shallow Tray process uses forced draft, countercurrent air stripping through baffled aeration trays to remove volatile organic compounds from water. Contaminated water is sprayed into the inlet chamber through a coarse mist spray nozzle. Air is blown through hundreds of holes in the bottom of the trays and form a froth of bubbles, generating a large mass transfer surface area where the contaminants are volatilized. The turbulent action of the froth scours the surfaces of the tray reducing build-up of scales. The schematic diagram of the single-tray stripper system coupled with the photolysis unit is shown in Figure 1. The photolysis system consists of xenon flashlamps (Model RC-600 pulsed UV System, Xenon Corporation).

Contaminated water samples for the stripping experiments were obtained by mixing known quantities of the liquid hydrocarbon (e.g., chlorinated volatile organic compounds or CVOCs) with water in a 300-gallon stainless steel tank (McMaster-Carr). Volatile and semi-volatile organic compounds used include BTX (Fisher Scientific), TCE (Fisher Scientific), MTBE (Sigma-Aldrich) and acetone (Fisher Scientific). Two turbine-blade electric mixers and the turbulence caused by the jet action of the water inlet hose were used to thoroughly disperse the CVOCs, allowing them to dissolve in water. During each run, the inlet and outlet water streams from the stripper were sampled periodically for about ten minutes or until the operation of the stripper reached steady state. Samples were collected in Napelene PE Amber sample bottles with care taken to avoid air bubbles. The samples were immediately refrigerated and later agitated with a Burrel Wrist-Action[R] shaker and analyzed spectrophotometrically for CVOCs at various wavelengths using a UV/VIS spectrophotometer (DU 7500, Beckman). Stripping efficiencies were determined by measuring the concentrations

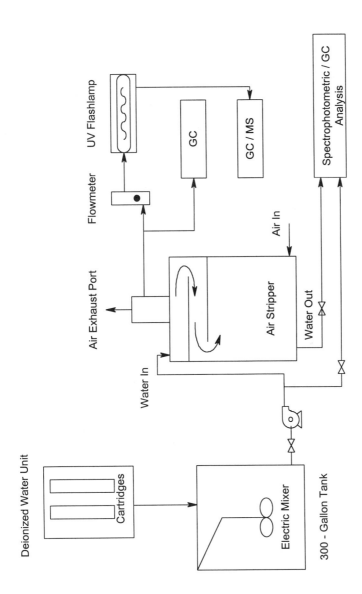

FIGURE 1. Schematic Diagram of the Combined Low - Profile Stripping / Photolysis Experimental Setup

of the pollutants in the inlet and outlet streams of the stripper. The steady-state stripping efficiency is calculated from: $E_s = 1 - C_{out}/C_{in}$.

RESULTS AND DISCUSSION

We have demonstrated the feasibility of effectively removing several organic pollutants in aqueous solutions using a Shallow-tray, low-profile air stripper. Experiments were conducted to determine the influence of air flowrate, water flowrate (2-20 GPM), volumetric air-to-water ratios (100-700), temperature, and inlet contaminants' concentrations (500-1500 ppm) on the stripping rate constants and steady-state removal efficiencies of BTX, TCE, MTBE, and acetone. The physical properties of these compounds and the stripping data obtained from these experiments are shown on Table 1. The relative effects of air and water flowrates on the stripping efficiencies are illustrated on Figure 2.

The strippability of a compound is a function of its volatility (or vapor pressure) and solubility. The easiest compounds to strip are those that are highly volatile and only slightly soluble such as BTX and TCE. These compounds also tend to have high Henry's Law constants. At room temperatures and varying air-to-water ratios, high stripping efficiencies of 97-100% were obtained for highly volatile compounds such as TCE and BTX at steady state conditions. However, the mass transfer rates of acetone and MTBE were limited by their high water solubilities, producing steady-state removal efficiencies of 30-65% and 70-80% respectively. The strippability of chemicals with high volatility and water solubility may be difficult to predict because of the opposite effects of volatility and solubility on stripping efficiency. Hydrophilic compounds, such as acetone and MTBE, are difficult to remove. However, higher volumetric air-to-water ratios significantly improve the removal efficiencies of these compounds. On the other hand, very hydrophobic compounds such as TCE and benzene are effectively stripped at room temperatures and relatively low air-water ratios.

The current efforts of this study involve the evaluation of optimal stripping conditions for more-difficult-to-strip compounds (e.g., acetone and naphthalene), development of a Shallow-tray, low-profile stripping model, destruction of stripped organics in the gas phase with UV photolysis (using advanced ultraviolet flashlamps) and catalytic reactions using novel solid catalysts (e.g., zeolites and activated carbon), characterization of reaction products with gas chromatograph/mass spectroscopy (GC/MS), and determination of mechanisms of decomposition. Our efforts also include the evaluation of the relative economics of our combined air stripping/UV photolysis system and comparison with conventional systems such as packed-tower air stripping with off-gas incineration, aqueous phase adsorption, steam stripping, direct UV-catalyzed oxidation in the liquid phase, biological and other developmental methods.

REFERENCES

Lamarre, B.L. and D.C. Shearouse. 1996. "Stripping Organics from Groundwater and Wastewater". Environ. Engin. World. March-April: 20-23.

LaBranche, D.F. and M.R. Collins. 1996. "Stripping Volatile Organics Compounds and Petroleum Hydrocarbons from Water". Water Environ. Res. 68(3): 348-358.

TABLE 1. The Physical Properties and Experimental Data for compounds Studied.

Compound	Physical Properties		Experimental Data		
	Solubility (ppm) at 25 °C	Henry's Law Constant at 25 °C	Temperature (°C)	Air / Water Ratio	Efficiency (%)
TCE	1,100	0.510	24.4	98	97
			26.7	132	99
			26.7	162	100
			26.7	176	100
			26.7	184	100
Benzene	1,780	0.225	26.7	98	97
			26.7	132	98
			26.7	162	98
			26.7	176	100
Toluene	560	0.195	26.7	98	99
Xylene	198	0.200	26.7	98	100
MTBE	51,000	0.045	26.7	323	70
			26.7	646	77
Acetone	Total	–	28.9	264	45
			27.8	323	64
			27.8	352	64
			22.2	264	33

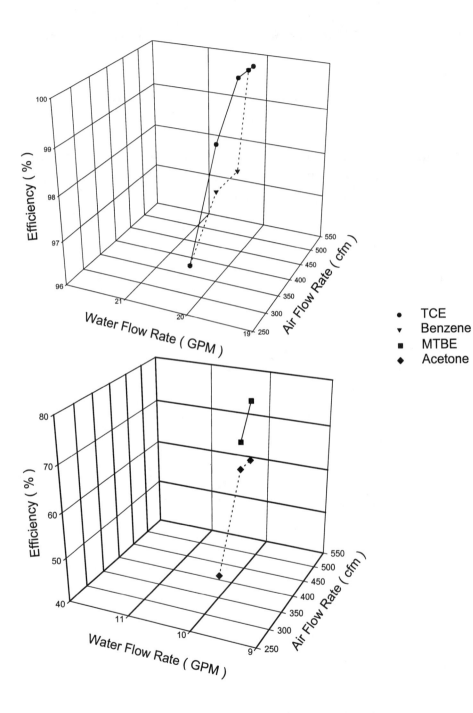

FIGURE 2. Effect of Air and Water Flowrates on the Removal Efficiencies .

ELECTROLYTIC REDUCTION OF CHLORINATED SOLVENTS —KINETICS, ELECTRODE SELECTIVITY, AND ELECTRODE MATERIALS

Zhijie Liu, Robert G. Arnold, and Eric A. Betterton
(University of Arizona; Tucson, AZ)

ABSTRACT: The electrolytic reductions of CCl_4 and other chlorinated solvents on electrodes consisting of elemental Ag, Au, Cu, Fe, Ni, Pd, and Zn were compared in terms of reaction kinetics, product distribution and rates of competitive reactions (predominantly H_2 (g) generation). Aside from the selection of cathode materials, conditions (electrode potential, bulk solution pH) in all reactors were identical. Reaction rates conformed to Butler-Volmer kinetics, modified to account for mass transfer limitation at the cathode surface. As expected, the limiting current was independent of cathode material selection. At higher (less negative) potentials that did not produce mass transfer limitation, rates of CCl_4 reduction were metal-dependent with Ni > Cu > Pd > Fe > Au > Ag > Zn. The extent of dehalogenation reactions at the same series of metal cathodes was established by measuring metal-dependent CCl_4 reaction products under otherwise identical reactor conditions. On that basis, Ni and Cu electrodes were superior to other metals tested. At $E_c=-1.0V$ (vs. SHE) and pH 7.0, methane production accounted for 80 percent of the CCl_4 transformed at the Ni electrode. Based on product identification of reduction of chlorinated ethanes and ethenes, dihalo-elimination and dehydrohalogenation were major reduction pathways for vicinal polychloroethanes and germinal polychloroethanes, respectively. For other chlorinated compounds, electrolytic reductions underwent hydrogenolysis to form less chlorinated or unchlorinated products. Based on selectivity for CCl_4 reduction, cathode materials were ordered Zn > Cu > Fe > Ni, suggesting that electrode selectivity was inversely related to the material-dependent exchange current density for hydrogen gas generation.

INTRODUCTION

Chlorinated solvents are of great environmental concern in groundwater due to their persistence and toxicity. There are environmental and economic incentives for improving extant technology or developing new treatment methods for destruction of such compounds. Reductive dehalogenation in electrolytic systems offers a promising remediation technology because it permits control of reaction kinetics and energetics via selection of electrode material and cathode potential.

Although electroreduction of chlorinated aliphatics has been studied for several decades, the application of this method for groundwater remediation received attention only recently. Electrolytic reduction of chlorinated compounds, e.g. carbon tetrachloride, 1,1,1-trichloroethane, and trichloroethylene, has been studied on metal and carbon electrodes (Criddle and McCarty, 1991; Nagaoka et al., 1994). To understand and better control electrolytic reactions of this nature, further investigation is needed to establish process feasibility, range of halogenated targets, extent of dehalogenation reactions, potential- and material-dependent kinetics, and electrode selectivity, i.e. avoidance of hydrogen gas generation. Here our objectives are to the explore dependence of reaction kinetics on cathode potential, electrode material, and target compounds; to identify electrolysis intermediates; and to investigate competition between dehalogenation and hydrogen generation reactions on metal cathodes. Kinetic findings are used to establish the potential utility of electrolytic strategies for the remediation of groundwater contaminated with chlorinated solvents.

MATERIALS AND METHODS

A glass electrolysis cell consisting of three compartments was used for all experiments. The 330-mL cathode compartment contained ports for gas- and liquid-phase sample withdrawal, addition of acid and purge gases, a pH probe and the working electrode. A side compartment was fitted with the Ag/AgCl reference electrode. A platinized platinum foil served as an anode. A cation-permeable membrane separated the cathode and anode compartments. Cathodes consisted of metal foils ($25 \times 21 \times 0.1$mm) or porous metal slabs (20 pores per inch). The cathode potential was maintained with a potentiostat. Current in the external circuit was determined indirectly by measuring the voltage drop across an in-line, 1Ω resistor. Bulk-phase pH was continuously measured and controlled using a pH-stat and addition of 1.0 M HCl.

For kinetic studies, the cathode compartment was filled with 0.1M Na_2SO_4. After purging the electrolyte with He for approximately an hour, 10-20 µL of the target compound was injected directly into the electrolyte. After the compound was completely dissolved, the cathode potential was established to initiate the experiments. A constant rate of stirring was maintained during the course of these experiments. Liquid-phase samples (20µL) were withdrawn at 10-20 minute intervals. Samples were extracted in pentane and analyzed via gas chromatograph with electron capture detector. Standards were prepared by injecting known volumes of pure liquid reagents into water-filled 165-mL vials. Liquid samples (1mL) were periodically withdrawn for Cl⁻ analysis using an ion chromatograph with an ion-exchange column and borate elution buffer.

In experiments designed to identify volatile products, 185mL of electrolyte was added to the 330-mL cathode compartment. After purging, 10-20 µL of pure target compound was injected into the liquid phase and stirred to achieve equilibrium between liquid and gas phases. The cathode potential was set per experimental objectives. Gas-phase samples (20µL) containing both chlorinated solvents and chlorine-free products were withdrawn at 10-20 min intervals for GC/FID analysis. Headspace $H_2(g)$ samples (20µL) were withdrawn at 10-15 min intervals for GC/TCD analysis. Gas standards were prepared by injecting known volume of the pure gases (99.5+%) into sealed glass vessels.

RESULTS AND DISCUSSION

Product Identification. Carbon tetrachloride was reduced on all metal electrodes tested (Ag, Au, Cu, Fe, Ni, Pd, and Zn). Intermediate included chloroform, which was further reduced although at a slower rate (Figure 1a). Major end products from both CCl_4 and $CHCl_3$ reduction were CH_4 and CH_2Cl_2. Trace amounts of unchlorinated hydrocarbons

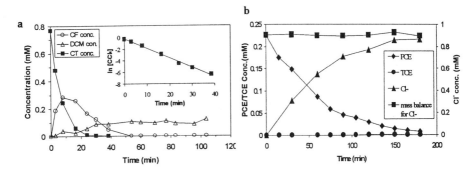

FIGURE 1. Reductive dehalogenation of carbon tetrachloride (a) and tetrachloroethene (b) in the electrochemical cell.

TABLE 1. Products and rate constants for reduction of chlorinated ethenes and ethanes at a porous nickel electrode (pH 7.0; E_c=-1.0V)

Compounds	rate constants $\times 10^4$ (1/min)	products
PCE	189	TCE; trans-DCE; ethane; ethene
TCE	77	TCE; trans-DCE; cis-DCE; 1,1-DCE; ethane; ethene
trans-DCE	48	ethane; ethene
cis-DCE	67	ethene; ethane
1,1-DCE	23	ethene; ethane
PCA	417	PCE; TCE; 1,1-DCA; 1,2-DCA; 1,1-DCE; ethane; ethene
1,1,1,2-TECA	581	PCE; TCE; 1,1-DCE; trans-DCE; cis-DCE; ethane; ethene
1,1,2,2-TECA	304	PCE; TCE; cis-DCE; trans-DCE
1,1,1-TCA	164	1,1-DCE; 1,1-DCA; ethane; ethene
1,1,2-TCA	79	1,1-DCA; 1,1-DCE; VC; ethane; ethene
1,1-DCA	6	ethane
1,2-DCA	1.5	ethane; ethene

were detected but not quantified. A reasonable mass balance on carbon (±10%) indicated that all major products were accounted for. CH_2Cl_2 was essentially a dead-end product, because its degradation was not observed during the course of the experiment (Figure 1a).

Tetrachloroethylene (PCE), another environmentally relevant target was successfully reduced on a porous Ni slab electrode (surface area ≈ $290 cm^2$) (Figure 1b). Trichloroethylene (TCE) was the primary reaction intermediate, but accumulated in only small amounts before it was further reduced to ethylene and ethane. Other expected intermediates, e.g. dichloroethylene isomers, were not detected. A mass balance on chlorine indicated that no major chlorinated products of PCE reduction were missed.

All the chlorinated solvents tested were reduced to less chlorinated or unchlorinated products on a Ni electrode (Table 1). Vogel et al.(1987) indicated that there are three major pathways for reductive transformation of chlorinated aliphatics: hydrogenolysis, dehydrohalogenation, and dihalo-elimination. All three pathways were evident in these experiments. The observed suite of reaction products suggested that (1) vicinal polychloroethanes undergo dihalo-elimination to yield the ethenes; (2) gem-polychloroethanes, e.g. 1,1,1-TCA, favor dehydrohalogenation and produce less chlorinated ethenes; and (3) hydrogenolysis is the major pathway for the electroreduction of other chlorinated compounds.

The extent of reaction was a function of cathode material and the applied cathode potential. For example, the ratio of CH_4 to CH_2Cl_2, both of which are final products of CCl_4 reduction, increased with decreasing cathode potential for every metal tested. At a single potential, Ni was superior to other metal cathodes and produced a higher CH_4/CH_2Cl_2 product ratio (Figure 3).

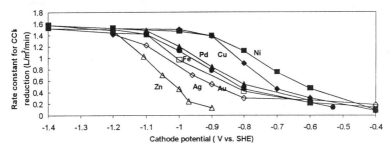

FIGURE 2. Dependence of first-order rate constants (k) for CCl_4 reduction on cathode potential and materials

Kinetics. At a given cathode potential, CCl_4 reduction was first-order reaction in the aqueous-phase concentration of CCl_4 (Figure 1a). The dependence of CCl_4 reduction kinetics on cathode metal selection and applied potential is illustrated in Figure 2. The relationship between the pseudo-first-order rate constant (k) and cathode potential (E_c) was well described by the modified Butler-Volmer equation with

$$k = k_0 e^{-\alpha F E_c / RT} \qquad (1)$$

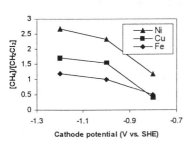

FIGURE 3. Major product distribution of CCl_4 reduction for various cathode material and potential after a 3-h reaction duration

where k_0 is the rate constant at E_c=0.0V vs. SHE, α is the transfer coefficient, F is the Faradaic constant (96,500 C/mole), and RT is the Boltzmann temperature (2480 J/mol). As predicted from Eqn (1), the rate constant for CCl_4 reduction increased exponentially with decreasing potential, E_c, in the reaction-controlled region. At more negative potentials, mass transfer limitations were encountered. Not surprisingly, the maximum rate constants were independent of cathode material. Metals were compared in terms of the magnitude of $E_{c,1/2}$, the potential at which $k=k_L/2$, where k_L is the rate constant corresponding to the transport-limited reaction. Based on $E_{c,1/2}$ values, the reactivity of metal cathodes followed the order: Ni > Cu > Pd > Fe > Au > Ag > Zn. The relative kinetics of reductions of chlorinated ethenes and ethanes at a fixed potential are compared in Table 1. Values indicate significant differences in reactivity among the compounds tested. In general, the more heavily substituted targets were more readily reduced, probably due to the electronegative character of chlorine substitutent. Vogel et al. (1987) suggested that rates of reductive dechlorinations are related to the standard redox potential of respective dehalogenation reactions. A satisfactory correlation was obtained between k, the first-order rate constant for reductive transformation, and the two-electron reduction potential for compound hydrogenolysis (Figure 4). Rate constants increase monotonically with increasing reduction potential. While useful for the purpose of general prediction, such correlations ignore the possibility of alternative or multiple reaction pathways for some halogenated targets.

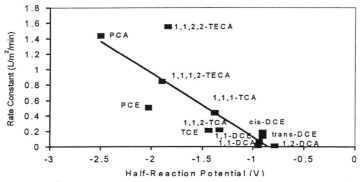

FIGURE 4. Relationship between reaction rate constants at –1.0V and the standard reduction potential of chlorinated solvents.

Selectivity. Hydrogen evolution is the most common reaction that competes with reductive dehalogenation in electrolytic systems. Here we define the instantaneous current efficiency as:

$$\varepsilon = \frac{i - \dfrac{dn_H}{dt}}{i} \qquad (2)$$

where i is the instantaneous current and n_H is the mass of hydrogen gas produced. Experimental results indicated that the instantaneous current efficiency for CCl_4 reduction depended on CCl_4 concentration, cathode potential, and cathode material (Figure 5). Current efficiencies decreased as disappeared due to (i) the dependence of dehalogenation of the concentrations of CCl_4 and $CHCl_3$, (ii) the displacement of water molecules from electrode surface by adsorbed CCl_4 at high CCl_4 activities, or both. The dependence of ε on E_c can be explained in terms of parameters that contribute to the electrolytic reductions of water and CCl_4. Shifting E_c below –0.8V generally changed the dehalogenation from kinetic control toward mass transfer control. Further decrease in E_c did little to improve the kinetics of CCl_4 reduction. However, the rate of hydrogen generation continued to increase at more negative potentials since proximity and abundance of water at cathode surface shift the mass transfer limitation for hydrogen production to very negative potentials. Results suggested that cathode material selection will be an important determinant of reactor efficiency. For example, hydrogen generation was slow on Zn electrode under the conditions of these experiments. At $E_c \geq -1.0V$, little or no hydrogen gas was produced for Zn electrodes. In order to compare electrode performances, we defined and average current efficiency $\hat{\varepsilon}$ as:

$$\hat{\varepsilon} = \frac{\int_0^\tau \varepsilon i dt}{\int_0^\tau i dt} \qquad (3)$$

where τ is the duration of the experiment. In order to obtain comparable $\hat{\varepsilon}$ for metals, the

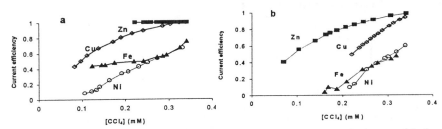

FIGURE 5. Selectivity of CCl_4 reduction in electrolytic systems at cathode potential of $-1.0V$ (a) and $-1.2V$ (b).

Table 2. Average current efficiency as a function of cathode material and potential.

Cathode material	$Logi_{0,H}$	$\hat{\varepsilon}$ at -0.8V	$\hat{\varepsilon}$ at −1.0V	$\hat{\varepsilon}$ at -1.2V
Fe	-5.5	0.84	0.56	0.3
Ni	-5.2	0.923	0.513	0.262
Cu	-8.5	1	0.895	0.7
Zn	-10.5	1	1	0.872

Note: Duration of reaction was 3 hrs, and initial CCl_4 concentration was 0.56mM.

initial CCl_4 concentrations and the durations of these experiments were identical. An inverse relationship between $\hat{\varepsilon}$ and $i_{0,H}$, the metal-dependent exchange current density for hydrogen evolution, was observed throughout the range $-1.2V \leq E_c \leq -0.8V$ (Table 2). The dependence of $\hat{\varepsilon}$ on $i_{0,H}$ was expected, particularly at more negative potentials, since the kinetics of hydrogen generation is directly related to $i_{0,H}$. In situations in which the generation of hydrogen gas is to be avoided, factors in cathode material selection should include $i_{0,H}$.

REFERENCES

Criddle, C. S., P. L. McCarty. 1991. *Environ. Sci. Technol.* 25: 973-978.

Nagaoka, T., J. Yamashita, M. Takase, and K. Ogura. 1994. *J. Electrochem. Soc.* 141: 1522-1526.

Vogel, T. M., C. S. Criddle, P. L. McCarty. 1987. *Environ. Sci. Technol.* 21: 722-736.

AUTHOR INDEX

This index contains names, affiliations, and book/page citations for all authors who contributed to the six books published in connection with the First International Conference on Remediation of Chlorinated and Recalcitrant Compounds, held in Monterey, California, in May 1998. Ordering information is provided on the back cover of this book.

The citations reference the six books as follows:

1(1): Wickramanayake, G.B., and R.E. Hinchee (Eds.). 1998. *Risk, Resource, and Regulatory Issues: Remediation of Chlorinated and Recalcitrant Compounds.* Battelle Press, Columbus, OH. 322 pp.

1(2): Wickramanayake, G.B., and R.E. Hinchee (Eds.). 1998. *Nonaqueous-Phase Liquids: Remediation of Chlorinated and Recalcitrant Compounds.* Battelle Press, Columbus, OH. 256 pp.

1(3): Wickramanayake, G.B., and R.E. Hinchee (Eds.). 1998. *Natural Attenuation: Chlorinated and Recalcitrant Compounds.* Battelle Press, Columbus, OH. 380 pp.

1(4): Wickramanayake, G.B., and R.E. Hinchee (Eds.). 1998. *Bioremediation and Phytoremediation: Chlorinated and Recalcitrant Compounds.* Battelle Press, Columbus, OH. 302 pp.

1(5): Wickramanayake, G.B., and R.E. Hinchee (Eds.). 1998. *Physical, Chemical, and Thermal Technologies: Remediation of Chlorinated and Recalcitrant Compounds.* Battelle Press, Columbus, OH. 512 pp.

1(6): Wickramanayake, G.B., and R.E. Hinchee (Eds.). 1998. *Designing and Applying Treatment Technologies: Remediation of Chlorinated and Recalcitrant Compounds.* Battelle Press, Columbus, OH. 348 pp.

KEYWORD INDEX

This index contains keyword terms assigned to the articles in the six books published in connection with the First International Conference on Remediation of Chlorinated and Recalcitrant Compounds, held in Monterey, California, in May 1998. Ordering information is provided on the back cover of this book.

In assigning the terms that appear in this index, no attempt was made to reference all subjects addressed. Instead, terms were assigned to each article to reflect the primary topics covered by that article. Authors' suggestions were taken into consideration and expanded or revised as necessary to produce a cohesive topic listing. The citations reference the six books as follows:

1(1): Wickramanayake, G.B., and R.E. Hinchee (Eds.). 1998. *Risk, Resource, and Regulatory Issues: Remediation of Chlorinated and Recalcitrant Compounds.* Battelle Press, Columbus, OH. 322 pp.

1(2): Wickramanayake, G.B., and R.E. Hinchee (Eds.). 1998. *Nonaqueous-Phase Liquids: Remediation of Chlorinated and Recalcitrant Compounds.* Battelle Press, Columbus, OH. 256 pp.

1(3): Wickramanayake, G.B., and R.E. Hinchee (Eds.). 1998. *Natural Attenuation: Chlorinated and Recalcitrant Compounds.* Battelle Press, Columbus, OH. 380 pp.

1(4): Wickramanayake, G.B., and R.E. Hinchee (Eds.). 1998. *Bioremediation and Phytoremediation: Chlorinated and Recalcitrant Compounds.* Battelle Press, Columbus, OH. 302 pp.

1(5): Wickramanayake, G.B., and R.E. Hinchee (Eds.). 1998. *Physical, Chemical, and Thermal Technologies: Remediation of Chlorinated and Recalcitrant Compounds.* Battelle Press, Columbus, OH. 512 pp.

1(6): Wickramanayake, G.B., and R.E. Hinchee (Eds.). 1998. *Designing and Applying Treatment Technologies: Remediation of Chlorinated and Recalcitrant Compounds.* Battelle Press, Columbus, OH. 348 pp.